Satellite Communications

About the Authors

Dennis Roddy (deceased) was professor emeritus of Electrical Engineering at Lakehead University in Thunder Bay, Ontario, Canada, and had more than 40 years of experience in both industrial and technical education. He was the single author of *Satellite Communications* (editions 1–4) as well as the author of *Radio and Line Transmission*, vols. 1 and 2; *Introduction to Microelectronics*; and *Microwave Technology*; and the coauthor of *Electronic Communications and Electronics*.

W. Linwood Jones, Ph.D. (Cocoa Beach, FL), is a professor in the Department of Electrical and Computer Engineering at the University of Central Florida in Orlando. He is the director of the Central Florida Remote Sensing Laboratory. He has over 40 years of experience in satellite communications and microwave remote sensing with over 30 years of teaching engineering. Dr. Jones is a life fellow of the IEEE and has received several outstanding educator awards.

David G. Long, Ph.D. (Provo, UT), is a professor in the Department of Electrical and Computer Engineering at Brigham Young University in Provo, UT. He is the director of BYU's Center for Remote Sensing and head of the Microwave Earth Remote Sensing laboratory. He has 40 years of experience in research in microwave remote sensing with over 30 years of teaching engineering. Dr. Long is fellow of IEEE and a coauthor of *Microwave Radar and Radiometric Sensing*.

Satellite Communications

Dennis Roddy

W. Linwood Jones

David G. Long

Fifth Edition

New York   Chicago   San Francisco
Athens   London   Madrid
Mexico City   Milan   New Delhi
Singapore   Sydney   Toronto

Satellite Communications, Fifth Edition

1 2 3 4 5 6 7 8 9   LCR   29 28 27 26 25 24

Library of Congress Control Number: 2023949511

ISBN 978-1-265-37254-5
MHID 1-265-37254-3

Sponsoring Editor
Lara Zoble

Project Manager
Tasneem Kauser,
KnowledgeWorks Global, Ltd.

Acquisitions Coordinator
Olivia Higgins

Copy Editor
Girish Sharma

Proofreader
Anshu Sinha

Indexer
W. Linwood Jones and David J. Long

Production Supervisor
Lynn M. Messina

Composition
KnowledgeWorks Global Ltd.

Illustration
KnowledgeWorks Global Ltd.

Art Director, Cover
Anthony Landi

In appreciation of the encouragement and support of my wife, Beverly Brown Jones.

—W. Linwood Jones

Contents

Preface

As with previous editions, this textbook provides a broad coverage of modern satellite communications, but with the passing of Professor Emeritus Dennis Roddy, this fifth edition has been significantly updated by Professors W. Linwood Jones and David G. Long, who individually have more than 40 years of experience in RF/microwave research and undergraduate and graduate education in satellite communications, radar, and microwave remote sensing.

Compared to the fourth edition, significant changes have been made to reflect the transition of current satellite systems to digital communications, while maintaining the presentation of analog and digital legacy systems in operation. The introduction of new material promotes better understanding of satellite link analysis with many new figures and numerical examples added. Of the 17 chapters, five, namely: Chap. 5—Polarization, Chap. 8—The Earth Segment, Chap. 9—Analog Signals, Chap. 11—Error Control Coding, and Chap. 16—Direct Broadcast Satellite (DBS) Television, are essentially unchanged. However, the remaining 12 chapters were revised as follows:

Chapter 1—Overview of Satellite Systems: provides updates to the communications satellite launches since 2000, and a presentation of the current satellite search and rescue system.

Chapter 2—Orbits and Launching Methods: removes outdated references to Space Shuttle satellite launches, expands the discussion of expendable rocket launches, introduces reusable first-stage boosters, and new numerical examples of GEO launch scenarios are given.

Chapter 3—The Geostationary Orbit: two new sections on antenna pointing geometry are introduced, with five additional examples and two new homework problems.

Chapter 4—Radio Wave Propagation: Increased emphasis on rain absorption effects with three new sections and two new homework problems.

Chapter 6—Antennas; three new sections concerning the calculation of satellite antenna line-of-sight gain, five new examples, and five new homework problems.

Chapter 7—The Space Segment: updated five sections (including two new) giving expanded satellite subsystem discussion, and added one homework problem.

Chapter 10—Digital Signals: updated and expanded the scope of four sections, with 10 new figures and two new examples.

Chapter 12—The Space Link: introduced an improved method of treating losses with major changes to all sections, including nine new sections, 14 new figures, seven new examples, and three new homework problems.

Chapter 13—Interference: minor changes with two new figures and one new example.

Chapter 14—Satellite Access: updated method of treating losses, addition of five new numerical examples and two new homework problems.

Chapter 15—Satellites in Networks: major update of presentation of satellite Internet services with three new sections added.

Chapter 17—Satellite Mobile and Specialized Services: updated five sections to remove outdated material and to include modern mobile SATCOM services.

Added two Appendices, G and H.

Acknowledgments

In this age of heightened security concerns, it has proved difficult to obtain technical information on current satellite systems and technology advancements. Special appreciation is expressed to the following individuals for their technical contributions:

Chapter 1 revisions were provided by Dr. Maria Marta Jacob of Techno-Sciences, Lanham, MD, and Dr. Zoubair Ghazi of Laplace Tech Solutions, Alexandria, VA.

Section 6.13.1 was provided by Dr. Alamgir Hossan of the University of Central Florida.

Chapter 15 revisions were provided by Mr. Wynn Rust of Cocoa Beach, FL.

Further, the authors appreciate the technical manuscript review of Dr. Maria Marta Jacob and Mr. Wynn Rust, and we appreciate the new graphics preparation and editorial review of Ms. Ashley Jones Sullivan.

Finally, the editorial team at McGraw Hill has contributed greatly to bring this project to fruition.

W. Linwood Jones
University of Central Florida

David G. Long
Brigham Young University

CHAPTER **1**

Overview of Satellite Systems

1.1 Introduction

The use of satellites in communications systems is very much a fact of everyday life, as is evidenced by the many homes equipped with antennas, or "dishes," used for reception of satellite television. What may not be so well known is that satellites form an essential part of telecommunications systems worldwide, carrying large amounts of data and telephone traffic in addition to television signals.

Satellites offer several features not readily available with other means of communications. Because very large areas of the earth are visible from a satellite, the satellite can form the star point of a communications net, simultaneously linking many users who may be widely separated geographically. The same feature enables satellites to provide communications links to remote communities in sparsely populated areas that are difficult to access by other means. Of course, satellite signals ignore political boundaries as well as geographic ones, which may or may not be a desirable feature.

To give some idea of cost, the construction and launch cost of the Canadian Anik-E1 satellite (in 1994 Canadian dollars) was $281.2 million, and that of the Anik-E2, $290.5 million. The combined launch insurance for both satellites was $95.5 million. In contrast, the SES 20 and SES 21 communications satellites launched in 2020 cost $400 million each with a combined launch cost of $140 million (2020 U.S. dollars).

A feature of any satellite system is that the cost is *distance insensitive*, meaning that it costs about the same to provide a satellite communications link over a short distance as it does over a large distance. Thus, a satellite communications system is economical only where the system is in continuous use and the costs can be reasonably spread over many users.

Satellites are also used for remote sensing, examples being the detection of water pollution and the monitoring and reporting of weather conditions. Some of these remote sensing satellites also form a vital link in search and rescue operations for downed aircraft, ships, and even hikers.

A good overview of the role of satellites is given by Kodheli et al. (2021), Evans et al. (2011), Pelton (2010), Huang and Cao (2020), Pritchard (1984), and Brown (1981). To provide a general overview of satellite systems here, three different types of applications are briefly described in this chapter: (1) the largest international system, INTELSAT, (2) the domestic satellite system in the United States, Domsat, and (3) U.S. *National Oceanographic and Atmospheric Administration* (NOAA) series of polar orbiting satellites used for environmental monitoring and search and rescue.

1.2   Frequency Allocations for Satellite Services

Allocating frequencies to satellite services is a complicated process that requires international coordination and planning. This is carried out under the auspices of the *International Telecommunication Union* (ITU).

To facilitate frequency planning, the world is divided into three regions:

Region 1: Europe, Africa, what was formerly the Soviet Union, and Mongolia
Region 2: North and South America and Greenland
Region 3: Asia (excluding region 1 areas), Australia, and the southwest Pacific

Within these regions, frequency bands are allocated to various satellite services, although a given service may be allocated different frequency bands in different regions. Some of the services provided by satellites are

Fixed satellite service (FSS)
Broadcasting satellite service (BSS)
Mobile satellite services
Navigational satellite services
Meteorological satellite services

There are many subdivisions within these broad classifications; for example, the FSS provides links for existing telephone networks as well as for transmitting television signals to cable companies for distribution over cable systems. Broadcasting satellite services are intended mainly for direct broadcast to the home, sometimes referred to as *direct broadcast satellite* (DBS) service (in Europe it may be known as *direct-to-home* [DTH] service). Mobile satellite services would include land mobile, maritime mobile, and aeronautical mobile. Navigational satellite services include *global positioning systems* (GPS), and satellites intended for the meteorological services often provide a search and rescue service.

Table 1.1 lists the frequency band designations in common use for satellite services. The Ku band signifies the band under the K band, and the Ka band is the band above the K band. The Ku band is the one used at present for DBS, and it is also used for certain FSS. The C band is used for FSS, and no DBS is allowed in this band. The very high-frequency (VHF) band is used for certain mobile and navigational services and for data transfer from weather satellites. The L band is used for mobile satellite services and navigation systems. For the FSS in the C band, the most widely used subrange is approximately 4 to 6 GHz. The higher frequency is nearly always used for the uplink to the satellite, for reasons that will be explained later, and common practice is to denote the C band by 6/4 GHz, giving the uplink frequency first. For the direct broadcast service in the Ku band, the most widely used range is approximately 12 to 14 GHz, which is denoted by 14/12 GHz. Although frequency assignments are made much more precisely and they may lie somewhat outside the values quoted here (an example of assigned frequencies in the Ku band is 14,030 and 11,730 MHz), the approximate values stated are quite satisfactory for use in calculations involving frequency, as are used later in the text.

Frequency Range (GHz)	Band Designation
0.1–0.3	VHF
0.3–1.0	UHF
1.0–2.0	L
2.0–4.0	S
4.0–8.0	C
8.0–12.0	X
12.0–18.0	Ku
18.0–27.0	K
27.0–40.0	Ka
40.0–75	V
75–110	W
110–300	mm
300–3000	μm

TABLE 1.1 Frequency Band Designations

Care must be exercised when using published references to frequency bands because the designations have been developed somewhat differently for radar and communications applications; in addition, not all countries use the same designations.

The official ITU frequency band designations are shown in Table 1.2 for completeness. However, in this text the designations given in Table 1.1 will be used, along with 6/4 GHz for the C band and 14/12 GHz for the Ku band.

Band Number	Symbols	Frequency Range (Lower Limit Exclusive, Upper Limit Inclusive)	Corresponding Metric Subdivision	Metric Abbreviations for the Bands
4	VLF	3–30 kHz	Myriametric waves	B.Mam
5	LF	30–300 kHz	Kilometric waves	B.km
6	MF	300–3000 kHz	Hectometric waves	B.hm
7	HF	3–30 MHz	Decametric waves	B.dam
8	VHF	30–300 MHz	Metric waves	B.m
9	UHF	300–3000 MHz	Decimetric waves	B.dm
10	SHF	3–30 GHz	Centimetric waves	B.cm
11	EHF	30–300 GHz	Millimetric waves	B.mm
12		300–3000 GHz	Decimillimetric waves	

Source: ITU Geneva.

TABLE 1.2 ITU Frequency Band Designations

1.3   INTELSAT

INTELSAT stands for *International Telecommunications Satellite*. The organization was created in 1964 and currently has over 140 member countries and more than 40 investing entities (see http://www.intelsat.com/ for more details). In July 2001, INTELSAT became a private company, and in May 2002 the company began providing end-to-end solutions through a network of teleports, leased fiber, and *points of presence* (PoPs) around the globe. Starting with the Early Bird satellite in 1965, a succession of satellites has been launched at intervals of a few years. Figure 1.1 illustrates the evolution of some of the INTELSAT satellites. As Fig. 1.1 shows, the capacity, in terms of number of voice channels, increased dramatically with each succeeding launch, as well as the design lifetime. These satellites are in *geostationary orbit*, meaning that they appear to be stationary in relation to the earth. The geostationary orbit is the topic of Chap. 3. At this point it may be noted that geostationary satellites orbit in the earth's equatorial plane and their position is specified by their longitude. For international traffic, INTELSAT covers three main regions—the *Atlantic Ocean Region* (AOR), the *Indian Ocean Region* (IOR), and the *Pacific Ocean Region* (POR) and what is termed *Intelsat America's Region*. For the ocean regions the satellites are positioned in geostationary orbit above the ocean, where they provide a transoceanic telecommunications route. For example, INTELSAT satellite 905 is positioned at 335.5° east longitude. The footprints for the C-band antennas are shown in Fig. 1.2*a*, and for the Ku-band spot beam antennas in Fig. 1.2*b* and *c*.

The INTELSAT VII-VII/A series was launched over a period from October 1993 to June 1996. The construction is like that for the V and VA/VB series, shown in Fig. 1.1, in that the VII series has solar sails rather than a cylindrical body. This type of construction is described in more detail in Chap. 7. The VII series was planned for service in the POR and for some of the less demanding services in the AOR. The antenna beam coverage is appropriate for that of the POR. Figure 1.3 shows the antenna beam footprints for the C-band hemispheric coverage and zone coverage, as well as the spot beam coverage possible with the Ku-band antennas (Lilly, 1990; Sachdev et al., 1990). When used in the AOR, the VII series satellite is inverted north for south (Lilly, 1990), minor adjustments then being needed only to optimize the antenna patterns for this region. The lifetime of these satellites ranges from 10 to 15 years depending on the launch vehicle. Figures from the INTELSAT Web site gives the capacity for the INTELSAT VII as 18,000 two-way telephone circuits and three TV channels; up to 90,000 two-way telephone circuits can be achieved with the use of "digital circuit multiplication." The INTELSAT VII/A has a capacity of 22,500 two-way telephone circuits and three TV channels; up to 112,500 two-way telephone circuits can be achieved with the use of digital circuit multiplication. In 1999, four satellites were in service over the AOR, one in the IOR, and two in the POR.

The INTELSAT VIII-VII/A series of satellites was launched over the period 1997 to 1998. Satellites in this series have similar capacity as the VII/A series, and the lifetime is 14 to 17 years. In the 2000s, several of the next generation of INTELSAT IX series have been launched forming a constellation of more than 50 communications satellites. Further, it is standard practice to have a spare satellite in orbit on high-reliability routes (which can carry preemptible traffic) and to have a ground spare in case of launch failure. Thus the cost for large international schemes can be high; for example, series IX, described later, represents a total investment of approximately $1 billion.

Table 1.3 summarizes the details of some of the more recent of the INTELSAT satellites. These satellites provide a much wider range of services than those available

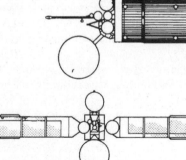

Designation: Intelsat	I	II	III	IV	IV A	V	V A/V B	VI
Year of first launch	1965	1966	1968	1971	1975	1980	1984/85	1986/87
Prime contractor	Hughes	Hughes	TRW	Hughes	Hughes	Ford Aerospace	Ford Aerospace	Hughes
Width (m)	0.7	1.4	1.4	2.4	2.4	2.0	2.0	3.6
Height (m)	0.6	0.7	1.0	5.3	6.8	6.4	6.4	6.4
Launch vehicles		Thor Delta		Atlas-Centaur		Atlas-Centaur and Ariane		STS and Ariane
Spacecraft mass in transfer orbit (kg)	68	182	293	1385	1489	1946	2140	12,100/3720
Communications payload mass (kg)	13	36	56	185	190	235	280	800
End-of-life (EOL) power of equinox (W)	40	75	134	480	800	1270	1270	2200
Design lifetime (years)	1.5	3	5	7	7	7	7	10
Capacity (number of voice channels)	480	480	2400	8000	12,000	25,000	30,000	80,000
Bandwidth (MHz)	50	130	300	500	800	2137	2480	3520

Figure 1.1 Evolution of INTELSAT satellites. (*From Colino 1985; courtesy of ITU Telecommunications Journal*).

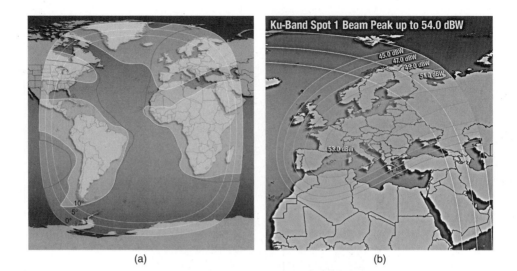

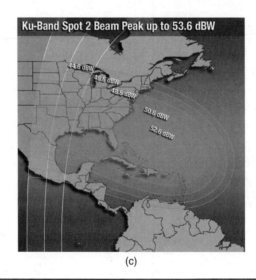

FIGURE 1.2 INTELSAT satellite 905 is positioned at 335.5° E longitude. (*a*) The footprints for the C-band antennas; (*b*) the Ku-band spot 1 beam antennas; and (*c*) the Ku-band spot 2 beam antennas.

previously, including such services as the Internet, DTH TV, tele-medicine, tele-education, and interactive video and multimedia. Transponders and the types of signals they carry are described in detail in later chapters, but for comparison purposes it may be noted that one 36 MHz transponder can carry about 9000 voice channels, or two analog TV channels, or about eight digital TV channels.

In addition to providing transoceanic routes, the INTELSAT satellites are also used for domestic services within any given country and regional services between countries. Two such services are Vista for telephone and Internet for data exchange. Figure 1.4 shows typical Vista applications.

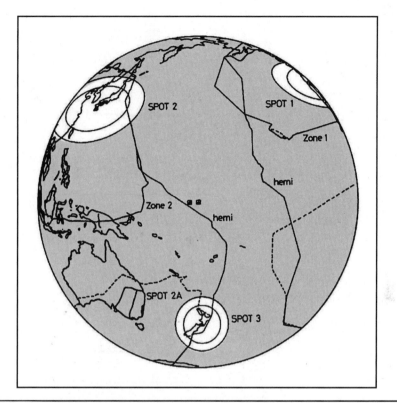

FIGURE 1.3   INTELSAT VII coverage (*Pacific Ocean Region; global, hemispheric, and spot beams*). (*From Lilly, 1990, with permission*).

Satellite	Location	Number of Transponders	Launch Date
21	58°E	Up to 36 @ 36 MHz in Ku-Band Up to 24 @ 36 MHz in C-Band	August 2012
23	53°E	Up to 27 @ 36 MHz in Ku Band Up to 72 @ 36 MHz in C-Band	October 2010
25	31.5°E	Up to 22 @ 34 MHz in Ku Band Up to 38 @ 46 MHz in C-Band	July 2008
32e	43°W	Up to 56 @ 36 MHz in Ku Band Up to 24 @ 36 MHz in Ku Band	February 2017
34	55.5°E	Up to 24 @ 36 MHz in C-Band Up to 39 @ 36 MHz in Ku Band	August 2015
35e	34.5°E	Up to 112 @ 36 MHz in C-Band Up to 247 @ 36 MHz in Ku Band	July 2017
37e	18°W	Up to 37 @ 36 MHz in Ka Band Up to 79 @ 36 MHz in C-Band	September 2017
39	62°E	Up to 72 @ 36 MHz in Ku Band Up to 56 @ 36 MHz in C-Band	August 2019

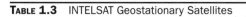

TABLE 1.3   INTELSAT Geostationary Satellites

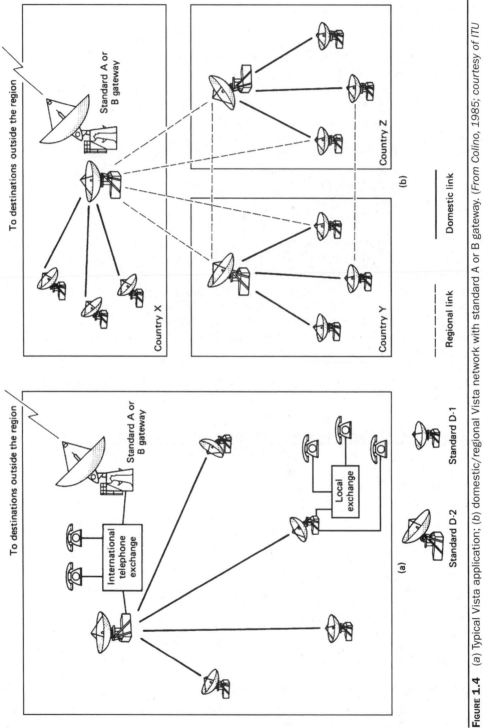

Figure 1.4 (a) Typical Vista application; (b) domestic/regional Vista network with standard A or B gateway. (*From Colino, 1985; courtesy of ITU Telecommunications Journal*).

1.4   U.S. Domsats

Domsat is an abbreviation for *domestic satellite.* Domestic satellites are used to provide various telecommunications services, such as voice, data, and video transmissions, within a country. In the United States, all domsats are situated in geostationary orbits. As is well known, they make available a wide selection of TV channels for the home entertainment market, in addition to carrying a large amount of commercial telecommunications traffic.

U.S. Domsats, which provide a DTH television service, can be classified broadly as high power, medium power, and low power (Reinhart, 1990). The defining characteristics of these categories are shown in Table 1.4.

The main distinguishing feature of these categories is the *equivalent isotropic radiated power* (EIRP). This is explained in more detail in Chap. 12, but for present purposes, it should be noted that the upper limit of EIRP is 60 dBW for the high-power category and 37 dBW for the low-power category, a difference of 23 dB. This represents an increase in received power of $10^{2.3}$ or about 200:1 in the high-power category, which allows much smaller antennas to be used with the receiver. As noted in the table, the primary purpose of satellites in the high-power category is to provide a DBS service. In the medium-power category, the primary purpose is point-to-point services, but space may be leased on these satellites for the provision of DBS services. In the low-power category, no official DBS services are provided. However, it was quickly discovered by home experimenters that a wide range of radio and TV programming could be received on this band, and it is now considered to provide a de facto DBS service, witness to which is the large number of *TV receive-only* (TVRO) dishes that have appeared in the yards and on the rooftops of homes in North America. TVRO reception of C-band signals in the home is prohibited in many other parts of the world, partly for aesthetic reasons, because of

	High Power	Medium Power	Low Power
Band	Ku	Ku	C
Downlink frequency allocation GHz	12.2–12.7	11.7–12.2	3.7–4.2
Uplink frequency allocation GHz	17.3–17.8	14–14.5	5.925–6.425
Space service	BSS	FSS	FSS
Primary intended use	DBS	Point-to-point	Point-to-point
Allowed additional use	Point-to-point	DBS	DBS
Terrestrial interference possible	No	No	Yes
Satellite spacing degrees	9	2	2–3
Satellite spacing determined by	ITU	FCC	FCC
Adjacent satellite interference possible?	No	Yes	Yes
Satellite EIRP range (dBW)	51–60	40–48	33–37

Notes: FCC, Federal Communications Commission; ITU, International Telecommunication Union.
Source: Reinhart (1990).

TABLE 1.4   Defining Characteristics of Three Categories of United States DBS Systems

the comparatively large dishes used, and partly for commercial reasons. Many North American C-band TV broadcasts are now encrypted, or scrambled, to prevent unauthorized access, although this also seems to be spawning a new underground industry in descramblers.

As shown in Table 1.4, true DBS service takes place in the Ku band. Figure 1.5 shows the components of a DBS system (Government of Canada, 1983). The television signal may be relayed over a terrestrial link to the uplink station. This transmits a very narrow beam signal to the satellite in the 14-GHz band. The satellite retransmits the television signal in a wide beam in the 12-GHz frequency band. Individual receivers within the beam coverage area will receive the satellite signal.

Table 1.5 shows the orbital assignments for domestic fixed satellites for the United States (FCC, 1996). These satellites are in geostationary orbit, which is discussed further in Chap. 3. Table 1.6 shows the U.S. Ka-band assignments. Broadband services, such as the Internet (see Chap. 15), can operate at Ka-band frequencies. In 1983, the U.S. FCC adopted a policy objective, setting 2° as the minimum orbital spacing for satellites operating in the 6/4-GHz band and 1.5° for those operating in the 14/12-GHz band (FCC, 1983). Interference between satellite circuits is likely to increase as satellites are positioned closer together. These spacings represent the minimum presently achievable in each band at acceptable interference levels. In fact, it seems likely that in some cases home satellite receivers in the 6/4-GHz band may be subject to excessive interference where 2° spacing is employed.

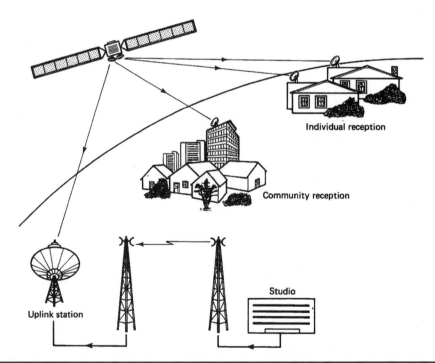

FIGURE 1.5 Components of a direct broadcasting satellite system. (*From Government of Canada, 1983, with permission.*)

Location, Degrees West Longitude	Satellite	Band/Polarization
139	Aurora II/Satcom C-5	4/6 GHz (vertical)
139	ACS-3K (AMSC)	12/14 GHz
137	Satcom C-1	4/6 GHz (horizontal)
137	Unassigned	12/14 GHz
135	Satcom C-4	4/6 GHz (vertical)
135	Orion O-F4	12/14 GHz
133	Galaxy 1-R(S)	4/6 GHz (horizontal)
133	Unassigned	12/14 GHz
131	Satcom C-3	4/6 GHz (vertical)
131	Unassigned	12/14 GHz
129	Loral 1	4/6 GHz (horizontal)/12/14 GHz
127	Galaxy IX	4/6 GHz (vertical)
127	Unassigned	12/14 GHz
125	Galaxy 5-W	4/6 GHz (horizontal)
125	GSTAR II/unassigned	12/14 GHz
123	Galaxy X	4/6 GHz (vertical)/12/14 GHz
121	EchoStar FSS-2	12/14 GHz
105	GSTAR IV	12/14 GHz
103	GE-1	4/6 GHz (horizontal)
103	GSTAR 1/GE-1	12/14 GHz
101	Satcom SN-4 (formerly Spacenet IV-n)	4/6 GHz (vertical)/12/14 GHz
99	Galaxy IV(H)	4/6 GHz (horizontal)/12/14 GHz
97	Telstar 401	4/6 GHz (vertical)/12/14 GHz
95	Galaxy III(H)	4/6 GHz (horizontal)/12/14 GHz
93	Telstar 5	4/6 GHz (vertical)
93	GSTAR III/Telstar 5	12/14 GHz
91	Galaxy VII(H)	4/6 GHz (horizontal)/12/14 GHz
89	Telestar 402R	4/6 GHz (vertical)/12/14 GHz
87	Satcom SN-3 (formerly Spacenet III-)/GE-4	4/6 GHz (horizontal)/12/14 GHz
85	Telstar 302/GE-2	4/6 GHz (vertical)
85	Satcom Ku-1/GE-2	12/14 GHz
83	Unassigned	4/6 GHz (horizontal)
83	EchoStar FSS-1	12/14 GHz
81	Unassigned	4/6 GHz (vertical)
81	Satcom Ku-2/unassigned	12/14 GHz

TABLE 1.5 FCC Orbital Assignment Plan (May 7, 1996)

Location, Degrees West Longitude	Satellite	Band/Polarization
79	GE-5	4/6 GHz (horizontal)/12/14 GHz
77	Loral 2	4/6 GHz (vertical)/12/14 GHz
76	Comstar D-4	4/6 GHz (vertical)
74	Galaxy VI	4/6 GHz (horizontal)
74	SBS-6	12/14 GHz
72	Unassigned	4/6 GHz (vertical)
71	SBS-2	12/14 GHz
69	Satcom SN-2/Telstar 6	4/6 GHz (horizontal)/12/14 GHz
67	GE-3	4/6 GHz (vertical)/12/14 GHz
64	Unassigned	4/6 GHz (horizontal)
64	Unassigned	12/14 GHz
62	Unassigned	4/6 GHz (vertical)
62	ACS-2K (AMSC)	12/14 GHz
60	Unassigned	4/6 GHz
60	Unassigned	12/14 GHz

TABLE 1.5   FCC Orbital Assignment Plan (May 7, 1996) (*Continued*)

Location	Company
147°WL	Morning Star Satellite Company, LLC
125°WL	PanAmSat Licensee Corporation
121°WL	Echostar Satellite Corporation
115°WL	Loral Space & Communications, Ltd.
113°WL*	VisionStar, Inc.
109.2°WL	KaStar Satellite Communications Corp.
105°WL†	GE American Communications, Inc.
103°WL	PanAmSat Corporation
101°WL	Hughes Communications Galaxy, Inc.
99°WL	Hughes Communications Galaxy, Inc.
97°WL	Lockheed Martin Corporation
95°WL	NetSat 28 Company, LLC
93°WL	Loral Space & Communications, Ltd.
91°WL	Comm, Inc.
89°WL	Orion Network Systems
87°WL	Comm, Inc.
85°WL	GE American Communications, Inc.
83°WL	Echostar Satellite Corporation
81°WL	Orion Network Systems

TABLE 1.6   Ka-Band Orbital Assignment Plan (FCC December 19, 1997)

Location	Company
77°WL	Comm, Inc.
75°WL	Comm, Inc.
73°WL	KaStar Satellite Corporation
67°WL	[under consideration]
62°WL	Morning Star Satellite Company, LLC
58°WL	PanAmSat Corporation
49°WL	Hughes Communications Galaxy, Inc.
47°WL	Orion Atlantic, LP
21.5°WL	Lockheed Martin Corporation
17°WL	GE American Communications, Inc.
2°EL	Lockheed Martin Corporation
25°EL	Hughes Communication Galaxy, Inc.
30°EL	Morning Star Satellite Company, LLC
36°EL	PanAmSat Corporation
40°EL	PanAmSat Corporation
48°EL	PanAmSat Corporation
54°EL	Hughes Communications Galaxy, Inc.
56°EL	GE American Communications, Inc.
78°EL	Orion Network Systems, Inc.
101°EL	Hughes Communications Galaxy, Inc.
105.5°EL	Loral Space & Communications, Ltd.
107.5°EL	Morning Star Satellite Company, LLC
111°EL	Hughes Communications Galaxy, Inc.
114.5°EL	GE American Communications, Inc.
124.5°EL	PanAmSat Corporation
126.5°EL	Orion Network Systems, Inc.
130°EL	Lockheed Martin Corporation
149°EL	Hughes Communications Galaxy, Inc.
164°EL	Hughes Communications Galaxy, Inc.
173°EL	PanAmSat Corporation
175.25°EL	Lockheed Martin Corporation

Notes: EL, east longitude; FCC, Federal Communications Commission; WL, west longitude.
*VisionStar will operate at a nominal orbit location of 113.05°WL, and with a station-keeping box of ±0.05°, thereby increasing the separation from the Canadian filing at 111.1°WL by 0.1° relative to the worst-case orbital spacing using the station keeping assumed in the ITU publications (±0.1°).
†The applicants in the range 95° to 105°WL have agreed to operate their satellites with a nominal 0.05° offset to the west, and with a station-keeping box of ±0.05°, thereby increasing the separation from the Luxembourg satellite at 93.2°WL by 0.1° relative to the worst-case orbital spacing using the station keeping assumed in the ITU publications (±0.1°).

TABLE 1.6 Ka-Band Orbital Assignment Plan (FCC December 19, 1997) (*Continued*)

1.5   Polar Orbiting Satellites

Polar orbiting satellites orbit the earth in such a way as to cover the north and south polar regions. (Note that the term *polar orbiting* does not mean that the satellite orbits around one or the other of the poles). Figure 1.6 shows a polar orbit in relation to the geostationary orbit. Whereas there is only one geostationary orbit, there are, in theory, an infinite number of polar orbits. The U.S. experience with weather satellites has led to the use of relatively low orbits, ranging in altitude between 800 and 900 km, compared with 36,000 km for the geostationary orbit. *Low earth orbiting* (LEO) satellites are known generally by the acronym LEOSATS.

In the United States, the *National Polar-orbiting Operational Environmental Satellite System* (NPOESS) was established in 1994 to consolidate the polar satellite operations of the Air Force, NASA (*National Aeronautics and Space Administration*) and NOAA (*National Oceanic and Atmospheric Administration*). NPOESS manages the *Integrated Program Office* (IPO), and the Web page can be found at http://www.ipo.noaa.gov/. As of 2005, a four-orbit system is in place, consisting of two U.S. Military orbits, one U.S. Civilian orbit and one EUMETSAT/METOP orbit. Here, METSAT stands for *meteorological satellite* and EUMETSAT stands for the *European organization for the exploration of the METSAT program*. METOP stands for *meteorological operations*. These orbits are sun synchronous, meaning that they cross the equator at the same local time each day. For example, the satellites in the NPOESS (civilian) orbit will cross the equator, going from south to north, at times 1:30 PM, 5:30 PM, and 9:30 PM sun-synchronous orbits are described in more detail in Chap. 2, but briefly, the orbit is arranged to rotate eastward at a rate of 0.9856°/day, to make it *sun synchronous*. In a sun-synchronous orbit the satellite crosses the same spot on the earth at the same local time each day, so that the same area of the earth can be viewed under approximately the same lighting conditions

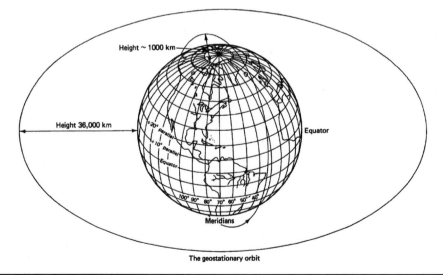

The geostationary orbit

FIGURE 1.6 Geostationary orbit and one possible polar orbit.

each day. A sun-synchronous orbit is inclined slightly to the west of the north pole. An orbital pass from south to north is referred to as an *ascending pass,* and from north to south as a *descending pass.*

The polar orbits are almost circular, and as previously mentioned they are at a height of between 800 and 900 km above earth. The polar orbiters can track weather conditions over the entire earth, and provide a wide range of data, including visible, infrared and millimeter wave radiometer data for imaging purposes, radiation measurements, and temperature and moisture atmospheric profiles. They carry ultraviolet sensors that measure ozone levels, and they can monitor the ozone hole over Antarctica. The polar orbiters carry a NOAA letter designation before launch, which is changed to a numeric designation once the satellite achieves orbit. For example, NOAA M, launched on June 24, 2002, became NOAA 17 when successfully placed in orbit. The series referred to as the *KLM satellites* carry much improved instrumentation. Some details are shown in Table 1.7.

Most of the polar orbiting satellites are used for weather and environmental monitoring and for search and rescue. As the satellite orbits the earth, the sensors view a swath on the earth's surface of about 6000 km (limb to limb) that passes over north and south poles. The orbital period of these satellites is about 102 min. Since a day has 1440 min, the number of orbits per day is 1440/102 or approximately 14. In the 102 min the earth rotates eastward $360° \times 102/1440$ or about 25°. Neglecting for the moment the small eastward rotation of the orbit required for sun synchronicity, the earth will rotate under the subsatellite path by this amount, as illustrated in Fig. 1.7.

Launch date	NOAA-K (NOAA-15): May 13, 1998
	NOAA-L: September 21, 2000
	NOAA-M: June 24, 2000
	NOAA-N: March 19, 2005 (tentative)
	NOAA-N': July 2007
Mission life	2 years minimum
Orbit	Sun synchronous, 833 ± 19 km or 870 ± 19 km
Mass	1478.9 kg on orbit; 2231.7 kg on launch
Length/Diameter	4.18 m/1.88 m
Sensors	Advanced very high-resolution radiometer (AVHRR/3)
	Advanced microwave sounding unit-A (AMSU-A)
	Advanced microwave sounding unit-B (AMSU-B)
	High resolution infrared radiation sounder (HIRS/3)
	Space environment monitor (SEM/2)
	Search and rescue (SAR) repeater and processor
	Data collection system (DCS/2)

TABLE 1.7   NOAA KLM Satellites

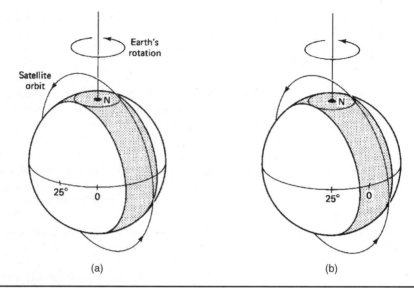

FIGURE 1.7 Polar orbiting satellite: (*a*) first pass; (*b*) second pass, earth having rotated 25°. Satellite period is 102 min.

1.6 Argos System

The Argos *data collection system* (DCS) collects environmental data radioed up from *platform transmitter terminals* (PTT) (Argos, 2005). The characteristics of the PTT are shown in Table 1.8.

The transmitters can be installed on many kinds of platforms, including fixed and drifting buoys, balloons, and animals. The physical size of the transmitters depends on the application. These can weigh as little as 17 g for transmitters fitted to birds, to track their migratory patterns. The PTTs transmit automatically at preset intervals, and those within the 6000 km swath are received by the satellite. As mentioned, the satellite completes about 14 orbits daily, and all orbits cross over the poles. A PTT located at the polar regions would therefore be able to deliver approximately 14 messages daily. At least two satellites are operational at any time, which doubles this number to 28. At the equator

Uplink frequency	401.75 MHz
Message length	Up to 32 bytes
Repetition period	45–200 s
Messages/pass	Varies depending on latitude and type of service
Transmission time	360–920 ms
Duty cycle	Varies
Power	Battery, solar, external

Source: www.argosinc.com/documents/sysdesc.pdf

TABLE 1.8 Platform Transmitter Terminals (PTT) Characteristics

the situation is different. The equatorial radius of the earth is approximately 6378 km, which gives a circumference of about 40,074 km. Relative to the orbital swath coverage, a given longitude at the equator will therefore rotate with the earth by $40{,}074 \times 102/1440$ or about 2839 km. This assumes a stationary orbital path, but as mentioned previously the orbit is sun synchronous, which means that it rotates eastward almost $1°$ per day (see Sec. 2.8.1), that is in the same direction as the earth's rotation. The overall result is that an equatorial PTT starting at the western edge of the footprint swath will "see" between three and four passes per day for one satellite. Hence the equatorial passes number between six and seven per day for two satellites. During any single pass the PTT is in contact with the satellite for 10 min on average. The messages received at the satellite are retransmitted in "real time" to one of several regional ground receiving stations whenever the satellite is within range. The messages are also stored aboard the satellites on tape recorders and are "dumped" to one of three main ground receiving stations. These are located at Wallops Island, VA, USA, Fairbanks, Alaska, USA, and Lannion, France. The Doppler shift in the frequency received at the satellite is used to determine the location of the PTT. This is discussed further in connection with the Cospas-Sarsat search and rescue satellites.

1.7 Cospas-Sarsat

COSPAS is an acronym from the Russian *Cosmicheskaya Sistyema Poiska Avariynich Sudov*, meaning space system for the search of vessels in distress and SARSAT stands for *Search and Rescue Satellite-Aided Tracking* (see http://www.equipped.com/cospas-sarsat_overview.htm). The initial Memorandum of Understanding that led to the development of the system was signed in 1979 by agencies from Canada, France, the USA, and the former USSR. There are (as of April 2022) 45 countries and organizations associated with the program. Canada, France, Russia, and the USA provide and operate the satellites and ground-segment equipment, and other countries provide ground-segment support. A full list of participating countries can be found in Cospas-Sarsat (2022). The system has now been developed to the stage where LEO satellites, geostationary earth orbiting (GEO) and medium-altitude earth orbiting (MEO) satellites are used, as shown in Fig. 1.8.

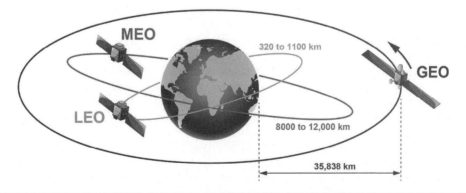

Figure 1.8 Geostationary orbit search and rescue (GEOSAR), medium earth orbit (MEOSAR) and low earth orbit search and rescue (LEOSAR) satellites. (Taken from: https://vita.mil-embedded.com/articles/spacevpx-enabling-the-next-generation-of-satellite-constellations/.)

The basic system requires users to carry distress radio beacons, which transmit a carrier signal when activated. Several different beacons are available: *emergency locator transmitter* (ELT) for aviation use; *emergency position indicating radio beacon* (EPIRB) for maritime use; and *personal locator beacon* (PLB) for personal use. The beacons can be activated manually or automatically (e.g., by a crash sensor). Depending on the transmitted signal power level, the signal is picked up by LEO, MEO, and/or GEO satellites. Because a LEO satellite is moving relative to the radio beacon, a *Doppler shift* in frequency is observed. In effect, if the line-of-sight distance between the transmitter and a satellite is shortened because of the relative motion, the wavelength of the emitted signal is also shortened. This in turn means the received frequency is increased. If the line-of-sight distance is lengthened, because of the relative motion, the wavelength is lengthened and therefore the received frequency decreased. It should be kept in mind that the radio-beacon emits a constant frequency, and the electromagnetic wave travels at constant velocity, that of light. Denoting the constant emitted frequency by f_0, the relative velocity between satellite and beacon, measured along the line of sight as v, and the velocity of light as c; then to a close approximation the received frequency is given by (assuming $v \ll c$):

$$f = \left(1 + \frac{v}{c}\right) f_0 \tag{1.1}$$

The relative velocity v is positive when the line-of-sight distance is decreasing, (satellite and beacon moving closer together) and negative when it is increasing (satellite and beacon moving apart). The relative velocity v is a function of the satellite motion and of the Earth's rotation. The frequency difference resulting from the relative motion is

$$\Delta f = f - f_0 = \frac{v}{c} f_0 \tag{1.2}$$

The fractional change is

$$\frac{\Delta f}{f_0} = \frac{v}{c} \tag{1.3}$$

When v is zero, the received frequency is the same as the transmitted frequency. When the beacon and satellite are approaching each other, v is positive, which results in a positive value of Δf. When the beacon and satellite are receding, v is negative, resulting in a negative value of Δf. The time at which Δf is zero is known as the *time of closest approach*.

Figure 1.9 shows how the beacon frequency, as received at the satellite, varies for different passes. In all cases, the received frequency goes from being higher to being lower than the transmitted value as the satellite approaches and then recedes from the beacon. The longest record and the greatest change in frequency are obtained if the satellite passes over the site, as shown for pass no. 2. This is so because the satellite is visible for the longest period during this pass. Knowing the orbital parameters for the satellite, the beacon frequency, and the Doppler shift for any one pass, the distance of the beacon relative to the projection of the orbit on the earth can be determined. However, whether the beacon is east or west of the orbit cannot be determined easily from a single pass. For two successive passes, the effect of the earth's rotation on the Doppler shift can be estimated more accurately, and from this it can be determined whether the

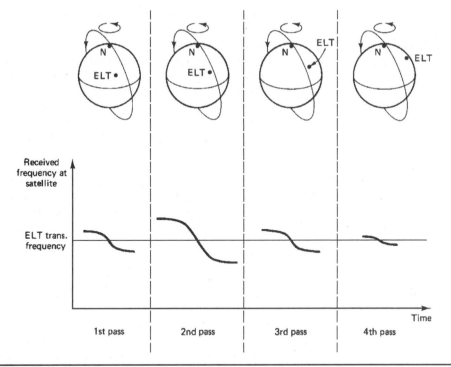

FIGURE 1.9 Showing the Doppler shift in received frequency on successive passes of the satellite. ELT—emergency locator transmitter.

orbital path is moving closer to or moving away from the beacon. In this way, the ambiguity in east-west positioning is resolved. The satellite must of course get the information back to an earth station so that the search and rescue operation can be completed, successfully one hopes. The SARSAT communicates on a downlink frequency of 1544.5 MHz to one of several LEO *local user terminals* (LUTs) established at various locations throughout the world.

In the original Cospas-Sarsat system, the signal from the emergency radio beacons was digital at a frequency of 121.5 MHz. It was found that over 98 percent of the alerts at this frequency were false, often being caused by interfering signals from other services and by inappropriate handling of the equipment. The 121.5-MHz system relied entirely on the Doppler shift, and the carrier does not carry any information about the beacon. The power is low, typically a few tenths of a watt, which limits locational accuracy to about 10 to 20 km. There were no signal storage facilities aboard the satellites for the 121.5-MHz signals, which therefore required that the distress site (the distress beacon) and the LUT must be visible simultaneously from the satellite. Because of these limitations, the 121.5-MHz beacons were phased out, and the 121.5-MHz service was terminated on February 1, 2009. Later satellites, Cospas-14, launched in 2019, and Sarsat-13, launched in 2012, do not carry 121.5-MHz beacons. However, all Cospas-Sarsat satellites launched prior to 2012 carry the 121.5-MHz processors. (Recall that Sarsat-7 is NOAA-15, Sarsat-8 is NOAA-L, Sarsat-9 is NOAA-M, and Sarsat-10 is NOAA-N.)

Newer beacons operating at a frequency of 406 MHz are being introduced. The power has been increased to 5 W, which should permit locational accuracy to 3 to 5 km (Scales and Swanson, 1984). The 406-MHz carrier is modulated with information such as an identifying code, the last known position, and the nature of the emergency. The satellite has the equipment for storing and forwarding the information from a continuous memory dump, providing complete worldwide coverage with 100 percent availability. The polar orbiters, however, do not provide continuous coverage. The mean time between a distress alert being sent and the appropriate search and rescue coordination center being notified is estimated at 27 min satellite storage time plus 44 min waiting time for a total delay of 71 min (Cospas-Sarsat, 1994a, b).

The nominal frequency is 406 MHz, and originally, a frequency of 406.025 MHz was used for operational coded beacons. Because of potential conflict with the GEOSAR system, new channels at 406.028 MHz and 406.037 MHz have been opened, and type approval for the 406.025 MHz channel ceased in January 2002. However, beacon types approved before the January 2001 date and still in production may continue to operate at 406.025 MHz. More details will be found at http://www.cospas-sarsat.org/Beacons/406Bcns.htm.

The status of the Cospas-Sarsat low earth orbiting search and rescue (LEOSAR) satellites as of August 11, 2022, is shown in Table 1.9.

The nominal space segment of LEOSAR consists of five satellites, although as shown in Table 1.9 more satellites may be in service at any one time. The status of the 406-MHz LEOSAR system as of December 2022 consists of repeaters on five polar

| Satellite | Repeater Instruments, MHz | Search and Rescue Processor | | Comments |
	406	Global	Local	
Sarsat 7	F	F	F	Reactivated August 11, 2022–13.00 UTC.
Sarsat 10	F	F	F	
Sarsat 12	F	F	F	
Sarsat 13	F	F	F	
Cospas 14	F	F	F	Subject to a periodic power-cycle reset occurring approximately every 14 days and resulting in a loss of SARP data in memory at the time of the reset. The procedure is scheduled to minimize the time between the latest memory download to a LEOLUT and the power-cycle reset.

Notes: F, fully operational; L, limited operation; NA, not applicable; NO, not operational.
Source: http://www.cospas-sarsat.int/en/system/space-segment-status-pro/current-space-segment-status-and-sar-payloads-pro

TABLE 1.9   Status of LEOSAR Payload Instruments

orbiters, 53 ground receiving stations (referred to as LEOSAR local user terminals, or LEOLUTs), 17 mission control centers (MCCs), and about 2,160,000 beacons operating at 406 MHz, carried mostly on aircraft and small vessels. The operational LEOLUTs provide at least 95 percent of locations accurate to within 5 km, and 98 percent to within 10 km. The MCC receives the messages from the LEOLUTs, processes them based on Cospas-Sarsat requirements, and alerts the *rescue coordination center* (RCC) nearest the location where the distress signal originated, and the RCC takes the appropriate action to effect a rescue.

The status of the GEOSAR segment of the Cospas-Sarsat system is shown in Table 1.10.

Since the geostationary satellites are stationary with respect to the earth, there is no Doppler shift of the received beacon carrier. The 406-MHz beacons for the *geostationary earth orbiting search and rescue* (GEOSAR) component carry position information obtained from the global navigation satellite systems such as the American GPS (see Sec. 17.5) system, the Russian *global navigation satellite system* (GLONASS), Galileo (European), and BeiDou (China).

Satellite	Status	Comments
Electro-L No.2 (14.5°W)	F	
Electro-L No.3 (76°E)	F	
GOES-13 (60°W)	OFF	In-orbit spare
GOES-14 (105°W)	OFF	In-orbit spare
GOES-15 (128°W)	OFF	In-orbit spare
GOES-16 (75.2°W)	F	GOES-East satellite. Downlink frequency 1544.55 MHz.
GOES-17 (137.2°W)	OFF	In-orbit spare
GOES-18 (137.2°W)	IOC	GOES-West satellite. Downlink frequency 1544.55 MHz. IOC since Jaunary 4, 2023
GSAT-17 (93.5°E)	IOC	
INSAT 3D (82°E)	F	
INSAT 3DR (74°E)	F	
Louch-5A (167°E)	F	Moving on an inclined orbit. Commissioned for active-tracking-capable antenna.
Louch-5V (95°E)	IOC	IOC since January 23, 2023.
MSG-2 (45.5°E)	F	
MSG-3 (9.5°E)	F	
MSG-4 (0°)	F	Reactivated August 11, 2022–13.00 UTC. Subject to periodic manoeuvers.

Notes: F, fully operational; GOES, geostationary operational environmental satellite (USA); INSAT, Indian satellite; IOC, initial operational capability; MSG, meteosat second generation (European).
Source: http://www.cospas-sarsat.int/en/system/space-segment-status-pro/current-space-segment-status-and-sar-payloads-pro

TABLE 1.10 Status of GEOSAR Payload Instruments

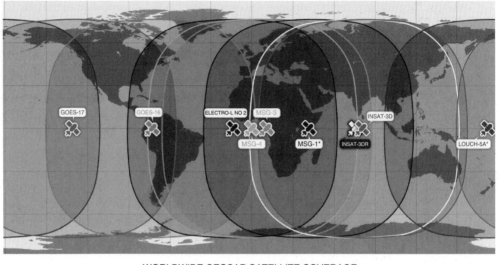

WORLDWIDE GEOSAR SATELLITE COVERAGE

*Indicates satellites moving on an elliptical orbit

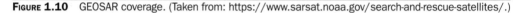

FIGURE 1.10 GEOSAR coverage. (Taken from: https://www.sarsat.noaa.gov/search-and-rescue-satellites/.)

Although the GEOSAR system provides wide area coverage it does not cover the polar regions, the antenna "footprint" being limited to latitudes of about 75° N and S. The coverage areas are shown in Fig. 1.10.

In 2004, the Cospas-Sarsat program initiated the development of MEO satellites for search and rescue missions. Since then, 406-MHz repeaters have been placed on Galileo, Glonass, GPS and, most recently, BDS (BeiDou) satellites. The medium earth orbiting search and rescue (MEOSAR) system complements the existing LEOSAR and GEOSAR systems and is gradually replacing the LEOSAR satellites as the primary architecture for Cospas-Sarsat.

The Cospas-Sarsat MEOSAR satellites carry the SAR instruments, which are radio repeaters that relay 406 MHz beacon signals, without onboard processing, to MEOLUTs for processing beacon identification. The SAR/Galileo, SAR/GLONASS, and SAR/Beidou payloads operate in the 1544–1545 MHz band, while SAR/GPS has a S-band, 2226 MHz, downlink assigned by ITU. The SAR instruments on board of the satellites simply translate the uplink band to the L-band or S-band signal, so that all measurements are made at the MEOLUTs. A single burst location of the transmitter beacon is derived by using time-of-arrival (TOA) and/or frequency-of-arrival (FOA), so that the RMS residuals of the estimated and measured value are minimal. A minimum of three satellites should detect the beacon signal to calculate position. The MEOSAR single-burst location accuracy are specified for three categories: (1) nearly static beacons, whose actual speed is between 0 and 0.5m/s, and the location accuracy shall be within 5 km, (2) slow moving beacons (0.5 m/s–10 m/s) with locations accurate to within 10 km, and (3) location accuracy for fast-moving beacons (speed above 10 m/s), which is still under development by Cospas-Sarsat community.

The status of the Cospas-Sarsat MEOSAR satellites is shown in Table 1.11.

Constellation	Satellites	Status	Comments
Galileo	GSAT201	F	GSAT0104 is not used for the Galileo Open Service but the SARR is active and used in operations.
	GSAT202	F	
	GSAT103	F	
	GSAT104	F	GSAT223 FOC since September 26, 2022.
	GSAT203	F	GSAT224 FOC since September 26, 2022.
	GSAT204	OFF	
	GSAT205	F	
	GSAT206	F	
	GSAT207	F	
	GSAT208	F	
	GSAT209	F	
	GSAT210	F	
	GSAT211	F	
	GSAT212	F	
	GSAT213	F	
	GSAT214	F	
	GSAT215	F	
	GSAT216	F	
	GSAT217	F	
	GSAT218	F	
	GSAT219	F	
	GSAT220	F	
	GSAT221	F	
	GSAT222	F	
	GSAT223	F	
	GSAT224	F	
GLONASS	Kosmos 2471	UT	Kosmos 2471 is not included in the Glonass navigation-satellite constellation and is not transmitting navigational signals nor ephemerides constantly and can be used for detection testing only.
	Kosmos 2501	F	Kosmos 2501 FOC since November 1, 2022.
GPS II DASS	NAVSTAR 51	F	
	NAVSTAR 54	F	
	NAVSTAR 56	F	
	NAVSTAR 57	F	
	NAVSTAR 59	F	
	NAVSTAR 60	F	
	NAVSTAR 61	F	
	NAVSTAR 66	F	

TABLE 1.11   Status of MEOSAR Payload Instruments

Constellation	Satellites	Status	Comments
GPS II DASS (*Continued*)	NAVSTAR 67	F	
	NAVSTAR 68	F	
	NAVSTAR 69	F	
	NAVSTAR 70	F	
	NAVSTAR 71	F	
	NAVSTAR 72	F	
	NAVSTAR 73	F	
	NAVSTAR 74	F	
	NAVSTAR 75	F	
	NAVSTAR 76	F	
GPS III DASS	NAVSTAR 50	F	
	NAVSTAR 77	F	
	NAVSTAR 79	F	
	NAVSTAR 80	F	
BeiDou	BeiDou-3 M13	A	
	BeiDou-3 M14	A	
	BeiDou-3 M21	A	
	BeiDou-3 M22	A	
	BeiDou-3 M23	A	
	BeiDou-3 M24	A	

Notes: A, available for ground segment testing; F, fully operational; OFF, repeater turned off; UT, under test.
Source: http://www.cospas-sarsat.int/en/system/space-segment-status-pro/current-space-segment-status-and-sar-payloads-pro

TABLE 1.11 Status of MEOSAR Payload Instruments (*Continued*)

1.8 Problems

1.1. Describe briefly the main advantages offered by satellite communications. Explain what is meant by a *distance-insensitive communications system.*

1.2. Comparisons are sometimes made between satellite and optical fiber communications systems. State briefly the areas of application for which you feel each system is best suited.

1.3. Briefly describe the development of INTELSAT starting from the 1960s through the present. Information can be found at Web site http://www.intelsat.com/.

1.4. From the Web site http://www.intelsat.com/, find the positions of the INTELSAT 901 and the INTELSAT 10-02 satellites, as well as the number of C-band and Ku-band transponders on each.

1.5. From Table 1.3, and by accessing the Intelsat Web site, determine which satellites provide service to each of the regions AOR, IOR, and POR.

1.6. Referring to Table 1.4, determine the power levels, in watts, for each of the three categories listed.

1.7. From Table 1.5, determine typical orbital spacing in degrees for (*a*) the 6/4-GHz band and (*b*) the 14/12-GHz band.

1.8. Give reasons why the Ku band is used for the DBS service.

1.9. An earth station is situated at longitude 91°W and latitude 45°N. Determine the range to the Galaxy VII satellite. A spherical earth of uniform mass and mean radius 6371 km may be assumed.

1.10. Given that the earth's equatorial radius is 6378 km, and the height of the geostationary orbit is 36,000 km, determine the intersatellite distance between the VisionStar Inc. satellite and the NetSat 28 Company LLC satellite, operating in the Ka band.

1.11. Explain what is meant by a *polar orbiting satellite*. A NOAA polar orbiting satellite completes one revolution around the earth in 102 min. The satellite makes a north to south equatorial crossing at longitude 90°W. Assuming that the orbit is circular and crosses exactly over the poles, estimate the position of the subsatellite point at the following times after the equatorial crossing: (*a*) 0 h, 10 min; (*b*) 1 h, 42 min; (*c*) 2 h, 0 min. A spherical earth of uniform mass may be assumed.

1.12. By accessing the NOAA Web page at http://www.noaa.gov/, find out how the GOES take part in weather forecasting. Give details of the GOES-12 characteristics.

1.13. The Cospas-Sarsat Web site is at http://www.cospas-sarsat.org. Access this site and find out the number and location of the LEOLUTs in current use.

1.14. Using information obtained from the Cospas-Sarsat Web site, find out which satellites carry (*a*) 406-MHz *SAR processors* (SARPs), (*b*) 406-MHz *SAR repeaters* (SARRs), and (*c*) 121.5-MHz SARRs. What is the basic difference between a SARP and a SARR?

1.15. Intelsat satellite 904 is situated at 60°E. Determine the land areas (markets) the satellite can service. The global EIRP is given as 31.0 up to 35.9 dBW, beam edge to beam peak. What are the equivalent values in watts? (See App. G for the definition of dBW.)

1.16. A satellite is in a circular polar orbit at a height of 870 km, the orbital period being approximately 102 min. Assuming an average value of earth's radius of 6371 km determine approximately the maximum period the satellite is visible from a beacon at sea level.

1.17. A satellite is in a circular polar orbit at a height of 870 km, the orbital period being approximately 102 min. The satellite orbit passes directly over a beacon at sea level. Assuming an average value of earth's radius of 6371 km determine approximately the fractional Doppler shift at the instant the satellite is first visible from the beacon.

References

Argos. 2005. General information to info@argosinc.com Customer support to DUS, at www.argosinc.com/documents. From the menu list select sysdesc.pdf

Brown, M. P., Jr. (ed.). 1981. *Compendium of Communication and Broadcast Satellites 1958 to 1980*. IEEE Press, New York.

Cospas-Sarsat, at http://www.cospas-sarsat.org/

Cospas-Sarsat, at http://www.cospas-sarsat.org/Beacons/406Bcns.htm

Cospas-Sarsat, at http://www.cospas-sarsat.org/Status/spaceSegmentStatus.htm

Cospas-Sarsat, at http://www.equipped.com/cospas-sarsat_overview.htm

Cospas-Sarsat. 1994a. *System Data No. 17*, February.

Cospas-Sarsat. 1994b. *Information Bulletin No. 8*, February.

Cospas-Sarsat. 2022. *Information Bulletin No.29*, April, at http://www.cospas-sarsat.org

Evans, B. G., P. T. Thompson, G. E. Corazza, A. Vanelli-Coralli, A., and E. A. Candreva. November 2011. "1945–2010: 65 Years of Satellite History from Early Visions to Latest Missions." In: *Proceedings of the IEEE*, vol. 99, no. 11, pp. 1840–1857. doi: 10.1109/JPROC.2011.2159467.

FCC. 1983. "Licensing of Space Stations in the Domestic Fixed-Satellite Service and Related Revisions of Part 25 of the Rules and Regulations, Report 83-184 33206, CC Docket 81-184." Federal Communications Commission, Washington, DC.

FCC. 1997. "Assignment of Orbital Locations to Space Stations in the Ka-Band." Adopted May 8, 1997, released May 9, at http://www.fcc.gov/Bureaus/International/Orders/1997/da970967.txt.

FCC. 1996. "Orbital Assignment Plan," at http://www.fcc.gov/Bureaus/International/Orders/1996/da960713.txt.

Government of Canada. 1983. "Direct-to-Home Satellite Broadcasting for Canada." *IEEE Spectrum*, March.

Huang, J., and J. Cao. 2020. Recent Development of Commercial Satellite Communications Systems. In: Liang, Q., Wang, W., Mu, J., Liu, X., Na, Z., Chen, B. (eds.), *Artificial Intelligence in China. Lecture Notes in Electrical Engineering*, vol. 572. Springer, Singapore. https://doi.org/10.1007/978-981-15-0187-6_63.

Information Services, Department of Communications. Intelsat, at http://www.intelsat.com/

Kodheli et al. 2021. "Satellite Communications in the New Space Era: A Survey and Future Challenges." In: IEEE Communications Surveys & Tutorials, vol. 23, no. 1, pp. 70–109. doi: 10.1109/COMST.2020.3028247.

Lilly, C. J. 1990. "INTELSAT's New Generation." *IEE Review*, vol. 36, no. 3, March.

NOAA, at http://www.ipo.noaa.gov/.

Pelton, J. N. March 2010. "The Start of Commercial Satellite Communications [History of Communications]." In: *IEEE Communications Magazine*, vol. 48, no. 3, pp. 24–31. doi: 10.1109/MCOM.2010.5434368.

Pritchard, W. L. 1984. "The History and Future of Commercial Satellite Communications." *IEEE Communications Magazine*, vol. 22, no. 5, May, pp. 22–37.

Reinhart, E. E. 1990. "Satellite Broadcasting and Distribution in the United States." *Telecommunication Journal*, vol. 57, no. V1, June, pp. 407–418.

Sachdev, D. K., P. Nadkarni, P. L. Neyret, R. Dest, K. Betaharon, and W. J. English. 1990. "INTELSAT V11: A Flexible Spacecraft for the 1990s and Beyond." *Proceedings of the IEEE*, vol. 78, no. 7, July, pp. 1057–1074.

Scales, W. C., and R. Swanson. 1984. "Air and Sea Rescue via Satellite Systems." *IEEE Spectrum*, March, pp. 48–52.

CHAPTER 2

Orbits and Launching Methods

2.1 Introduction

Satellites (spacecrafts) orbiting the Earth follow the same laws that govern the motion of the planets around the sun. From early times much has been learned about planetary motion through careful observations. Johannes Kepler (1571–1630) was able to derive empirically three laws describing planetary motion. Later, in 1665, Sir Isaac Newton (1642–1727) derived Kepler's laws from his own laws of mechanics and developed the theory of gravitation (for very readable accounts of much of the work of these two great men, see Arons [1965] and Bate et al. [1971]).

Kepler's laws apply quite generally to any two bodies in space which interact through gravitation. The more massive of the two bodies is referred to as the *primary*, the other, the *secondary* or *satellite*.

2.2 Kepler's First Law

Kepler's first law states that the path followed by a satellite around the primary is an ellipse. An ellipse has two focal points, shown as F_1 and F_2 in Fig. 2.1. The center of mass of the two-body system, termed the *barycenter*, is always centered on one of the foci. In our specific case, because of the enormous difference between the masses of the Earth and the satellite, the center of mass coincides with the center of the Earth, which is therefore always at one of the foci.

The *semi-major axis* of the ellipse is denoted by a, and the semi-minor axis by b. The *eccentricity e* is given by

$$e = \frac{\sqrt{a^2 - b^2}}{a} \tag{2.1}$$

The eccentricity and the semi-major axis are two of the orbital parameters specified for satellites orbiting the Earth. For an elliptical orbit, $0 < e < 1$. When $e = 0$, the orbit is circular. The geometrical significance of eccentricity, along with some of the other geometrical properties of the ellipse, is developed in App. B.

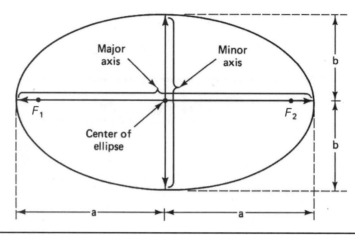

FIGURE 2.1   The foci F_1 and F_2, the semi-major axis a, and the semi-minor axis b of an ellipse.

2.3   Kepler's Second Law

Kepler's second law states that, for equal time intervals, a satellite sweeps out equal areas in its orbital plane, centered at the barycenter. Referring to Fig. 2.2, assuming the satellite travels distances S_1 and S_2 meters in 1 s, then the areas A_1 and A_2 are equal. The average velocity in each case is V_1 and V_2 m/s, and because of the equal area law, it follows that the velocity at S_2 is less than that at S_1. An important consequence of this is that the satellite takes longer to travel a given distance when it is farther away from the Earth. Use is made of this property to increase the length of time a satellite can be seen from geographic regions of the Earth. Also, based upon the second law, the velocity can be derived at each location on the orbit path.

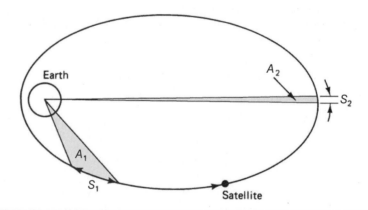

FIGURE 2.2   Kepler's second law: any two areas A_1 and A_2 swept out in unit time are equal.

2.4   Kepler's Third Law

Kepler's third law states that the square of the periodic time (*P*) of the orbit is proportional to the cube of the mean distance between the two bodies, which is equal to the semi-major axis *a*. For the artificial satellites orbiting the Earth, Kepler's third law can be written in the form

$$a^3 = \frac{\mu}{n^2} \tag{2.2}$$

where *n* is the mean angular motion of the satellite in radians per second and μ is the Earth's geocentric gravitational constant. Its value is (see Wertz, 1984, Table L3)

$$\mu = 3.986005 \times 10^{14} \ \text{m}^3/\text{s}^2 \tag{2.3}$$

Equation (2.2) applies only to the ideal situation of a satellite orbiting a perfectly spherical Earth of uniform mass, with no perturbing forces acting, such as atmospheric drag. Later, in Sec. 2.8, the effects of the Earth's oblateness and atmospheric drag are considered.

With *n* in radians per second, the orbital period in seconds is given by

$$P = \frac{2\pi}{n} \tag{2.4}$$

Finally, the familiar form of Kepler's third law becomes

$$P^2 = \left(\frac{4\pi^2}{\mu}\right) a^3 \tag{2.5a}$$

or

$$P = \frac{2\pi}{\sqrt{\mu}} a^{3/2} \ \text{s} \tag{2.5b}$$

The importance of Kepler's third law is that it shows there is a fixed relationship between period and semi-major axis. One very important orbit, known as *geostationary orbit*, is determined by the rotational period of the Earth, as described in Chap. 3. In anticipation of this, the approximate radius of the geostationary orbit is determined in the following example.

Example 2.1   Calculate the radius of a circular orbit for which the period is 1 day.

Solution   There are 86,400 s in 1 day, and therefore the mean motion is

$$n = \frac{2\pi}{86400} = 7.272 \times 10^{-5} \ \text{rad/s}$$

From Kepler's third law:

$$a = \left[\frac{3.986005 \times 10^{14}}{(7.272 \times 10^{-5})^2}\right]^{1/3} = 42{,}241 \ \text{km}$$

Since the orbit is circular the semi-major axis is also the radius.

2.5   Definitions of Terms for Earth-Orbiting Satellites

As mentioned previously, Kepler's laws apply in general to satellite motion around a primary body. For the case of Earth-orbiting satellites, certain terms are used to describe the position of the orbit with respect to the Earth.

Subsatellite path. This is the path traced out on the Earth's surface directly below the satellite.

Orbital radius. This is the distance between the centers of mass of the Earth and the satellite.

Apogee. This is the point farthest from Earth (maximum orbit radius, r_a). Apogee height, measured from the surface of the Earth, is shown as h_a in Fig. 2.3.

Perigee. This is the point of closest approach to Earth (minimum orbit radius, r_p). The perigee height, measured from the surface of the Earth, is shown as h_p in Fig. 2.3.

Eccentricity. This is a measure of the ellipticity of the orbit that ranges between 0 (circular) to 1 (highly elliptical). It is calculated using Eq. (2.6) in terms of the apogee and perigee radial distances

$$e = \frac{r_a - r_p}{r_a + r_p} \tag{2.6}$$

Line of apsides. This is the line joining the perigee and apogee through the center of the Earth.

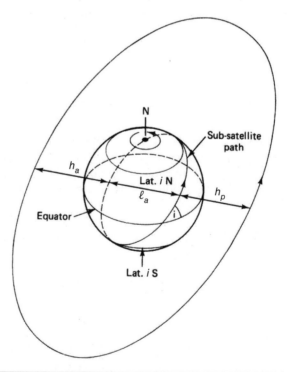

Figure 2.3   Apogee height h_a, perigee height h_p, and inclination i. l_a is the line of apsides.

Ascending node. This is the point where the orbit *subsatellite path* crosses the equatorial plane moving from south to north.

Descending node. This is the point where the orbit *subsatellite path* crosses the equatorial plane moving from north to south.

Line of nodes. This is the line joining the ascending and descending nodes through the center of the Earth.

Inclination. This is the angle between the orbital plane and the Earth's equatorial plane shown as i in Fig. 2.3. It is measured at the ascending node from the equator to the orbit plane, going from east to north.

Prograde orbit. This is an orbit in which the satellite moves in the same direction as the Earth's rotation, as shown in Fig. 2.4. The inclination of a prograde orbit always lies between 0° and 90°.

Retrograde orbit. This is an orbit in which the satellite moves in a direction counter to the Earth's rotation, as shown in Fig. 2.4. The inclination of a retrograde orbit always lies between 90° and 180°.

Argument of perigee. This is the angle from ascending node to perigee, measured in the orbital plane at the Earth's center, in the direction of satellite motion. The argument of perigee is shown as ω in Fig. 2.5.

Right ascension of the ascending node. This is the angle measured eastward, in the equatorial plane, from the first point of Aries (symbol ♈) to the ascending node, shown as Ω in Fig. 2.5. The position of the ascending node is specified to define the position of the orbit in inertial space (with reference to the fixed stars).

Mean anomaly. This is the average value of the angular position of the satellite with reference to the perigee. For a circular orbit, M gives the angular position of the satellite

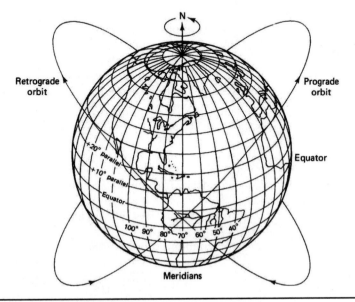

Figure 2.4 Prograde and retrograde orbits.

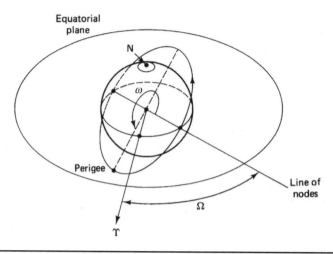

Equatorial
plane

N

ω

Perigee

Line of
nodes

Ω

Υ

FIGURE 2.5   The argument of perigee ω and the right ascension of the ascending node Ω, where the first point of Aries is the symbol Υ.

in the orbit. For elliptical orbit, the position is much more difficult to calculate, and M is used as an intermediate step in the calculation as described in Sec. 2.9.5.

True anomaly. This is the angle from the perigee to the satellite position, measured at the Earth's center. This gives the true angular position of the satellite in the orbit as a function of time. A method of determining the true anomaly is described in Sec. 2.9.5.

2.5.1   Satellite Velocity

Occasionally, it is necessary to know the satellite instantaneous velocity on the orbital path, which can be derived from Keppler's second and third laws. Consider first a satellite in a circular orbit (an ellipse with eccentricity = 0). In a circular orbit, the velocity magnitude is constant, and the direction is always perpendicular to the orbit radius, as shown in Fig. 2.6a. Since the velocity magnitude is constant, its value is the ratio of the orbit length (circle circumference) to the orbit period calculated by Kepler's third law as

$$V_{\text{circ}} = \sqrt{\frac{\mu}{r_o}} \qquad (2.7)$$

where is the Earth's gravitational constant Eq. (2.3) and r_o is the orbit radial distance.

For an elliptical orbit, the velocity (magnitude and direction) varies with orbit position (as shown in Fig. 2.6b); however, at the apogee and the perigee the velocity direction is perpendicular to the orbit radius. Thus, at these locations the velocity magnitude can be easily calculated.

The minimum velocity magnitude V_a occurs at the orbit apogee

$$V_a = \sqrt{\frac{\mu(1-e)}{a(1+e)}} \text{ m/s} \qquad (2.8)$$

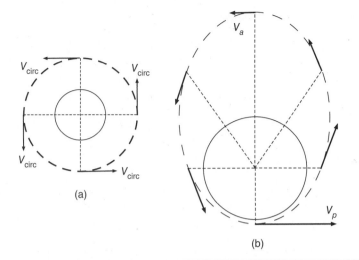

FIGURE 2.6   Satellite orbital velocity for (a) circular orbit and (b) elliptical orbit.

where a is the orbit semi-major axis

$$a = \frac{(r_a + r_p)}{2} \text{ m}$$ (2.9)

and e is the orbit eccentricity, Eq. (2.1).

The maximum velocity magnitude V_p occurs at the orbit perigee

$$V_p = \sqrt{\frac{\mu(1+e)}{a(1-e)}} \text{ m/s}$$ (2.10)

Example 2.2   Calculate the satellite velocity magnitude for circular orbit at 450 km altitude, given $\mu = 3.986005 \times 10^{14}$ m³/s².

Solution   First, calculate the orbit radius $r_o = h + R = 450 + 6378.14 = 6,828.14$ km
The velocity magnitude, from Eq. (2.7), is

$$V_{\text{circ}} = \sqrt{\frac{\mu}{r_o}} = \sqrt{\frac{3.986005 \times 10^{14}}{6,828.14 \times 1000}} = 7.64 \text{ km/s}$$

Example 2.3   Calculate the satellite velocity magnitude at apogee and perigee for an elliptical orbit at 450 km × 35,786.1 km altitude, given $\mu = 3.986005 \times 10^{14}$ m³/s².

Solution   First, calculate the perigee orbit radius:

$$r_p = h + R = 450 + 6378.14 = 6,828.14 \text{ km}$$

and at apogee the orbit radius:

$$r_a = 35,786.1 + 6378.14 = 42,164.2 \text{ km}$$

from Eq. (2.9), the semi-major axis is

$$a = \frac{(r_a + r_p)}{2} = \frac{(42,164.2 + 6,828.14)}{2} = 24,496.2 \text{ km}$$

From Eq. (2.6), calculate the orbit eccentricity

$$e = \frac{r_a - r_p}{r_a + r_p} = \frac{(42,164.2 - 6,828.14)}{42,164.2 + 6,828.14} = 0.7213$$

From Eq. (2.8), at apogee

$$V_a = \sqrt{\frac{\mu(1-e)}{a(1+e)}} = \sqrt{\frac{3.986005 \times 10^{14}(1-0.7213)}{24,496.2 \times 10^3 \times (1+0.7213)}} = 1.623 \text{ km/s}$$

From Eq. (2.10), at perigee

$$V_p = \sqrt{\frac{\mu(1+e)}{a(1-e)}} = \sqrt{\frac{3.986005 \times 10^{14}(1+0.7213)}{24,496.2 \times 10^3 \times (1-0.7213)}} = 10.024 \text{ km/s}$$

Example 2.4   Calculate the satellite velocity magnitude for circular geostationary orbit at 35,786.063 km altitude, given $\mu = 3.986005 \times 10^{14}$ m³/s².

Solution   First, calculate the orbit radius

$$r_0 = h + R = 35,786.06 \text{ km} + 6378.14 = 42,164.20 \text{ km}$$

The velocity magnitude from Eq. (2.7)

$$V_{\text{circ}} = \sqrt{\frac{\mu}{r_0}} = \sqrt{\frac{3.986005 \times 10^{14}}{42,164.24 \times 1000}} = 3.075 \text{ km/s}$$

2.6   Orbital Elements

Earth-orbiting artificial satellites are defined by six orbital elements referred to as the *Keplerian element set*. Two of these, the semi-major axis a and the eccentricity e, described in Sec. 2.2, give the shape of the ellipse. A third, the mean anomaly M_0, gives the position of the satellite in its orbit at a reference time known as the *epoch*. A fourth, the argument of perigee ω, gives the rotation of the orbit's perigee point relative to the orbit's line of nodes in the Earth's equatorial plane. The remaining two elements, the inclination i and the right ascension of the ascending node Ω, relate the orbital plane's position to the Earth. These four elements are described in Sec. 2.5.

Because the equatorial bulge causes slow variations in ω and Ω, and because other perturbing forces may alter the orbital elements slightly, the values are specified for the reference time or epoch, and thus the epoch also must be specified.

Appendix C describes the commonly used two-line element orbit descriptions. The two-line elements may be downloaded from Celestrak at http://celestrak.com/NORAD/elements/. Figure 2.7 shows how to interpret the NASA two-line elements.

The two-line elements do not include the semi-major axis, but this can be calculated from the data given. An example calculation is presented below.

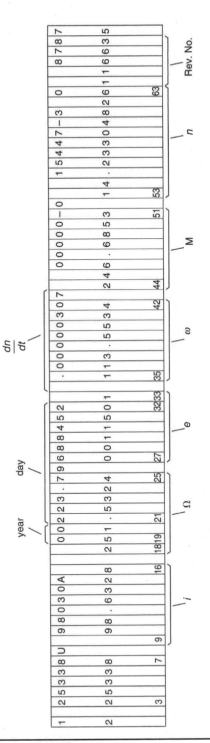

Figure 2.7 Two-line elements for NOAA-15.

Value	Description
25338	*Satellite number*
00	*Epoch year* (last two digits of the year)
223.79688452	*Epoch day* (day and fractional day of the year) (see Sec. 2.9.2)
0.00000307	*First-time* derivative of the mean motion (rev/day^2)
98.6328	*Inclination* (degrees)
251.5324	*Right ascension of the ascending node* (degrees)
0011501	*Eccentricity* (leading decimal point assumed)
113.5534	*Argument of perigee* (degrees)
246.6853	*Mean anomaly* (degrees)
14.23304826	*Mean motion* (rev/day)
11,663	*Revolution number at epoch* (rev)

TABLE 2.1 Example Parameters from a Two-Line Element (see App. C)

Example 2.5 Calculate the semi-major axis for the satellite parameters given in Table 2.1.

Solution The mean motion is given in Table 2.1 as NN = 14.23304826 day^{-1}. In rad/s this is

$$n_0 = 2 \times \pi \times \text{NN} = 0.00104 \text{ s}^{-1}$$

Kepler's third law gives

$$a = \left[\frac{\mu}{n_0^2} \right]^{1/3} = 7192.335 \text{ km}$$

2.7 Apogee and Perigee Heights

Also not specified as orbital elements, the apogee height and perigee height are often required. As shown in App. B, the length of the radius vectors at apogee and perigee can be obtained from the geometry of the ellipse:

$$r_a = a(1+e) \qquad (2.11)$$

$$r_p = a(1-e) \qquad (2.12)$$

To find the apogee and perigee heights, the radius of the Earth must be subtracted from the radii lengths, as shown in the following example.

Example 2.6 Calculate the apogee and perigee heights for the orbital parameters given in Table 2.1. Assume a mean Earth radius of 6378.137 km.

Solution From Table 2.1, $e = .0011501$ and from Example 2.1, $a = 7192.335$ km. Using Eqs. (2.11) and (2.12):

$$r_a = 7192.335(1 + 0.0011501) = 7200.607 \text{ km}$$

$$r_p = 7192.335(1 - 0.0011501) = 7184.063 \text{ km}$$

The corresponding heights (orbit altitudes) are:

$$h_a = r_a - R = 829.6 \text{ km}$$

$$h_p = r_p - R = 813.1 \text{ km}$$

2.8   Orbit Perturbations

The *Keplerian orbit* described so far is ideal in the sense that it assumes that the Earth is a uniform spherical mass and that the only force acting is the centrifugal force resulting from satellite motion balancing the gravitational pull of the Earth. In practice, other forces, which can be significant, are the gravitational forces of the sun and the moon and atmospheric drag. The gravitational pulls of sun and moon have negligible effect on low-orbiting satellites, but they do affect satellites in the geostationary orbit as described in Sec. 3.5. Atmospheric drag, on the other hand, has negligible effect on geostationary satellites but does affect low-orbiting Earth satellites below about 1000 km.

2.8.1   Effects of a Nonspherical Earth

For a spherical Earth of uniform mass, Kepler's third law (Eq. 2.2) gives the nominal mean motion n_0 as

$$n_0 = \sqrt{\frac{\mu}{a^3}} \tag{2.13}$$

The 0 subscript is included as a reminder that this result applies for a perfectly spherical Earth of uniform mass. However, it is known that the Earth is not perfectly spherical, there being an equatorial bulge and a flattening at the poles, a shape described as an *oblate spheroid*. When the Earth's oblateness is considered, the mean motion, denoted in this case by symbol n, is modified to (Wertz, 1984).

$$n = n_0 \left[1 + \frac{K_1(1 - 1.5 \sin^2 i)}{a^2(1 - e^2)^{1.5}} \right] \tag{2.14}$$

K_1 is a constant which evaluates to 66,063.1704, km². The Earth's oblateness has negligible effect on the semi-major axis a, and if a is known, the mean motion is readily calculated. The orbital period considering the Earth's oblateness is termed the *anomalistic period* (e.g., from perigee to perigee). The mean motion specified in the NASA bulletins is the reciprocal of the anomalistic period. The anomalistic period is

$$P_A = \frac{2\pi}{n} \text{ s} \tag{2.15}$$

where n is in radians per second.

If the known quantity is n (e.g., as is given in the NASA bulletins), one can solve Eq. (2.8) for a, keeping in mind that n_0 is also a function of a. Equation (2.8) may be solved for a by finding the root of the following equation:

$$n - \sqrt{\frac{\mu}{a^3}} \left[1 + \frac{K_1(1 - 1.5 \sin^2 i)}{a^2(1 - e^2)^{1.5}} \right] = 0 \tag{2.16}$$

This is illustrated in the following example.

Example 2.7   A satellite is orbiting in the equatorial plane with a period from perigee to perigee of 12 h. Given that the eccentricity is 0.002, calculate the semi-major axis. The Earth's average equatorial radius is 6378.137 km.

Solution   Given data: $e = 0.002$; $i = 0°$; $P = 12$ h; $K_1 = 66063.1704$ km²; $a_E = 6378.137$ km; $\mu = 3.986005 \times 1014$ m³/s²

The mean motion is:

$$n = \frac{2\pi}{P} = 1.454 \times 10^{-4} \text{ s}^{-1}$$

Assuming this is the same as n_0, Kepler's third law gives

$$a = \left(\frac{\mu}{n^2}\right)^{1/3} = 26{,}610 \text{ km}$$

Solving the root equation yields a value of 26,612 km.

The oblateness of the Earth also produces two rotations of the orbital plane. The first of these, known as *regression of the nodes,* is where the nodes appear to slide along the equator. In effect, the line of nodes, which is in the equatorial plane, rotates about the center of the Earth. Thus Ω, the right ascension of the ascending node, shifts its position.

If the orbit is prograde (see Fig. 2.4), the nodes slide westward, and if retrograde, they slide eastward. As seen from the ascending node, a satellite in prograde orbit moves eastward, and in a retrograde orbit, westward. The nodes therefore move in a direction opposite to the direction of satellite motion, hence the term *regression of the nodes.* For a polar orbit ($i = 90°$), the regression is zero.

The second effect is rotation of apsides in the orbital plane, described below. Both effects depend on the mean motion n, the semi-major axis a, and the eccentricity e. These factors can be grouped into one factor K given by

$$K = \frac{nK_1}{a^2(1-e^2)^2} \tag{2.17}$$

K has the same units as n. Thus, with n in rad/day, K is in rad/day, and with n in degrees/day, K is in degrees/day. An approximate expression for the rate of change of Ω with respect to time is (Wertz, 1984)

$$\frac{d\Omega}{dt} = -K \cos i \tag{2.18}$$

where i is the inclination. The rate of regression of the nodes has the same units as n.

When the rate of change given by Eq. (2.18) is negative, the regression is westward, and when the rate is positive, the regression is eastward. Therefore, eastward regression, i must be greater than 90°, or the orbit must be retrograde. It is possible to choose values of a, e, and i such that the rate of rotation is 0.9856°/day eastward. Such an orbit is said to be *sun synchronous* and is described further in Sec. 2.10.

The other major effect produced by the equatorial bulge is a rotation of the line of apsides. This line rotates in the orbital plane, resulting in the argument of perigee changing with time. The rate of change is given by (Wertz, 1984)

$$\frac{d\omega}{dt} = K(2 - 2.5 \sin^2 i) \tag{2.19}$$

Again, the units for the rate of rotation of the line of apsides are the same as those for n (incorporated in K). When the inclination i is equal to $63.435°$, the term within the parentheses is equal to zero, and hence no rotation takes place. Use is made of this fact in the orbit chosen for the Russian Molniya satellites (see Probs. 2.23 and 2.24).

Denoting the epoch time by t_0, the right ascension of the ascending node by Ω_0, and the argument of perigee by ω_0 at epoch gives the new values for Ω and ω at time t as

$$\Omega = \Omega_0 + \frac{d\Omega}{dt}(t-t_0) \tag{2.20}$$

$$\omega = \omega_0 + \frac{d\omega}{dt}(t-t_0) \tag{2.21}$$

Keep in mind that the orbit is not a physical entity. Rather, it is the result of forces from an oblate Earth, which act on the satellite to produce the changes in the orbital parameters. Thus, rather than follow a closed elliptical path in a fixed plane, the satellite drifts because of the regression of the nodes, and the latitude of the point of closest approach (the perigee) changes because of the rotation of the line of apsides. It is permissible to visualize the satellite as following a closed elliptical orbit but with the orbit itself moving relative to the Earth because of the changes in Ω and ω. Thus, as stated earlier, the period P_A is the time required to go around the orbital path from perigee to perigee, even though the perigee has moved relative to the Earth.

Suppose, for example, that the inclination is $90°$ so that the regression of the nodes is zero (from Eq. 2.18), and the rate of rotation of the line of apsides is $-K/2$ (from Eq. 2.19), and further, imagine the situation where the perigee at the start of observations is exactly over the ascending node. One period later, the perigee is at an angle $-KP_A/2$ relative to the ascending node or, in other words, is south of the equator. The time between crossings at the *ascending node* is P_A $(1 + K/2n)$, which is the period observed from the Earth. Recall that K has the same units as n, for example, rad/s.

Example 2.8 Determine the rate of regression of the nodes and the rate of rotation of the line of apsides for the satellite parameters specified in Table 2.1. The value for a obtained in Example 2.2 may be used.

Solution From Table 2.1 and Example 2.2 $i = 98.6328°$; $e = 0.0011501$; NN = 14.23304826 day^{-1}; $a = 7192.335$ km, and the known constant: $K_1 = 66063.1704$ km^2

Converting n to rad/s:

$$n = 2\pi NN = 0.00104 \text{ rad/s}$$

From Eq. (2.17):

$$K = \frac{nK_1}{a^2(1-e^2)^2} = 6.544 \text{ deg/day}$$

From Eq. (2.18):

$$\frac{d\Omega}{dt} = -K\cos i = 0.981 \text{ deg/day}$$

From Eq. (2.19):

$$\frac{d\omega}{dt} = -K(2 - 2.5 \sin^2 i) = -2.904 \text{ deg/day}$$

Example 2.9   For the satellite in Example 2.8, calculate the new values for Ω and ω one period after epoch.

Solution   From Table 2.1:

$$NN = 14.23304826 \text{ day}^{-1}; \ \omega_0 = 113.5534°; \ \Omega_o = 251.5324°$$

The anomalistic period is

$$P_A = \frac{1}{NN} = 0.070259 \text{ day}$$

This is also the time difference $(t - t_0)$ since the satellite has completed one revolution from perigee to perigee. Hence:

$$\Omega = \Omega_0 + \frac{d\Omega}{dt}(t - t_0) = 251.5324 + 0.981(0.070259) = 251.601°$$

$$\omega = \omega_0 + \frac{d\omega}{dt}(t - t_0) = 113.5534 + (-2.903)(0.070259) = 113.349°$$

In addition to the equatorial bulge, the Earth is not perfectly circular in the equatorial plane; it has a small eccentricity of the order of 10^{-5}. This is referred to as the *equatorial ellipticity*. The effect of the equatorial ellipticity is to set up a gravity gradient, which has a pronounced effect on satellites in geostationary orbit (Sec. 7.4). Very briefly, a satellite in geostationary orbit ideally should remain fixed relative to the Earth. However, the gravity gradient resulting from the equatorial ellipticity causes the satellites in geostationary orbit to drift to one of two stable points, which coincide with the minor axis of the equatorial ellipse. These two points are separated by 180° on the equator and are at approximately 75° E longitude and 105° W longitude. Satellites in service are prevented from drifting to these points through station-keeping maneuvers, described in Sec. 7.4. Because old, out-of-service satellites eventually do drift to these points, they are referred to as "satellite graveyards." It may be noted that the effect of equatorial ellipticity is negligible on most other satellite orbits.

2.8.2   Atmospheric Drag

For near-Earth satellites, those below about 1000 km, the effects of atmospheric drag are significant. Because the drag is greatest at the perigee, the drag acts to reduce the velocity at this point, with the result that the satellite does not reach the same apogee height on successive revolutions.

The result is that the semi-major axis and the eccentricity are both reduced. Drag does not noticeably change the other orbital parameters, including perigee height. In the program used for generating the orbital elements given in the NASA bulletins, a pseudo-drag term is included, which is equal to one-half the rate of change of mean motion (ADC USAF, 1980). An approximate expression for the change of major axis is

$$a \cong a_0 \left[\frac{n_0}{n_0 + n_0'(t - t_0)} \right]^{2/3} \tag{2.22}$$

where the "$_0$" subscripts denote values at the reference time t_0, and n_0' is the first derivative of the mean motion. The mean anomaly is also changed, with an approximate value for the change being:

$$\delta M = \frac{n_0'}{2}(t-t_0)^2 \qquad (2.23)$$

From Table 2.1 it is seen that the first-time derivative of the mean motion is listed in columns 34 to 43 of line 1 of the NASA bulletin. For the example shown in Fig. 2.6, the first-time derivative of the mean motion is 0.00000307 rev/day^2. Thus, the changes resulting from the drag term are significant only for long time intervals, and for present purposes are ignored. For a more accurate analysis, suitable for long-term predictions, the reader is referred to ADC USAF (1980).

2.9   Inclined Orbits

A study of the general situation of a satellite in an inclined elliptical orbit is complicated by the fact that different parameters relate to different reference frames. The orbital elements are known with reference to the plane of the orbit, the position of which is fixed (or slowly varying) in space, while the location of the Earth station is usually given in terms of the local geographic coordinates which rotate with the Earth. Rectangular coordinate systems are generally used in calculations of satellite position and velocity in space, while the Earth station quantities of interest may be the azimuth and elevation angles and range. Transformations between coordinate systems are therefore required.

Here, to illustrate the method of calculation for elliptical inclined orbits, the problem of finding the Earth station look angles and range is considered. It should be kept in mind that with inclined orbits the satellites are not geostationary, and therefore, the required look angles and range change with time. Detailed and very readable treatments of orbital properties in general can be found, for example, in Bate et al. (1971) and Wertz (1984). Much of the explanation and the notation in this section is based on these two references.

Determination of the look angles and range involves the following quantities and concepts:

1. The *orbital elements*, as published in the NASA bulletins and described in App. C

2. Various measures of *time*

3. The perifocal *coordinate system*, which is based on the orbital plane

4. The *geocentric-equatorial coordinate system*, which is based on the Earth's equatorial plane

5. The *topocentric-horizon coordinate system*, which is based on the observer's horizon plane

The two major coordinate transformations needed are:

- The satellite position measured in the perifocal system is transformed to the geocentric-horizon system in which the Earth's rotation is measured, thus enabling the satellite position and the Earth station location to be coordinated.

- The satellite-to-Earth station position vector is transformed to the topocentric-horizon system, which enables the look angles and range to be calculated.

2.9.1 Calendars

A calendar is a time-keeping device in which the year is divided into months, weeks, and days. Calendar days are units of time based on the Earth's motion relative to the sun. Of course, it is more convenient to think of the sun moving relative to the Earth. This motion is not uniform, and so a fictitious sun, termed the *mean sun*, is introduced.

The mean sun does move at a uniform speed but otherwise requires the same time as the real sun to complete one orbit of the Earth, this time being the *tropical year*. A day measured relative to this mean sun is termed a *mean solar day*. Calendar days are mean solar days, and generally they are just referred to as days.

A tropical year contains 365.2422 days. To make the calendar year, also referred to as the *civil year*, more easily usable, it is normally divided into 365 days. The extra 0.2422 of a day is significant, and for example, after 100 years, there is a discrepancy of 24 days between the calendar year and the tropical year. Julius Caesar made the first attempt to correct the discrepancy by introducing the *leap year*, in which an extra day is added to February whenever the year number is divisible by 4. This gave the *Julian calendar*, in which the civil year was 365.25 days on average, a reasonable approximation to the tropical year.

By the year 1582, an appreciable discrepancy once again existed between the civil and tropical years. Pope Gregory XIII took matters in hand by abolishing the days October 5 through October 14, 1582, to bring the civil and tropical years into line and by placing an additional constraint on the leap year in that years ending in two zeros must be divisible by 400 without remainder to be reckoned as leap years. This dodge was used to miss out 3 days every 400 years. To see this, let the year be written as $X00$ where X stands for the hundreds. For example, for 1900, $X = 19$. For $X00$ to be divisible by 400, X must be divisible by 4. Now a succession of 400 years can be written as $X + (n - 1)$, $X + n$, $X + (n + 1)$, and $X + (n + 2)$, where n is any integer from 0 to 9. If $X + n$ is evenly divisible by 4, then the adjoining three numbers are not, since some fraction from $-1/4$ to $2/4$ remains, so these three years must be omitted. The resulting calendar is the *Gregorian calendar*, which is the one in use today.

Example 2.10   Calculate the average length of the civil year in the Gregorian calendar.

Solution   The nominal number of days in a 400-year period is $400 \times 365 = 146,000$.
The nominal number of leap years is $400/4 = 100$, but as shown earlier, this must be reduced by 3, and therefore, the number of days in 400 years of the Gregorian calendar is $146,000 + 100 - 3 = 146,097$. This gives a yearly average of $146,097/400 = 365.2425$.

In calculations requiring satellite predictions, it is necessary to determine whether a year is a leap year or not, and the simple rule is: If the year number ends in two zeros and is divisible by 400 without remainder, it is a leap year. Otherwise, if the year number is divisible by 4 without remainder, it is a leap year.

Example 2.11   Determine which of the following years are leap years: (*a*) 1987, (*b*) 1988, (*c*) 2000, (*d*) 2100.

Solution
(*a*) $1987/4 = 496.75$ (therefore, 1987 is not a leap year)
(*b*) $1988/4 = 497$ (therefore, 1988 is a leap year)
(*c*) $2000/400 = 5$ (therefore, 2000 is a leap year)
(*d*) $2100/400 = 5.25$ (therefore, 2100 is not a leap year, even though 2100 is divisible by 4 without remainder)

2.9.2 Universal Time

Universal time coordinated (UTC) is the time used for all civil time–keeping purposes, and it is the time reference which is broadcast by the National Bureau of Standards as a standard for setting clocks. It is based on an atomic time-frequency standard. The fundamental unit for UTC is the mean solar day (see App. J in Wertz, 1984). In terms of "clock time," the mean solar day is divided into 24 h, an hour into 60 min, and a minute into 60 s. Thus, there are 86,400 "clock seconds" in a mean solar day. Satellite-orbit epoch time is given in terms of UTC.

Example 2.12   Calculate the time in days, hours, minutes, and seconds for the epoch day 324.95616765.

Solution   This represents the 324th day of the year plus 0.95616765 mean solar day. The decimal fraction in hours is $24 \times 0.95616765 = 22.9480236$; the decimal fraction of this expressed in minutes is $0.9480236 \times 60 = 56.881416$; the decimal fraction of this expressed in seconds is $0.881416 \times 60 = 52.88496$. Thus, the epoch is day 324, at 22 h, 58 m, 52.88 s.

Universal time coordinated is equivalent to *Greenwich mean time* (GMT), as well as *Zulu* (Z) time. There are several other "universal time" systems, all interrelated (Wertz, 1984) and all with the mean solar day as the fundamental unit. For present purposes, the distinction between these systems is not critical, and the term *universal time* (UT) is used in this text.

For computations, UT may be required in two forms: (1) as a fraction of a day and (2) in degrees. Given UT in the normal form of hours, minutes, and seconds, it is converted to fractional days as

$$UT_{day} = \frac{1}{24} \left(hours + \frac{minutes}{60} + \frac{seconds}{3600} \right) \tag{2.24}$$

In turn, this may be converted to degrees as

$$UT° = 360° \times UTday \tag{2.25}$$

2.9.3 Julian Dates

Calendar times are expressed in UT, and although the time interval between any two events may be measured as the difference in their calendar times, the calendar time notation is not suited to computations where the timing of many events must be computed. What is required is a reference time to which all events can be related in decimal days. Such a reference time is provided by the Julian zero-time reference, which is 12 noon (12:00 UT) on January 1 in the year 4713 b.c.! Of course, this date is a hypothetical starting point, which can be established by counting backward according to a certain formula. For details of this intriguing time reference, see Wertz (1984, p. 20). The important point is that ordinary calendar times are easily converted to Julian dates, measured on a continuous time scale of Julian days. To do this, first determine the day of the year, keeping in mind that day zero, denoted as Jan 0.0 is midnight between December 30 and 31 of the previous year. For example, noon on December 31 is January 0.5, and noon on January 1 is January 1.5. It may seem strange that the last day of December can be denoted as "day zero in January," but this makes the day count correspond to the actual calendar day.

Code for calculating the Julian day for any date and time is given in Wertz (1984, p. 20), and a general method is given in Duffett-Smith (1986, p. 9). Once the Julian day is known for a given reference date and time, the Julian day for any other time can be easily calculated by adding or subtracting the required day difference. Some "reference times" are listed in Table 2.2.

For convenience in calculations the day number of the year is given in Table 2.3.

January 0.0	Julian Day	January 0.0	Julian Day	January 0.0	Julian Day
1999	2451178.5	2010	2455196.5	2021	2459214.5
2000	2451543.5	2011	2455561.5	2022	2459579.5
2001	2451909.5	2012	2455926.5	2023	2459944.5
2002	2452274.5	2013	2456292.5	2024	2460309.5
2003	2452639.5	2014	2456657.5	2025	2460675.5
2004	2453004.5	2015	2457022.5	2026	2461040.5
2005	2453370.5	2016	2457387.5	2027	2461405.5
2006	2453735.5	2017	2457753.5	2028	2461770.5
2007	2454100.5	2018	2458118.5	2029	2462136.5
2008	2454465.5	2019	2458483.5	2030	2462501.5
2009	2454831.5	2020	2458848.5	2031	2462866.5

TABLE **2.2**   Some Reference Julian Dates

Date	Day Number for Start of Day (Midnight) (Numbers in Parentheses are for Leap Years)
January 31	31
February 28 (29)	59 (60)
March 31	90 (91)
April 30	120.5 (121.5)
May 31	151 (152)
June 30	181 (182)
July 31	212 (213)
August 31	243 (244)
September 30	273 (274)
October 31	304 (305)
November 30	334 (335)
December 31	365 (366)

TABLE **2.3**   Day Number for the Last Day of the Month

Example 2.13   Find the Julian day for 13 h UT on December 18, 2000.

Solution   The year 2000 is a leap year, and from Table 2.3, December 18 is day number $335 + 18 = 353$. This is for midnight December 17/18. UT = 13 h as a fraction of a day is $13/24 = 0.5416667$. From Table 2.2, the Julian date for January 0.0, 2000, is 2451543.5, and therefore the required Julian date is $2451543.5 + 353 + 0.5416667 = 2451897.0417$.

In Sec. 2.9.7, certain calculations require a time interval measured in *Julian centuries*, where a Julian century consists of 36,525 mean solar days. The time interval is reckoned from a reference time of January 0.5, 1900, which corresponds to 2,415,020 Julian days.

Denoting the reference time as JD_{ref}, the Julian century by JC, and the time in question by JD, then the interval in Julian centuries from the reference time to the time in question is given by

$$T = \frac{JD - JD_{ref}}{JC} \tag{2.26}$$

This is illustrated in the following example.

Example 2.14   Find the time in Julian centuries from the reference time January 0.5, 1900, to 13 h UT on December 18, 2000.

Solution JD_{ref} = 2415020 days; JC = 36525 days. From Example 2.10: JD = 2451897.0417 days. Equation (2.20) gives

$$T = \frac{2451897.0417 - 2415020}{36525} = 1.00963838$$

Note that the time units are days and T is dimensionless.

2.9.4   Sidereal Time

Sidereal time is time measured relative to the fixed stars (Fig. 2.8). Note that one complete rotation of the Earth relative to the fixed stars is not a complete rotation relative to the sun. This is because the Earth moves in its orbit around the sun.

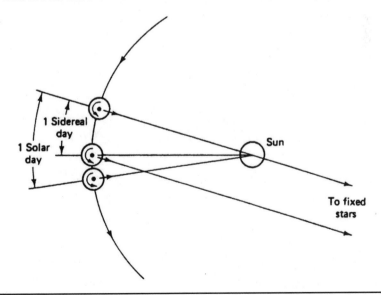

FIGURE 2.8   A sidereal day, or one rotation of the Earth relative to fixed stars, is shorter than a solar day.

The *sidereal day* is defined as one complete rotation of the Earth relative to the fixed stars. One sidereal day has 24 sidereal hours, 1 sidereal hour has 60 sidereal minutes, and 1 sidereal minute has 60 sidereal seconds. Care must be taken to distinguish between sidereal times and mean solar times, which use the same basic subdivisions. The relationships between the two systems, given in Bate et al. (1971), are

$$1 \text{ mean solar day} = 1.0027379093 \text{ mean sidereal days}$$
$$= 24 \text{ h } 3 \text{ m } 56.55536 \text{ s sidereal time}$$
$$= 86{,}636.55536 \text{ mean sidereal seconds}$$

(2.27)

$$1 \text{ mean sidereal day} = 0.9972695664 \text{ mean solar days}$$
$$= 23 \text{ h } 56 \text{ m } 04.09054 \text{ s mean solar time}$$
$$= 86{,}164.09054 \text{ mean solar seconds}$$

(2.28)

Measurements of longitude on the Earth's surface require the use of sidereal time (discussed further in Sec. 2.9.7). The use of 23 h, 56 min as an approximation for the mean sidereal day is used later in determining the height of the geostationary orbit.

2.9.5 The Orbital Plane

In the orbital plane, the position vector **r** and the velocity vector **v** specify the motion of the satellite, as shown in Fig. 2.9. For present purposes, only the magnitude of the position vector is required. From the geometry of the ellipse (see App. B), this is found to be

$$r = \frac{a(1 - e^2)}{1 + e \cos v}$$

(2.29)

The true anomaly v is a function of time, and determining it is one of the more difficult steps in the calculations.

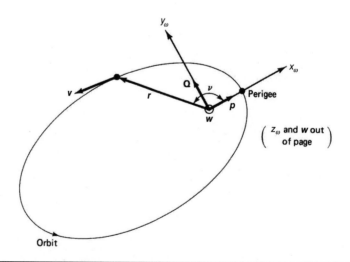

FIGURE 2.9 Perifocal coordinate system (**PQW** frame).

The vector components in the **IJK** frame are

$$
\begin{bmatrix} r_I \\ r_J \\ r_K \end{bmatrix} = \begin{bmatrix} 0.567 & -0.25 \\ -0.25 & 0.856 \\ 0.785 & 0.453 \end{bmatrix} \begin{bmatrix} -6500 \\ 4000 \end{bmatrix} = \begin{bmatrix} -4685.3 \\ 5047.7 \\ -3289.1 \end{bmatrix} \text{ km}
$$

The magnitude of the **r** vector in the **IJK** frame is

$$
\mathbf{r} = \sqrt{(-4685.3)^2 + (5047.7)^2 + (-3289.1)^2} = 7632.2 \text{ km}
$$

This is the same as that was obtained from the **P** and **Q** components.

2.9.7 Earth Station Referred to the IJK Frame

The Earth station's position is given by the geographic coordinates of latitude λ_E and longitude ϕ_E. Unfortunately, there does not seem to be any standardization of the symbols used for latitude and longitude. In some texts, as here, the Greek λ is used for latitude and the Greek ϕ for longitude. In other texts, the reverse of this happens. One minor advantage of the former is that latitude and λ both begin with the same "la" which makes the relationship easy to remember.

Care also must be taken regarding the sign conventions used for latitude and longitude because different systems are sometimes used, depending on the application. In this book, north latitudes are written as positive numbers and south latitudes as negative numbers, zero latitude, of course, being the equator. Longitudes east of the Greenwich meridian are positive numbers, and longitudes west are negative numbers.

The position vector of the Earth station relative to the **IJK** frame is **R** as shown in Fig. 2.11, and the angle between **R** and the equatorial plane, denoted by Ψ_E, is closely related, but not quite equal to, the Earth station latitude. More is said about this angle

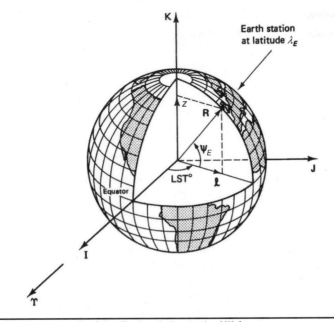

FIGURE 2.11 Position vector **R** of the Earth relative to the **IJK** frame.

shortly. **R** is obviously a function of the rotation of the Earth, and so first it is necessary to find the position of the Greenwich meridian relative to the **I** axis as a function of time. The angular distance from the **I** axis to the Greenwich meridian is measured directly as *Greenwich sidereal time* (GST), also known as the *Greenwich hour angle*, or GHA. Sidereal time is described in Sec. 2.9.4.

GST may be found using values tabulated in some almanacs (see Bate et al., 1971), or it may be calculated using formulas given in Wertz (1984). In general, sidereal time may be measured in time units of sidereal days, hours, minutes, seconds, or it may be measured in angular units (degrees, minutes, seconds, or radians). Conversion is easily accomplished, since 2π radians or 360° correspond to 24 sidereal hours. The formula for GST in degrees is

$$GST = 99.9610° + 36000.7689° \times T + 0.0004° \times T^2 + UT° \qquad (2.40)$$

Here, $UT°$ is universal time expressed in degrees, as given by Eq. (2.19). T is the time in Julian centuries, given by Eq. (2.26).

Once GST is known, the *local sidereal time* (LST) is found by adding the east longitude of the station in degrees. East longitude for the Earth station is denoted as EL. Recall that previously longitude was expressed in positive degrees east and negative degrees west. For east longitudes, EL = $\phi_{E'}$ while for west longitudes, $WL = -360° + EL$. For example, for an Earth station at east longitude 40°, EL = 40°. For an Earth station at west longitude 40°, EL = 360 + (−40) = 320°. Thus the LST in degrees is given by

$$LST = GST + EL \qquad (2.41)$$

The procedure is illustrated in the following examples.

Example 2.20   Find the GST for 13 h UT on December 18, 2000.

Solution   From Example 2.11: $T = 1.00963838$. The individual terms of Eq. (2.40) are:

$$X = 36000.7689° \times T = 347.7578° \text{ (mod 360°)}$$

$$Y = 0.0004° \times T^2 = 0.00041° \text{ (mod 360°)}$$

$$UT = \frac{13}{24} \times 360° = 195°$$

$$GST = 99.6910° + X + Y + UT = 282.4493° \text{ (mod 360)}$$

Example 2.21   Find the LST for Thunder Bay, longitude 89.26° W for 13 h UT on December 18, 2000.

Solution   Expressing the longitude in degrees west: WL = −89.26°
In degrees east this is EL = 360° + (−89.26°) = 270.74°

$$LST = GST + EL = 282.449 + 270.74 = 93.189° \text{ (mod 360°)}$$

Knowing the LST enables the position vector **R** of the Earth station to be located with reference to the **IJK** frame as shown in Fig. 2.11. However, when **R** is resolved into its rectangular components, account must be taken of the oblateness of the Earth. The Earth may be modeled as an *oblate spheroid*, in which the equatorial plane is circular, and any *meridional plane* (i.e., any plane containing the Earth's polar axis) is elliptical, as illustrated in Fig. 2.12.

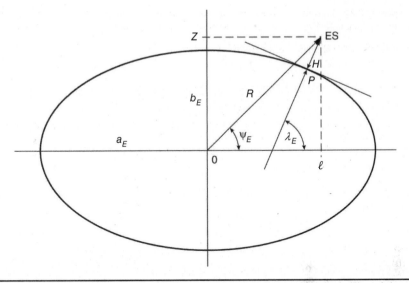

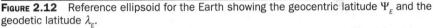

Figure 2.12 Reference ellipsoid for the Earth showing the geocentric latitude Ψ_E and the geodetic latitude λ_E.

For one model, known as a *reference ellipsoid,* the semi-major axis of the ellipse is equal to the equatorial radius, the semi-minor axis is equal to the polar radius, and the surface of the ellipsoid represents the *mean sea level.* Denoting the semi-major axis by a_E and the semi-minor axis by b_E and using the known values for the Earth's radii gives

$$a_E = 6378.1414 \text{ km} \tag{2.42}$$

$$b_E = 6356.755 \text{ km} \tag{2.43}$$

From these values the eccentricity of the Earth is seen to be

$$e_E = \frac{\sqrt{a_E^2 - b_E^2}}{a_E} = 0.08182 \tag{2.44}$$

In Figs. 2.11 and 2.12, what is known as the *geocentric latitude* is shown as Ψ_E. This differs from what is normally referred to as latitude. An imaginary plumb line dropped from the Earth station makes an angle λ_E with the equatorial plane, as shown in Fig. 2.12. This is known as the *geodetic latitude,* and for all practical purposes here, this can be taken as the geographic latitude of the Earth station.

With the height of the Earth station above mean sea level denoted by H, the geocentric coordinates of the Earth station position are given in terms of the geodetic coordinates by (Thompson, 1966)

$$N = \frac{a_E}{\sqrt{1 - e_E^2 \sin^2 \lambda_E}} \tag{2.45}$$

$$R_I = (N+H)\cos \lambda_E \cos \text{LST} = l \cos \text{LST} \tag{2.46}$$

$$R_J = (N+H)\cos \lambda_E \sin \text{LST} = l \sin \text{LST} \tag{2.47}$$

$$R_K = \left[N\left(1 - e_E^2\right) + H \right] \sin \lambda_E = z \tag{2.48}$$

Example 2.22 Find the components of the radius vector to the Earth station at Thunder Bay, given that the latitude is 48.42°, the height above sea level is 200 m, and the LST is 167.475°.

Solution With all distances in km, $H = 0.2$

$$N = \frac{a_E}{\sqrt{1 - e_E^2 \sin^2 \lambda_E}} = \frac{6378.1414}{\sqrt{1 - 0.08182^2 \ \sin^2 48.42}} = 6390.121 \text{ km}$$

$$l = (N + H) \cos \lambda_E = 6390.321 \times 0.66366 = 4241.033 \text{ km}$$

$$R_I = l \cos(\text{LST}) = 4241.033 \times (-0.9762) = -4140.103 \text{ km}$$

$$R_J = l \sin(\text{LST}) = 4241.033 \times (0.216865) = 919.734 \text{ km}$$

$$R_K = [N(1 - e_E^2) + H] \sin \lambda_E = [6390.121 \times (1 - 0.08182^2) + 0.2] \times 0.74803$$

$$= 4748.151 \text{ km}$$

At this point, both the satellite radius vector **r** and the Earth station radius vector **R** are known in the **IJK** frame for any position of satellite and Earth. From the vector diagram shown in Fig. 2.13*a*, the range vector ρ is obtained as

$$\boldsymbol{\rho} = \mathbf{r} - \mathbf{R} \tag{2.49}$$

This gives ρ in the **IJK** frame. It then remains to transform ρ to the observer's frame, known as the *topocentric-horizon frame*, shown in Fig. 2.13*b*.

2.9.8 The Topocentric-Horizon Coordinate System

The position of the satellite, as measured from the Earth station, is usually given in terms of the azimuth and elevation angles and the range ρ. These are measured in the *topocentric-horizon coordinate system* illustrated in Fig. 2.13*b*. In this coordinate system, the fundamental plane is the observer's horizon plane. In the notation given in Bate et al. (1971), the positive x axis is taken as south, the unit vector being denoted by **S**. The positive y axis points east, the unit vector being **E**. The positive z axis is "up," pointing to the observer's zenith, the unit vector being **Z**. (*Note*: This is not the same z as that used in Sec. 2.9.7.) The frame is referred to as the **SEZ** frame, which of course rotates with the Earth.

As shown in Sec. 2.9.7, the range vector ρ is known in the **IJK** frame, and it is now necessary to transform this to the **SEZ** frame. Again, this is a standard transformation procedure. See Bate et al. (1971).

$$\begin{bmatrix} \rho_S \\ \rho_E \\ \rho_Z \end{bmatrix} = \begin{bmatrix} \sin \psi_E \cos \text{LST} & \sin \psi_E \sin \text{LST} & -\cos \psi_E \\ -\sin \text{LST} & \cos \text{LST} & 0 \\ \cos \psi_E \cos \text{LST} & \cos \psi_E \sin \text{LST} & \sin \psi_E \end{bmatrix} \begin{bmatrix} \rho_I \\ \rho_J \\ \rho_K \end{bmatrix} \tag{2.50}$$

From Fig. 2.12, the geocentric angle Ψ_E is seen to be given by

$$\psi_E = \arctan \frac{z}{l} \tag{2.51}$$

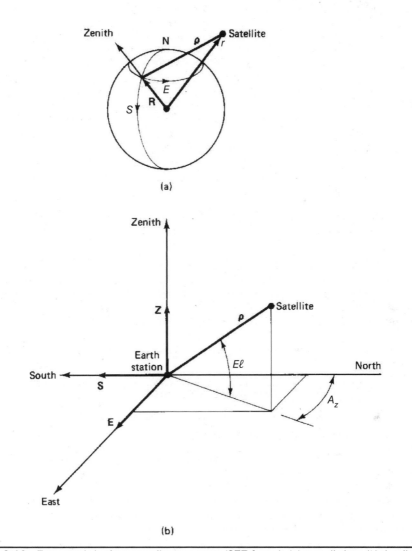

FIGURE 2.13 Topocentric-horizon coordinate system (**SEZ** frame): (*a*) overall view; (*b*) detailed view.

The coordinates l and z given in Eqs. (2.46) and (2.48) are known in terms of the Earth station height and latitude, and hence the range vector is known in terms of these quantities and the LST. As a point of interest, for zero height, the angle Ψ_E is related to λ_E by

$$\tan \psi_{E(H=0)} = \left(1 - e_E^2\right)\tan \lambda_E \tag{2.52}$$

Here, e_E is the Earth's eccentricity, equal to 0.08182. The difference between the geodetic and geocentric latitudes reaches a maximum at a geocentric latitude of 45°, when the geodetic latitude is 45.192°.

The magnitude of the range is

$$\rho = \sqrt{\rho_S^2 + \rho_E^2 + \rho_Z^2} \qquad (2.53)$$

The antenna elevation angle is

$$El = \arcsin \frac{\rho_z}{\rho} \qquad (2.54)$$

The antenna azimuth angle is found from

$$\alpha = \arctan \frac{|\rho_E|}{|\rho_S|} \qquad (2.55)$$

The azimuth depends on which quadrant α is in. With α in degrees the azimuth is as given in Table 2.4.

Example 2.23   The IJK range vector components for a certain satellite, at GST = 240°, are as given below. Calculate the corresponding range and the look angles for an Earth station the coordinates for which are—latitude 48.42° N, longitude 89.26° W, height above mean sea level 200 m.

Solution   Given data: $\rho_I = -1280$ km; $\rho_J = -1278$ km; $\rho_K = 66$ km; GST = 240°; $\lambda_E = 48.42°$; $\phi_E = -89.26°$; $H = 200$ m

The required Earth constants are $a_E = 6378.1414$ km; $e_E = 0.08182$

$$N = \frac{a_E}{\sqrt{1 - e_E^2 \sin^2 \lambda_E}} = \frac{6378.1414}{\sqrt{1 - 0.08182^2 \, \sin^2 48.42}} = 6390.121 \text{ km}$$

$$l = (N + H)\cos \lambda_E = 6390.321 \times 0.66366 = 4241.033 \text{ km}$$

$$z = \left[N(1 - e_E^2) + H \right] \sin \lambda_E$$

$$= [6390.121 \times (1 - 0.08182^2) + 0.2] \times 0.74803 = 4748.151 \text{ km}$$

$$\psi_E = \arctan \frac{z}{l} = 42.2289°$$

Substituting the known values in Eq. (2.50), and with all distances in km:

$$\begin{bmatrix} \rho_S \\ \rho_E \\ \rho_Z \end{bmatrix} = \begin{bmatrix} -.6507 & .3645 & -.6662 \\ -.4888 & -.8724 & 0 \\ -.5812 & .3256 & .7458 \end{bmatrix} \begin{bmatrix} -1280 \\ -1278 \\ 66 \end{bmatrix} = \begin{bmatrix} 322.9978 \\ 1740.571 \\ 376.9948 \end{bmatrix} \text{ km}$$

ρ_S	ρ_E	Azimuth (Degrees)
−	+	α
+	+	$180° - \alpha$
+	−	$180° + \alpha$
−	−	$360° - \alpha$

TABLE 2.4   Azimuth Angles

The magnitude is

$$\rho = \sqrt{322.9978^2 + 1740.571^2 + 376.9948^2} \cong 1810 \text{ km}$$

The antenna angle of elevation is

$$El = \arcsin \frac{376.9948}{1810} \cong 12°$$

The angle α is

$$\alpha = \arctan \frac{1740.571}{322.9978} = 79.487°$$

Since both ρ_E and ρ_S are positive, Table 2.4 gives the azimuth as

$$Az = 180° - \alpha = 100.5°$$

2.9.9 The Subsatellite Point

The point on the Earth vertically under the satellite is referred to as the *subsatellite point*. The latitude and longitude of the subsatellite point, and the height of the satellite above the subsatellite point, can be determined from knowledge of the radius vector **r**. Figure 2.14 shows the meridian plane which contains the subsatellite point. The height of the terrain above the reference ellipsoid at the subsatellite point is denoted by H_{SS}, and the height of the satellite above this, by h_{SS}. Thus, the total height of the satellite above the reference ellipsoid is

$$h = H_{SS} + h_{SS} \tag{2.56}$$

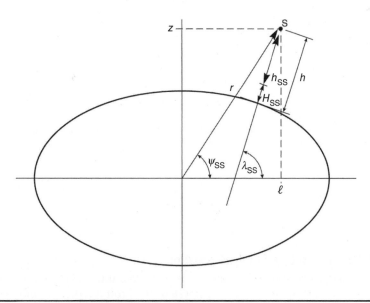

FIGURE 2.14 Geometry for determining the subsatellite point.

Now the components of the radius vector **r** in the **IJK** frame are given by Eq. (2.39). Figure 2.14 is similar to Fig. 2.12, with the difference that r replaces R, the height to the point of interest is h rather than H, and the subsatellite latitude λ_{SS} is used. Thus, Eqs. (2.45) through (2.48) may be written as

$$N = \frac{a_E}{\sqrt{1 - e_E^2 \sin^2 \lambda_{SS}}} \tag{2.57}$$

$$r_I = (N + h)\cos\lambda_{SS}\cos \text{LST} \tag{2.58}$$

$$r_J = (N + h)\cos\lambda_{SS}\sin \text{LST} \tag{2.59}$$

$$r_K = [N(1 - e_E^2) + h]\sin\lambda_{SS} \tag{2.60}$$

We now have three equations in three unknowns, LST, λ_E, and h, which can be solved. In addition, by analogy with the situation shown in Fig. 2.11, the east longitude is obtained from Eq. (2.41) as

$$EL = LST - GST \tag{2.61}$$

where GST is the Greenwich sidereal time.

Example 2.24 Determine the subsatellite height, latitude, and LST for the satellite in Example 2.19.

Solution From Example 2.19, the known components of the radius vector **r** in the **IJK** frame can be substituted in the left-hand side of Eqs. (2.58) through (2.60). The known values of a_E and e_E can be substituted in the right-hand side to give

$$-4685.3 = \left(\frac{6378.1414}{\sqrt{1 - 0.08182^2 \sin^2 \lambda_{SS}}} + h\right)\cos\lambda_{SS}\cos \text{LST}$$

$$5047.7 = \left(\frac{6378.1414}{\sqrt{1 - 0.08182^2 \sin^2 \lambda_{SS}}} + h\right)\cos\lambda_{SS}\sin \text{LST}$$

$$-3289.1 = \left(\frac{6378.1414 \times (1 - 0.08182^2)}{\sqrt{1 - 0.08182^2 \sin^2 \lambda_{SS}}} + h\right)\cos\lambda_{SS}\cos \text{LST}$$

Each equation contains the unknowns LST, λ_{SS}, and h. Unfortunately, these unknowns cannot be separated out in the form of explicit equations. The following values were obtained by a computer solution.

$$\lambda_{SS} \cong -25.654°$$

$$h \cong 1258.012 \text{ km}$$

$$LST \cong 132.868°$$

2.9.10 Predicting Satellite Position

The basic factors affecting satellite position are outlined in the previous sections. The NASA two-line elements are generated by orbit prediction models contained in Spacetrack report No. 3 (ADC USAF, 1980), which also contains computer code for the models. Readers desiring highly accurate prediction methods are referred to this report. Spacetrack report No. 4 (ADC USAF, 1983) gives details of the models used for atmospheric density analysis.

2.10   Local Mean Solar Time and Sun-Synchronous Orbits

The *celestial sphere* is an imaginary sphere of infinite radius, where the points on the surface of the sphere represent stars or other celestial objects. The points represent directions, and distance has no significance for the sphere. The orientation and center of the sphere can be selected to suit the conditions being studied, and in Fig. 2.15, the sphere is centered on the geocentric-equatorial coordinate system (see Sec. 2.9.6). What this means is that the celestial equatorial plane coincides with the Earth's equatorial plane, and the direction of the north celestial pole coincides with the Earth's polar axis. For clarity the **IJK** frame is not shown, but from the definition of the line of Aries in Sec. 2.9.6, the point for Aries lies on the celestial equator and on the x-axis, and the z-axis passes through the north celestial pole.

Also shown in Fig. 2.15 is the sun's meridian. The angular distance along the celestial equator, measured eastward from the point of Aries to the sun's meridian is the *right ascension of the sun*, denoted by α_s. In general, the right ascension of a point P is the angle measured eastward along the celestial equator from the point of Aries to the meridian passing through P. This is shown as α_p in Fig. 2.15. The *hour angle* of a star is the angle measured westward along the celestial equator from the meridian to meridian of the star. Thus for point P the hour angle of the sun is $(\alpha_p - \alpha_s)$ measured westward (the hour angle is measured in the opposite direction to the right ascension).

Now the *apparent solar time* of point P is the local hour angle of the sun, expressed in hours, plus 12 h. The 12 h is added because zero-hour angle corresponds to midday, when the P meridian coincides with the sun's meridian. Because the Earth's path around the sun is elliptical rather than circular, and because the plane containing the path of the Earth's orbit around the sun (the *ecliptic plane*) is inclined at an angle of

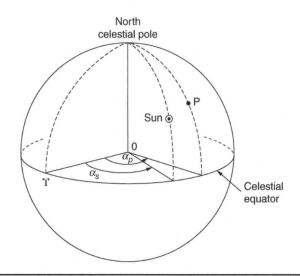

FIGURE 2.15   Sun-synchronous orbit.

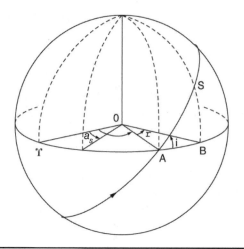

Figure 2.16 The condition for sun synchronicity is that the local solar time should be constant.

approximately 23.44°, the apparent solar time does not measure out uniform intervals along the celestial equator, in other words, the length of a solar day depends on the position of the Earth relative to the sun. To overcome this difficulty a fictitious mean sun is introduced, which travels in uniform circular motion around the sun (this is similar in many ways to the mean anomaly defined in Sec. 2.5). The time determined in this way is the *mean solar time*. Tables are available in various almanacs, which give the relationship between mean solar time and apparent solar time through the *equation of time*.

The relevance of this to a satellite orbit is illustrated in Fig. 2.16. This shows the trace of a satellite orbit on the celestial sphere (again keeping in mind that directions and not distances are shown). Point A corresponds to the ascending node. The hour angle of the sun from the ascending node of the satellite is $\Omega - \alpha_s$ measured westward. The hour angle of the sun from the satellite (projected to S on the celestial sphere) is $\Omega - \alpha_s + \beta$ and thus the local mean (solar) time is

$$t_{\text{SAT}} = \frac{1}{15}(\Omega - \alpha_s + \beta) + 12 \tag{2.62}$$

To find β requires solving the spherical triangle defined by the points ASB. This is a right spherical triangle because the angle between the meridian plane through S and the equatorial plane is a right angle. The triangle also contains the inclination i (the angle between the orbital plane and the equatorial plane) and the latitude λ (the angle measured at the center of the sphere going north along the meridian through S). The inclination i and the latitude λ are the same angles already introduced in connection with orbits. The solution of the right spherical triangle (see Wertz, 1984) yields for β

$$\beta = \arcsin\left(\frac{\tan \lambda}{\tan i}\right) \tag{2.63}$$

The local mean (solar) time for the satellite is therefore

$$t_{\text{SAT}} = \frac{1}{15}\left[\Omega - \alpha_s + \arcsin\left(\frac{\tan \lambda}{\tan i}\right)\right] + 12 \tag{2.64}$$

Notice that as the inclination i approaches 90° angle β approaches zero.

Accurate formulas are available for calculating the right ascension of the sun, but a good approximation to this is

$$\alpha_s = \frac{\Delta d}{365.24} 360° \qquad (2.65)$$

where Δd is the time in days from the vernal equinox. This is so because in 1 year of approximately 365.24 days the Earth completes a 360° orbit around the sun.

For a *sun-synchronous orbit* the local mean time must remain constant. The advantage of a sun-synchronous orbit for weather satellites and environmental satellites is that each time the satellite passes over a given latitude, the lighting conditions are approximately the same. This is particularly important for Earth remote sensing satellites. Equation (2.64) shows that for a given latitude and fixed inclination, the only variables are α_s and Ω. In effect, the angle $(\Omega - \alpha_s)$ must be constant for a constant local mean time. Let Ω_0 represent the right ascension of the ascending node at the vernal equinox and Ω' the time rate of change of Ω then

$$t_{SAT} = \frac{1}{15}\left[\Omega_0 + \Omega'\Delta d - \frac{\Delta d}{365.24} 360° + \arcsin\left(\frac{\tan\lambda}{\tan i}\right) + 12\right]$$

$$= \frac{1}{15}\left[\Omega_0 + \left(\Omega' - \frac{360}{365.24}\right)\Delta d + \arcsin\left(\frac{\tan\lambda}{\tan i}\right) + 12\right] \qquad (2.66)$$

For this to be constant the coefficient of Δd must be zero, or

$$\Omega' = \frac{360°}{365.24}$$

$$= 0.9856 \text{ degrees/day} \qquad (2.67)$$

Use is made of the regression of the nodes to achieve sun synchronicity. As shown in Sec. 2.8.1 by Eqs. (2.18) and (2.20), the rate of regression of the nodes and the direction are determined by the orbital elements a, e, and i. These can be selected to give the required regression of 0.9856° east per day. The orbital parameters for the Tiros-N satellites are listed in Table 2.5. These satellites follow near-circular, near-polar orbits.

	833-km Orbit	870-km Orbit
Inclination	98.739°	98.899°
Nodal period	101.58 min	102.37 min
Nodal regression	25.40°/day E	25.59°/day E
Nodal precession	0.986°/day E	0.986°/day E
Orbits per day	14.18	14.07

Source: Schwalb (1982a and b).

TABLE 2.5 Tiros-N Series Orbital Parameters

2.11   Standard Time

Local mean time is not suitable for civil time–keeping purposes because it changes with longitude (and latitude), which make it difficult to order day-to-day affairs. The approach taken internationally is to divide the world into 1-h time zones, the zonal meridians being 15° apart at the equator. The Greenwich meridian is used as zero reference and in the time zone that is ~7.5° about the Greenwich meridian the civil time is the same as the GMT. Care must be taken, however, since in the spring the clocks are advanced by 1 h, leading to *British summer time* (BST), also known as daylight saving time. Thus, BST is equal to GMT plus 1 h.

In the first zone east of the GMT zone, the basic civil time is GMT +1 h, and in the first zone west of the GMT zone, the basic civil time is GMT –1 h. One hour is added or subtracted for each additional zone east or west. Again, care must be taken to allow for summertime if it is in force (not all regions have the same summertime adjustment, and some regions may not use it at all). Also, in some instances the zonal meridians are adjusted where necessary to suit regional or country boundaries.

Orbital elements are normally specified in relation to GMT (or as noted in Sec. 2.9.2, UTC), but results (such as times of equatorial crossings) usually need to be known in the standard time for the zone where observations are being made. Care must be taken therefore to allow for the zone change, and for daylight saving time if in force. Many useful time zone maps and other information can be obtained from the Internet through a general search for "time zones."

2.12   Problems

2.1.   State Kepler's three laws of planetary motion. Illustrate in each case their relevance to artificial satellites orbiting the Earth.

2.2.   Using the results of App. B, show that for any point P, the sum of the focal distances to S and S' is equal to $2a$.

2.3.   Show that for the ellipse the differential element of area $dA = r^2 dv/2$, where dv is the differential of the true anomaly. Using Kepler's second law, show that the ratio of the speeds at apoapsis and periapsis (or apogee and perigee for an Earth-orbiting satellite) is equal to

$$(1-e)/(1+e)$$

2.4.   A satellite orbit has an eccentricity of 0.2 and a semi-major axis of 10,000 km. Find the values of (*a*) the latus rectum; (*b*) the minor axis; (*c*) the distance between foci.

2.5.   For the satellite in Prob. 2.4, find the length of the position vector when the true anomaly is 130°.

2.6.   The orbit for an Earth-orbiting satellite has an eccentricity of 0.15 and a semi-major axis of 9000 km. Determine (*a*) its periodic time; (*b*) the apogee height; (*c*) the perigee height. Assume a mean value of 6371 km for the Earth's radius.

2.7.   For the satellite in Prob. 2.6, at a given observation time during a south to north transit, the height above ground is measured as 2000 km. Find the corresponding true anomaly.

2.8.   The semi-major axis for the orbit of an Earth-orbiting satellite is found to be 9500 km. Determine the mean anomaly 10 min after passage of perigee.

2.9. The exact conversion factor between feet and meters is 1 ft = 0.3048 m. A satellite travels in an unperturbed circular orbit of semi-major axis a = 27,000 km. Determine its tangential speed in (*a*) km/s, (*b*) ft/s, and (*c*) mi/h.

2.10. Explain what is meant by *apogee height* and *perigee height*. The Cosmos 1675 satellite has an apogee height of 39,342 km and a perigee height of 613 km. Determine the semi-major axis and the eccentricity of its orbit. Assume a mean Earth radius of 6371 km.

2.11. The Aussat 1 satellite in geostationary orbit has an apogee height of 35,795 km and a perigee height of 35,779 km. Assuming a value of 6378 km for the Earth's equatorial radius, determine the semi-major axis and the eccentricity of the satellite's orbit.

2.12. Explain what is meant by the ascending and descending nodes. In what units should these be measured, and in general, should you expect them to change with time?

2.13. Explain what is meant by (*a*) line of apsides and (*b*) line of nodes. Is it possible for these two lines to be coincident?

2.14. Make a figure to explain what is meant by each of the angles: *inclination; argument of perigee; right ascension of the ascending node.* Which of these angles do you expect, in general, to change with time?

2.15. The inclination of an orbit is 67°. What is the greatest latitude, north and south, reached by the subsatellite point? Is this orbit retrograde or prograde?

2.16. Briefly describe the main effects of the Earth's equatorial bulge on a satellite orbit. Given that a satellite is in a circular equatorial orbit for which the semi-major axis is equal to 42,164.2 km, calculate (*a*) the mean motion, (*b*) the rate of regression of the nodes, and (*c*) the rate of rotation of argument of perigee.

2.17. A satellite in polar orbit has a perigee height of 600 km and an apogee height of 1200 km. Calculate (*a*) the mean motion, (*b*) the rate of regression of the nodes, and (*c*) the rate of rotation of the line of apsides. The mean radius of the Earth may be assumed equal to 6371 km.

2.18. What is the fundamental unit of universal coordinated time? Express the following times in (*a*) days and (*b*) degrees: 0 h, 5 min, 24 s; 6 h, 35 min, 20 s; your present time.

2.19. Determine the Julian days for the following dates and times: midnight March 10, 1999; noon, February 23, 2000; 16:30 h, March 1, 2003; 3 p.m., July 4, 2010.

2.20. Find, for the times and dates given in Prob. 2.19, (*a*) T in Julian centuries and (*b*) the corresponding GST in degrees.

2.21. Find the month, day, and UT for the following epochs: (*a*) day 3.00, year 1999; (*b*) day 186.125, year 2000; (*c*) day 300.12157650, year 2001; (*d*) day 3.29441845, year 2004; (*e*) day 31.1015, year 2010.

2.22. Find the GST corresponding to the epochs given in Prob. 2.21.

2.23. The Molnya 3-(25) satellite has the following parameters specified: perigee height 462 km; apogee height 40,850 km; period 736 min; inclination 62.8°. Using an average value of 6371 km for the Earth's radius, calculate (*a*) the semi-major axis and (*b*) the eccentricity. (*c*) Calculate the nominal mean motion n_0. (*d*) Calculate the mean motion. (*e*) Using the calculated value for a, calculate the anomalistic period and compare with the specified value. Calculate (*f*) the rate of regression of the nodes, and (*g*) the rate of rotation of the line of apsides.

2.24. Repeat the calculations in Prob. 2.23 for an inclination of 63.435°.

2.25. Determine the orbital condition necessary for the argument of perigee to remain stationary in the orbital plane. The orbit for a satellite under this condition has an eccentricity of 0.001 and a semi-major axis of 27,000 km. At a given epoch the perigee is exactly on the line of Aries. Determine the satellite position relative to this line after a period of 30 days from epoch.

2.26. For a given orbit, K as defined by Eq. (2.17) is equal to 0.112 rev/day. Determine the value of inclination required to make the orbit sun synchronous.

2.27. A satellite has an inclination of 90° and an eccentricity of 0.1. At epoch, which corresponds to time of perigee passage, the perigee height is 2643.24 km directly over the north pole. Determine (*a*) the satellite mean motion. For 1 day after epoch determine (*b*) the true anomaly, (*c*) the magnitude of the radius vector to the satellite, and (*d*) the latitude of the subsatellite point.

2.28. The following elements apply to a satellite in inclined orbit:
$\Omega_o = 0°$; $\omega_o = 90°$; $M_o = 309°$; $i = 63°$; $e = 0.01$; $a = 7130$ km. An Earth station is situated at 45°N, 80°W, and at zero height above sea level. Assuming a perfectly spherical Earth of uniform mass and radius 6371 km and given that epoch corresponds to a GST of 116°, determine at epoch the orbital radius vector in the (*a*) **PQW** frame; (*b*) **IJK** frame; (*c*) the position vector of the Earth station in the **IJK** frame; (*d*) the range vector in the **IJK** frame; (*e*) the range vector in the **SEZ** frame; and (*f*) the Earth station look angles.

2.29. A satellite moves in an inclined elliptical orbit, the inclination being 63.4°. State with explanation the maximum northern and southern latitudes reached by the subsatellite point. The nominal mean motion of the satellite is 14 rev/day, and at epoch the subsatellite point is on the ascending node at 100°W. Calculate the longitude of the subsatellite point 1 day after epoch. The eccentricity is 0.01.

2.30. A "no name" satellite has the following parameters specified: perigee height 197 km; apogee height 340 km; period 88.2 min; inclination 64.6°. Using an average value of 6371 km for the Earth's radius, calculate (*a*) the semi-major axis and (*b*) the eccentricity. (*c*) Calculate the nominal mean motion n_o. (*d*) Calculate the mean motion. (*e*) Using the calculated value for *a*, calculate the anomalistic period and compare with the specified value. Calculate (*f*) the rate of regression of the nodes, and (*g*) the rate of rotation of the line of apsides.

2.31. Given that $\Omega_o = 250°$; $\omega_o = 85°$; $M_o = 30°$; for the satellite in Prob. 2.30, calculate, for 65 min after epoch ($t_0 = 0$), the new values of Ω, ω, and M. Also find the true anomaly and radius.

2.32. From the NASA bulletin given in App. C, determine the date and the semi-major axis.

2.33. Determine, for the satellite listed in the NASA bulletin of App. C, the rate of regression of the nodes, the rate of change of the argument of perigee, and the nominal mean motion n_o.

2.34. From the NASA bulletin in App. C, verify that the orbital elements specified are for a nominal S–N equator crossing.

2.35. A satellite in exactly polar orbit has a slight eccentricity (just sufficient to establish the idea of a perigee). The anomalistic period is 110 min. Assuming that the mean motion is $n = n_0$ calculate the semi-major axis. Given that at epoch the perigee is exactly over the north pole, determine the position of the perigee relative to the north pole after one anomalistic period and the time taken for the satellite to make one complete revolution relative to the north pole.

2.36. A satellite is in an exactly polar orbit with apogee height 7000 km and perigee height 600 km. Assuming a spherical Earth of uniform mass and mean radius 6371 km, calculate (*a*) the semi-major axis, (*b*) the eccentricity, and (*c*) the orbital period. (*d*) At a certain time the satellite is observed ascending directly overhead from an Earth station on latitude 49°N. Give that the argument of perigee is 295° calculate the true anomaly at the time of observation.

2.37. The two-line elements for satellite NOAA 18 are as follows:

> NOAA 18
> 1 28654U 05018A 05154.51654998-.00000093 00000-0-28161-4 0 189
> 2 28654 98.7443 101.8853 0013815 210.8695 149.1647 14.10848892 1982

Determine the approximate values of (*a*) the semi-major axis, and (*b*) the latitude of the subsatellite point at epoch.

2.38. Using the two-line elements given in Prob. 2.37, determine the longitude of the subsatellite point and the LST at epoch.

2.39. Equation (2.34) gives the GST in degrees as

$$GST = 99.9610° + 36000.7689° \times T + 0.0004° \times T^2 + UT°$$

where *T* is the number of Julian centuries that have elapsed since noon, January 0, 1900. The GST equation is derived from (Wertz, 1984) $GST = \alpha_s - 180° + UT°$ where α_s is the right ascension of the mean sun. Determine the right ascension of the mean sun for noon on June 5, 2005.

2.40. If the orbits detailed in Table 2.5 are circular and using Eq. (2.2) to find the semi-major axis, calculate the regression of the nodes for these orbits.

2.41. Determine the standard zone time in the following zones, for 12 noon GMT: (*a*) 285°E, (*b*) 255°E, (*c*) 45°E, (*d*) 120°E.

2.42. Determine the GMT for the following local times and locations: (*a*) 7 a.m. Los Angeles, USA; (*b*) 1 p.m. Toronto, Canada; (*c*) 12 noon Baghdad, Iraq; (*d*) 3 p.m. Tehran, Iran.

References

ADC USAF. 1980. *Model for Propagation of NORAD Element Sets*. Spacetrack Report No. 3. Aerospace Defense Command, U.S. Air Force, December.

ADC USAF. 1983. *An Analysis of the Use of Empirical Atmospheric Density Models in Orbital Mechanics*. Spacetrack Report No. 3. Aerospace Defense Command, U.S. Air Force, February.

Arons, A. B. 1965. *Development of Concepts of Physics*. Addison-Wesley, Reading, MA.

Bate, R. R., D. D. Mueller, and J. E. White. 1971. *Fundamentals of Astrodynamics*. Dover, New York.

Celestrak, at http://celestrak.com/NORAD/elements/noaa.txt.

Duffett-Smith, P. 1986. *Practical Astronomy with Your Calculator*. Cambridge University Press, New York.

Schwalb, A. 1982a. The TIROS-N/NOAA-G Satellite Series. NOAA Technical Memorandum NESS 95, Washington, DC.

Schwalb, A. 1982b. Modified Version of the TIROS-N/NOAA A-G Satellite Series (NOAA E-J): Advanced TIROS N (ATN). NOAA Technical Memorandum NESS 116, Washington, DC.

Thompson, Morris M. (editor-in-chief). 1966. *Manual of Photogrammetry*, 3d ed., Vol. 1. American Society of Photogrammetry, New York.

Wertz, J. R. (ed.). 1984. *Spacecraft Attitude Determination and Control*. D. Reidel, Holland.

CHAPTER 3

The Geostationary Orbit

3.1   Introduction

A satellite in a geostationary orbit appears to be stationary with respect to the Earth, hence the name *geostationary*. Three conditions are required for an orbit to be geostationary:

1. The satellite must travel eastward at the same rotational speed as the Earth.

2. The orbit must be circular.

3. The inclination of the orbit must be zero.

The first condition is obvious. If the satellite is to appear stationary, it must rotate at the same speed as the Earth, which is constant. The second condition follows from this and from Kepler's second law (Sec. 2.3). Constant speed means that equal areas must be swept out in equal times, and this can only occur with a circular orbit (see Fig. 2.2). The third condition, that the inclination must be zero, follows from the fact that any inclination has the satellite moving north and south (see Sec. 2.5 and Fig. 2.3), and hence it is not geostationary. North and south movements can be avoided only with zero inclination, which means that the orbit lies in the Earth's equatorial plane.

Kepler's third law may be used to find the radius of the orbit (for a circular orbit, the semi-major axis is equal to the radius). Denoting the radius by α_{GSO}, then from Eqs. (2.2) and (2.4),

$$\alpha_{\text{GSO}} = \left(\frac{\mu P}{4\pi^2} \right)^{1/3}\tag{3.1}$$

The period P for the geostationary is 1 sidereal day or 23 h, 56 min, 4 s mean solar time (ordinary clock time). This is the time taken for the Earth to complete one revolution about its N–S axis, measured relative to the fixed stars (see Sec. 2.9.4). Substituting this value along with the value for μ given by Eq. (2.3) results in

$$\alpha_{\text{GSO}} = 42,164.20 \text{ km}\tag{3.2}$$

The average equatorial radius of the Earth specified in the World Geodetic System (WGS84) is

$$\alpha_E = 6,378.137 \text{ km}\tag{3.3}$$

and hence the geostationary height is

$$h_{GSO} = a_{GSO} - a_E = 42,164.2 - 6,378.137 = 35,786.063 \text{ km} \qquad (3.4)$$

This value is often rounded up to 36,000 km for approximate calculations.

In practice, a precise geostationary orbit cannot be attained because of disturbance forces in space and the effects of the Earth's equatorial bulge. The gravitational fields of the sun and the moon produce a shift of about 0.85°/year in inclination. Also, the Earth's *equatorial ellipticity* causes the satellite to drift eastward along the orbit. In practice, station-keeping maneuvers must be performed periodically to correct for these shifts, as described in Sec. 7.4.

An important point to grasp is that there is only one geostationary orbit because there is only one value of a that satisfies Eq. (2.3) for a periodic time of 23 h, 56 min, 4.1 s. Communications authorities throughout the world regard the geostationary orbit as a natural resource, and its use is carefully regulated through national and international agreements.

3.2 Earth Station Antenna Look Angles

The *look angles* for the ground station antenna are the azimuth and elevation angles required at the antenna, so that it points directly at the satellite. In Sec. 2.9.8 the look angles are determined in the general case of an elliptical orbit, and there the angles vary in time to track the satellite as it passes from horizon to horizon. With a geostationary orbit, the situation is much simpler because the satellite is stationary with respect to the Earth. Although in general no tracking is necessary, with the large earth stations used for commercial communications, the antenna beamwidth is very narrow (see Chap. 6), and a tracking mechanism is required to compensate for the movement of the satellite about the nominal geostationary position. On the other hand, with the types of antennas used for small earth terminals, the beamwidth is quite broad, and no tracking is necessary. This allows the antenna to be fixed in position, as evidenced by the small antennas commonly used for reception of satellite TV or the slightly larger dishes used for convenience stores, gasoline stations, restaurants, etc. that can be seen fixed to the sides of buildings.

The three pieces of information that are needed to determine the look angles for the geostationary orbit are

1. The earth-station latitude, denoted here by λ_E
2. The earth-station longitude, denoted here by ϕ_E
3. The longitude of the subsatellite point, denoted here by ϕ_{ss} (often this is just referred to as the satellite longitude)

The convention used is that north latitudes are positive angles and south latitudes are negative angles. Longitudes east of the Greenwich meridian are positive angles and longitudes west are negative angles. For example, if a latitude of 40°S is specified, this is −40°, while a longitude of 35°W is specified as −35°.

As discussed in Chap. 2, when calculating the look angles for *low-earth-orbit* (LEO) satellites, it was necessary to include the variation in the local Earth's radius, which depends on latitude. Fortunately, with the geostationary orbit, this variation has negligible effect on the look angles.

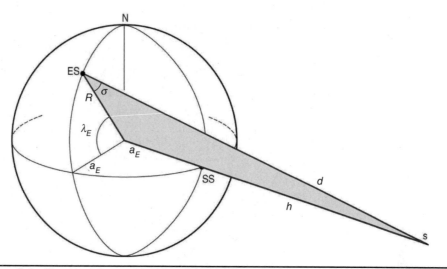

Figure 3.1 The geometry used in determining the look angles for a geostationary satellite.

The GEOSAT/earth station geometry is shown in Fig. 3.1. Here, ES denotes the position of the earth station, SS the subsatellite point, S the satellite, and d is the range from the earth station to the satellite. The shaded triangle contains the line-of-sight vector d between the earth station and the satellite, and the angle σ is the angle used to calculate the antenna elevation pointing angle.

There are two types of triangles involved in this geometry illustrated in Fig. 3.1, namely, the spherical triangle (ABC) shown in heavy outline in Fig. 3.2a and the plane triangle in Fig. 3.1 that is given in a planar view in Fig. 3.2b.

Considering first the spherical triangle ABC, in Fig. 3.2a, where there are six angles that define this figure. The three angles ($\angle A$, $\angle B$, and $\angle C$) are the vertex angles between three planes, and three spherical triangle sides (a, b, c) are defined by central angles that subtend arcs of great circles. Great circles are the intersection of the sphere with a plane that passes through the center of the sphere. For example, side a is the central angle between the radius to the north pole and the radius to the subsatellite point in the equatorial plane, which has a value $a = \pi/2$, radians. Side b is the central angle between the radius to the earth station and the radius to the subsatellite point, and side c is the central angle between the radius to the earth station and the radius to the north pole, which from Fig. 3.2a has a value $c = \pi/2 - \lambda_E$, radians.

Concerning the vertex angles, angle A is the angle between the plane containing side c and the plane containing side b. Angle B is the angle between the plane containing c and the plane containing a. From Fig. 3.2a, $B = \phi_E - \phi_{SS}$, and the maximum value of $B = 1.419$ radians (81.3°). Finally, angle C is the angle between the plane containing b and the plane containing a. To summarize, the information known about the spherical triangle is

$$a = \frac{\pi}{2} \tag{3.5}$$

$$c = \frac{\pi}{2} - \lambda_E \tag{3.6}$$

$$B = \phi_E - \phi_{SS} \tag{3.7}$$

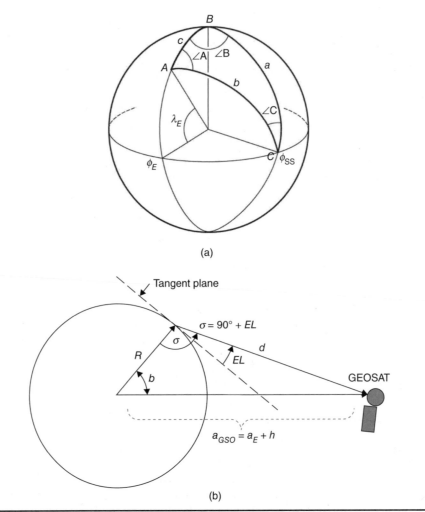

FIGURE 3.2   (a) The spherical triangle geometry related to Fig. 3.1, and (b) the earth station antenna incident plane with the line-of-sight to the GEOSAT.

Note that when the earth station is west of the subsatellite point, B is negative, and when east, B is positive. When the earth-station latitude is north, c is less than $\pi/2$, and when south, c is greater than $\pi/2$. Special rules, known as *Napier's rules*, are used to solve the spherical triangle (see Wertz, 1984), and these have been modified here to account for the signed angles B and λ_E. Only the result stated here gives the earth central angle b as

$$b = \arccos(\cos B \cos \lambda_E) \tag{3.8}$$

and the vertex angle A as

$$A = \arcsin\left(\frac{\sin |B|}{\sin b}\right) \tag{3.9}$$

Because of the harmonic nature the sine function, there are two solutions that satisfy Eq. (3.9), namely, A and $180° - A$. The correct solution must be determined by inspection of the four cases of spherical triangles as shown in Fig. 3.3. Note that the earth station azimuth pointing angle is measured, in the clockwise direction from north, in the local horizontal plane.

In Fig. 3.3a, when the earth station resides in the Southern Hemisphere and west of the satellite longitude, the vertex angle A is acute (less than 90°), whereas the azimuth angle, measured from North in a clockwise direction, is $A_z = A$. In Fig. 3.3b, where the earth station resides in the Southern Hemisphere and east of the satellite longitude, the angle A is acute, and the azimuth is, by inspection, $A_z = 360° - A$.

Next, consider Fig. 3.3c, where the earth station resides in the Northern Hemisphere and west of the satellite longitude. Here the angle A_c is obtuse and is given by $A_c = 180° - A$, by inspection, $A_z = A_c - 180° - A$.

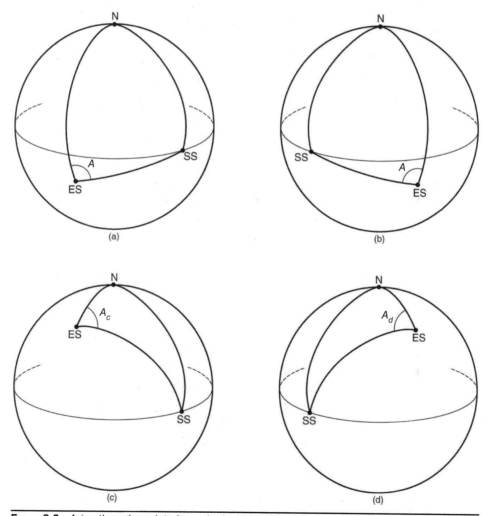

(a)   (b)

(c)   (d)

Figure 3.3   Azimuth angles related to spherical triangle vertex angle A (see Table 3.1).

Fig. 3.3	λ_E	B	A_z, Degrees
a	<0	<0	A
b	<0	>0	360° − A
c	>0	<0	180° − A
d	>0	>0	180° + A

TABLE 3.1   Azimuth Angles A_z from Fig. 3.3

Finally, in Fig. 3.3d, the vertex angle is obtuse and is given by $A_d = 180° − A$, and by inspection, $A_z = 360° − A_d = 180° + A$. In all the above cases, A is the acute angle returned by Eq. (3.9). These conditions are summarized in Table 3.1.

Example 3.1   A geostationary satellite is located at 90°W. Calculate the azimuth angle for an earth-station antenna at latitude 35°N and longitude 100°W.

Solution   The given quantities are: $\phi_{SS} = −90°$, $\phi_E = −100°$, $\lambda_E = 35°$

From Eq. (3.7) $B = \phi_E − \phi_{SS} = −10°$

and Eq. (3.8) $b = \arccos(\cos B \cos \lambda_E) = 36.23°$

and Eq. (3.9) $A = \arcsin\left(\dfrac{\sin |B|}{\sin b}\right) = 17.1°$

By inspection, $\lambda_E > 0$ and $B < 0$; therefore, Fig. 3.3c applies, and

$$A_z = 180° − A = 162.9°$$

Now, consider Fig. 3.2b for the plane geometry used to calculate the range and antenna elevation angle to the satellite. By applying the law of cosines for plane triangles, the range d is

$$d = \sqrt{R^2 + a_{GSO}^2 − 2Ra_{GSO}\cos b}$$

$$= 42,164.20\sqrt{1.02288 − 0.30254\cos b} \tag{3.10}$$

Next, we construct a tangent plane at the earth station location, which serves as the local horizontal plane for the elevation angle calculation. Applying the law of sines for plane triangles enables the angle of elevation to be found:

$$El = \arccos\left(\frac{a_{GSO}}{d}\sin b\right) \tag{3.11}$$

Example 3.2   Find the range and antenna elevation angle for the case specified in Example 3.1.

Solution   Given: $R = 6378.137$ km; $a_{GSO} = 42164.2$ km, and $b = 36.23°$. Equation (3.10) gives:

$$d = \sqrt{6378.137^2 + 42164.2^2 − 2 \times 6378.137 \times 42164.2 \times \cos 36.23°} = 37,210.2 \text{ km}$$

Equation (3.11) gives:

$$El = \arccos\left(\frac{42164.2}{37210.2}\sin 36.23°\right) \cong 48°$$

Figure 3.4 shows the look angles (El and Az) for Ku-band satellites as seen from Thunder Bay, Ontario, Canada.

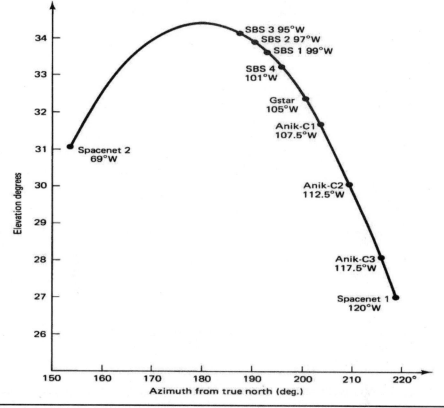

FIGURE 3.4 Azimuth-elevation angles for several GEOSAT's from an earth-station location of 48.42°N, 89.26°W (Thunder Bay, Ontario). Ku-band satellites are shown.

The preceding results do not consider the case when the earth station is on the equator. Obviously, when the earth station is directly under the satellite, the elevation is 90° and the azimuth is irrelevant. When the subsatellite point is east of the equatorial earth station ($B < 0$), the azimuth is ~90°, and when west ($B > 0$), the azimuth is ~270°. Also, the range as determined by Eq. (3.10) is approximate, and where more accurate values are required, as, for example, where propagation times need to be known accurately, the range is determined by measurement (see Sec. 14.7).

For a typical home satellite installation, practical adjustments are typically made to align the antenna to a known satellite for maximum signal. Thus, the look angles need not be determined with great precision, but they are calculated to give the expected values for a satellite longitude, which is close to the earth-station longitude. In some cases, especially with *direct broadcast satellites* (DBS), the home antenna is aligned to one particular cluster of satellites, as described in Chap. 16, and no further adjustments are necessary.

3.2.1 Earth Station Antenna Off-Boresight Angles

In Sec. 13.2, we will discuss interference between two satellites, and for this calculation, it is necessary to calculate the earth station antenna off-boresight angles to/from adjacent GEOSAT's. For this scenario consider Fig. 3.5 that shows the geometry required to

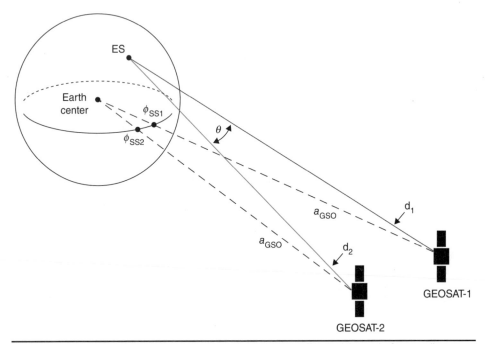

FIGURE 3.5 Geometry for calculating earth station antenna off-boresight angle for an adjacent GEOSAT-2.

calculate the earth station antenna off-boresight angle to an adjacent GEOSAT-2. Assume that the earth station antenna is boresighted at the desired GEOSAT-1, which is located at longitude ϕ_{ss1}. The adjacent GEOSAT-2, which is located at longitude ϕ_{ss2}, appears as an off-boresight angle θ in the antenna gain pattern. Based upon the spherical trigonometry discussed earlier the off-boresight angle is calculated as

$$\theta = \arccos\left(\frac{\left(d_1^2 + d_2^2 - 2a_{\mathrm{GSO}}^2(1 - \cos(\phi_{ss1} - \phi_{ss2}))\right)}{2d_1 d_2} \right) \tag{3.12}$$

Example 3.3 For the earth station and satellite given in Example 3.1, find the off-boresight angle for an adjacent GEOSAT-2 located at an orbital location of 92° west.

Solution From Example 3.2, R = 6378.137 km; a_{GSO} = 42164.2 km, and b = 36.23°. Eq. (3.10) gives:

$$d_1 = \sqrt{6378.137^2 + 42,164.2^2 - 2 \times 6378.137 \times 42,164.2 \times \cos 36.23°}$$

$$= 37{,}210.2 \text{ km}$$

Next, we calculate the range (d_2) from the earth station to GEOSAT-2

The given quantities are: $\phi_{ss2} = -92°$, $\phi_E = -100°$, $\lambda_E = 35°$

From Eq. (3.7) $B = \phi_E - \phi_{ss} = -8°$

and Eq. (3.8) $b = \arccos(\cos B \, \cos \lambda_E) = 35.79°$

Using Eq. (3.10), for R = 6371 km; a_{GSO} = 42164.2 km, and b = 35.79° gives:

$$d_2 = \sqrt{6378.137^2 + 42164.2^2 - 2 \times 6378.137 \times 42164.2 \times \cos 35.79°} = 37{,}177.9 \text{ km}$$

Next using Eq. (3.12),

$$\theta = \arccos\left(\frac{(d_1^2 + d_2^2 - 2\alpha_{GSO}^2(1 - \cos(\phi_{ss1} - \phi_{ss2}))}{2d_1 d_2}\right)$$

Calculate the argument of the arccos

$$= \frac{(37,210.2^2 + 37,177.9^2 - 2 \times 42,164.2^2(1 - \cos 2°))}{(2 \times 37,210.2 \times 37,177.9)}$$

$$= 0.99922$$

and $\theta = \arccos(0.99922) = 0.03962$ radians $= 2.27°$

3.3   The Polar Mount Antenna

When a small satellite antenna needs to be steerable, expense usually precludes the use of separate azimuth and elevation actuators. Instead, a single actuator is used which moves the antenna in a circular arc, which is known as a *polar mount antenna*. The antenna pointing is only accurate for one satellite, and some pointing error must be accepted for satellites on either side of this. With a polar mount antenna, the dish is mounted on an axis termed the *polar axis* such that the antenna boresight is normal to this axis, as shown in Fig. 3.6a. The polar mount is aligned along a true north line, as shown in Fig. 3.6, with the boresight pointing due south. The angle between the polar mount and the local horizontal plane is set equal to the earth-station latitude λ_E; simple geometry shows that this makes the boresight lie parallel to the equatorial plane. Next, the dish is tilted at an angle δ relative to the polar mount until the boresight is pointing at a satellite position due south of the earth station. Note that there does not need to be an actual satellite at this position. (The angle of tilt is often referred to as the *declination*, which must not be confused with the magnetic declination used in correcting compass readings. The term *angle of tilt* is used for δ in this text.)

The required angle of tilt is found as follows: From the geometry of Fig. 3.6b,

$$\delta = 90° - El_0 - \lambda_E \tag{3.13}$$

where El_0 is the angle of elevation required for the satellite position due south of the earth station. But for the due south situation, angle B in Eq. (3.7) is equal to 0; hence, from Eq. (3.8), $b = \lambda_E$. Hence, from Eq. (3.11), or Fig. 3.6c.

$$\cos El_0 = \frac{a_{GSO}}{d} \sin \lambda_E \tag{3.14}$$

Combining Eqs. (3.13) and (3.14) gives the required angle of tilt as

$$\delta = 90° - \arccos\left(\frac{a_{GSO}}{d}\right)\sin \lambda_E - \lambda_E \tag{3.15}$$

In the calculations leading to d, a spherical earth of mean radius 6378.137 km may be used, and the earth-station elevation may be ignored. The value obtained for δ is sufficiently accurate for initial alignment and fine adjustments can be made, if necessary. Calculation of the angle of tilt is illustrated next.

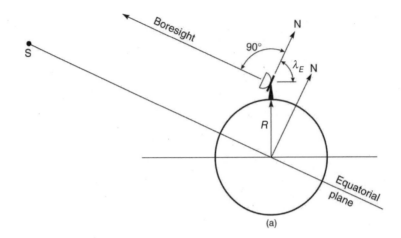

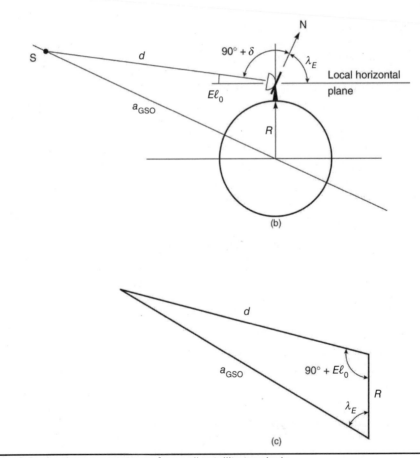

Figure 3.6   A polar mount antenna for small satellite terminals.

Example 3.4 Determine the angle of tilt required for a polar mount used with an earth station at latitude 49° north. Assume a spherical earth of mean radius 6378.137 km and ignore the earth-station altitude.

Solution Given data:

$\lambda_E = 49°$; $a_{GSO} = 42{,}164.2$ km; $R = 6371$ km; and $b = \lambda_E = 49°$.

Equation (3.10) gives:

$$d = \sqrt{6378.137^2 + 42{,}164.2^2 - 2 \times 6378.137 \times 42{,}164.2 \times \cos 49°} \cong 38{,}287 \text{ km}$$

From Eq. (3.11):

$$El = \arccos\left(\frac{42{,}164.2}{38{,}287} \sin 49°\right) \cong 33.8°$$

$$\delta = 90° - 33.8° - 49° = 7.2°$$

3.4 Limits of Visibility

There are east and west limits on the geostationary arc visible from any given earth station. The limits are determined by the geographic coordinates of the earth station and the antenna elevation. The lowest elevation in theory is 0°, when the antenna is pointing along the horizontal. A quick estimate of the longitudinal limits can be made by considering an earth station at the equator, with the antenna pointing either west or east along the horizontal, as shown in Fig. 3.7.

The limiting angle is given by

$$\theta = \arccos\frac{a_E}{a_{GSO}} = \arccos\frac{6378.137}{42{,}164.20} = 81.3° \tag{3.16}$$

Thus, for this situation, an earth station can see satellites over a geostationary arc bounded by ±81.3° about the earth-station longitude.

In practice, to avoid reception of excessive noise from the earth, some finite minimum value of elevation is used, here denoted by El_{min}. A typical value is 5°. The limits

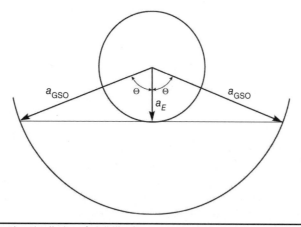

Figure 3.7 Illustrating the limits of visibility for an earth station on the equator.

of visibility also depend on the earth-station latitude. As in Fig. 3.2b, let S represent the angle subtended at the satellite when the angle $\sigma_{min} = 90° + El_{min}$. Applying the spherical trigonometry sine rule gives

$$S = \arcsin\left(\frac{R}{a_{GSO}} \sin \sigma_{min}\right) \tag{3.17}$$

A sufficiently accurate estimate is obtained by assuming a spherical earth of mean radius 6378 km as was done previously. Once angle S is known, angle b is found from

$$b = 180 - \sigma_{min} - S \tag{3.18}$$

From Eq. (3.8):

$$B = \arccos\left(\frac{\cos b}{\cos \lambda_E}\right) \tag{3.19}$$

Once angle B is found, the satellite longitude can be determined from Eq. (3.7). This is illustrated next.

Example 3.5   Determine the limits of visibility for an earth station situated at mean sea level, at latitude 48.42° north, and longitude 89.26° west. Assume a minimum angle of elevation of 5°.

Solution   Given data:

$\lambda_E = 48.42°$; $\phi_E = -89.26°$; $El_{min} = 5°$; $a_{GSO} = 42164.2$ km; $R = 6378.137$ km

$$\sigma_{min} = 90° + El_{min}$$

Equation (3.17) gives:

$$S = \arcsin\left(\frac{6378.137}{42,164.2} \sin 95°\right) = 8.67°$$

Equation (3.18) gives:

$$b = 180 - 95° - 8.67° = 76.33°$$

Equation (3.19) gives:

$$B = \arccos\left(\frac{\cos 76.34°}{\cos 48.42°}\right) = 69.14°$$

The satellite limit east of the earth station is at

$$\phi_E + B = -20° \text{ approx.}$$

and west of the earth station at

$$\phi_E - B = -158° \text{ approx.}$$

3.5   Near Geostationary Orbits

As mentioned in Sec. 2.8, there are many perturbing forces that cause an orbit to depart from the ideal Keplerian orbit. For the geostationary case, the most important of these are the gravitational fields of the moon and the sun, and the nonspherical shape of the Earth. Other significant forces are solar radiation pressure and reaction of the satellite itself to motor movement within the satellite. As a result, station-keeping maneuvers must be carried out to maintain the satellite within set limits of its nominal geostationary position. Station keeping is discussed in Sec. 7.4.

An exact geostationary orbit therefore is not attainable in practice, and the orbital parameters vary with time. The two-line orbital elements are published at regular intervals, and an example is given in Table 3.2 showing typical values. The period for a geostationary satellite is 23 h, 56 min, 4 s, or 86,164 s. The reciprocal of this is 1.00273896 rev/day, which is about the value for most of the satellites in this table. Thus, these satellites are *geosynchronous*, in that they rotate in synchronism with the rotation of the earth. However, they are not geostationary. The term *geosynchronous satellite* is used in many cases instead of *geostationary* to describe these near-geostationary satellites. It should be noted, however, that in general a geosynchronous satellite does not have to be near-geostationary, and there are several geosynchronous satellites that are in highly elliptical orbits with comparatively large inclinations (e.g., the Tundra satellites).

Although in principle the two-line elements can be used as described in Chap. 2 to determine orbital motion, the small inclination makes it difficult to locate the position of the ascending node, and the small eccentricity makes it difficult to locate the position of the perigee. However, because of the small inclination, the angles ω and Ω can be assumed to be in the same plane.

Referring to Fig. 2.10 with this assumption, the subsatellite point is $\Omega + \omega + v$ east of the line of Aries. The longitude of the subsatellite point (the satellite longitude) is the easterly rotation from the Greenwich meridian. The *Greenwich sidereal time* (GST) gives the eastward position of the Greenwich meridian relative to the line of Aries, and hence the subsatellite point is at longitude

$$\phi_{SS} = \omega + \Omega + v - \text{GST} \tag{3.20}$$

```
INTELSAT 901
1 26824U 01024A   05122.92515626 -.00000151 00000-0  10000-3 0  7388
2 26824   0.0158 338.7780 0004091 67.7508 129.4375  1.00270746 14318
INTELSAT 902
1 26900U 01039A   05126.99385197 .00000031  00000-0  10000-3 0  6260
2 26900   0.0156 300.5697 0002640 112.8823 231.2391 1.00271845 13528
INTELSAT 903
1 27403U 02016A 05125.03556931    .00000000  00000-0  10000-3 0  6249
2 27403   0.0362 171.6123 0002986 232.5077 157.1571 1.00265355 11412
INTELSAT 904
1 27380U 02007A   05125.62541657 .00000043 00000-0  00000+0 0  5361
2 27380   0.0202   0.0174 0003259 40.3723 108.3316  1.00272510 11761
INTELSAT 905
1 27438U 02027A   05125.03693822 .00000000  00000-0  10000-3 0  5812
2 27438   0.0205 164.2424 0002820 218.0675 189.4691 1.00265924 10746
INTELSAT 906
1 27513U 02041A   05126.63564565 .00000012  00000-0 00000+0 0  4817
2 27513   0.0111 324.7901 0003200 99.2828   93.4848 1.00272600  9803
INTELSAT 907
1 27683U 03007A   05124.32309516 .00000000  00000-0  10000-3 0  3108
2 27683   0.0206  13.5522 0009594 61.6856 235.7624  1.00266570  8131
INTELSAT 1002
1 28358U  4022A   05124.94126775 -.00000018 00000-0 00000+0 0  1527
2 28358   0.0079 311.0487 0000613  59.4312 190.2817 1.00271159  3289
```

TABLE 3.2 Typical Two-Line Orbital Elements for Selected Geostationary Satellites

and the mean longitude of the satellite is given by

$$\phi_{SSmean} = \omega + \Omega + M - GST \qquad (3.21)$$

Equation (2.31) can be used to calculate the true anomaly, and because of the small eccentricity, this can be approximated as

$$v = M + 2e\sin M \qquad (3.22)$$

The two-line elements for the Intelsat series, obtained from Celestrak at http://celestrak.com/NORAD/elements/intelsat.txt, are shown in Table 3.2.

Example 3.6   Using the data given in Table 3.2, calculate the longitude for INTELSAT 10-02.

Solution   From Table 3.2 the inclination is seen to be 0.0079°, which makes the orbit almost equatorial. Also the revolutions per day are 1.00271159, or approximately geosynchronous. Other values taken from this table are:

epoch day = 124.94126775 days; year = 2005; Ω = 311.0487°; ω = 59.4312°; M = 190.2817°; and e = 0.0000613

From Table 2.2 the Julian day for Jan 2005 is JD = 2453370.5 days. The Julian day for epoch is JD = 2453370.5 + 124.94126775 = 2453495.44126775 days. The reference value is (see Eq. 2.26) JD_{ref} = 2415020 days. Hence T in Julian centuries is:

$$T = \frac{JD - JD_{ref}}{36525}$$

$$= \frac{38475.442}{36525}$$

$$= 1.05340017$$

The decimal fraction of the epoch gives the UT as a fraction of a day, and in degrees this is:

$$UT° = 0.94126775 \qquad 360° = 338.85637°$$

Substituting these values in Eq. (2.34) gives, for the GST:

$$GST = 99.9610° + 36000.7689° \times T + 0.0004° \times T^2 + UT°$$
$$= 201.764° \;(\text{mod } 360°)$$

Equation (3.22) gives:

$$v = M + 2e\sin M = 3.32104 \text{ rad} + 2 \times .0000613 \sin(190.2817°) = 190.28°$$

Equation (3.20) then gives:

$$\phi_{SS} = \omega + \Omega + v - GST = 59.4313° + 311.0487° + 190.2804° - 201.764° = 359.00°$$

and Eq. (3.21) gives:

$$\phi_{SSmean} = \omega + \Omega + M - GST$$
$$= 59.4313° + 311.0487° + 190.2804° - 201.764° = 359.00°$$

From Table 1.3 the assigned spot for INTELSAT 10-02 is 359° east.

Modified inclination and eccentricity parameters can be derived from the specified values of inclination i, the eccentricity e, and the angles ω and Ω. Details of these are found in Maral and Bousquet (1998).

3.6   Earth Eclipse of Satellite

If the Earth's equatorial plane coincides with the plane of the Earth's orbit around the sun (the ecliptic plane), geostationary satellites are eclipsed by the Earth once each day. As it is, the equatorial plane is tilted at an angle of 23.4° to the ecliptic plane, and this keeps the satellite in full view of the sun for most days of the year, as illustrated by position A in Fig. 3.8. Around the spring and autumnal equinoxes, when the sun is crossing the equator, the satellite passes into the Earth's shadow at certain periods, these being periods of eclipse as illustrated in Fig. 3.8. The spring equinox is the first day of spring, while the autumnal equinox is the first day of autumn.

Eclipses begin 23 days before equinox and end 23 days after equinox. The eclipse lasts about 10 min at the beginning and end of the eclipse period and increases to a maximum duration of about 72 min at full eclipse (Spilker, 1977). During an eclipse, the solar cells are not illuminated by the sun so that operating power must be supplied from batteries. This is discussed further in Sec. 7.2, and Fig. 7.3 shows eclipse time as a function of days of the year.

Where the satellite longitude is east of the earth station, the satellite enters eclipse during daylight (and early evening) hours for the earth station, as illustrated in Fig. 3.9. This can be undesirable if the satellite must operate on reduced battery power. Where the satellite longitude is west of the earth station, eclipse does not

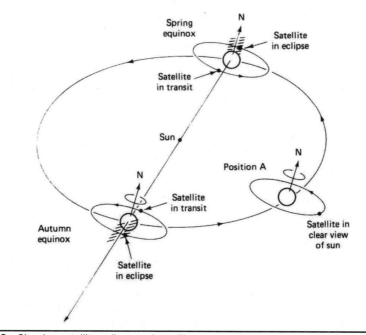

FIGURE 3.8   Showing satellite eclipse and satellite sun transit around spring and autumn equinoxes.

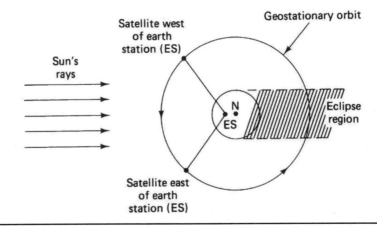

FIGURE 3.9   A satellite east of the earth station enters eclipse during daylight and early evening (high activity) hours at the earth station. A satellite west of the earth station enters eclipse during night and early morning (low activity) hours.

occur until the earth station is in darkness (or early morning) when usage is likely to be low. Thus satellite longitudes that are west, rather than east, of the earth station are more desirable.

3.7   Sun Transit Outage

Another event that must be considered during the equinoxes is the transit of the satellite between Earth and sun (see Fig. 3.8), such that the sun comes within the beamwidth of the earth-station antenna. When this happens, the sun appears as an extremely noisy source that completely blanks out the signal from the satellite. This effect is termed *sun transit outage*. It lasts for short periods—each day for about 6 days around the equinoxes. The occurrence and duration of the sun transit outage depends on the latitude of the earth station, a maximum outage time of 10 min being typical.

3.8   Launching Orbits

Satellites may be *directly injected* into low-altitude orbits, up to about 200 km altitude, from a launch vehicle. The United States has two principal space launch facilities at Vandenberg Space Force Base (retrograde orbits) and Cape Canaveral SSB (prograde orbits) for launching both governmental and civilian payloads into space using both expendable and reusable liquid propellant rockets. The major US launch vehicles include: Titan, Atlas-Centaur, and Delta, and a single reusable booster the Falcon 9.

Also, within the commercial SATCOM community there is much competition for launches among the following countries that have such facilities. The major suppliers are the European Space Agency using the Ariane rocket launched from Kourou, French Guiana; Russia using the Proton and Zenit rockets from Baikonur, Kazakhstan and a floating ocean platform in the Pacific; Japan using the H-II rocket from Tanegashima, Japan; and China using the Long March 5 rocket launched from both Wenchan, Hainan and Xichang, Sichuan, China.

As mentioned above, during the past several years, the SpaceX corp. has developed the Falcon 9 reusable first-stage rocket, which offers significant lost cost savings. This has revolutionized the satellite launch market, and now many such re-usable boosters are under development within the international launcher community.

3.8.1 GEO Launching Scenario

For orbit injection, the spacecraft must achieve a selected altitude and velocity combination, in a desired orbit plane, and for orbital altitudes >200 km, the *direct injection* approach is not the most energy-efficient approach in terms of launch vehicle thrust energy required. Thus, the satellite is usually placed into an initial LEO *parking orbit*, which is followed by an elliptical transfer orbit to reach the final high-altitude orbit. In most cases, the transfer orbit is selected to minimize the energy required for transfer, and such an orbit is known as a *Hohmann transfer* orbit. However, it should be noted that the time required for transfer is longer for this orbit than all other possible transfer obits.

For example, consider the case for GEO transfer, which is required between two circular orbits in the same plane as illustrated in Fig. 3.10. The Hohmann elliptical orbit is tangent to the low-altitude orbit at its perigee and tangent to the high-altitude orbit at its apogee. The process begins with the firing of a *perigee kick-motor* rocket at the desired perigee location to accelerate the satellite and to provide the desired delta-V. The resulting orbit is elliptical in altitude and the high point (apogee) is determined by the magnitude of the delta-V provided at perigee. Once the desired orbit apogee is achieved, an *apogee kick motor* rocket is fired at this location to impart additional delta-V that circularizes the orbit and becomes geostationary.

Traditionally, the satellites carried two solid propellant (single use) rockets for perigee and apogee kick motors; however, modern technology often uses a single multi-use liquid motor for this purpose. A typical example of a GEO launch, using an expendable Atlas-Centaur vehicle, is shown in Fig. 3.11. During the first-stage Atlas booster burn, the satellite is placed into the parking orbit, and the transfer orbit injection is initiated at launch time +27 min, with the Centaur second-stage burn for transfer orbit injection.

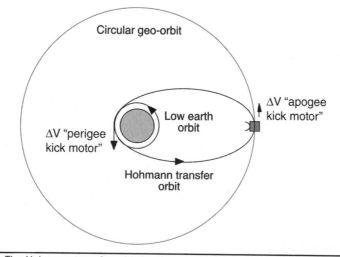

FIGURE 3.10 The Hohmann transfer orbit for a typical GEO launch scenario.

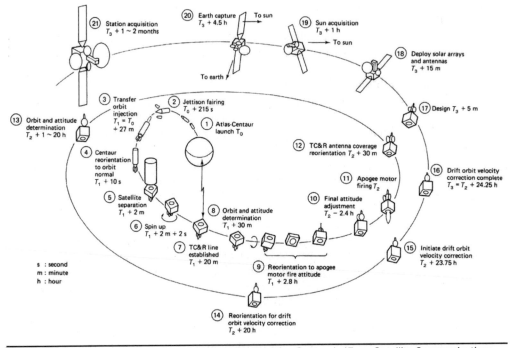

FIGURE 3.11   From launch to station of INTELSAT V (by Atlas-Centaur). (*From Satellite Communications Technology, edited by K. Miya, 1981; courtesy of KDD Engineering & Consulting, Inc., Tokyo.*)

After about 2 min of the rocket burn, the satellite has reached the required velocity, and the motor is shutdown. Thus, having fulfilled its launch requirements, the Centaur stage is separated from the satellite (to remove useless mass).

During the next 2.4 h, the satellite travels in the transfer orbit to the apogee, at which time the apogee kick motor is fired to circularize the orbit. Normally, the satellite is not at the assigned longitude, but it slowly drifts toward that point for the next several weeks, while the satellite undergoes comprehensive on-orbit check-out and testing before becoming fully operational. Throughout the launch and acquisition phases, a network of ground stations, spread across the Earth, is used to perform the *tracking, telemetry, and command* (TT&C) functions.

Now, consider the usual case where the orbits do not occur in the same plane. Since satellite velocity changes in the same plane change the ellipticity of the orbit (but not its inclination), a velocity change is required normal to the orbital plane. These changes are made at either the apogee or perigee, without affecting the other orbital parameters. Because energy must be expended to make any orbital changes, a geostationary satellite is initially launched into an orbit with as low an orbital inclination as possible, which is equal to the latitude of the launch site. Thus, GEOSAT launching from sites not on the equator are disadvantaged, because the satellite must carry extra fuel to affect a change in inclination.

Prograde (direct) orbits (Fig. 2.4) have an easterly component of velocity, so prograde launches gain from the Earth's rotational velocity. For a given launcher size, a significantly larger payload can be launched in an easterly direction than is possible with a retrograde (westerly) launch. In particular, easterly launches are used for the initial launch into the geostationary orbit.

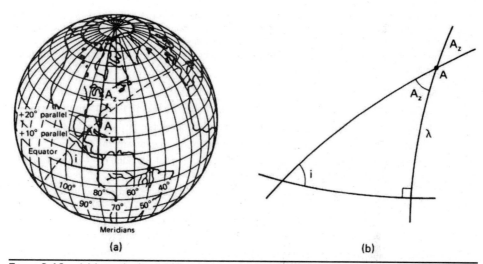

Figure 3.12 (a) Launch site A @ Cape Canaveral, FL, showing launch azimuth A_z; (b) enlarged version of the spherical triangle shown in (a). λ is the latitude of the launch site.

The relationship between inclination, latitude, and azimuth may be seen as follows (this analysis is based on that given in Bate et al. [1971]). Figure 3.12a shows the geometry at the launch site A at latitude λ (the slight difference between geodetic and geocentric latitudes may be ignored here). The dotted line shows the satellite earth track, the satellite having been launched at some azimuth angle A_z, where the angle i is the resulting inclination.

The spherical triangle of interest is shown in more detail in Fig. 3.12b. This is a right spherical triangle, and Napier's rule for this gives

$$\cos i = \cos \lambda \sin A_z \tag{3.23}$$

For a prograde orbit (see Fig. 2.4 and Sec. 2.5), $0 \le i \le 90°$ and hence $\cos i$ is positive. Also, $-90° \le \lambda \le 90°$, and hence $\cos \lambda$ is also positive. It follows therefore from Eq. (3.23) that $0 \le A_z \le 180°$, or the launch azimuth must be easterly to obtain a prograde orbit, confirming what was already known.

For a fixed λ, Eq. (3.23) also shows that to minimize the inclination i, $\cos i$ should be a maximum, which requires $\sin A_z$ to be maximum, or $A_z = 90°$. Equation (3.23) shows that under these conditions

$$\cos i_{min} = \cos \lambda \tag{3.24}$$

or

$$i_{min} = \lambda \tag{3.25}$$

Thus the *lowest* inclination possible on initial launch is equal to the latitude of the launch site. This result confirms the converse statement made in Sec. 2.5 under *inclination* that the greatest latitude north or south is equal to the inclination. From Cape Kennedy, which is located at a latitude of approximately 28°, the smallest initial inclination that can be achieved for easterly launches is approximately 28°.

Example 3.7 Consider an ideal GEOSAT launch scenario from a circular parking orbit of 150 km, shown in Fig. 3.10. Calculate the delta-V required for the perigee kick motor (PKM) to place the satellite in the desired Hohmann transfer orbit of 150 km × 35,786.06 km altitude. Also, calculate the delta-time from the PKM firing until the apogee kick motor (AKM) firing, and calculate the AKM delta-V required to circularize the orbit.

Solution The solution has the following steps:

Step 1: Using the earth radius $R = 6378.137$ km, calculate the orbit radius for the parking orbit.

Step 2: Using Eq. (2.7), calculate the satellite orbital speed for the parking orbit.

Step 3: Using Eq. (2.9), calculate the semi-major axis for the transfer orbit.

Step 4: For the desired Hohmann orbit, using Eq. (2.6) calculate the orbit eccentricity, and using Eq. (2.10) calculate the orbit speed @ perigee.

Step 5: Calculate the orbital speed difference between step 4 and step 2, which is the required PKM delta-V.

Step 6: Set T_o as the time of the start of the PKM firing and using Keppler's third law Eq. (2.5b), calculate the period of the Hohmann transfer orbit. The time to AKM firing is the period/2.

Step 7: Using Eq. (2.8) calculate the orbit speed @ apogee for the transfer orbit.

Step 8: using Eq. (2.7), calculate the satellite orbital speed for the circular GEO orbit.

Step 9: Calculate the orbital speed difference between step 8 and step 7, which is the required AKM delta-V.

The results are tabulated below.

Alt @ Perigee, km	Radius @ Perigee, km	Alt @ Apogee, km	Radius @ Apogee, km	Semi-Major Axis, km	Eccentricity, e	Speed Perigee km/s	Speed Apogee km/s	Period, s
450.00	6,828.14	450.00	6,828.14	6,828.14	0.0000	7.640	7.640	5,615.2
450.00	6,828.14	35,786.06	42,164.20	24,496.17	0.7213	10.024	1.623	38,155.6
35,786.06	42,164.20	35,786.06	42,164.20	42,164.20	0.0000	3.075	3.075	86,164.1

Example 3.8 Consider a GEOSAT launch scenario from a parking orbit of 450 km, in the equatorial plane. At time 0 s, the solid propellant perigee kick motor (PKM) was fired, but the kick motor thrusting was greater than expected. This resulted in a transfer orbit of 450 km × 40,347.8 km (eccentricity = 0.7450 see plot attached in Fig. 3-13).

Fortunately, we may use the satellite thrusters to maneuver the satellite into its desired geosynchronous orbit. The properly aligned thruster may be fired at perigee to reduce the velocity to the required value for the Hohmann orbit of 450 × 35,786.1 km. Assume that the thruster can be fired in 0.1 s increments using a linear relationship between delta-V and thruster burn time, with a slope of 1 m/s/s burn. After reaching the proper transfer orbit, an apogee kick motor (AKM) is fired to circularize the orbit.

Find the firing time for the first thruster firing relative to the initial perigee kick motor firing and find the required thruster burn duration. Also find the delta-V requirement for the apogee kick motor to circularize the new transfer orbit to GEO.

Solution Following the steps of Example 3.6 the following table is presented:

Alt @ Perigee, km	Radius @ Perigee, km	Alt @ Apogee, km	Radius @ Apogee, km	Semi-Major Axis, km	Eccentricity, e	Speed Perigee, km/s	Speed Apogee, km/s	Period, s
450.0	6,828.1	40347.8	46,726.0	26,777.1	0.7450	10.093	1.475	43,607
450.0	6,828.1	35786.1	42,164.2	24,496.2	0.7213	10.024	1.623	38,156
35,786.1	42,164.2	35786.1	42,164.2	42,164.2	0.0000	3.075	3.075	86,164

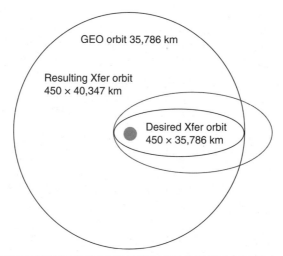

Figure 3.13 Failed GEO launch scenario in Example 3.8.

The first row represents the failed transfer orbit, the second row is the desired Hohmann transfer orbit, and the last row is the desired circular geostationary orbit.

Consider the first row, which started with the PKM firing at perigee T_o s). After one orbit (43,607 s) the satellite will be located at the perigee starting point, and its speed will be 10.093 km/s. However, the desired speed @ perigee is given by the second row (10.024 km/s). So, we desire that the thruster reduce the speed @ perigee by

$$\text{delta-V}_p = 10.024 - 10.093 = -0.069 \text{ km/s} = -69.0 \text{ m/s}$$

which means that the thruster burn will be 69.0 s duration @ $T_o + 43,607$ s

For simplicity, assume that the delta-V_p occurred instantaneously, and at this time the satellite is on the desired transfer orbit. After 1/2 of the new orbit period = 38,155.6/2 = 19,077.8 s, the satellite will be at the orbit apogee. At this time, its speed is 1.623 km/s, but for the desired GEO orbit, the speed must be 3.075 km/s. Therefore, the AKM must deliver a delta-Va = 3.075 − 1.623 = 1.452 km/s and afterwards the satellite will be on the desired GEO orbit.

3.9 Problems

3.1. Explain what is meant by the geostationary orbit. How do the geostationary orbit and a geosynchronous orbit differ?

3.2. (*a*) Explain why there is only one geostationary orbit. (*b*) Show that the range *d* from an earth station to a geostationary satellite is given by

$$d = \sqrt{(R\sin El)^2 + h(2R + h)} - R\sin El,$$

where *R* is the Earth's radius (assumed spherical), *h* is the height of the geostationary orbit above the equator, and *El* is the elevation angle of the earth station antenna.

3.3. Determine the latitude and longitude of the farthest north earth station which can link with any given geostationary satellite. The longitude should be given relative to the satellite longitude, and a minimum elevation angle of 5° should be assumed for the earth station antenna. A spherical Earth of mean radius 6378.137 km may be assumed.

3.4.   An earth station at latitude 30°S is in communication with an earth station on the same longitude at 30°N, through a geostationary satellite. The satellite longitude is 20° east of the earth stations. Calculate the antenna-look angles for each earth station and the round-trip time, assuming this consists of propagation delay only.

3.5.   Determine the maximum possible longitudinal separation which can exist between a geostationary satellite and an earth station while maintaining line-of-sight communications, assuming the minimum angle of elevation of the earth station antenna is 5°. State also the latitude of the earth station.

3.6.   An earth station is located at latitude 35°N and longitude 100°W. Calculate the antenna-look angles for a satellite at 67°W.

3.7.   An earth station is located at latitude 12°S and longitude 52°W. Calculate the antenna-look angles for a satellite at 70°W.

3.8.   An earth station is located at latitude 35°N and longitude 65°E. Calculate the antenna-look angles for a satellite at 19°E.

3.9.   An earth station is located at latitude 30°S and longitude 130°E. Calculate the antenna-look angles for a satellite at 156°E.

3.10.   Calculate for your home location the look angles required to receive from the satellite (*a*) immediately east and (*b*) immediately west of your longitude.

3.11.   CONUS is the acronym used for the 48 contiguous states. Allowing for a 5° elevation angle at earth stations, verify that the geostationary arc required to cover CONUS is 55° to 136°W.

3.12.   Referring to Prob. 3.11, verify that the geostationary arc required for CONUS plus Hawaii is 85° to 136° W and for CONUS plus Alaska is 115° to 136°W.

3.13.   By considering the Mississippi River as the dividing line between east and west, verify that the western region of the United States can be covered by satellites in the geostationary arc from 136° to 163°W and the eastern region by 25° to 55°W. Assume a 5° angle of elevation.

3.14.   (*a*) An earth station is located at latitude 35°N. Assuming a polar mount antenna is used, calculate the angle of tilt. (*b*) Would the result apply to polar mounts used at the earth stations specified in Probs. 3.6 and 3.8?

3.15.   Repeat Prob. 3.14 for an earth station located at latitude 12°S. Would the result apply to a polar mount used at the earth station specified in Prob. 3.7?

3.16.   Repeat Prob. 3.14 for an earth station located at latitude 30°S. Would the result apply to a polar mount used at the earth station specified in Prob. 3.9?

3.17.   Calculate the angle of tilt required for a polar mount antenna used at your home location.

3.18.   The borders of a certain country can be roughly represented by a triangle with coordinates 39°E, 33.5°N; 43.5°E, 37.5°N; 48.5°E, 30°N. If a geostationary satellite must be visible from *any point* in the country, determine the limits of visibility (i.e., the limiting longitudinal positions for a satellite on the geostationary arc). Assume a minimum angle of elevation for the earth station antenna of 5°. Show which geographic location fixes which limit.

3.19.   Explain what is meant by the *earth eclipse* of an earth-orbiting satellite. Why is it preferable to operate with a satellite positioned west, rather than east, of earth station longitude?

3.20. Explain briefly what is meant by *sun transit outage.*

3.21. Using the data given in Fig. 3.7, calculate the longitude for INTELSAT 904.

3.22. Calculate the semi-major axis for INTELSAT 901.

3.23. Calculate the apogee and perigee heights for INTELSAT 906.

3.24. Calculate the rate of regression of the nodes and the rate of rotation of the line of apsides for INTELSAT 907.

3.25 A satellite was launched into a 0° inclination LEO parking orbit of 150 km × 250 km, and the desire is to reach a GEO orbit. Determine the least PKM delta-V_p to reach the GEO orbit (i.e., determine which location (apogee or perigee) of the parking orbit will require the minimum PKM performance to achieve the desired Hohmann Xfer orbit to GEO?

3.26 Using Example 3.8 as a guide consider the following failed GEOSAT launch scenario. A satellite was launched into a 0° inclination LEO parking orbit of 150 km × 250 km, and the desire is to reach a GEO orbit. Assume that the PKM under-performs and delivers 85% of the required delta-V_p. What are the new perigee and apogee altitudes of the failed transfer orbit? Fortunately, we may use the satellite thrusters to maneuver the satellite into its desired geosynchronous orbit. The properly aligned thruster may be fired in 0.1 s increments up to a maximum of 100 s, which corresponds to a delta-V of 100 m/s (assume a linear relationship between delta-V and thruster burn time of 1 m/s/s burn). After a burn, there is a minimum delay of 3 h before the next thruster firing to allow for cooling of the thruster. For minimum energy considerations, the thruster firings must occur at the transfer orbit perigee. Determine the minimum number of thruster firings and the delta-V_p for each to reach the desired Hohmann transfer orbit.

References

Bate, R. R., D. D. Mueller, and J. E. White. 1971. *Fundamentals of Astrodynamics.* Dover, New York.

Celestrak, at http://celestrak.com/NORAD/elements/intelsat.txt.

Mahon, J., and J. Wild. 1984. "Commercial Launch Vehicles and Upper Stages." *Space Commun. Broadcast.*, Vol. 2, pp. 339–362.

Maral, G., and M. Bousquet. 1998. *Satellite Communications Systems.* Wiley, New York.

Spilker, J. J. 1977. *Digital Communications by Satellite.* Prentice-Hall, Englewood Cliffs, NJ.

Wertz, J. R. (ed.). 1984. *Spacecraft Attitude Determination and Control.* D. Reidel, Holland.

WGS84, [https://nssdc.gsfc.nasa.gov/planetary/factsheet/earthfact.html].

CHAPTER 4

Radio Wave Propagation

4.1 Introduction

A signal traveling between an earth station and a satellite must pass through the Earth's atmosphere, including the ionosphere, as shown in Fig. 4.1, and this can introduce certain impairments, which are summarized in Table 4.1. Some of the more important of these impairments are described in this chapter.

4.2 Atmospheric Losses

Losses occur in the Earth's atmosphere because of energy absorption by atmospheric gases. These losses are treated quite separately from those which result from adverse weather conditions, which of course are also atmospheric losses.

The atmospheric absorption loss varies with frequency, as shown in Fig. 4.2. The figure is based on statistical data (CCIR Report 719-1, 1982). Two absorption peaks occur, the first one at a frequency of 22.225 GHz, resulting from resonance absorption by water vapor (H_2O), and the second one is a series of resonant oxygen (O_2) lines over the range of 50 to 70 GHz. However, at frequencies outside of these peaks, the absorption is quite low (typically > 97% transmissivity). The graph in Fig. 4.2 is for total atmospheric transmissivity at zenith incidence, that is, for an elevation angle of 90° at the earth-station antenna. Denoting this value of absorption loss as $[AA]_{90}$ dB, then for elevation angles down to 10°, an approximate formula for the absorption loss in decibels is (CCIR Report 719-1, 1982):

$$[AA] = [AA]_{90} \operatorname{cosec}(El) \tag{4.1}$$

where El is the angle of elevation.

In addition, an effect known as *atmospheric scintillation* can occur, which is a fading phenomenon that results in loss or significant attenuation of the signal, with the fading period being several tens of seconds (Miya, 1981). It is caused by differences in the atmospheric refractive index, which in turn results in focusing and defocusing of the radio waves, which follow different ray paths through the atmosphere. It may be necessary to make an allowance for atmospheric scintillation, through the introduction of a fade margin in the link power-budget calculations.

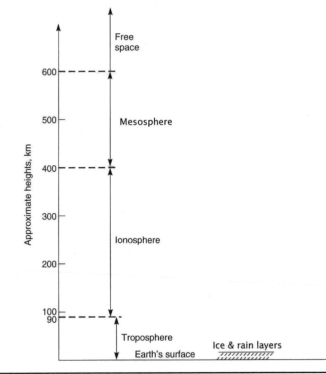

Figure 4.1   Simplified diagram of layers in the Earth's atmosphere.

Propagation Impairment	Physical Cause	Prime Importance
Attenuation and sky noise increases	Atmospheric gases, cloud, rain	Frequencies above about 10 GHz
Signal depolarization	Rain, ice crystals	Dual-polarization systems at C and Ku bands (depends on system configuration)
Refraction, atmospheric multipath	Atmospheric gases	Communication and tracking at low elevation angles
Signal scintillations	Tropospheric and ionospheric refractivity fluctuations	Tropospheric at frequencies above 10 GHz and low-elevation angles; ionospheric at frequencies below 10 GHz
Reflection multipath, blockage	Earth's surface, objects	Mobile satellite services on surface
Propagation delays, variations	Troposphere, ionosphere	Precise timing and location systems; *time division multiple access* (TDMA) systems
Intersystem interference	Ducting, scatter, diffraction	Mainly at C-band; rain scatter may be significant at higher frequencies

Source: Brussard and Rogers (1990).

Table 4.1   Propagation Concerns for Satellite Communications Systems

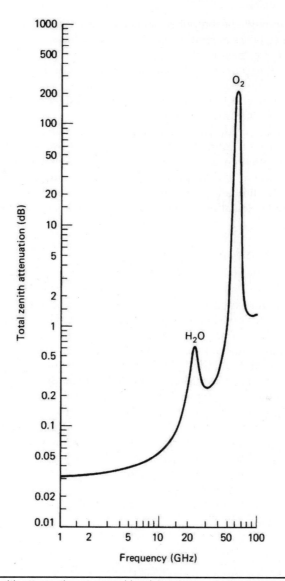

FIGURE 4.2 Total zenith attenuation at ground level: pressure = 1 atm, temperature = 20°C, and water vapor = 7.5 g/m³. (*Adapted from CCIR Report 719-2, with permission from International Telecommunication Union.*)

4.3 Ionospheric Effects

Radio waves traveling between satellites and earth stations must pass through the ionosphere. The ionosphere is the upper region of the Earth's atmosphere, which has been ionized, mainly by high-energy solar radiation. The free electrons in the ionosphere are not uniformly distributed; rather, they form in layers, which exhibit diurnal thickness variations, with increased plasma density during the day and reduced during the night. Furthermore, clouds of electrons (known as *traveling ionospheric disturbances*) may travel through the ionosphere and give rise to fluctuations in the signal that can only be determined on a statistical basis. The effects include *scintillation, absorption, variation in the direction of arrival, propagation delay, dispersion, frequency change,* and *polarization rotation* (CCIR Report 263-5, 1982). All these effects decrease as the frequency increases, most in inverse proportion to the frequency squared, and only the polarization rotation and scintillation effects are of major concern for satellite communications. Polarization rotation is described in Sec. 5.5.

Ionospheric scintillations are variations in the amplitude, phase, polarization, or angle of arrival of radio waves. They are caused by irregularities in the ionosphere that change with time. The main effect of scintillations is fading of the signal. The fades can be quite severe, and they may last up to several minutes. As with fading caused by atmospheric scintillations, it may be necessary to include a fade margin in the link power-budget calculations to allow for ionospheric scintillation.

4.4   Rain Attenuation

Rainfall attenuates the propagation of microwaves primarily by absorption of energy, but in addition, it also attenuates by scattering from raindrops, and the combined effect is known as rain extinction. Furthermore, rain extinction is a function of *rain rate* (R), which is the rate at which the rainwater accumulates in a container that is measured in units of millimeters of depth per hour. When the rain rate increases, there are more rain drops, but most importantly, their diameters increase by a factor > 2 from low to high rain rates, which dramatically increases the total liquid volume. The degree of microwave scattering is proportional to the ratio of raindrop diameter to the wavelength raised to a power > 1; therefore, for frequencies ($f > 10\,\text{GHz}$) and rain rates $R_p > 10\,\text{mm/h}$, scattering becomes a significant part of the total rain attenuation.

The occurrence of rainfall is a random variable, and its intensity is approximately logarithmically distributed in time (high rain rates much less frequent). Of interest is the percentage of time that specified values are exceeded, and this time percentage is usually for a 1-year period. For example, a rain rate exceedance of 0.001% means that the given rain rate would be exceeded for 0.001 percent of a year, or about 5.3 min randomly distributed during any 1 year. In this case the rain rate R_p is denoted by "subscript p" as R_{001}.

The *specific rain attenuation (extinction) coefficient* α is

$$\alpha = -k_e(f)R_p^{b(f)}, \text{dB/km} \tag{4.2}$$

where

$$k_e = k_a + k_s, \text{dB/km} \tag{4.3}$$

where "subscript e" is the total extinction, which compromises an absorption component "subscript a" and a scattering component "subscript s." Note these coefficients are additive negative dBs, and by convention the negative sign is omitted, but it is included in Eq. (4.2). *Throughout this text, the reader should be careful to include the sign of the dB and to note that losses are gains < 1 that are negative dBs, such as Eq. (4.2). Further since our equations for power ratios are defined as products of terms, when expressed as dB, the terms are always added (see App. G for further discussion).*

A simple analytical expression for the rain extinction coefficients for satellite communications have been provided from regression analysis (Olsen et al., 1978 and CCIR report, 1981) as

$$k_e(f) = (4.21e - 05)f^{2.42} \qquad \text{for } 2.9 \leq f \leq 54 \text{ GHz} \qquad (4.4)$$

and the exponent
$$b(f) = 0.851\, f^{0.158} \qquad \text{for } f < 8.5 \text{ GHz}$$
$$= 1.41\, f^{-0.0779} \qquad \text{for } 8.5 \leq f \leq 25 \text{ GHz}$$
$$= 2.63\, f^{-0.272} \qquad \text{for } 25 \leq f < 164 \text{ GHz} \qquad (4.5)$$

Also, since rain attenuation is polarization-dependent, values for $k_e(f)$ and $b(f)$ are available in tabular form in several publications, and the values in Table 4.2 have been abstracted from Table 4.3 of Ippolito (1986). The subscripts h and v refer to horizontal and vertical polarizations respectively.

Frequency, GHz	$k_{e,h}$	$k_{e,v}$	b_h	b_v
1	0.0000387	0.0000352	0.912	0.880
2	0.000154	0.000138	0.963	0.923
4	0.00065	0.000591	1.121	1.075
6	0.00175	0.00155	1.308	1.265
7	0.00301	0.00265	1.332	1.312
8	0.00454	0.00395	1.327	1.310
10	0.0101	0.00887	1.276	1.264
12	0.0188	0.0168	1.217	1.200
15	0.0367	0.0335	1.154	1.128
20	0.0751	0.0691	1.099	1.065
25	0.124	0.113	1.061	1.030
30	0.187	0.167	1.021	1.000

Source: Ippolito (1986, p. 46).

TABLE 4.2   Specific Attenuation Coefficients

For $p = 0.001\%$	$r_{0.001} = \dfrac{10}{10 + L_G}$
For $p = 0.01\%$	$r_{0.01} = \dfrac{90}{90 + 4L_G}$
For $p = 0.1\%$	$r_{0.1} = \dfrac{180}{180 + L_G}$
For $p = 1\%$	$r_1 = 1$

Source: Ippolito (1986).

TABLE 4.3   Reduction Factors

Once the specific attenuation coefficient α is found, then the total attenuation is

$$A = -\alpha L_{eff}, \text{ dB} \qquad (4.6)$$

where L_{eff} is the *effective slant path length* of the signal through the rain. Because the rain density is unlikely to be uniform over the actual path length, a statistical effective path length is used rather than the actual (geometric) length L_S given in Fig. 4.3. This depends on the antenna angle of elevation θ and the *rain height* h_R, which usually occurs at the height of the freezing level (273 K) in the atmosphere. It should be noted that rain effects are highly location-dependent; therefore, curves of h_R are shown in Fig. 4.4 for three climate zones labeled: Method 1—*maritime climates*; Method 2—*tropical climates*; Method 3—*continental climates*. For the last zone, curves are shown for p values of 0.001, 0.01, 0.1, and 1 percent.

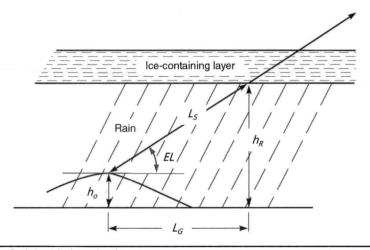

Figure 4.3 Path length through rain.

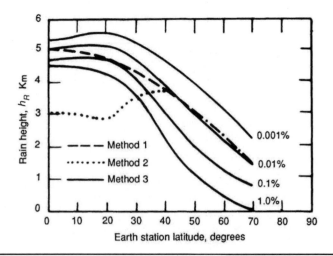

Figure 4.4 Rain height as a function of Earth-station latitude for different climatic zones.

For small angles of elevation ($El < 10°$), the determination of L_S is complicated by the Earth curvature (see CCIR Report 564-2, 1982). However, for $El \geq 10°$ a flat Earth approximation may be used, and from Fig. 4.3 it is seen that

$$L_S = \frac{(h_R - h_0)}{\sin(El)} \tag{4.7}$$

Then, the effective path length is given in terms of the slant length by

$$L_{eff} = L_s r_p, \text{ km} \tag{4.8}$$

where r_p is a *reduction factor*, which is a function of the percentage time p and L_G, the horizontal projection of L_S. From Fig. 4.3 the horizontal projection is

$$L_G = L_s \cos(El) \tag{4.9}$$

The reduction factors are given in Table 4.3.

Combining these factors together into one equation, the rain attenuation in decibels is given by

$$A_p = -\alpha R_p^b L_s r_p, \text{ dB} \tag{4.10}$$

Detailed computation of rain attenuation requires information about the drop size density and rain height (e.g., Ulaby and Long, 2014). However, for the purposes of computing rain attenuation, simplified models and assumptions can be used, but different climates and seasons have different mean heights. Thus, interpolation formulas, which depend on the climate zone being considered (see Fig. 16.8), are available for values of p other than those quoted earlier (see, e.g., Ippolito, 1986), and polarization shifts resulting from rain are described in Sec. 5.6.

For a given rain rate, the rain attenuation increases exponentially with the carrier frequency as shown in Fig. 4.5. For example, at the C- (6 GHz) uplink frequency, for a medium rain rate of 10 mm/h, a typical rain attenuation at is $A = -0.25$ dB, and the corresponding value for the Ku-band (14 GHz) is $A = -2.04$ dB, and this ratio of the attenuation in dB diverges with increasing rain rate.

In Sec. 12.9, we will consider the impacts of rain on the satellite link, and sometimes it will be necessary to consider rain absorption and scattering separately. Therefore, in Fig. 4.6, we characterize the frequency response of rain single scattering albedo defined as

$$AB = \frac{k_s}{k_e}, \text{ (dB/km)/(dB/km), ratio} \tag{4.11}$$

This figure, for homogeneous rain rates of 12, 24, and 48 mm/h, is an empirical characterization based upon published calculations of Mie extinction and absorption coefficients in rain (Ulaby and Long, 2014, Section 8-8.2). While, this figure is qualitatively correct, the detailed values are extrapolations of published material, and therefore only represents a typical case. It is included for only educational purposes, to demonstrate how to perform link calculations in rain with significant scattering loss.

At a frequency of 10 GHz, the intercept of the 12 mm/h rain rate curve is 0.1 albedo, which represents 10% scattering and 90% (absorption) loss. As the rain rate doubles to 24 mm/h, the 10 GHz albedo increases to about 15%, and in the limit of increasing rain rate, the albedo asymptotically approached about 60% scattering contribution for high frequencies. However, even at the highest rain rates (>100 mm/h), the scattering never

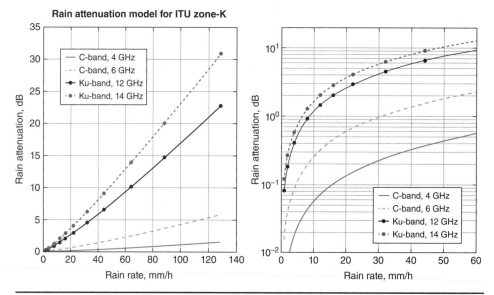

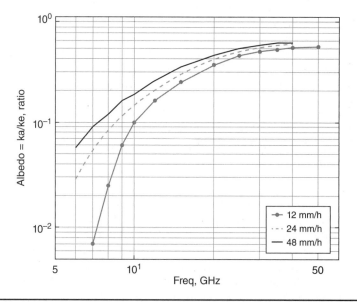

Figure 4.5 Typical rain attenuation for C-band (3 and 6 GHz) and Ku-band (12 and 14 GHz) for ITU climate zone-K. Left panel is a linear plot of [A] in dB as a function of R in mm/h, and the right panel is the same data presented in a semilog format.

Figure 4.6 Single scattering albedo (k_s/k_e) for three rain rates, where $k_{rain} = k_a + k_s$, in units of dB/km.

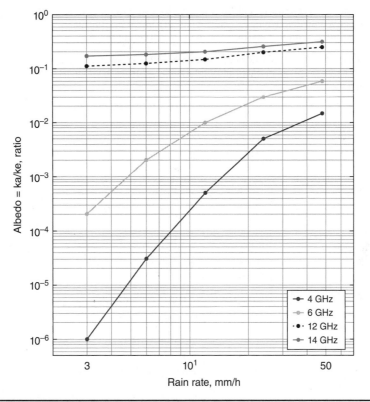

FIGURE 4.7   Typical rain albedo (k_s/k_e) for C- and Ku-bands.

becomes an issue for the C-band frequencies of 4 and 6 GHz, so the approximation of neglecting scattering for $f < 10$ GHz is well justified. Further, for the convenience of the reader, the C- and Ku-band rain albedo is presented in Fig. 4.7 as a function of rain rate, and the results are given in Table 4.4.

Finally, studies have shown (CCIR Report 338-3, 1978) that the rain attenuation for horizontal polarization is considerably greater than for vertical polarization. Therefore, rain produces significant performance impacts on the satellite circuit, which are discussed in detail in Sec. 12.9.

(Rain Rate)/ Freq	4 GHz	6 GHz	12 GHz	14 GHz
3	1.00E-06	2.00E-04	0.10819	0.17017
6	3.00E-05	2.00E-03	0.12298	0.18314
12	5.00E-04	1.00E-02	0.15021	0.20756
24	5.00E-03	3.06E-02	0.19534	0.25024
48	1.50E-02	5.99E-02	0.24825	0.31104

TABLE 4.4   Rain Scattering Albedo for C- and Ku-Bands (from Fig. 4.7)

Example 4.1   For a carrier frequency of 12 GHz and for horizontal and vertical polarizations, calculate the rain attenuation which exceeds for 0.01 percent of the time in any year, for a $R_{01} = 10$ mm/h. The earth station altitude is 600 m, the antenna elevation angle is 50°, and assume the rain height is 3 km.

Solution   Because the CCIR formula contains hidden conversion factors, units are not attached to the data, and it is understood that all lengths and heights are in kilometers, and rain rate is in millimeters per hour.

$$El = 50°;\ h_0 = 0.6;\ h_r = 3;\ R_{01} = 10$$

From Eq. (4.7):

$$L_S = \frac{(h_R - h_0)}{\sin(El)} = \frac{(3 - 0.6)}{\sin(50°)} = 3.13 \text{ km}$$

From Eq. (4.9):

$$L_G = L_S\cos(50°) = 3.133\ \cos(50°) = 2.014 \text{ km}$$

From Table 4.3, the reduction factor is

$$r_{01} = \frac{90}{(90 + 4L_G)} = 0.9178$$

For horizontal polarization, from Table 4.2 at $f = 12$ GHz; $a_h = 0.0188$; $b_h = 1.217$

From Eq. (4.10):

$$A_p = -\alpha_h R_{01}{}^{b_h} L_s r_{01}, = -0.0188 \times 10^{1.217} \times 3.133 \times 0.9178, = -0.891 \text{ dB}$$

For vertical polarization, from Table 4.2 at $f = 12$ GHz; $a_v = 0.0168$; $b_v = 1.2$

$$A_p = -0.0168 \times 10^{1.2} \times 3.133 \times 0.9178 = -0.766 \text{ dB}$$

The corresponding equations for circular polarization are

$$a_c = \frac{a_h + a_v}{2} \tag{4.12a}$$

$$b_c = \frac{a_h b_h + a_v b_v}{2a_c} \tag{4.12b}$$

The attenuation for circular polarization is compared with that for linear polarization in the following example.

Example 4.2   Repeat Example 4.1 for circular polarization.

Solution   From Eq. (4.12a):

$$a_c = \frac{(a_h + a_v)}{2} = \frac{(0.0188 + 0.0168)}{2} = 0.0178$$

From Eq. (4.12b):

$$b_c = \frac{a_h b_h + a_v b_v}{2a_c} = \frac{0.0188 \times 1.217 + 0.0168 \times 1.2}{2 \times 0.0178} = 1.209$$

From Eq. (4.10):

$$A_p = -a_c R_{01}^{b_c} L_S r_{01}$$
$$= -0.0178 \times 10^{1.209} \times 3.133 \times 0.9178 = -0.83 \text{ dB}$$

4.5 Rain Exceedance

As was mentioned in Sec. 4.4, the occurrence of rainfall is a random variable, and its intensity is approximately logarithmically distributed in time (high rain rates much less frequent). Therefore, the statistics of rain attenuation, denoted as A_p in Eq. (4.10), are usually available in the form of curves or tables showing the fraction of time that a given rain rate or attenuation is exceeded, or equivalently, the probability that a given attenuation will be exceeded (see Hogg and Chu, 1975; Crane, R. K., 1980; Lin et al., 1980, and Pratt et al., 2003). An example of typical exceedance tables, for ITU climate zone-K (see Fig. 16.8) over Southern Canada and Northern United States, is given for C- and Ku-band in Fig. 4.8 and in tabular form in Table 4.4. These data are developed using the simple attenuation model given by Eqs. (4.4) and (4.5), with the effective rain path length, $L_{eff} = 4.7$ km (Tables 4.5 and 4.6).

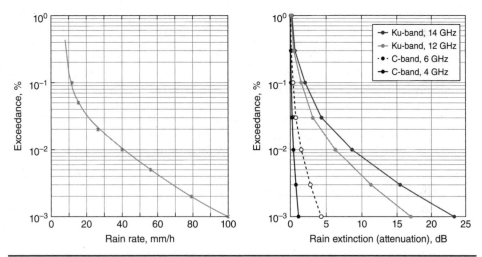

Figure 4.8 Rain exceedance curve (left panel) for the ITU climate zone-K and corresponding rain attenuation exceedance curves (right panel) for C-band and Ku-band.

Exceedance %	0.440	0.140	0.086	0.062	0.042	0.040	0.33	0.029	0.025
R_p, mm/h	8	10	12	14	16	18	20	22	24

Exceedance %	0.025	0.0223	0.0198	0.0177	0.0159	0.0143	0.0129	0.0117	0.0106
R_p, mm/h	24	26	28	30	32	34	36	38	40

Exceedance %	0.0106	0.00959	0.00871	0.00793	0.00722	0.00658	0.00423	0.00281	0.00193
R_p, mm/h	40	42	44	46	48	50	60	70	80

Table 4.5 Exceedance Rain Rates (from Fig. 4.8)

Exceedance	R_p	A_p, dB	A_p, dB	A_p, dB	A_p, dB
%	mm/h	4 GHz	6 GHz	12 GHz	14 GHz
0.100	12	−0.13	−0.32	−1.45	−2.04
0.05	15.4	−0.18	−0.43	−1.94	−2.71
0.02	26.5	−0.36	−0.84	−3.64	−5.06
0.010	42	−0.64	−1.48	−6.22	−8.58
0.005	56	−0.93	−2.10	−8.69	−11.94
0.002	79	−1.43	−3.20	−12.96	−17.72
0.001	100	−1.93	−4.29	−17.05	−23.23

TABLE 4.6   Exceedance Rain Rates and Associated Rain Extinctions (from Fig. 4.8)

The left panel of Fig. 4.8 is the R_p exceedance curve (cumulative distribution function), where the y axis is the cumulative probability in %, and the x axis is the R_p in mm/h. The cumulative probability is the % of 1 year that the abscissa value is exceeded. For example, a y value of 10^2 (0.01%) has a corresponding x value of 42 mm/h, which means that the R_p exceeds 42 mm/h only 0.01% of the time (~52.6 min/yr). Alternatively, one could interpret this as 99.99% of the time, the $R_p < 42$ mm/h.

The right panel in Fig. 4.8 is the corresponding A_p exceedance curve for C-band (4 and 6 GHz) and Ku-band (12 and 14 GHz) that corresponds to the R_p given in the left panel. For example, consider the 12 GHz curve in this figure, where for 0.01% the $R_{01} = 42$ mm/h (from left panel) and the corresponding $A_{01} = -6.22$ dB.

4.6   Other Propagation Impairments

Because of the large difference in dielectric constant of liquid and frozen water, hail, ice, and snow have little effect on attenuation because of the low liquid water content (Ulaby and Long, 2014), but ice can cause depolarization, described briefly in Chap. 5. On the other hand, clouds are formed by very small water droplets, and thereby contain small volumes of liquid water. As a result, clouds are a weak source of signal attenuation that varies with cloud liquid water density, and can be calculated like that for rain (Ippolito, 1986, p. 56; Ulaby and Long, 2014). For example, at a frequency of 10 GHz and for light clouds (water content of 0.25 g/m³), the specific attenuation is about −0.05 dB/km, and for heavy clouds (water content of 2.5 g/m³) the corresponding attenuation is about −0.2 dB/km. Furthermore, cloud thicknesses are typically of order 1–2 km, which results in a range of attenuations of −0.05 to −0.4 dB. Finally, high altitude clouds are formed of ice crystals and have negligible attenuation (Ulaby and Long, 2014).

4.7   Problems and Exercises

4.1. With reference to Table 4.1, identify the propagation impairments which most affect transmission in the C-band.

4.2. Repeat Prob. 4.1 for Ku-band transmissions.

4.3. Calculate the approximate value of atmospheric attenuation for a satellite transmission at 14 GHz, for an angle of elevation of the Earth-station antenna equal to 15°.

4.4. Calculate the approximate value of atmospheric attenuation for a satellite transmission at 6 GHz, for an angle of elevation of the Earth-station antenna equal to 30°.

4.5. Describe the major effects that the ionosphere has on the transmission of satellite signals at frequencies of (*a*) 4 GHz and (*b*) 12 GHz.

4.6. Explain what is meant by *rain rate* and how this is related to specific attenuation.

4.7. Compare the specific attenuations for vertical and horizontal polarization at a frequency of 4 GHz and a point rain rate of 8 mm/h which is exceeded for 0.01 percent of the year.

4.8. Repeat Prob. 4.7 for a frequency of 12 GHz.

4.9. Explain what is meant by *effective path length* in connection with rain attenuation.

4.10. For a satellite transmission path, the angle of elevation of the earth station antenna is 35°, and the earth station is situated at mean sea level. The signal is vertically polarized at a frequency of 18 GHz. The rain height is 1 km, and a rain rate of 10 mm/h is exceeded for 0.001% of the year. Calculate the rain attenuation under these conditions.

4.11. Repeat Prob. 4.10 for the rain rate that is exceeded (*a*) 0.01% and (*b*) 0.1% of the year.

4.12. Given a satellite transmission: $El = 22°$, $R_{0.01} = 15$ mm/h, $h_0 = 600$ m, $h_R = 1500$ m, and horizontal polarization, calculate the rain attenuation for a signal frequency of 14 GHz.

4.13. Determine the specific attenuation for a circularly polarized satellite signal at a frequency of 4 GHz, where a point rain rate of 8 mm/h is exceeded for 0.01% of the year.

4.14. A circularly polarized wave at a frequency of 12 GHz is transmitted from a satellite. The point rain rate for the region is $R_{0.01} = 13$ mm/h. Calculate the specific attenuation.

4.15. Given that for Prob. 4.13 the earth station is situated at altitude 500 m and the rain height is 2 km, calculate the rain attenuation, when the angle of elevation of the path is 35°.

4.16. Given that for Prob. 4.14 the earth station is situated at altitude 200 m and the rain height is 2 km, calculate the rain attenuation, when the angle of elevation of the path is 25°.

4.17. Given a Ka-band uplink operating at 20 GHz, in the ITU climate zone-K. Assuming an effective rain path distance of 3 km, calculate the expected rain attenuation for 99.99% link availability.

4.18. For Prob. 4.17, calculate the scattering and absorption components of the rain attenuation.

References

Brussard, G., and D. V. Rogers. 1990. "Propagation Considerations in Satellite Communication Systems." *Proceedings of the IEEE*, vol. 78, no. 7, July, pp. 1275–1282.

CCIR ITU Draft Report 721 (MOD F). 1981. "Attenuation by Precipitation and Other Atmospheric Particles." Document 5/5046-E, Geneva.

CCIR Report 263-5. 1982. "Ionospheric Effects upon Earth-Space Propagation." *15th Plenary Assembly*, vol. VI, Geneva, pp. 124–146.

CCIR Report 564-2. 1982. "Propagation Data Required for Space Telecommunication System." *15 Plenary Assembly*, vol. IX, part 1, Geneva.

CCIR Report 719-1. 1982. "Attenuation by Atmospheric Gases." *15th Plenary Assembly*, vol. V, Geneva, pp. 138–150.

Crane, R. K. 1980. "Prediction of Attenuation by Rain." *IEEE Transactions on Communications*, vol. COM-28, no. 9, pp. 1717–1733.

Hogg, D. C., and T. Chu. 1975. "The Role of Rain in Satellite Communications." *Proceedings of the IEEE*, vol. 63, no. 9, pp. 1308–1331.

Ippolito, L. J. 1986. *Radiowave Propagation in Satellite Communications*. Van Nostrand Reinhold, New York.

Lin, S. H., H. J. Bergmann, and M. V. Pursley. 1980. "Rain Attenuation on Earth-Satellite Paths - Summary of 10-Year Experiments and Studies." *Bell System Technical Journal*, vol. 59, no. 2, pp. 183–228.

Miya, K. (ed.). 1981. *Satellite Communications Technology*. KDD Engineering and Consulting, Japan.

Olsen, R., D. V. Rogers, and D. B. Hodge. 1978. "The aR^b Relation in the Calculation of Rain Attenuation." *IEEE Transactions on Antennas and Propagation*, vol. 26, no. 2, pp. 318–329.

Pratt, T., C. Bostian, and J. Allnutt. 2003. *Satellite Communications*, 2nd ed. John Wiley & Sons Inc.

Tsang, L,. J. A. Kong, E. G. Njoku, D. H. Staelin, and J. W. Waters. 977). "Theory for Microwave Thermal Emission from a Layer of Cloud or Rain." *IEEE Transactions on Antennas and Propagation*, vol. 25, no. 5, pp. 650–657.

Ulaby, F., and D. G. Long. 2014. *Microwave Radiometric and Radar Remote Sensing*. University of Michigan Press, Ann Arbor, Michigan.

CHAPTER 5

Polarization

5.1 Introduction

In the *far field zone* of a transmitting antenna, the radiated wave takes on the characteristics of a *transverse electromagnetic* (TEM) wave. Far field zone refers to distances greater than $2D^2/\lambda$ from the antenna, where D is the largest linear dimension of the antenna and λ is the wavelength. For a parabolic antenna of 3 m diameter transmitting a 6-GHz wave ($\lambda = 5$ cm), the far field zone begins at approximately 360 m. The TEM designation is illustrated in Fig. 5.1, where both the magnetic field **H** and the electric field **E** are transverse to the direction of propagation, denoted by the propagation vector **k**.

E, H, and **k** represent vector quantities, and it is important to note their relative directions. When one looks along the direction of propagation, the rotation from **E** to **H** is in the direction of rotation of a right-hand-threaded screw and the vectors are said to form a *right-hand set*. The wave always retains the directional properties of the right-hand set, even when reflected, for example. One way of remembering how the right-hand set appears is to note that the letter **E** comes before **H** in the alphabet and rotation is from **E** to **H** when looking along the direction of propagation.

At great distances from the transmitting antenna, such as are normally encountered in radio systems, the TEM wave can be considered to be plane. This means that the **E** and **H** vectors lie in a plane, which is at right angles to the vector **k**. The vector **k** is said to be normal to the plane. The magnitudes are related by $E = HZ_0$, where $Z_0 = 120\pi\,\Omega$.

The direction of the line traced out by the tip of the electric field vector determines the *polarization* of the wave. Keep in mind that the electric and magnetic fields are varying as functions of time. The magnetic field varies exactly in phase with the electric field, and its amplitude is proportional to the electric field amplitude, so it is only necessary to consider the electric field in this discussion. The tip of the **E** vector may trace out a straight line, in which case the polarization is referred to as *linear*. Other forms of polarization, specifically elliptical and circular, are introduced later.

In the early days of radio, there was little chance of ambiguity in specifying the direction of polarization in relation to the surface of the earth. Most transmissions utilized linear polarization and were along terrestrial paths. Thus, *vertical polarization* meant that the electric field was perpendicular to the earth's surface, and *horizontal polarization* meant that it was parallel to the earth's surface. Although the terms vertical and horizontal are used with satellite transmissions, the situation is not quite so clear. A linear polarized wave transmitted by a geostationary satellite may be designated vertical if its electric field is parallel to the earth's polar axis, but even so the electric field is parallel to the earth at the equator. This situation is clarified shortly.

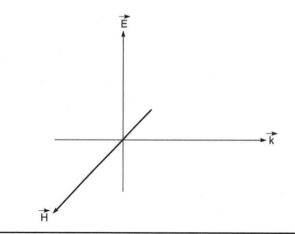

FIGURE 5.1   Vector diagram for a transverse electromagnetic (TEM) wave.

Suppose for the moment that horizontal and vertical are taken as the x and y axes of a right-hand set, as shown in Fig. 5.2a. A vertically polarized electric field can be described as

$$\mathbf{E}_y = \hat{a}_y E_y \sin \omega t \tag{5.1}$$

where $\hat{a}_y$ is the unit vector in the vertical direction and E_y is the *peak value* or *amplitude* of the electric field. Likewise, a horizontally polarized wave could be described by

$$\mathbf{E}_x = \hat{a}_x E_x \sin \omega t \tag{5.2}$$

These two fields would trace out the straight lines shown in Fig. 5.2b. Now consider the situation where both fields are present simultaneously. These would add vectorially, and the resultant would be a vector $\mathbf{E}$ (Fig. 5.2c) of amplitude $\sqrt{E_x^2 + E_y^2}$, at an angle to the horizontal given by

$$\alpha = \arctan \frac{E_y}{E_x} \tag{5.3}$$

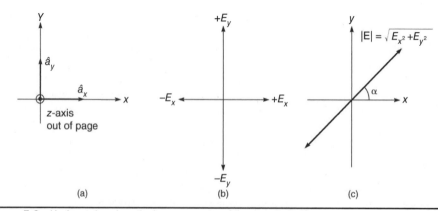

(a)   (b)   (c)

FIGURE 5.2   Horizontal and vertical components of linear polarization.

E varies sinusoidally in time in the same manner as the individual components. It is still linearly polarized but cannot be classified as simply horizontal or vertical. Arguing back from this, it is evident that **E** can be resolved into vertical and horizontal components, a fact which is of great importance in practical transmission systems. The power in the resultant wave is proportional to the voltage $\sqrt{E_x^2 + E_y^2}$, squared, which is $E_x^2 + E_y^2$. In other words, the power in the resultant wave is the sum of the powers in the individual waves, which is to be expected.

More formally, $\mathbf{E}_y$ and $\mathbf{E}_x$ are said to be *orthogonal.* The dictionary definition of orthogonal is at *right angles,* but a wider meaning will be attached to the word later.

Consider now the situation where the two fields are equal in amplitude (denoted by E), but one leads the other by 90° in phase. The equations describing these are

$$\mathbf{E}_y = \hat{a}_y E \sin \omega t \qquad (5.4a)$$

$$\mathbf{E}_x = \hat{a}_x E \cos \omega t \qquad (5.4b)$$

Applying Eq. (5.3) in this case yields $\alpha = \omega t$. The tip of the resultant electric field vector traces out a circle, as shown in Fig. 5.3a, and the resultant wave is said to be *circularly polarized.* The amplitude of the resultant vector is E. The resultant field in this case does not go through zero. At $\omega t = 0$, the y component is zero and the x component is E. At $\omega t = 90°$, the y component is E and the x component is zero. Compare this with the linear polarized case where at $\omega t = 0$, both the x and y components are zero, and at $\omega t = 90°$, both components are maximum at E. Because the resultant does not vary in time, the power must be found by adding the powers in the two linear polarized, sinusoidal waves. This gives a resultant proportional to $2E^2$.

The direction of circular polarization is defined by the sense of rotation of the electric vector, but this also requires that the way the vector is viewed must be specified.

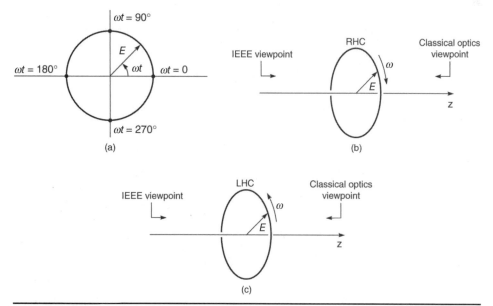

FIGURE 5.3 Circular polarization.

The *Institute of Electrical and Electronics Engineers* (IEEE) defines *right-hand circular* (RHC) *polarization* as a rotation in the clockwise direction when the wave is viewed along the direction of propagation, that is, when viewed from "behind," as shown in Fig. 5.3*b*. *Left-hand circular* (LHC) *polarization* is when the rotation is in the counterclockwise direction when viewed along the direction of propagation, as shown in Fig. 5.3*c*. LHC and RHC polarizations are orthogonal. The direction of propagation is along the +z axis.

As a caution it should be noted that the classical optics definition of circular polarization is just the opposite of the IEEE definition. The IEEE definition is used throughout this text.

For a right-hand set of axes (Fig. 5.1) and with propagation along the +z axis, then when viewed along the direction of propagation (from "behind") and with the +y axis directed upward, the +x axis is directed toward the left. Consider now Eq. (5.4). At $\omega t = 0$, E_y is 0 and E_x is a maximum at E along the +x axis. At $\omega t = 90°$, E_x is zero and E_y is a maximum at E along the +y axis. In other words, the resultant field of amplitude E has rotated from the +x axis to the +y axis, which is a clockwise rotation when viewed along the direction of propagation. Equation (5.4) therefore represents RHC polarization.

Given that Eq. (5.4) represents RHC polarization, it is left as an exercise to show that the following equations represent LHC polarization:

$$E_y = \hat{a}_y E \sin \omega t \tag{5.5a}$$

$$E_x = -\hat{a}_x E \cos \omega t \tag{5.5b}$$

In the more general case, a wave may be *elliptically polarized*. This occurs when the two linear components are

$$E_y = \hat{a}_y E_y \sin \omega t \tag{5.6a}$$

$$E_x = \hat{a}_x E_x \sin(\omega t + \delta) \tag{5.6b}$$

Here, E_y and E_x are not equal in general, and δ is a fixed phase angle. It is left as an exercise for the student to show that when $E_y = 1$, $E_x = 1/3$, and $\delta = 30°$, the polarization ellipse is as shown in Fig. 5.4.

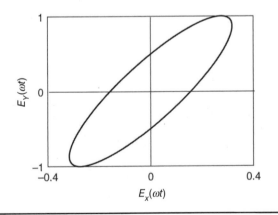

Figure 5.4   Elliptical polarization.

The *axial ratio* of an elliptical polarized wave is the ratio of major axis to minor axis of the ellipse. Orthogonal elliptical polarization occurs when a wave has the same value of axial ratio but opposite sense of rotation.

Satellite communications links use linear polarization and circular polarization, but transmission impairments can change the polarization to elliptical in each case. Some of these impairments, relating to the transmission medium, are described in Secs. 5.5, 5.6, and 5.7, and the influence of the antenna structure on polarization is described in Chap. 6. Antennas are covered in detail in Chap. 6, but at this stage the relationship of the antenna to the polarization type is defined.

5.2 Antenna Polarization

The polarization of a transmitting antenna is defined by the polarization of the wave it transmits. Thus, a horizontal dipole would produce a horizontally polarized wave. Two dipoles mounted close together symmetrically and at right angles to each other would produce a circularly polarized wave if fed with currents equal in amplitude but differing in phase by 90°. This is shown by Eqs. (5.4) and (5.5). Note that because of the symmetry of the circular polarization, the dipoles need not lie along the horizontal and vertical axes; they just need to be spatially at right angles to each other. The terms *horizontal* and *vertical* are used for convenience.

The polarization of a receiving antenna must be aligned to that of the wave for maximum power transfer. Taking again the simple dipole as an example, a vertical dipole receives maximum signal from a vertically polarized wave. Figure 5.5 illustrates this. In Fig. 5.5a the dipole is parallel to the electric field E, and hence the induced voltage V is at a maximum, denoted by V_{max}. In Fig. 5.5b the dipole is at right angles to the electric field, and the induced voltage is zero. In Fig. 5.5c the dipole lies in the plane of polarization (the wavefront) but is at some angle α to the electric field. The induced voltage is given by

$$V = V_{max} \cos \alpha \tag{5.7}$$

Note that for Eq. (5.7) to apply, the dipole must lie in the same plane as E (the wavefront). If the dipole is inclined at some angle $\theta°$ to the wavefront, the received

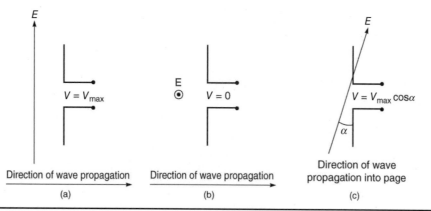

Figure 5.5 Linear polarization relative to a receiving dipole.

signal is reduced further by the radiation pattern of the antenna. This is described more fully in Sec. 6.6. The reciprocity theorem for antennas (see Sec. 6.2) ensures that an antenna designed to transmit in a given polarization receives maximum power from a wave with that polarization. An antenna designed for a given sense of polarization receives no energy from a wave with the orthogonal polarization. Figures 5.5*a* and *b* illustrate the specific case where the desired signal is vertically polarized and the orthogonal signal is horizontally polarized. However, as mentioned above, certain impairments can result in a loss of polarization discrimination, discussed in later sections.

The combined power received by the two crossed dipoles is a maximum when the incoming wave is circularly polarized. The average power received from a sinusoidal wave is proportional to the square of the amplitude. Thus, for a circularly polarized wave given by either of Eq. (5.4) or Eq. (5.5), the power received from each component is proportional to E^2, and the total power is twice that of one component alone. The crossed dipoles would receive this total. A single dipole receives a signal from a circularly polarized wave, but at a loss of −3 dB. This is so because the single dipole responds only to one of the linear components, and hence the received power is half that of the crossed dipoles. Again, because of the symmetry of the circularly polarized wave, the dipole need only lie in the plane of polarization; its orientation with respect to the x and y axes is not a factor.

A grid of parallel wires reflects a linear polarized wave when the electric field is parallel to the wires, and it transmits the orthogonal wave. This is illustrated in Fig. 5.6, and this is used in one type of *dual polarized antenna*, illustrated in Fig. 5.7. Here, the grid allows the wave, the electric field of which is transverse to the wires to pass through, whereas it reflects the parallel (E_v) wave. The reflector behind the grid reflects the wave that passes through. Thus two orthogonal, linearly polarized waves, having high polarization isolation (see Sec. 5.4) are transmitted from the antenna system. Some details of the construction of this type of antenna are found in Maral and Bousquet (1998).

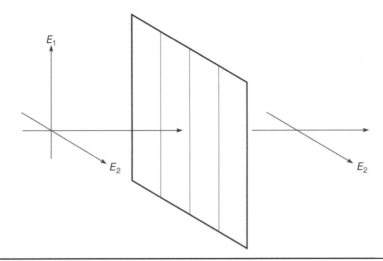

Figure 5.6 A wire grid polarizer.

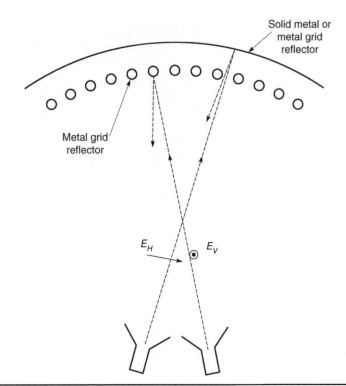

FIGURE 5.7   A wire grid polarizer used in a dual-polarized antenna.

5.3   Polarization of Satellite Signals

As mentioned above, the directions "horizontal" and "vertical" are easily visualized with reference to the earth. Consider, however, the situation where a geostationary satellite is transmitting a linear polarized wave. In this situation, the usual definition of horizontal polarization is where the electric field vector is parallel to the equatorial plane, and vertical polarization is where the electric field vector is parallel to the Earth's polar axis. At the subsatellite point on the equator, both polarizations result in electric fields that are parallel to the local horizontal plane, and care must be taken therefore not to use "horizontal" as defined for terrestrial systems. For other points on the Earth's surface within the footprint of the satellite beam, the polarization vector (the unit vector in the direction of the electric field) is at some angle relative to a reference plane. Following the work of Hogg and Chu (1975), the reference plane is taken to be that which contains the direction of propagation and the local gravity direction (a "plumb line"), which is shown in Fig. 5.8.

With the propagation direction denoted by **k** and the local gravity direction at the ground station by **r,** the direction of the normal to the reference plane is given by the vector cross-product:

$$\mathbf{f} = \mathbf{k} \times \mathbf{r} \tag{5.8}$$

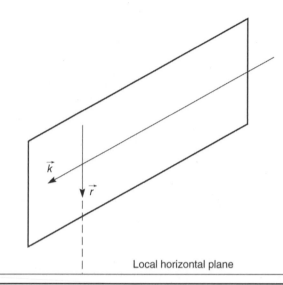

Local horizontal plane

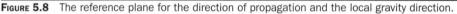

FIGURE 5.8 The reference plane for the direction of propagation and the local gravity direction.

With the unit polarization vector at the earth station denoted by **p,** the angle between it and **f** is obtained from the vector dot product as

$$\eta = \arccos \frac{\mathbf{P} \cdot \mathbf{f}}{|\mathbf{f}|} \tag{5.9}$$

Since the angle between a normal and its plane is 90°, the angle between **p** and the reference plane is $\xi = |90° - \eta|$ and

$$\xi = \arcsin \frac{\mathbf{P} \cdot \mathbf{f}}{|\mathbf{f}|} \tag{5.10}$$

This is the desired angle. Keep in mind that the polarization vector is always at right angles to the direction of propagation.

The next step is to relate the polarization vector **p** to the defined polarization at the satellite. Let unit vector **e** represent the defined polarization at the satellite. For vertical polarization, **e** lies parallel to the earth's N-S axis. For horizontal polarization, **e** lies in the equatorial plane at right angles to the geostationary radius a_{GSO} to the satellite. A cross-product vector can be formed,

$$\mathbf{g} = \mathbf{k} \times \mathbf{e} \tag{5.11}$$

where **g** is normal to the plane containing **e** and **k,** as shown in Fig. 5.9. The cross-product of **g** with **k** gives the direction of the polarization in this plane. Denoting this cross-product by **h** gives

$$\mathbf{h} = \mathbf{g} \times \mathbf{k} \tag{5.12}$$

The unit polarization vector at the earth station is therefore given by

$$\mathbf{p} = \frac{\mathbf{h}}{|\mathbf{h}|} \tag{5.13}$$

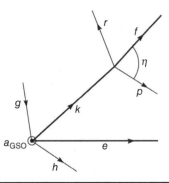

Figure 5.9   Vectors **g** = **k** × **e** and **h** = **g** × **h**.

All these vectors can be related to the known coordinates of the earth station and satellite shown in Fig. 5.10. With the longitude of the satellite as the reference, the satellite is positioned along the positive x axis at

$$x_s = a_{GSO} \tag{5.14}$$

The coordinates, for the earth-station position vector **R**, are (ignoring the slight difference between geodetic and geocentric latitudes and assuming the earth station to be at mean sea level)

$$R_x = R\cos\lambda \, \cos B \tag{5.15a}$$

$$R_y = R\cos\lambda \, \sin B \tag{5.15b}$$

$$R_z = R\sin\lambda \tag{5.15c}$$

where $B = \phi_E - \phi_{SS}$ as defined in Eq. (3.8).

The local gravity direction is $\mathbf{r} = -\mathbf{R}$. The coordinates for the direction of propagation **k** are

$$k_x = R_x - a_{GSO} \tag{5.16a}$$

$$k_y = R_y \tag{5.16b}$$

$$k_z = R_z \tag{5.16c}$$

Calculation of the polarization angle is illustrated in the following example.

Example 5.1   A geostationary satellite is stationed at 105°W and transmits a vertically polarized wave. Determine the angle of polarization at an earth station at latitude 18°N longitude 73°W.

Solution   Given data:

$\lambda = 18°$; $\phi_E = -73°$; $\phi_{SS} = -105°$; $a_{GSO} = 42164$ km; $R = 6371$ km (spherical earth of mean radius R assumed).
Eq. (3.8) gives: $B = \phi_E - \phi_{SS} = 32°$

Applying Eq. (5.15), the geocentric-equatorial coordinates for the earth station position vector are:

$$R_x = R\cos\lambda \, \cos B = 6371\cos18° \, \cos32° = 5138.48 \text{ km}$$

$$R_y = R\cos\lambda \, \sin B = 6371\cos18° \, \sin32° = 3210.88 \text{ km}$$

$$R_z = R\sin\lambda = 6371 \sin18° = 1968.75 \text{ km}$$

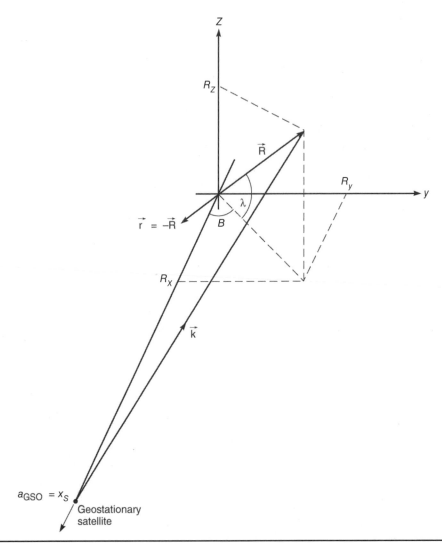

FIGURE 5.10 Vectors **k** and **R** in relation to satellite and earth station positions.

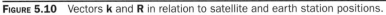

The coordinates for the local gravity direction, obtained from $\mathbf{r} = -\mathbf{R}$ are

$$\mathbf{r} = -\begin{bmatrix} 5138.48 \\ 3210.88 \\ 1968.75 \end{bmatrix} \text{km}$$

From Eq. (5.16), the geocentric-equatorial coordinates for the propagation direction are

$$\mathbf{k} = \begin{bmatrix} R_x - a_{\text{GSO}} \\ R_y \\ R_z \end{bmatrix} = \begin{bmatrix} -37025.5 \\ 3210.88 \\ 1968.75 \end{bmatrix} \text{km}$$

For vertical polarization at the satellite, the geocentric-equatorial coordinates for the polarization vector are $x = 0$, $y = 0$, and $z = 1$:

$$\mathbf{e} = \begin{bmatrix} 0 \\ 0 \\ 1 \end{bmatrix}$$

The vector cross products can be written in determinant form, where $\mathbf{a}_x$, $\mathbf{a}_y$, $\mathbf{a}_z$, are the unit vectors along the x, y, z, axes. Thus, Eq. (5.8) is

$$\mathbf{f} = \mathbf{k} \times \mathbf{r}$$

$$= -\begin{bmatrix} \mathbf{a}_x & \mathbf{a}_y & \mathbf{a}_z \\ -37025.5 & 3210.88 & 1968.75 \\ 5138.48 & 3210.88 & 1968.75 \end{bmatrix}$$

$$= \mathbf{a}_x 0 - \mathbf{a}_y 8.3 \cdot 10^7 + \mathbf{a}_z 1.35 \cdot 10^8 \ \mathrm{km}^2$$

From Eq. (5.11):

$$\mathbf{g} = \mathbf{k} \times \mathbf{e}$$

$$= \begin{bmatrix} \mathbf{a}_x & \mathbf{a}_y & \mathbf{a}_z \\ -37025.5 & 3210.88 & 1968.75 \\ 0 & 0 & 1 \end{bmatrix}$$

$$= \mathbf{a}_x 3210.88 + \mathbf{a}_y 37025.5 + \mathbf{a}_z 0 \ \mathrm{km}$$

From Eq. (5.12):

$$\mathbf{h} = \mathbf{g} \times \mathbf{k}$$

$$= \begin{bmatrix} \mathbf{a}_x & \mathbf{a}_y & \mathbf{a}_z \\ 3210.88 & 37025.5 & 0 \\ -37025.5 & 3210.88 & 1968.75 \end{bmatrix}$$

$$= \mathbf{a}_x 7.2894 \cdot 10^7 + \mathbf{a}_y 6.3214 \cdot 10^6 + \mathbf{a}_z 71.3812 \cdot 10^9 \ \mathrm{km}^2$$

The magnitude of $\mathbf{h}$ is

$$|\mathbf{h}| = \sqrt{(7.2894 \cdot 10^7)^2 + (-6.3214 \cdot 10^6)^2 + (1.3812 \cdot 10^9)^2} = 1.383 \cdot 10^9$$

From Eq. (5.13):

$$\mathbf{p} = \frac{\mathbf{h}}{|\mathbf{h}|} = \mathbf{a}_x 0.0527 - \mathbf{a}_y 0.0046 + \mathbf{a}_z 0.9986$$

The dot product of $\mathbf{p}$ and $\mathbf{f}$ is

$$\mathbf{p} \cdot \mathbf{f} = 0.0527 \times 0 + (-0.0046) \times 8.3 \times 10^7 + 0.9986 \times 1.35 \times 10^8$$

$$= 1.356 \times 10^8 \ \mathrm{km}^2$$

The magnitude of $\mathbf{f}$ is

$$|\mathbf{f}| = \sqrt{(-8.3 \times 10^7)^2 + (1.35 \times 10^8)^2} = 1.588 \times 10^8 \ \mathrm{km}^2$$

and from Eq. (5.10)

$$\xi = \arcsin \frac{1.356}{1.588} = 58.64°$$

5.4   Cross-Polarization Discrimination

The propagation path between a satellite and earth station passes through the iono-sphere, and possibly through layers of ice crystals in the upper atmosphere and rain, all of which can alter the polarization of the wave being transmitted. An orthogonal com-ponent may be generated from the transmitted polarization, an effect referred to as *depolarization*. This can cause interference where orthogonal polarization is used to pro-vide isolation between signals, as in the case of frequency reuse.

Two measures are in use to quantify the effects of polarization interference. The most widely used measure is called *cross-polarization discrimination* (XPD). Figure 5.11*a* shows how this is defined. The transmitted electric field is shown having a magnitude E_1 before it enters the medium that causes depolarization. At the receiving antenna the electric field may have two components, a *copolar* component, having magnitude E_{11}, and a *cross-polar* component, having magnitude E_{12}.

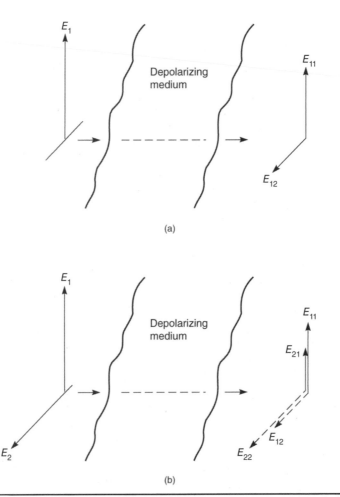

Figure 5.11   Vectors defining (*a*) cross-polarization discrimination (XPD), and (*b*) polarization isolation (I).

The cross-polarization discrimination in decibels is defined as

$$[\text{XPD}] = 10 \log \left(\frac{E_{11}}{E_{12}} \right)^2, \text{dB} \tag{5.17}$$

The second situation is shown in Fig. 5.11b. Here, two orthogonally polarized signals, with magnitudes E_1 and E_2, are transmitted. After traversing the depolarizing medium, copolar and cross-polar components exist for both waves. The *polarization isolation* is defined by the ratio of received copolar power to received cross-polar power and thus considers any additional depolarization introduced by the receiving system (Ippolito, 1986). Since received power is proportional to the square of the electric field strength, the polarization isolation in decibels is defined as

$$[I] = 10 \log \left(\frac{E_{11}}{E_{21}} \right)^2, \text{dB} \tag{5.18}$$

When the transmitted signals have the same magnitudes ($E_1 = E_2$) and where the receiving system introduces negligible depolarization, then I and XPD give identical results. For clarity, linear polarization is shown in Fig. 5.11, but the same definitions for XPD and I apply for any other system of orthogonal polarization.

5.5   Ionospheric Depolarization

The ionosphere is the upper region of the earth's atmosphere that has been ionized, mainly by solar radiation. The free electrons in the ionosphere are not uniformly distributed but form altitude-stratified layers. Furthermore, clouds of electrons (known as *traveling ionospheric disturbances*) may travel through the ionosphere and give rise to fluctuations in the signal. One of the effects of the ionosphere is to produce a rotation of the polarization of a signal, an effect known as *Faraday rotation*.

When a linearly polarized wave traverses the ionosphere, it sets in motion the free electrons in the ionized layers. These electrons move in the earth's magnetic field, and therefore, they experience a force (like that which a current-carrying conductor experiences in the magnetic field of a motor). The direction of electron motion is no longer parallel to the electric field of the wave, and as the electrons react back on the wave, the net effect is to shift the polarization. The angular shift in polarization (the Faraday rotation) is dependent on the length of the path in the ionosphere, the strength of the earth's magnetic field in the ionized region, and the electron density in the region. Faraday rotation is inversely proportional to frequency squared and is not considered to be a serious problem for frequencies above about 10 GHz.

Suppose a linearly polarized wave produces an electric field E at the receiver antenna when no Faraday rotation is present. The received power is proportional to E^2. A Faraday rotation of θ_F degrees results in the copolarized component (the desired component) of the received signal being reduced to $E_{co} = E \cos \theta_F$, the received power in this case being proportional to E_{co}^2. The *polarization loss* (PL) in decibels is

$$[\text{PL}] = 10 \log \left(\frac{E_{co}}{E} \right)^2$$

$$= 10 \log(\cos \theta_F)^2 \tag{5.19}$$

At the same time, a cross-polar component $E_x = E \sin \theta_F$ is created, and hence the XPD is

$$[\text{XPD}] = 10 \log\left(\frac{E_{co}}{E_x}\right)^2$$

$$= 10 \log(\cot \theta_F)^2 \ \text{dB} \tag{5.20}$$

Maximum values quoted by Miya (1981) for Faraday rotation are 9° at 4 GHz and 4° at 6 GHz. To counter the depolarizing effects of Faraday rotation, circular polarization may be used. With circular polarization, a Faraday shift simply adds to the overall rotation and does not affect the copolar or cross-polar components of electric field. Alternatively, if linear polarization is to be used, polarization tracking equipment may be installed at the antenna.

5.6   Rain Depolarization

The ideal shape of a raindrop is spherical, because this minimizes the energy (the surface tension) required to hold the raindrop together. The shape of small raindrops is close to spherical, but larger drops are better modeled as oblate spheroids with some flattening underneath, because of the air resistance. These are sketched in Fig. 5.12a and b. For vertically falling rain, the axis of symmetry of the raindrops is parallel to the local vertical as shown in Fig. 5.12b, but more realistically, aerodynamic forces cause some canting, or tilting, of the drops. Thus, there is a certain randomness in the angle of tilt as sketched in Fig. 5.12c.

As shown earlier, a linearly polarized wave can be resolved into two component waves, one vertically polarized and the other horizontally polarized. Consider a wave with its electric vector at some angle τ relative to the major axis of a raindrop, which for clarity is shown horizontal in Fig. 5.13. The vertical component of the electric field lies parallel to the minor axis of the raindrop, and therefore it experiences less interaction than the horizontal component. Thus, there is a difference in the attenuation and phase shift experienced by each of the electric field components. These differences are termed as the *differential attenuation and differential phase shift*, and they result in depolarization of the wave. For the situation shown in Fig. 5.13, the angle of polarization of the wave emerging from the rain is altered relative to that of the wave entering the rain. Experience has shown that the depolarization resulting from the differential phase shift is more significant than that resulting from differential attenuation.

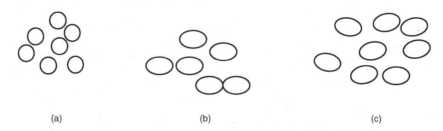

(a)   (b)   (c)

Figure 5.12   Raindrops: (a) small spherical, (b) flattening resulting from air resistance, and (c) angle of tilt randomized through aerodynamic force.

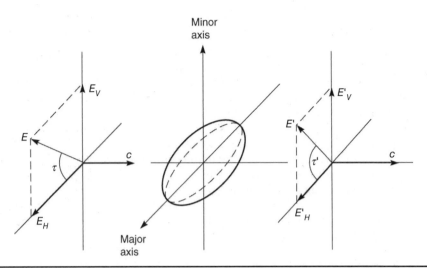

Figure 5.13   Polarization vector relative to the major and minor axes of a raindrop.

The cross-polarization discrimination in decibels associated with rain is given to a good approximation by the empirical relationship (CCIR Report 564-2, 1982)

$$\text{XPD} = U - V \log A \tag{5.21}$$

where U and V are empirically determined coefficients and A is the rain attenuation. U, V, and A must be in decibels in this equation. The attenuation A is as determined in Sec. 4.4. The following formulas are given in the CCIR reference for U and V for the frequency range 8 to 35 GHz:

$$V = \begin{cases} 20 & \text{for } 8 \leq f \leq 15 \text{ GHz} \\ 23 & \text{for } 15 \leq f \leq 35 \text{ GHz} \end{cases} \tag{5.22a}$$

and

$$U = 30 \log f - 10 \log(0.5 - 0.4697 \cos 4\tau) - 40 \log(\cos\theta) \tag{5.22b}$$

where f is the frequency in gigahertz, θ is the angle of elevation of the propagation path at the earth station, and τ is the tilt angle of the polarization relative to the horizontal. For circular polarization $\tau = 45°$. As shown earlier, for a satellite transmission, the angle ξ between the reference plane containing the direction of propagation and the local vertical is a complicated function of position, but the following general points can be observed. When the electric field is parallel to the ground (horizontal), $\tau = 0$, the second term on the right-hand side of the equation for U contributes a +15-dB amount to the XPD, whereas with circular polarization the contribution is only about +0.13 dB. With the electric field vector in the reference plane containing the direction of propagation and the local vertical, $\tau = 90° - \theta$ (all angles in degrees), and the $\cos 4\tau$ term becomes $\cos 4\theta$.

5.7   Ice Depolarization

As shown in Fig. 4.3 an ice layer is present at the top of a rain region, and as noted in Table 4.1, the ice crystals can result in depolarization. The experimental evidence suggests that the chief mechanism producing depolarization in ice is differential phase shift, with little differential attenuation present. This is because ice is a low-loss dielectric, unlike water, which has considerable absorptive losses. Ice crystals tend to be needle-shaped or platelike and, if randomly oriented, have little effect, but depolarization occurs when they become aligned. Sudden increases in XPD that coincide with lightning flashes are thought to be a result of the lightning producing alignment. An *International Radio Consultative Committee* (CCIR) recommendation for taking ice depolarization into account is to add a fixed decibel value to the XPD value calculated for rain. Values of –2 dB are suggested for North America and –4 to –5 dB for maritime regions, and it is further suggested that the effects of ice can be ignored for time percentages less than 0.1 percent (Ippolito, 1986).

5.8   Problems and Exercises

5.1.   Explain what is meant by a plane TEM wave.

5.2.   Two electric fields, in time phase and each of unity amplitude, act at right angles to one another in space. On a set of x-y axes draw the path traced by the tip of the resultant electric field vector. Given that the total power developed across a 50 Ω load is 5 W, find the peak voltage corresponding to the unity amplitude.

5.3.   Two electric fields with an amplitude ratio of 3:1 and in time phase, act at right angles to one another in space. On a set of x-y axes draw the path traced by the tip of the resultant. Given that the total power developed across a 50 Ω load is 10 W, find the peak voltage corresponding to the unity amplitude.

5.4.   Two electric field vectors of equal amplitude are 90° out of time phase with one another. On a set of x-y axes draw the path traced by the tip of the resultant vector.

5.5.   Two electric field vectors of amplitude ratio 3:1 are 90° out of time phase with one another. On a set of x-y axes draw the path traced by the tip of the resultant vector. If the peak voltages are 3 V and 1 V, determine the average power developed in a 10 Ω load.

5.6.   With reference to a right-hand set of rectangular coordinates, and given that Eq. (5.4) applies to a plane TEM wave, the horizontal component being directed along the x axis and the vertical component along the y axis, determine the sense of polarization of the wave.

5.7.   With $\delta = -45°$ and equal amplitude components, determine the sense of polarization of a wave represented by Eq. (5.6).

5.8.   A plane TEM wave has a horizontal (+x directed) component of electric field of amplitude 3 V/m and a vertical (+y directed) component of electric field of amplitude 5 V/m. The horizontal component leads the vertical component by a phase angle of 20°. Determine the sense of polarization.

5.9.   A plane TEM wave has a horizontal (+x directed) component of electric field of amplitude 3 V/m and a vertical (+y directed) component of electric field of amplitude 5 V/m. The horizontal component lags the vertical component by a phase angle of 20°. Determine the sense of polarization.

5.10.   Given that the plane TEM wave of Prob. 5.8 propagates in free space, determine the magnitude of the magnetic field.

5.11.   Explain what is meant by *orthogonal polarization* and the importance of this in satellite communications.

5.12.   The TEM wave represented by Eq. (5.4) is received by a linearly polarized antenna. Determine the reduction in emf induced in the antenna compared to what would be obtained with polarization matching.

5.13.   A plane TEM wave has a horizontal (+x-directed) component of electric field of amplitude 3 V/m and a vertical (+y-directed) component of electric field of amplitude 5 V/m. The components are in time phase with one another. Determine the angle a linearly polarized antenna must be at with reference to the x axis to receive maximum signal.

5.14.   For Prob. 5.13, what would be the reduction in decibels of the received signal if the antenna is placed along the x axis?

5.15.   Explain what is meant by *vertical polarization* of a satellite signal. A vertically polarized wave is transmitted from a geostationary satellite and is received at an earth station that is west of the satellite and in the northern hemisphere. Is the wave received at the earth station vertically polarized? Give reasons for your answer.

5.16.   Explain what is meant by *horizontal polarization* of a satellite signal. A horizontally polarized wave is transmitted from a geostationary satellite and is received at an earth station that is west of the satellite and in the northern hemisphere. Is the wave received at the earth station horizontally polarized? Give reasons for your answer.

5.17.   A geostationary satellite stationed at 90°W transmits a vertically polarized wave. Determine the polarization of the resulting signal received at an earth station situated at 70°W, 45°N.

5.18.   A geostationary satellite stationed at 10°E transmits a vertically polarized wave. Determine the polarization of the resulting signal received at an earth station situated at 5°E, 45°N.

5.19.   Explain what is meant by *cross-polarization discrimination* and briefly describe the factors which militate against good cross-polarization discrimination.

5.20.   Explain the difference between *cross-polarization discrimination* and *polarization isolation*.

5.21.   A linearly polarized wave traveling through the ionosphere suffers a Faraday rotation of 9°. Calculate (*a*) the polarization loss and (*b*) the cross-polarization discrimination.

5.22.   Why is Faraday rotation of no concern with circularly polarized waves?

5.23.   Explain how depolarization is caused by rain.

5.24.   A transmission path between an earth station and a satellite has an angle of elevation of 32° with reference to the earth. The transmission is circularly polarized at a frequency of 12 GHz. Given that rain attenuation on the path is 1 dB, calculate the cross-polarization discrimination.

5.25.   Repeat Prob. 5.24 for a linearly polarized signal where the electric field vector is parallel to the earth at the earth station.

5.26.   Repeat Prob. 5.24 for a linearly polarized signal where the electric field vector lies in the plane containing the direction of propagation and the local vertical at the earth station.

5.27.   Repeat Prob. 5.24 for a signal frequency of 18 GHz and an attenuation of 1.5 dB.

References

CCIR Report 564-2. 1982. "Propagation Data Required for Space Telecommunication System." *15 Plenary Assembly*, Vol. IX, Part 1, Geneva.

Hogg, D. C., and T. Chu. 1975. "The Role of Rain in Satellite Communications." *Proc. IEEE*, Vol. 63, No. 9, pp. 1308–1331.

Ippolito, L. J. 1986. *Radiowave Propagation in Satellite Communications*. Van Nostrand Reinhold, New York.

Maral, G., and M. Bousquet. 1998. *Satellite Communications Systems*. Wiley, New York.

Miya, K. (ed.). 1981. *Satellite Communications Technology*. KDD Engineering and Consulting, Japan.

CHAPTER 6

Antennas

6.1 Introduction

Antennas can be broadly classified according to function—as *transmitting antennas* and *receiving antennas*. Although the requirements for each function or mode of operation, can be markedly different, a single antenna may be, and frequently is, used for transmitting and receiving signals simultaneously. Many of the properties of an antenna, such as its directional characteristics, apply equally to both modes of operation, this being a result of the *reciprocity theorem* described in Sec. 6.2.

Certain forms of interference (see Chap. 13) can present problems for satellite systems, which are not encountered in other radio systems and minimizing these requires special attention to those features of the antenna design which control interference.

Another way in which antennas for use in satellite communications can be classified is into *earth station* antennas and *satellite* or *spacecraft* antennas. Although the general principles of antennas may apply to each type, the constraints set by the physical environment lead to quite different designs in each case.

Before looking at antennas specifically for use in satellite systems, some of the general properties and definitions for antennas are given in this and the next few sections. As already mentioned, antennas form the link between transmitting and receiving equipment, and the space propagation path. Figure 6.1a shows the antenna as a radiator. The power amplifier in the transmitter is shown as generating P_TW. A feeder connects this to the antenna with the net power reaching the antenna P_T minus the losses in the feeder. These losses include ohmic losses and mismatch losses. The power is further reduced by losses in the antenna so that the power radiated, shown as P_{rad}, is less than that generated at the transmitter.

The antenna as a receiver is shown in Fig. 6.1b. Power P_{rec} is transferred to the antenna from a passing radio wave. Losses in the antenna reduce the power available for the feeder so that the amount P_R reaching the receiver is less than that received by the antenna.

6.2 Reciprocity Theorem for Antennas

The reciprocity theorem for antennas states that if a current I is induced in an antenna B, operated in the receive mode, by an emf applied at the terminals of antenna A operated in the transmit mode, then the same emf applied to the terminals of B induces the same current at the terminals of A. This is illustrated in Fig. 6.2. For a proof of the reciprocity theorem, see for example, Glazier and Lamont (1958) or Balanis (2005).

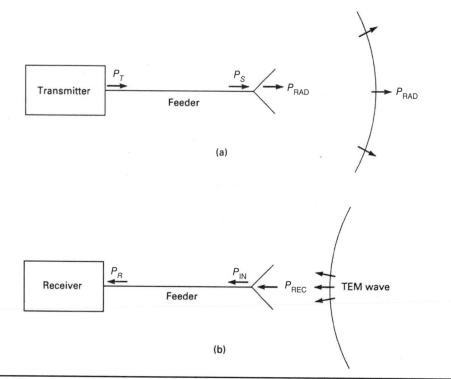

FIGURE 6.1 (a) Transmitting antenna. (b) Receiving antenna.

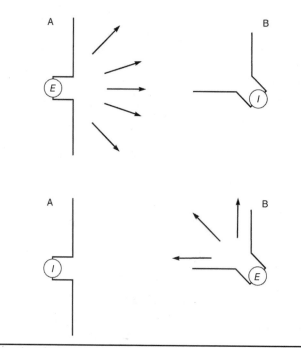

FIGURE 6.2 Illustration of the reciprocity theorem for B receive (top) and B transmit (bottom) operation.

Several important consequences result from the reciprocity theorem. All practical antennas have directional patterns; that is, they transmit more energy in some directions than others, and they receive more energy when pointing in some directions than others. The reciprocity theorem requires that *the directional pattern for an antenna operating in the transmit mode is the same as that when operating in the receive mode.*

Another important consequence of the reciprocity theorem is that *the antenna impedance is the same for both modes of operation.*

6.3   Antenna Pattern Coordinate System

To discuss the directional patterns of an antenna, it is necessary to set up a coordinate system to which these can be referred. The system in most common use is the *spherical coordinate system* illustrated in Fig. 6.3 that comprises a right-hand rectangular (X, Y, Z) coordinate system with two *polar coordinate systems* in orthogonal planes (XY and ZY'), that define the *r*, θ, and ϕ *spherical coordinates*. Thus, the Z-axis defines the north pole and the XY plane defines the equatorial plane. Note that this is the same as the right-hand set introduced in Sec. 5.1 and that this becomes particularly significant when the polarization of the radio waves associated with antennas is described.

Usually, the antenna is located at the origin, with the direction of maximum electric field intensity coaligned with the z-axis, and a distant observation point P in space is related to the origin by the coordinates r, θ, and ϕ. Thus, r is the radius vector, the magnitude of which gives the distance between the observation point P and the antenna; ϕ is the angle measured from the x axis clockwise to the projection of r in the xy plane; and θ is the angle measured from the z axis to r.

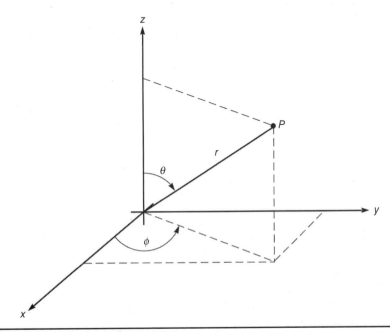

Figure 6.3   The spherical coordinate system.

6.4   The Radiated Fields

There are three main regions to the radiated electromagnetic fields surrounding an antenna: two near-field regions and a far-field region. The field strengths of the near-field components decrease rapidly with increasing distance from the antenna, one component being inversely related to distance squared, and the other to the distance cubed. At comparatively short distances these components are negligible compared with the radiated component used for radio communications, the field strength of which decreases in proportion to distance. Estimates for the distances at which the fields are significant are shown in Fig. 6.4a. Here, D is the largest dimension of the antenna (e.g., the diameter of a parabolic dish reflector), and λ is the wavelength. Only the far-field region is of interest here, which applies for distances greater than about $2D^2/\lambda$.

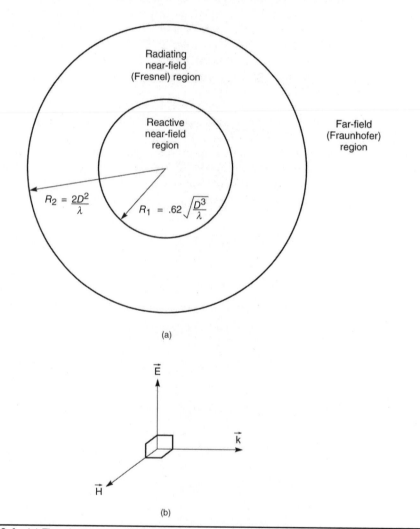

(a)

(b)

FIGURE 6.4   (a) The electromagnetic-field regions surrounding an antenna. (b) Vector diagrams in the far-field region.

In the far-field region, the radiated fields form a *transverse electromagnetic* (TEM) wave in which the electric field is at right angles to the magnetic field, and both are at right angles (transverse) to the direction of propagation. The vector relationship is shown in Fig. 6.4*b*, where **E** represents the electric field, **H** the magnetic field, and **k** the direction of propagation. These vectors form a right-hand set in the sense that when one looks along the direction of propagation, a clockwise rotation is required to go from **E** to **H**. An important practical point is that the wavefront can be assumed to be plane; that is, **E** and **H** lie in a plane to which **k** is the normal.

In the far field, the electric field vector can be resolved into two components, which are shown in relation to the coordinate system in Fig. 6.5*a*. The component labeled E_θ is tangent at point P to the circular arc of radius r. The component labeled E_ϕ is tangent at

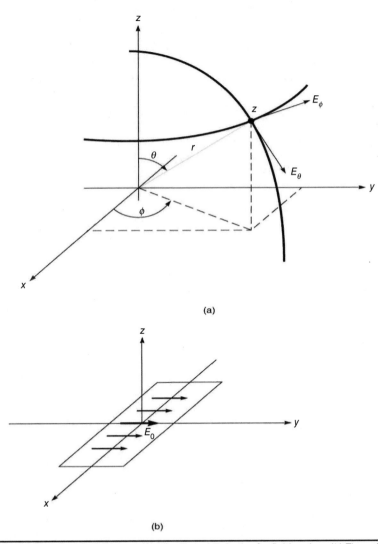

(a)

(b)

FIGURE 6.5 (a) The electric field components E_θ and E_ϕ in the far-field region. (b) The reference vector E_0 at the origin.

point P to the circle of radius $r \sin\theta$ centered on the z axis (that is like a circle of latitude on the earth's surface). Both these components are functions of θ and ϕ and in functional notation are written as $E_\theta(\theta, \phi)$ and $E_\phi(\theta, \phi)$. The resultant magnitude of the electric field is given by

$$E = \sqrt{E_\theta^2 + E_\phi^2} \tag{6.1}$$

If E_θ and E_ϕ are peak values, E is the peak value of the resultant, and if they are rms values, E is the rms value of the resultant.

The vector $\mathbf{E}_0$ shown at the origin of the coordinate system represents the principal electric vector of the antenna itself. For example, for a horn antenna, this is the electric field vector across the aperture as shown in Fig. 6.5b. For definiteness, the $\mathbf{E}_0$ vector is shown aligned with the y axis, since this allows two important planes to be defined:

The H plane is the xz plane, for which $\phi = 0$

The E plane is the yz plane, for which $\phi = 90°$

Magnetic field vectors are associated with these electric field components. Thus, following the right-hand rule, the magnetic vector associated with the E_θ component lies parallel with E_ϕ and is normally denoted by H_ϕ, while that associated with E_ϕ lies parallel (but pointing in the opposite direction) to E_θ and is denoted by H_θ. For clarity, the **H** fields are not shown in Fig. 6.5. The magnitudes of the fields are related through the *wave impedance* Z_W. For radio waves in free space, the value of the wave impedance is (in terms of field magnitudes)

$$Z_W = \frac{E_\phi}{H_\theta} = \frac{E_\theta}{H_\phi} = 377, \text{Ohms} \tag{6.2}$$

The same value can be used with negligible error for radio waves in the Earth's atmosphere.

6.5 Power Flux Density

The *power flux density* of a radio wave is a quantity used in calculating the performance of satellite communications links. The concept can be understood by imagining the transmitting antenna to be at the center of a sphere. The power from the antenna radiates outward, normal to the surface of the sphere, and the power flux density is the power flow per unit surface area. Power flux density is a vector quantity, and its magnitude is given by

$$\psi = \frac{E^2}{Z_W} \tag{6.3}$$

Here, E is the rms value of the field given by Eq. (6.1). The units for ψ are watts per square meter with E in volts per meter and Z_W in Ohms. Because the E field is inversely proportional to distance (in this case the radius of the sphere), the power density is inversely proportional to the square of the distance.

6.6 The Isotropic Radiator and Antenna Gain

The word *isotropic* means, rather loosely, equally in all directions. Thus an *isotropic radiator* is one which radiates equally in all directions. No real antenna can radiate equally in all directions, and so an isotropic radiator is only hypothetical. However, an isotropic radiator provides a very useful theoretical standard against which real antennas can be compared. Being hypothetical, it can be made 100 percent efficient, meaning that it radiates all the power fed into it. Thus, referring to Fig. 6.1a, $P_{rad} = P_S$. By imagining the isotropic radiator to be at the center of a sphere of radius r, the power flux density, which is the power flow through unit area, is

$$\Psi_i = \frac{P_S}{4\pi r^2} \tag{6.4}$$

The flux density from a real antenna varies with direction, but with most antennas a well-defined maximum occurs. The power *gain* of the antenna is the ratio of this maximum flux to that for the isotropic radiator at the same radius r:

$$G_o = \frac{\Psi_M}{\Psi_i} \tag{6.5}$$

A very closely related gain metric is the antenna *directivity*. This differs from the power gain in that in determining the isotropic flux density, the actual power P_{rad} radiated by the real antenna is used, rather than the power P_S supplied to the antenna. These two values are related as $P_{rad} = \eta_A P_S$, where η_A is the *antenna efficiency*. Denoting the directivity by D gives $G = \eta_A D$.

Often, the directivity is the parameter which can be calculated, and the efficiency is assumed to be equal to unity so that the power gain is also known. Note that η_A does not include feeder mismatch or polarization losses, which are accounted for separately.

The power gain G as defined by Eq. (6.5) is called the *isotropic power gain*, sometimes denoted by G_i. The power gain of an antenna also may be referred to some standard other than isotropic. For example, the gain of a reflector-type antenna may be stated relative to the antenna illuminating the reflector. Care must be taken therefore to know what reference antenna is being used when gain is stated. The isotropic gain is the most used reference and is assumed throughout this text (without use of a subscript) unless otherwise noted.

6.7 Radiation Pattern

The *radiation pattern* shows how the gain of an antenna varies with direction. Referring to Fig. 6.3, at a fixed distance r, the gain varies with θ and ϕ and may be written generally as $G(\theta, \phi)$. The radiation pattern is the gain normalized to its maximum value. Denoting the maximum value simply by G_o (as given by Eq. 6.5) the radiation pattern is

$$g(\theta, \phi) = \frac{G(\theta, \phi)}{G_o} \tag{6.6}$$

The radiation pattern gives the directional properties of the antenna normalized to the maximum value, in this case the maximum gain. The same function gives the power density normalized to the maximum power density. For most satellite antennas, the three-dimensional plot of the radiation pattern shows a well-defined main lobe, as sketched in Fig. 6.6a. In this diagram, the length of a radius line to any point on the surface of the lobe gives the value of the radiation function at that point. Frequently the maximum value is normalized to unity, and or convenience, this is shown as pointing along the positive z axis. Be very careful to observe that the axes shown in Fig. 6.6 *do not represent distance*. The distance r is assumed to be fixed at some value in the far field. What is shown is a plot of normalized gain as a function of angles θ and ϕ.

The main lobe represents a *beam* of radiation, and the beamwidth is specified as the angle subtended by the -3 dB $\Rightarrow 0.5$ power ratio lines. Because in general the beam may not be symmetrical, it is usual practice to give the beamwidth in the H plane ($\phi = 0°$), as shown in Fig. 6.6b, and in the E plane ($\phi = 90°$), as shown in Fig. 6.6c, and we define the half-power beamwidth to be β_H and β_E, respectively. Because the radiation pattern is defined in terms of radiated power, the normalized electric field strength pattern is given by $\sqrt{g(\theta, \phi)}$.

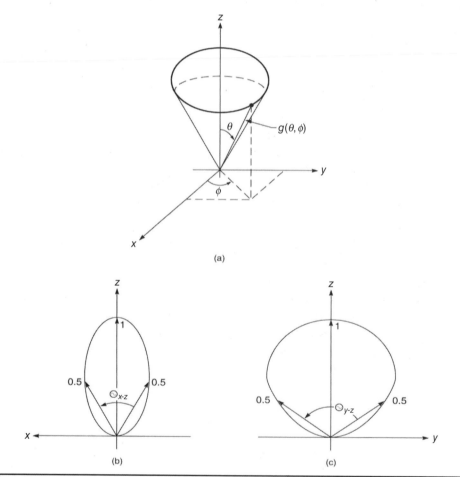

Figure 6.6 (a) A radiation pattern. (b) The beamwidth in the H-plane. (c) The beamwidth in the E-plane.

6.8   Beam Solid Angle and Directivity

Plane angles are measured in radians, and an arc of length R, which is equal to the radius, subtends an angle of one radian at the center of a circle. An angle of θ radians defines an arc length of $R\theta$ on the circle. This is illustrated in Fig. 6.7a. The circumference of a circle is given by $2\pi R$, and hence the total angle subtended at the center of a circle is 2π rad. All this should be familiar to the reader, but what may not be so familiar is the concept of solid angle. A surface area of R^2 on the surface of a sphere of radius R subtends unit solid angle at the center of the sphere that is shown in Fig. 6.7b. The unit for the solid angle is the *steradian*, abbreviated sr. A solid angle of Ω steradians defines a surface area on the sphere (a spherical cap) of $R^2\Omega$. Looking at this another way, a surface area A subtends a solid angle A/R^2 at the center of the sphere. Since the total surface area of a sphere of radius R is $4\pi R^2$, the total solid angle subtended at the center of the sphere is 4π sr.

The *radiation intensity* is the power radiated per unit solid angle. For a power P_{rad} radiated, the average radiation intensity (which is also the isotropic value) taken over a sphere is

$$U_i = \frac{P_{rad}}{4\pi}, \text{W/sr} \tag{6.7}$$

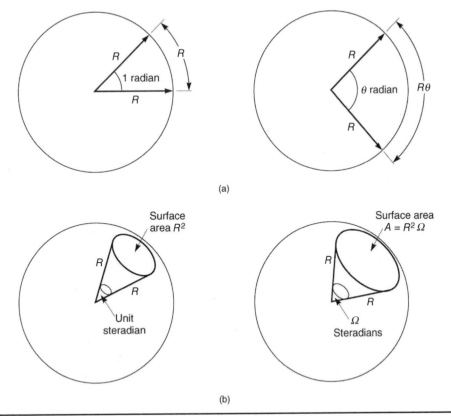

(a)

(b)

Figure 6.7   (a) Defining the radian. (b) Defining the steradian.

From the definition of directivity D, the maximum radiation intensity is

$$U_{max} = DU_i \tag{6.8}$$

The *beam solid angle*, Ω_A, for an actual antenna is defined as the solid angle through which all the power flows to produce a constant radiation intensity equal to the maximum value. Thus

$$U_{max} = \frac{P_{rad}}{\Omega_A} \tag{6.9}$$

Combining Eqs. (6.7), (6.8), and (6.9) yields the important result

$$D = \frac{4\pi}{\Omega_A} \tag{6.10}$$

This is important because for narrow-beam antennas such as used in many satellite communications systems, a good approximation to the solid angle is

$$\Omega_A \cong \beta_E \times \beta_H \tag{6.11}$$

where β_E is the half-power beamwidth in the E plane and β_H is the half-power beamwidth in the H plane, as shown in Fig. 6.6. This equation requires the half-power beamwidths to be expressed in radians, and the resulting solid angle is in steradians.

The usefulness of this relationship is that the half-power beamwidths can be measured, and hence the directivity can be found. When the half-power beamwidths are expressed in degrees, the equation for the directivity becomes

$$D = \frac{(180/\pi)^2 * 4\pi}{(\beta_E^o * \beta_H^o)} = \frac{41253}{(\beta_E^o * \beta_H^o)}, \text{ power ratio} \tag{6.12}$$

6.9   Effective Aperture

So far, the properties of antennas have been described in terms of their radiation characteristics. A receiving antenna has directional properties also described by the radiation pattern, but in this case it refers to the ratio of received power normalized to the maximum value.

An important concept used to describe the reception properties of an antenna is that of *effective aperture*. Consider a TEM wave of a given power density ψ at the receiving antenna. Let the load at the *antenna terminals* be a complex conjugate match so that maximum power transfer occurs and power P_{rec} is delivered to the load. With the receiving antenna aligned for maximum reception (including polarization alignment), the received power is proportional to the power density of the incoming wave. The constant of proportionality is the effective aperture A_{eff} which is defined by the equation

$$P_{rec} = A_{eff} \psi \tag{6.13}$$

However for practical systems, because of system losses (see Sec. 12.4), the power delivered to the actual receiver may be less than this; therefore, for these cases, there is a multiplicative gain < 1 to account for these losses. This method is applied to losses throughout this text that we believe that this is more intuitive than the alternative approach of treating losses as power ratios > 1.0 and then dividing into the loss-free quantity. This topic is discussed further in App. G.

For antennas that have easily identified physical apertures, such as horns and parabolic reflector types, the effective aperture is related in a direct way to the physical aperture. If the wave could uniformly illuminate the physical aperture, then this is equal to the effective aperture. However, the presence of the antenna in the field of the incoming wave alters the field distribution, thereby preventing uniform illumination. Thus, the effective aperture is smaller than the physical aperture by a factor known as the *illumination efficiency*. Denoting the illumination efficiency by η_I the effective and physical apertures are related according to

$$A_{\text{eff}} = \eta_I A_{\text{physical}} \tag{6.14}$$

The illumination efficiency is usually a specified number, and it typically ranges between about 0.5 and 0.8. It cannot exceed unity. A conservative value often used in calculations is 0.55 for a communications link antenna.

A fundamental relationship exists between the power gain of an antenna and its effective aperture. This is

$$\frac{A_{\text{eff}}}{G} = \frac{\lambda^2}{4\pi} \tag{6.15}$$

where λ is the wavelength of the TEM wave, assumed sinusoidal (for practical purposes, this is the wavelength of the radio wave carrier). The importance of this equation is that the gain is normally the known (measurable) quantity, but once this is known, the effective aperture is also known.

6.10 The Half-Wave Dipole

The half-wave dipole is a basic antenna type which finds limited but essential use in satellite communications. It enables some radiation in all directions except along the dipole axis itself. It is this near-omnidirectional property which makes it useful for telemetry and command signals to and from the satellite, which is essential during the launch phase when highly directional antennas cannot be deployed.

The half-wave dipole is shown in Fig. 6.8a, and its radiation pattern in the *xy* plane and in any one meridian plane in Fig. 6.8b and c. Because the phase velocity of the radio wave along the wire is somewhat less than the free-space velocity, the wavelength is also slightly less, and the antenna is cut to about 95 percent of the free-space half-wavelength. This tunes the antenna correctly to resonance.

The main properties of the half-wave dipole are:

Impedance: 73 Ω
Directivity: 1.64 (2.15 dB)
Effective aperture: 0.13 λ^2
−3 dB beamwidth: 78°

Assuming an antenna efficiency of unity ($\eta_A = 1$), the power gain is also 1.64, or 2.15 dB. This is the gain referred to an *isotropic radiator*, which is expressed in decibel units as dBi.

As shown in Fig. 6.8b, the radiation is a maximum in the *xy* plane, the normalized value being unity. The symmetry of the dipole means that the radiation pattern in this plane is a circle of unit radius. Symmetry also means that the pattern is the same for any

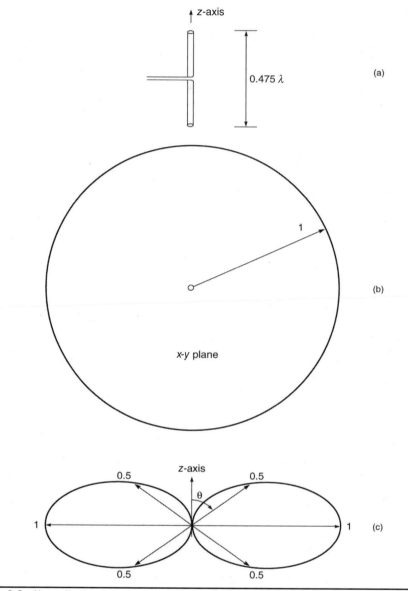

Figure 6.8 Normalized antenna patterns for the half-wave dipole.

plane containing the dipole axis (the z axis). Thus the radiation pattern is a function of θ only and is given by

$$g(\theta) = \frac{\cos^2\left(\frac{\pi}{2}\cos\theta\right)}{\sin^2\theta} \qquad (6.16)$$

A plot of this function is shown in Fig. 6.8c. It is left as an exercise for the reader to show that the −3 dB beamwidth obtained from this pattern is 78°.

6.11 Aperture Antennas

The open end of a waveguide is an example of a simple aperture antenna. It is capable of radiating energy being carried by the guide, and it can receive energy from a wave impinging on it. In satellite communications, the most encountered aperture antennas are horn and reflector antennas. Before describing some of the practical aspects of these, the radiation pattern of an idealized aperture is considered to illustrate certain features that are important in satellite communications.

An idealized aperture is shown in Fig. 6.9. It consists of a rectangular aperture of sides a and b cut in an infinite ground plane. A uniform electric field exists across the aperture parallel to the side b, and the aperture is centered on the coordinate system shown in Fig. 6.3, with the electric field parallel to the y axis. Radiation from different parts of the aperture adds constructively in some directions and destructively in others, with the result that the radiation pattern exhibits a main lobe and several sidelobes. Mathematically, this is shown as follows:

At some fixed distance r in the far-field region, the electric field components described in Sec. 6.4 are given by

$$E_\theta(\theta, \phi) = C \sin\phi \frac{\sin X}{X} \frac{\sin Y}{Y} \tag{6.17}$$

$$E_\phi(\theta, \phi) = C \cos\theta \cos\phi \frac{\sin X}{X} \frac{\sin Y}{Y} \tag{6.18}$$

Here, C is a constant which depends on the distance r, the lengths a and b, the wavelength λ, and the electric field strength E_0. For present purposes, E_0 can be set equal to unity. X and Y are variables given by

$$X = \frac{\pi a}{\lambda} \sin\theta \cos\phi \tag{6.19}$$

$$Y = \frac{\pi b}{\lambda} \sin\theta \sin\phi \tag{6.20}$$

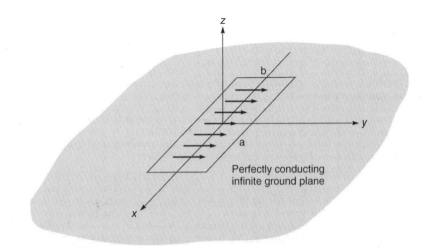

FIGURE 6.9 An idealized aperture radiator.

Even for the idealized, and hence simplified, aperture situation, the electric field equations are quite complicated. The two principal planes of the coordinate system are defined as the H plane, which is the xz plane, for which $\phi = 0$, and the E plane, which is the xy plane, for which $\phi = 90°$. It simplifies matters to examine the radiation pattern in these two planes. Consider first the H plane. With $\phi = 0$, it is seen that $Y = 0$, $E_\theta = 0$, and

$$X = \frac{\pi a}{\lambda} \sin \theta \tag{6.21}$$

and with C set equal to unity,

$$E_\phi(\theta) = \cos \theta \frac{\sin X}{X} \tag{6.22}$$

The radiation pattern is given by:

$$g_H(\theta) = \left| E_\phi(\theta) \right|^2 = \cos^2 \theta \left| \frac{\sin X}{X} \right|^2 \tag{6.23}$$

A similar analysis may be applied to the E plane resulting in $X = 0$, $E_\phi = 0$, and

$$Y = \frac{\pi b}{\lambda} \sin \theta \tag{6.24}$$

$$E_\theta(\theta) = \frac{\sin Y}{Y} \tag{6.25}$$

$$g_E(\theta) = \left| E_\theta(\theta) \right|^2 = \left| \frac{\sin Y}{Y} \right|^2 = \text{Sinc}(Y) \tag{6.26}$$

The Sinc or sampling function defined as $\text{Sinc}(x) = \sin x / x$ occurs frequently in communications engineering. This is available as a library function in many computer applications such as MatLab, Python, IDL, C++, etc. A point to bear in mind when evaluating this function is that x must be in radians. $\text{Sinc}(x) = 1$ for $x = 0$, and $\text{Sinc}(x) = 0$ for $x = n\pi$, where n is the integer. It is seen that the H plane pattern contains the function $\text{Sinc}(Y)$ and the E plane, the function $\text{Sinc}(X)$. These radiation patterns are illustrated in Example 6.1.

Example 6.1   Plot the E-plane and H-plane radiation patterns for the uniformly illuminated aperture for which $a = 3\lambda$, and $b = 2\lambda$.

Solution   Looking first at the H plane, for $a = 3\lambda$, $X = 3\pi \sin \theta$. As noted here, the sampling function has well-defined zeros, occurring in this case when $\sin \theta = 1/3, 2/3$, or 1. The $g_H(\theta)$ function has correspondingly zeros or nulls at these points. (The $\cos \theta$ term also has zeros at $\theta = n\pi/2$, where n is any odd integer.

For the E plane and $b = 2\lambda$, $Y = 2\pi \sin \theta$. Again, the sampling function has well-defined zeros occurring in this case when $\sin \theta = 1/2$ or 1, and the $g_E(\theta)$ function shows corresponding zeros, or nulls. Plots of the radiation functions are shown. The curves are symmetrical about the vertical axis and so only one-half of the curves are shown.

The results of Example 6.1 show the main lobe and the sidelobes. These are general features of aperture antennas. The sidelobes can produce in interference to adjacent channels. As a result, maximum allowable levels are specified to minimize this (see Fig. 6.19). The nulls in the radiation pattern can be useful in some situations where these can be aligned with an interfering source.

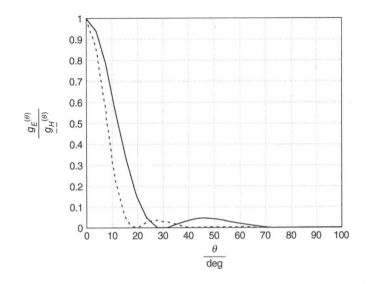

The uniform field distribution across the ideal aperture (Fig. 6.9) cannot be realized in practice; rather, the actual distribution depends on the way the aperture is energized. The radiation pattern is also influenced by the physical construction of the antenna. With reflector-type antennas, for example, the position of the primary feed can change the pattern in important ways.

Another important practical consideration with real antennas is the *cross-polarization isolation*. This refers to the antenna in the transmit mode radiating, and in the receive mode responding to, an unwanted signal with polarization orthogonal to the desired polarization (see Sec. 5.2). As mentioned in Chap. 5, frequency reuse makes use of orthogonal polarization, and any unwanted cross-polarized component results in interference. The cross-polarization characteristics of some practical antennas are considered in the following sections.

The aperture shown in Fig. 6.9 is linearly polarized, the **E** vector being directed along the y axis. At some arbitrary point in the far-field region, the wave remains linearly polarized, the magnitude E being given by Eq. (6.1). It is necessary only for the receiving antenna to be oriented so that E induces maximum signal, with no component orthogonal to E so that cross-polarization contamination is absent. Care must be taken, however, in how cross-polarization is defined. The linearly polarized field **E** can be resolved into two vectors, one parallel to the plane containing the aperture vector E_0, referred to as the *co-polar* component, and a second component orthogonal to this, referred to as the *cross-polarized* component. The way in which these components are used in antenna measurements is detailed in Balanis (2005), Kraus (1988), Chang (1989), and Rudge et al. (1982).

6.12   Horn Antennas

The horn antenna is an example of an aperture antenna that provides a smooth transition from a waveguide to a larger aperture that couples more effectively into space. Horn antennas are used directly as radiators aboard satellites to illuminate comparatively large areas of the Earth, and they are also widely used as primary feeds for reflector type antennas both in transmitting and receiving modes. The three most used types of horns are illustrated in Fig. 6.10.

6.12.1   Conical Horn Antennas

The *smooth-walled* conical antenna shown in Fig. 6.10 is the simplest horn structure. The term smooth-walled refers to the inside wall. The horn may be fed from a rectangular waveguide, but this requires a rectangular-to-circular transition at the junction. Feeding

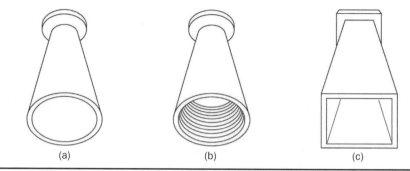

FIGURE 6.10   Horn antennas: (a) smooth-walled conical, (b) corrugated, and (c) pyramidal.

from a circular guide is direct and is the preferred method, with the guide operating in the TE_{11} mode. The conical horn antenna may be used with linear or circular polarization, but to illustrate some of the important features, linear polarization is assumed.

The electric field distribution at the horn mouth is sketched in Fig. 6.11 for vertical polarization. The curved field lines can be resolved into vertical and horizontal components as shown. The TEM wave in the far field is linearly polarized, but the horizontal components of the aperture field give rise to cross-polarized waves in the far-field region. Because of the symmetry, the cross-polarized waves cancel in the principal planes (the E and H planes); however, they produce four peaks, one in each quadrant around the main lobe. Referring to Fig. 6.5, the cross-polarized fields peak in the $\phi = \pm 45$ planes. The peaks are about -19 dB relative to the peak of the main (copolar) lobe (Olver, 1992).

The smooth-walled horn does not produce a symmetrical main beam, even though the horn itself is symmetrical. The radiation patterns are complicated functions of the horn dimensions, and details can be found in Balanis (2005) or Chang (1989), where it is shown that the beamwidths in the principal planes can differ widely. This lack of symmetry is a disadvantage for satellite antennas, where wide coverage is required.

By operating a conical horn in what is termed a *hybrid mode*, which is a nonlinear combination of transverse electric (TE) and transverse magnetic (TM) modes, the pattern symmetry is improved, the cross-polarization is reduced, and a more efficient main beam is produced with low sidelobes. It is especially important to reduce the cross-polarization in satellite applications, where frequency reuse is employed, as described in Sec. 5.2.

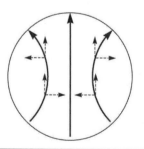

FIGURE 6.11   Aperture field in a smooth-walled conical horn.

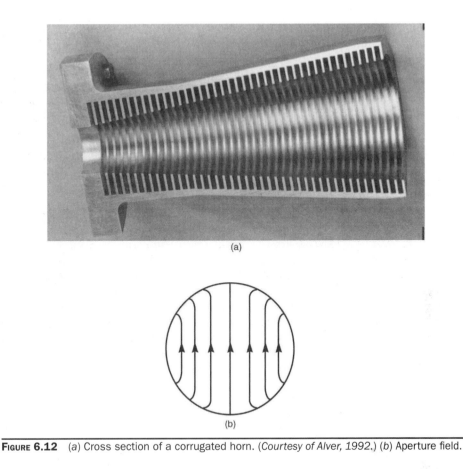

Figure 6.12 (a) Cross section of a corrugated horn. (*Courtesy of Alver, 1992.*) (b) Aperture field.

One method of achieving a hybrid mode is to corrugate the inside wall of the horn, thus giving rise to the *corrugated horn* antenna. The cross section of a corrugated horn is shown in Fig. 6.12a. The aperture electric field is shown in Fig. 6.12b, where it is seen to have a much lower cross-polarized component. This field distribution is sometimes referred to as a *scalar field* and the horn as a *scalar horn*. A development of the scalar horn is the scalar feed, Fig. 6.13, which can be seen on most domestic receiving systems. Here, the flare angle of the horn is 90°, and the corrugations are in the form of a flange surrounding the circular waveguide. The corrugated horn is obviously more difficult to make than the smooth-walled version, and close manufacturing tolerances must be maintained, especially in machining the slots or corrugations, all of which contribute to increased costs. A comprehensive description of the corrugated horn is found in Olver (1992), and design details can be found in Balanis (2005) or Chang (1989).

A hybrid mode also can be created by including a dielectric rod along the axis of the smooth-walled horn, this being referred to as a *dielectric-rod-loaded antenna* (see Miya, 1981).

A *multimode* horn is one which is excited by a linear combination of transverse electric and transverse magnetic fields, the most common type being the *dual-mode horn*, which combines the TE_{11} and TM_{11} modes. The advantages of the dual-mode horn are like those of the hybrid-mode horn, that is, better main lobe symmetry, lower

FIGURE 6.13   A scalar feed.

cross-polarization, and a more efficient main beam with low sidelobes. Dual-mode horns have been installed aboard various satellites (see Miya, 1981).

Horns that are required to provide earth coverage from geostationary satellites must maintain low cross-polarization and high gain over a cone angle of ±9°. This is achieved most simply and economically with dual-mode horns (Hwang, 1992).

6.12.2 Pyramidal Horn Antennas

The pyramidal horn antenna, illustrated in Fig. 6.14, is primarily designed for linear polarization. In general, it has a rectangular cross section $a \times b$ and operates in the TE_{10} waveguide mode, which has the electric field distribution shown in Fig. 6.14.

In general, the beamwidths for the pyramidal horn differ in the E and H planes, but it is possible to choose the aperture dimensions to make these equal. The pyramidal horn can be operated in horizontally and vertically polarized modes simultaneously, giving rise to dual-linear polarization. According to Chang (1989), the cross-polarization characteristics of the pyramidal horn have not been studied to a great extent, and if required, they should be measured.

For any of the aperture antennas discussed, the isotropic gain can be found in terms of the area of the physical aperture by using the relationships given in Eqs. (6.14) and (6.15). For accurate gain determinations, the difficulties lie in determining the illumination efficiency η_I, which can range from 35 to 80 percent for horns and from 50 to 80 percent for circular reflectors (Balanis, 2005). Circular reflectors are discussed in Sec. 6.13.

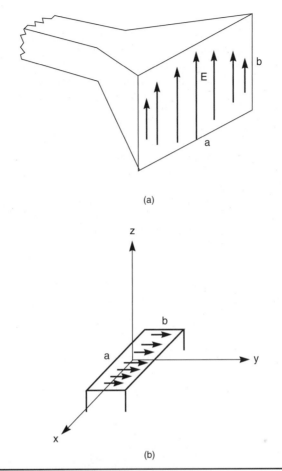

(a)

(b)

FIGURE 6.14   The pyramidal horn and its equivalent aperture.

6.13   The Parabolic Reflector

Parabolic reflectors are widely used in microwave communications systems, and the principal design criteria is to provide the maximum transfer of power from the transmitter to the receiver, which maximizes the on-axis gain. However, for Earth stations, there are strong ITU restrictions that limit the sidelobe levels to reduce interference to/from neighboring GEOSATs. For satellites, the antenna design often employs offset parabolic reflector to prevent satellite structure blockage and scattering from lowering the gain and from distorting the desired spatial coverage pattern including increased sidelobes. Often, because of frequency reuse restrictions, satellite antennas may use more sophisticated multi-reflector designs to provide shaped beam radiation patterns to restrict signals from causing interference in certain geographic areas.

For SATCOM, the trade is between the use of horns and reflector antennas, and for nearly all cases, reflectors are the design of choice. The issue of using horns is that typical designs require horn lengths to be 5 to 10 times the aperture diameter. Since most of the Earth station and space antennas require aperture diameters > 1 m, this favors the folded optics design of reflector antennas that typically have lengths/diameters ratios between 0.5 and 1.0. The only space application for horns is for Earth coverage antennas from GEO orbit, which requires a half-power beamwidth of 17.4° and a corresponding $D/\lambda \sim 3.3$. The reflector provides a focusing mechanism which concentrates the energy in a desired direction, and the most used form is the parabolic reflector with a circular aperture, as shown in Fig. 6.15. We note that reflector antennas often use horns as their antenna feeds.

The main property of the paraboloidal reflector is its focusing property, associated with a plane electromagnetic (EM) wave traveling in a direction that is parallel to the axis of symmetry for the 3D dish. When the rays, perpendicular to the aperture plane wave, strike the reflector's surface, they are specularly reflected and converge on a single point known as the *focus*. Conversely, spherical rays originating at the focus, are reflected as a parallel-rays, with equal path lengths to the focal plane, which is illustrated in Fig. 6.16*b*. The path equality means that a wave originating from an isotropic point source, located at the focus, has a uniform phase distribution over the aperture plane. This property, along with the parallel-beam property, means that the wavefront is plane. Thus, radiation from the paraboloidal reflector appears to originate as a plane wave from the aperture plane, normal to the axis (see App. B for more details). Although the characteristics of the reflector antenna are more readily described in terms of radiation (transmission), it should be kept in mind that the reciprocity theorem makes these applicable to the receiving mode as well.

For communication links, only the far-field needs to be considered. The reflected wave is a plane wave, while the wave originating from the isotropic source and striking the reflector has a spherical wavefront. The power density in the plane wave is independent of distance. However, for the spherical wave reaching the reflector from the

FIGURE 6.15 A medium size Earth station parabolic reflector antenna.

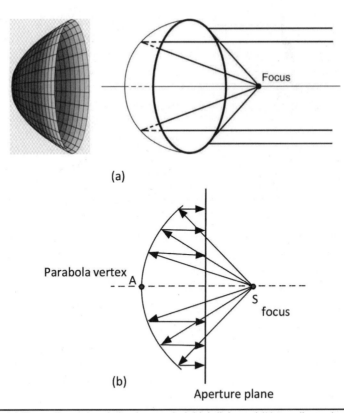

(a)

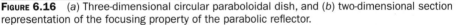

(b)

Aperture plane

Figure 6.16   (a) Three-dimensional circular paraboloidal dish, and (b) two-dimensional section representation of the focusing property of the parabolic reflector.

source, the power density of the far-field component decreases in inverse proportion to the distance squared, and therefore, the illumination at the edge of the reflector is less than that at the vertex. This gives rise to a nonuniform amplitude distribution across the aperture plane, which in effect means that the illumination efficiency is reduced. Nevertheless, there is a benefit in that sidelobe levels are reduced.

In satellite applications, the primary antenna is usually a horn (or an array of horns, as shown later) pointed toward the reflector. The radiation from the horn is a spherical wave, with the *phase center* being the center of curvature of the wavefront. Thus, the horn is positioned so that the phase center lies on the focus. Also, a high illumination efficiency is desirable that requires that the radiation pattern of the primary antenna should approximate the inverse of the space attenuation factor (see App. B Eq. B.35). Further, the space attenuation effect can be mitigated by adding higher-order modes to the horn feed so that the horn-radiation pattern approximates the inverse of the space attenuation function (Chang, 1989).

For example, consider the position of the feed (S) in relation to the reflector for various values of f/D as shown in Fig. 6.17. For $f/D < 0.25$, the primary antenna lies in the space between the reflector and the aperture plane, and the illumination tapers toward the edge of the reflector. For $f/D > 0.25$, the primary antenna lies outside the aperture plane, which results in more nearly uniform illumination, but *spillover*

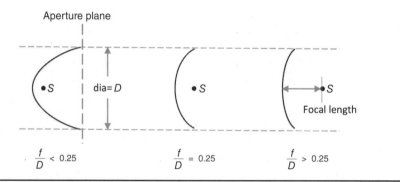

Figure 6.17   Position of the focus for various f/D values.

increases. In the transmitting mode, spillover is the radiation from the primary antenna, which is directed toward the reflector but misses the reflector surface.

The gain and beamwidths of the paraboloidal antenna are as follows. The antenna aperture area is defined as physical area of the projected circular reflector in the aperture plane.

$$\text{Area} = \frac{\pi D^2}{4}, \text{m}^2 \tag{6.27}$$

From the relationships given by Eqs. (6.14) and (6.15), the peak (or boresight) gain is

$$G_o = \left(\frac{4\pi}{\lambda^2}\right)\eta_I \text{ area}$$

$$= \eta_I \left(\frac{\pi}{\lambda/D}\right)^2 \tag{6.28}$$

The radiation pattern for the paraboloidal reflector is like that developed in Example 6.1 for the rectangular aperture, in that there is a main lobe and several sidelobes, although there are differences in detail. In practice, the sidelobes are accounted for by an envelope function as described in Sec. 13.2.4. Useful approximations for the half-power beamwidth (HPBW, β) and the beamwidth *between the first nulls* (BWFN) are

$$\beta \cong 1.2\left(\frac{\lambda}{D}\right), \text{ radians} \tag{6.29}$$

$$\text{BWFN} \cong 2\beta, \text{ radians} \tag{6.30}$$

In these relationships, the beamwidths are given in radians, but the conversion to degrees is to multiply by $(180°/\pi)$. Furthermore, the paraboloidal antenna described is *center-fed*, in that the primary horn is located at the focus and pointed toward the center of the reflector (apex of the parabola). With this arrangement the primary horn and its supports present a partial blockage to the reflected wave. The energy scattered by the blockage is lost from the main lobe, and it can create additional sidelobes.

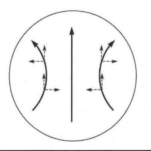

FIGURE 6.18 Current paths in a paraboloidal reflector for linear polarization.

One effective solution is to use an *offset feed* as described in Sec. 6.14. The wave from the primary radiator induces surface currents in the reflector. The curvature of the reflector causes the currents to follow curved paths so that both horizontal and vertical components are present, even where the incident wave is linearly polarized in one or other of these directions. The situation is sketched for the case of vertical polarization in Fig. 6.18. The resulting radiation consists of co-polarized (co-pol) and cross-polarized (x-pol) fields. The symmetry of the arrangement means that the cross-polarized component is zero in the principal planes (the E and H planes), and the cross-polarization peaks in the $\phi = \pm 45$ planes, assuming a coordinate system as shown in Fig. 6.5a. Sketches of the co-pol and x-pol radiation patterns for the 45 planes are shown in Fig. 6.19.

6.13.1 Simple Analytical Approximations for Main Bean Gain Patterns

As described in Chap. 12, communication engineers perform satellite link-power budget analysis to determine the signal to noise power ratio for the space links. A key parameter

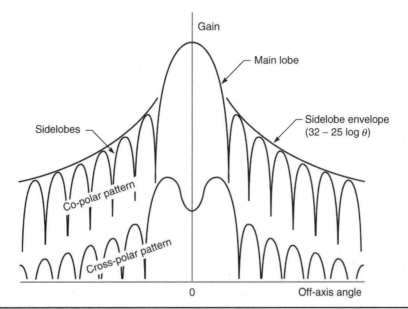

FIGURE 6.19 Co-polar and cross-polar radiation patterns. (*Courtesy of FCC Report FCC/OST R83-2, 1983.*)

in these calculations is the antenna line-of-sight (LOS) gains for Earth station and satellite antennas. In practice, these antenna patterns are measured; and given the pointing vector for the antenna boresight, the off-boresight angle θ can be determined by simple geometry. For Geostationary satellites, the calculation of antenna look angles is presented in Sec. 3.2, and a detailed description of antenna misalignment losses is presented in Sec. 12.4.1.

Often, neither the real antenna patterns nor reflector antenna-pattern codes are available; therefore, we present two simple analytical functions that provide realistic approximations of the main-beam, normalized, parabolic reflector antenna pattern, given the reflector diameter in wavelengths.

These functions are used in examples and homework problems for both earth stations and satellite antennas to provide relative gain $G(\theta)$ for circular beams from parabolic reflectors. We assume that the E-plane and H-plane patterns are equal, which is a reasonable assumption for the main beam. With this assumption, the normalized pattern is circularly symmetric, and the off-boresight gain is only a function of the spherical angle θ, which greatly simplifies the LOS gain calculation, i.e., the off-boresight gain is independent of the spherical angle, ϕ.

Bessel Function Approximation

The normalized antenna pattern (relative gain) is defined by

$$G(\theta)_{\text{Bessel}} = 64 \left| \frac{J_2(u)}{u^2} \right|^2 \tag{6.31}$$

where J_2 is the Bessel function of the 1st kind of 2nd order, and the argument is

$$u = \frac{\pi}{\lambda/D} \sin \theta$$

which, for small off-boresight angles < 0.2 radians, becomes

$$u \cong \frac{\pi}{\lambda/D} \theta = \pi \Theta_n \tag{6.32}$$

where λ/D and θ are given in radians.

If we replace theta in Eq. (6.32) by the normalized off-boresight angle ratio Θ_n defined as

$$\Theta_n = \frac{\theta}{\lambda/D}, \text{ ratio} \tag{6.33}$$

then the normalized gain pattern becomes independent of the λ/D. Thus, it can be used to calculate the absolute LOS gain is given by

$$g(\theta)_{\text{LOS}} = G_0 G(\theta)_{\text{Bessel}}, \text{ power ratio} \tag{6.34a}$$

or, expresses in decibels, is

$$\left[g(\theta)_{\text{LOS}} \right] = \left[G_0 \right] + \left[G(\theta)_{\text{Bessel}} \right], \text{ dB} \tag{6.34b}$$

where G_0 is the peak (bore-sight) gain that is given by Eq. (6.28), and because $G(\theta)_{\text{Bessel}} \leq 1$, the corresponding $\left[G(\theta)_{\text{Bessel}} \right]$ is a negative dB.

Example 6.2 Given an antenna of diameter 20 wavelengths, calculate the normalized gain pattern to the −10 dB level using the Bessel function approximation. Assuming an aperture efficiency = 65%, determine the peak gain and the absolute line-of-sight (LOS) gain at an off-boresight angle = 1.82°.

Solution Using Eq. (6.28) calculate the peak (boresight gain)

$$G_o = \eta_I \left(\frac{\pi}{\lambda/D}\right)^2 = 0.65\left(\frac{\pi}{\lambda/20\lambda}\right)^2 = 0.65(20\pi)^2 = 2.57e + 03 \Rightarrow 34.09 \text{ dB}$$

Using Eq. (6.32), the argument of the Bessel function, at an off-boresight angle = 1.82°, is

$$u = \frac{\pi}{\lambda/D}\sin\theta = \left(\frac{\pi D}{\lambda}\right)\sin\theta = 20\pi \, \sin(1.82°) = 1.996$$

Finally, using Eq. (6.31) for $u = 1.996$

$$G(\theta)_{\text{Bessel}} = 64\left|\frac{J_2(u)}{u^2}\right|^2 = 0.50 \Rightarrow -3.0 \text{ dB}$$

and the corresponding LOS gain is

$$\left[g(\theta)_{\text{LOS}}\right] = \left[G_0\right] + \left[G(\theta)\right] = 34.1 + (-3.0) = 31.1, \text{ dB}$$

and Fig. 6.20 shows an extended plot of the Bessel approximation normalized antenna gain pattern.

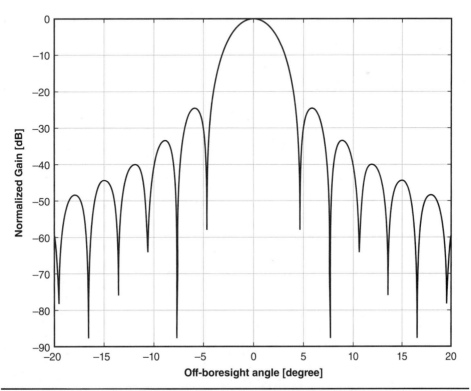

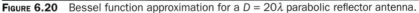

FIGURE 6.20 Bessel function approximation for a $D = 20\lambda$ parabolic reflector antenna.

Part-B Alternative Solution

As an alternative, the LOS gain can also be calculated from Table 6.1, where the normalized antenna gain is given as a function of the normalized off-boresight angle ratio Eq. (6.33)., which makes the main-beam pattern independent of (λ/D).

The first step is to express theta in radians, theta = $1.82° \Rightarrow 0.0318$ radians, and then using Eq. (6.33) to calculate the normalized off-boresight ratio

$$\Theta_n = \frac{\theta}{\left(\dfrac{\lambda}{D}\right)} = \frac{0.0318}{\dfrac{\lambda}{20\lambda}} = 0.635$$

entering the table and performing linear interpolation, yields

$$\Delta \text{gain} = \frac{(0.540 - 0.483)(0.60 - 0.635)}{(0.60 - 0.65)} = 0.040$$

$$= G(\theta)_{\text{norm}} = 0.54 - \Delta\text{gain} = 0.540 - 0.040 = 0.50 \Rightarrow -3.0 \text{ dB}$$

Θ_norm Ratio	Bessel Approx Power Ratio	Bessel Approx dB
0.00	1.00E+00	0.00
0.05	9.96E-01	−0.02
0.10	9.84E-01	−0.07
0.15	9.64E-01	−0.16
0.20	9.36E-01	−0.29
0.25	9.02E-01	−0.45
0.30	8.61E-01	−0.65
0.35	8.15E-01	−0.89
0.40	7.65E-01	−1.16
0.45	7.12E-01	−1.48
0.50	6.55E-01	−1.83
0.55	5.98E-01	−2.23
0.60	5.40E-01	−2.68
0.65	4.83E-01	−3.16
0.70	4.27E-01	−3.70
0.75	3.73E-01	−4.28
0.80	3.22E-01	−4.92
0.85	2.74E-01	−5.62
0.90	2.30E-01	−6.38
0.95	1.90E-01	−7.20
1.00	1.55E-01	−8.10
1.05	1.24E-01	−9.08
1.10	9.66E-02	−10.15

TABLE 6.1   Bessel Normalized Main Beam Gain Approximation

Part-C

Suppose the reflector diameter is 30 wavelengths, what is the off-boresight gain?

Solution: Because the λ/D is now $(1/30)$, we must recalculate the normalized off-boresight ratio, using theta expressed in radians, theta = $1.82° \Rightarrow 0.0318$ radians, and using Eq. (6.33)

$$\Theta_n = \frac{\theta}{\left(\dfrac{\lambda}{D}\right)} = \frac{0.0318}{\dfrac{\lambda}{30\lambda}} = 0.954$$

entering the table, yields ~ −7.2 dB.

To assess the effectiveness of the Bessel approximation, the gain pattern of an on-axis fed parabolic reflector antenna was calculated using a computational electromagnetic (CEM) software tool for simulation of radiation patterns of realistic antennas (FEKO, Altair Engineering Inc., 2023). Selected results, given in Fig. 6.21, compare E and H principal planes gain patterns, with equivalent Bessel

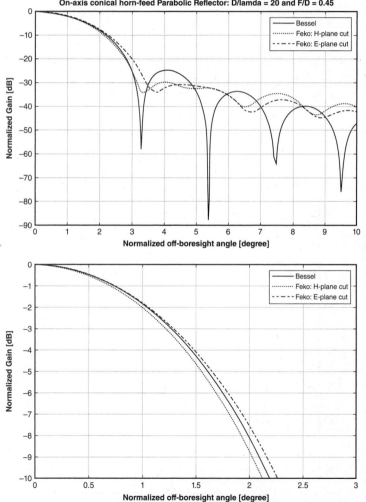

FIGURE 6.21   Comparison of Bessel function approximation with CEM antenna pattern simulation for a center-fed parabolic reflector antenna. All patterns are plotted in normalized off-boresight angle ratio (Eq. 6.33), which makes the pattern independent of λ/D.

results for the same reflector diameter. In this simulation, a frequency domain (method of moments) solution was used to simulate the primary pattern for the conical feed horn, and the asymptotic large element physical optics approach was used to simulate the reflector secondary radiation patterns.

To evaluate the quality of the Bessel approximation, a series of CEM calculations were made using an operating frequency = 12.0 GHz, reflector diameter $D = 20\ \lambda$, and for several focal lengths, $0.4 \le f/D \le 0.75$. The closest main beam pattern agreement resulted from the $f/D = 0.45$ case, and this result is shown in Fig. 6.21. For this case, the Bessel pattern provided an excellent match with the average of the E-plane and H-plane normalized gain patterns over the main beam.

Sinc² Function Approximation

The second approximation for the normalized main beam antenna pattern is square of the Sinc function given by

$$G(\theta)_{\text{sinc}} = \left(\frac{\sin(x)}{x}\right)^2 \tag{6.35}$$

where θ is the off-boresight angle in radians, and the argument of the Sinc is

$$x = 0.690\,u \tag{6.36}$$

where the coefficient 0.690 was empirically derived to match the Bessel normalized gain pattern over the main beam, and $u = \dfrac{\pi}{\lambda/D}\sin\theta$, Eq. (6.32)

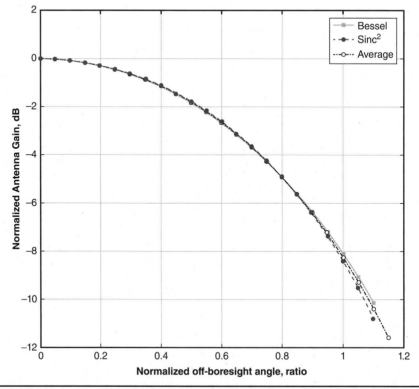

Figure 6.22 Comparison of the Bessel and Sinc² approximations used for earth station and satellite antennas. All patterns are plotted using normalized off-boresight angle ratio, which makes the pattern independent of the λ/D ratio.

Θ_norm Ratio	Sinc² Approx Power Ratio	Sinc² Approx Db
0.00	1.00E+00	0.00
0.05	9.96E-01	−0.02
0.10	9.84E-01	−0.07
0.15	9.65E-01	−0.15
0.20	9.39E-01	−0.27
0.25	9.06E-01	−0.43
0.30	8.67E-01	−0.62
0.35	8.22E-01	−0.85
0.40	7.73E-01	−1.12
0.45	7.20E-01	−1.42
0.50	6.65E-01	−1.77
0.55	6.07E-01	−2.17
0.60	5.49E-01	−2.60
0.65	4.91E-01	−3.09
0.70	4.33E-01	−3.63
0.75	3.77E-01	−4.23
0.80	3.24E-01	−4.90
0.85	2.73E-01	−5.63
0.90	2.27E-01	−6.45
0.95	1.84E-01	−7.36
1.00	1.46E-01	−8.37
1.05	1.12E-01	−9.51
1.10	8.30E-02	−10.81

TABLE 6.2   Sinc² Normalized Main Beam Gain Approximation

As with the Bessel approximation, when we replace theta by the normalized off-boresight angle ratio (Θ_{norm} Eq. (6.33), then the pattern is also independent of the λ/D, and this Sinc² approximation pattern is tabulated in Table 6.2. The objective is that these patterns should be two different analytical approximations that are essentially identical, and an examination of Tables 6.1 and 6.2 shows that the differences are very small, <0.1 dB, within the half-power beamwidth, and gradually increase to <0.5 dB at the −10 dB level. Therefore, the use of the Bessel and Sinc² approximations are fully equivalent without any preference.

6.14   The Offset Feed

An example of a small Earth station antenna with an off-set feed configuration is shown in Fig. 6.23. The upper panel (a) shows the ray traces for a horn feed at the focus of a virtual parent parabolic reflector (shown in dashed lines). For this configuration, the primary radiation pattern of the horn is offset such that it illuminates only the upper

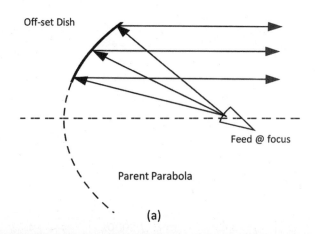

FIGURE 6.23   (a) Ray paths for an offset reflector. (b) Typical small Earth terminal antennas.

portion of the parent reflector. As a result, the feed horn and its support are well clear of the secondary reflector pattern so that no blockage occurs. With the center-fed arrangement described in the previous section, the blockage results typically in a 10 percent reduction in efficiency (Brain and Rudge, 1984) and increased radiation in the sidelobes.

This offset arrangement avoids this blockage and results an excellent main beam secondary pattern, but, because of the reflector asymmetry, there is a slight disadvantage compared with the center-fed antenna concerning the cross-polarization with a linear polarized feed. However, polarization compensation can be introduced into the primary feed to correct for the cross-polarization, or a *polarization-purifying grid* can be incorporated into the antenna structure (Brain and Rudge, 1984). Thus, this is a minor issue for the antenna RF design that is easily accommodated. Figure 6.23*b* shows two ~1.5 m dishes that are viewing different satellites in geostationary orbit.

Also, for spacecraft use, there are significant mechanical advantages that make the off-set reflector configuration, the choice for the vast majority of GEOSATs, where

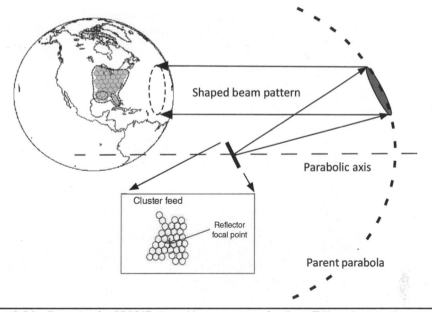

Figure 6.24 Example of a GEOSAT shaped beam antenna for direct TV broadcast to the eastern and central time zones.

large apertures are required (see, e.g., Figs. 7.6, 7.21, and 7.26). This is especially true for the three-axis stabilized spacecraft, where multiple large reflectors can be folded against the spacecraft cube structure for launch, and then once on orbit released with a single hinge mechanical joint to swing into proper configuration without the feed assembly moving. Therefore, the cost and risk associated with antenna stowage for launch and on-orbit deployments are significantly reduced by this configuration, and it also provides superior antenna patterns without spacecraft structure blockage issues and fields of view conflicts between multiple antennas.

An example of a *direct broadcast satellite* (DBS) television antenna is shown in Fig. 6.24, where a cluster of feeds provides the primary illumination pattern on a large offset parabolic reflector. Because the RF signals to the individual feeds are coherent, their electric fields are spatially combined as an array, which produces a shaped beam specifically tailored for subscriber coverage for the United States eastern and central time zones.

Further, off-set reflectors are also used with double-reflector antennas, as discussed next.

6.15 Double-Reflector Antennas

With reflector-type antennas, the transmission lines connecting the feed horn to the transmit/receive equipment must be kept as short as possible to minimize losses. This is particularly important with large earth stations where the transmit power is large and where very low receiver noise is required. The single-reflector system described in Sec. 6.13 does not lend itself very well to achieving this, however more satisfactory, but more costly, arrangements are possible with a double-reflector system.

Figure 6.25 Tracking Cassegrain satellite dish (15 m diameter).

For this configuration, the feed horn is mounted at the rear of the main reflector through an opening at the vertex, as illustrated in Figs. 6.25 and 6.26 for two earth station antennas of 15 m and 5 m diameter respectively. The rear mount makes for a compact feed, which is an advantage where steerable (tracking) antennas must be used, and this makes access for servicing much easier. The subreflector, which is mounted at the front of the main reflector, is generally smaller than the feed horn assembly and causes

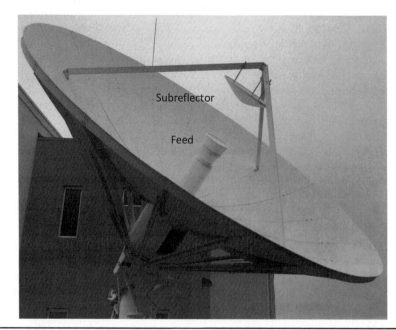

Figure 6.26 Fixed Cassegrain satellite dish (5 m diameter).

less blockage. Two main types are in use, namely, the Cassegrain antenna and the Gregorian antenna, named after the astronomers who first developed them.

6.15.1 Cassegrain Antenna

The basic Cassegrain configuration, which is a common design for optical telescopes, consists of a main paraboloid and a subreflector, which is a hyperboloid (see App. B). The subreflector has two focal points, one of which is made to coincide with that of the main reflector and the other with the phase center of the feed horn, as shown in Fig. 6.27a. The Cassegrain system is equivalent to a single paraboloidal reflector of focal length

$$f_e = \frac{e_h + 1}{e_h - 1} f \qquad (6.37)$$

where e_h is the eccentricity of the hyperboloid (see App. B) and f is the focal length of the main reflector. The eccentricity of the hyperboloid is always greater than unity and typically ranges from about 1.4 to 3. The equivalent focal length, therefore, is greater

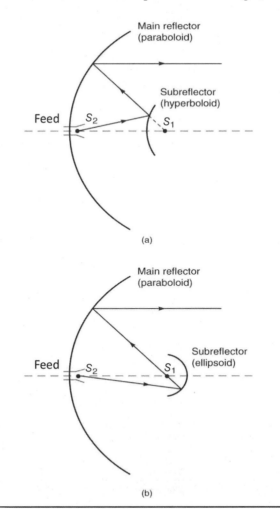

FIGURE 6.27 Ray paths for (a) Cassegrain antenna. (b) Gregorian antenna.

than the focal length of the main reflector. The diameter of the equivalent paraboloid is the same as that of the main reflector, and hence the f/D ratio is increased.

As shown in Fig. 6.17, a large f/D ratio leads to more uniform illumination, and in the case of the Cassegrain, this is achieved without the spillover associated with the single-reflector system, which results in a significantly improved aperture efficiency. Also, the larger f/D ratio results in lower cross-polarization (Miya, 1981). The Cassegrain system is widely used in large earth-station installations.

6.15.2 Gregorian Antenna

The basic configuration of the Gregorian is like the Cassegrain form, which comprises a main paraboloid and a subreflector, which is an ellipsoid (see App. B). As with the hyperboloid, the subreflector has two focal points, one of which is made to coincide with that of the main reflector and the other with the phase center of the feed horn, as shown in Fig. 6.27*b*. The performance of the Gregorian system is similar in many respects to the Cassegrain. An offset Gregorian antenna is illustrated in Fig. 6.28.

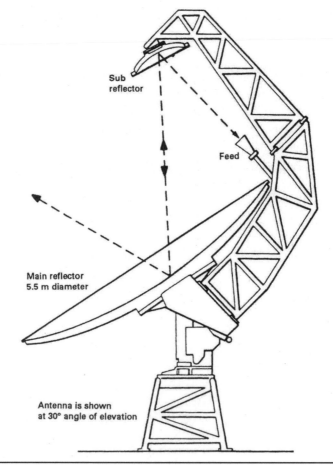

Figure 6.28 Offset Gregorian antenna. (*Courtesy of Radio Electr. Eng., vol. 54, No. 3, Mar. 1984, p. 112.*)

6.16   Shaped Reflector Systems

With the double-reflector systems described, the illumination efficiency of the main reflector can be increased, while avoiding the problem of increased spillover, by shaping the surfaces of the subreflector and main reflector. For example, with the Cassegrain system, altering the curvature of the center section of the subreflector to be greater than that of the hyperboloid allows it to reflect more energy toward the edge of the main reflector, which makes the amplitude distribution more uniform. At the same time, the curvature of the center section of the main reflector is made smaller than that required for the single reflector paraboloid. This compensates for the reduced path length so that the constant phase condition across the aperture is maintained. The edge of the subreflector surface is shaped in a manner to reduce spillover, and of course, the overall design must consider the radiation pattern of the primary feed. The process, referred to as *reflector shaping*, employs computer-aided design methods. Further details can be found in Miya (1981) and Rusch (1992).

With the Hughes Space and Communications Company's shaped reflector (Fig. 6.29), dimples and/or ripples are created on the surface. The depth of these is no more than a

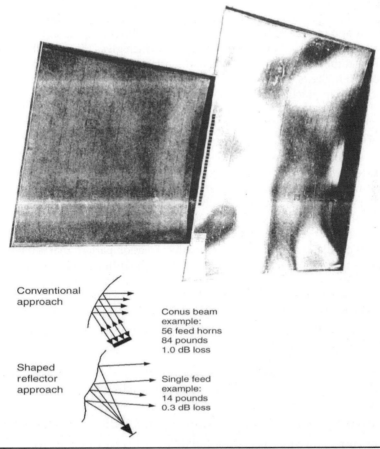

FIGURE 6.29   Shaped-beam reflector, showing ray paths. (*Courtesy of Hughes Space and Communications Company. Reproduced from Vectors XXXV(3):14, 1993. © Hughes Aircraft Co.*)

wavelength, which makes them rather difficult to see, especially at the Ka-band. Reflections from the uneven surface reinforce radiation in some directions and reduce it in others. The design steps start with a map of the ground coverage area desired. A grid is overlaid on the map, and at each grid intersection a weighting factor is assigned, which corresponds to the antenna gain desired in that direction. The intersection points on the coverage area also can be defined by the azimuth and elevation angles required at the ground stations, which enables the beam contour to be determined.

The beam-shaping stage starts by selecting a smooth parabolic reflector that forms an elliptical beam encompassing the coverage area. The reflector surface is computer modeled as a series of mathematical functions that are changed or perturbed until the model produces the desired coverage. On a first pass the computer analyzes the perturbations and translates these into surface ripples. The beam footprint computed for the rippled surface is compared with the coverage area. The perturbation analysis is performed iteratively until a satisfactory match is obtained.

As an example, the conventional approach to producing a CONUS beam requires 56 feed horns with the feed assembly weighs 38 kg and having a −1 dB loss. With a shaped reflector, a single-feed horn is used, at a weight of 6.4 kg and having a −0.3 dB loss (see Vectors, 1993).

Shaped reflectors also have been used to compensate for rainfall attenuation, and this has application in DBS systems (see Chap. 16). In this case, the reflector design is based on a map like that shown in Fig. 16.8, which gives the rainfall intensity as a function of latitude and longitude. The attenuation resulting from the rainfall is calculated as shown in Sec. 4.4, and the reflector is shaped to redistribute the radiated power to match, within practical limits, the attenuation.

6.17 Arrays

Beam shaping can be achieved by using an array of basic elements. The elements are arranged so that their radiation patterns provide mutual reinforcement in certain directions and cancellation in others. Although most arrays used in satellite communications are two-dimensional horn arrays, the principle is most easily explained with reference to an in-line array of dipoles (Fig. 6.30a and b).

As shown previously (Fig. 6.8), the radiation pattern for a single dipole in the xy plane is circular, and it is this aspect of the radiation pattern that is altered by the array configuration. Two factors contribute to this: the difference in distance from each element to some point in the far field and the difference in the current feed to each element. For the coordinate system shown in Fig. 6.30b, the xy plane, the difference in distance is given by $(s \cos \phi)$. Although this distance is small compared with the range between the array and point P, it plays a crucial role in determining the phase relationships between the radiation from each element.

It should be kept in mind that at any point in the far field the array appears as a point source, the situation being as sketched in Fig. 6.30c. For this analysis, the point P is taken to lie in the xy plane. Since one wavelength corresponds to a phase difference of 2π, the phase lead of element n relative to $(n-1)$ resulting from the difference in distance is $(2\pi \frac{s}{\lambda} \cos \phi)$. To illustrate the array principles, it is assumed that each element is fed by currents of equal magnitude but differing in phase progressively by some angle α. Positive values of α mean a phase lead and negative values a phase lag.

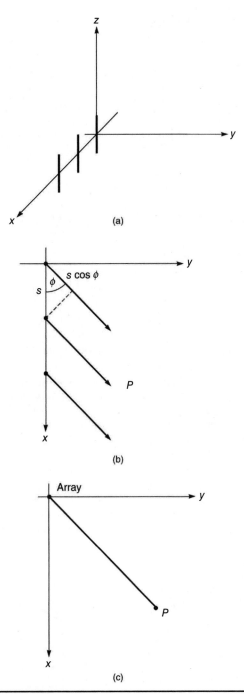

FIGURE 6.30 An in-line array of dipoles.

The total phase lead of element n relative to $(n-1)$ is given as

$$\psi = \alpha + 2\pi \frac{S}{\lambda} \cos \phi \qquad (6.38)$$

The Argand diagram for the phasors is shown in Fig. 6.31. The magnitude of the resultant phasor can be found by first resolving the individual phasors into horizontal (real axis) and vertical (imaginary axis) components, adding these, and finding the resultant. The contribution from the first element is E, and from the second element, $E \cos \psi + jE \sin \psi$. The third element contributes $E \cos 2\psi + jE \sin \psi$, and in general the Nth element contributes $E \cos(N-1)\psi + jE \sin (N-1)\psi$. These contributions can be added to get:

$$E_R = E + E \cos \psi + jE \, \sin\psi + E \cos 2\psi + jE \sin 2\psi + \cdots$$

$$= \sum_{n=0}^{N-1} E \cos n\psi + jE \, \sin n\psi = E \sum_{n=0}^{N-1} e^{jn\psi} \qquad (6.39)$$

Here, N is the total number of elements in the array. A single element results in a field E, and the array is seen to modify this by the summation factor. The magnitude of summation factor is termed the *array factor* (AF):

$$\mathrm{AF} = \left| \sum_{n=0}^{N-1} e^{jn\psi} \right| \qquad (6.40)$$

The AF has a maximum value of N when $\psi = 0$, and hence the maximum value of E_R is $E_{Rmax} = NE$. Recalling that ψ as given by Eq. (6.38) is a function of the current phase angle, α, and the angular coordinate, ϕ, it is possible to choose the current phase to make the AF show a peak in some desired direction ϕ_0. The required relationship is

$$\alpha = -2\pi \frac{S}{\lambda} \cos \phi_0 \qquad (6.41)$$

Combining this with Eq. (6.38) gives

$$\psi = \frac{2\pi S}{\lambda} (\cos \phi - \cos \phi_0) \qquad (6.42)$$

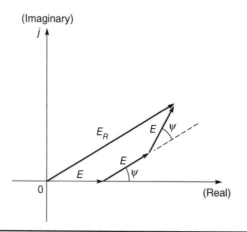

FIGURE 6.31   Phasor diagram for the in-line array of dipoles.

This can be substituted into Eq. (6.39) to give the AF as a function of ϕ relative to maximum.

Example 6.3   A dipole array has two elements equispaced at 0.25 wavelength. The AF is required to have a maximum along the positive axis of the array. Plot the magnitude of the AF as a function of ϕ.

Solution   Given data are $N = 2$; $s = 0.25\lambda$; $\phi_0 = 0$.
From Eq. (6.42):

$$\psi = \frac{\pi}{2}(\cos\phi - 1)$$

The AF is

$$AF = \left| \sum_{n=0}^{1} e^{jn\psi} \right| = |1 + \cos\Psi + j\sin\Psi| = \sqrt{2(1 + \cos\psi)}$$

When $\phi = 0$, $\psi = 0$, and hence the AF is 2 (as expected for two elements in the $\phi = 0$ direction). When $\phi = \pi$, $\psi = -\pi$, and hence the AF is zero in this direction. The plot on polar graph paper is shown below.

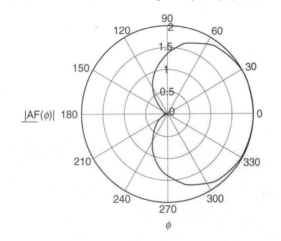

For this example, the values are purposely chosen to illustrate what is termed an *end-fire array*, where the main beam is directed along the positive axis of the array. Keep in mind that a single dipole has a circular pattern. An example of a five-element end fire array is given in Prob. 6.32.

The current phasing can be altered to make the main lobe appear at $\phi = 90°$, giving rise to a *broadside array*. The symmetry of the dipole array means that two broadside lobes occur, one on each side of the array axis. This is illustrated in Example 6.4.

Example 6.4   Repeat Example 6.3 for $\phi = 90$ and $s = 0.5\lambda$

Solution   The general expression for the AF for the two-element array is not altered and is given by

$$AF = \sqrt{2(1 + \cos\psi)}$$

However, from Eq. (6.42), the phase angle for $s = 0.5\lambda$ and $\phi_0 = 90°$ becomes

$$\psi = \pi\cos\phi$$

With $\phi = 0$, $\psi = \pi$, and hence the AF is zero. Also, with $\phi = 180°$, $\psi = -\pi$ and once again the AF is zero. With $\phi = \pm 90°$, $\psi = 0$ and the AF is 2 in each case. The plot on polar graph paper is as shown below. An example of a five-element broadside array is given in Prob. 6.37.

As these examples show, the current phasing controls the position of the main lobe, and a continuous variation of current can be used to produce a *scanning array*. With the simple dipole array, the shape of the beam changes drastically with changes in the current phasing, and in practical scanning arrays, steps are taken to avoid this. A detailed discussion of arrays is found in Kummer (1992).

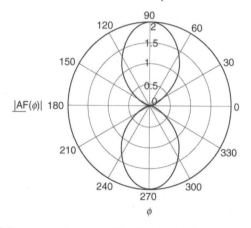

Arrays may be used directly as antennas, and details of a nine-horn array used to provide an earth coverage beam are given in Hwang (1992). Arrays are also used as feeds for reflector antennas, and such a horn array is shown in Fig. 6.32.

FIGURE 6.32   A multi-feed contained-beam reflector antenna. (*Courtesy of Brain and Rudge, 1984.*)

6.18   Planar Antennas

A *microstrip antenna* is an antenna etched in one side of a printed circuit board, a basic *patch antenna* being as sketched in Fig. 6.33*a*. A variety of construction techniques are in use, both for the antenna pattern itself, and the feed arrangement, but the principles of operation can be understood from a study of the basic patch radiator. In Fig. 6.33*a* the antenna input is a microstrip line connecting to the patch, and the copper on the underside of the board forms a ground plane. The dielectric substrate is thin (less than about one-tenth of a wavelength) and the field under the patch is concentrated in the dielectric. At the edges of the patch the electromagnetic fields are associated with surface waves and radiated waves, the radiation taking place from the "apertures" formed in the substrate between the edges of the patch and the ground plane. The radiated fields are sketched in Fig. 6.33*b*.

Figure 6.34 shows the patch of sides *a* and *b* situated at the origin of the coordinate system of Fig. 6.3. Approximate expressions for the radiation pattern in the principal planes at $\phi = 0$ and $\phi = 90$ are (see James et al., 1981, Eqs. 4.26a and b):

$$g(\theta, \phi = 90°) = \cos^2\left(\frac{\pi b}{\lambda_0}\sin\theta\right) \tag{6.43}$$

$$g(\theta, \phi = 0) = \cos^2\theta\left[\frac{\sin X}{X}\right]^2 \tag{6.44}$$

where $X = (\pi a / \lambda_0)\sin\theta$, and λ_0 is the free space wavelength. Equation (6.44) is similar to Eq. (6.23). A plot of these functions, for a half wavelength patch is shown in Fig. 6.35. In practice, the length of each side of the patch is less than half the free space wavelength because the phase velocity v_p of the wave is less than the free space value. Recall that

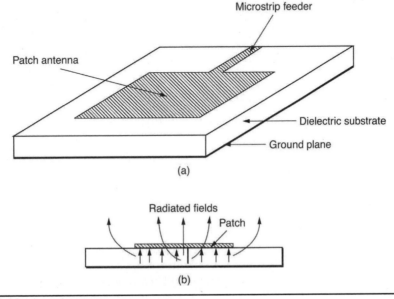

(a)

(b)

FIGURE 6.33   A patch antenna.

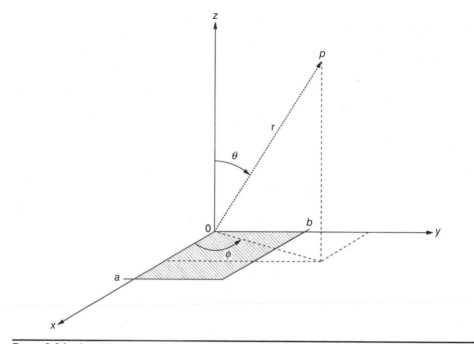

Figure 6.34 A patch antenna and its coordinate system.

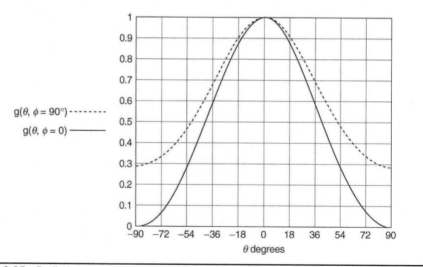

$g(\theta, \phi = 90°)$ ‑ ‑ ‑ ‑ ‑ ‑

$g(\theta, \phi = 0)$ ———

Figure 6.35 Radiation patterns for a patch antenna.

$\lambda f = v_p$, where λ is the wavelength and f the frequency, and the phase velocity of a wave in a dielectric medium of relative permittivity ε_r is $c/\sqrt{\varepsilon_r}$, where c is the free space velocity of an electromagnetic wave. For a microstrip board of thickness 1.59 mm and a relative permittivity of 2.32, at a frequency of 10 GHz, the side length is $0.32\lambda_0$ where λ_0 is the free space wavelength (see James et al., 1981 [Table 5.3]).

Other geometries are in use, for example circular patches (disc patches), and coplanar boards and stripline boards are also used. The patch dimensions are usually half or quarter board wavelength. Figure 6.36a shows a disc element with a balanced coaxial feed. Figure 6.36b shows a *copolanar waveguide* construction. Here the board has ground planes on both sides, which are bonded together. The antenna element is etched into one of the ground planes. Figure 6.36c shows a *triplate* construction, where a *stripline*,

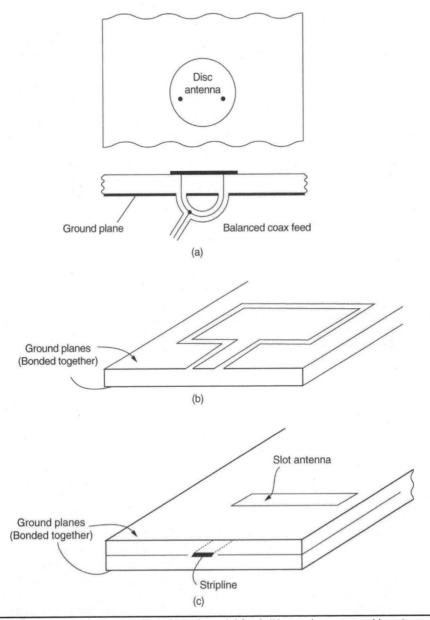

FIGURE 6.36 (a) a disc antenna with balanced coaxial feed; (b) a coplanar waveguide antenna; (c) a triplate slot antenna.

which is etched on the inner layer of one of the boards, forms a central conductor which passes under the slot in the upper ground plane, the slot forming an aperture antenna. The two dielectrics are glued tightly together, and the ground planes are bonded together too.

The basic microstrip patch is a linear polarized antenna, but various feed arrangements are in use to convert it to a circularly polarized antenna (see, e.g., James et al., 1981).

6.19   Planar Arrays

The patch antenna is widely used in *planar arrays*. These are arrays of basic antenna elements etched on one side of a printed circuit board. A multilayer board is normally used so that associated connections and circuitry can be accommodated, as shown in Fig. 6.37. Flat panels are used, and these may be circular (as shown) or rectangular. The use of phase shifters to provide the tracking (beam scanning) is a key feature of planar arrays. The most economical method of beam forming and scanning is mechanical as described in the earlier sections. The beam is formed by a shaped reflector and azimuth and elevation motors provide the scanning. Such motors can also be used with flat panel arrays, as shown in Fig. 6.37, although the beam is formed by phasing of the elements, as described in Sec. 6.17, rather than by means of a mechanical reflector. The beam can be made to scan by introducing a progressive phase shift to the driving voltage applied to the various patch elements.

Figure 6.38 shows two basic configurations. In the active configuration, each antenna element has its own amplifier and phase shifter, while in the passive arrangement, a single amplifier drives each element through the individual phase shifters. The active arrangement has the advantage that the failure of one amplifier causes degradation but not complete loss of signal, but this must be offset against the greater cost and complexity incurred.

Electronic scanning can be achieved in one of several ways. The phase shift coefficient of a transmission line is given by $\beta = 2\pi/\lambda$, where λ is the wavelength of the signal passing along the line. A section of transmission line of length l introduces a phase lag (the output lags the input) of amount

$$\varphi = \beta l \tag{6.45}$$

Phase shift can be achieved by changing either l or β. The concept of changing the length l is illustrated in Fig. 6.39. Making the shorter length the reference line, the phase shift obtained in switching from one to the other is

$$\Delta\varphi = \beta(l_1 - l_2) \tag{6.46}$$

Note that the switched line phase shifter requires a *double pole single throw* (DPST) switch at each end. Several types of switches have been utilized in practical designs, including PIN diodes, *field effect transistors* (FETs) and *micro-electro-mechanical* (MEM) switches. In a PIN diode, the p-type semiconductor region is separated from the n-type region by an intrinsic region (hence the name PIN). At frequencies below about 100 MHz, the diode behaves as a normal rectifying diode. Above this frequency, the stored charge in the intrinsic region prevents rectification from occurring and the diode conducts in both directions. The diode resistance is inversely related to the stored charge, which in turn is controlled by a steady bias voltage. With full forward bias the diode appears as

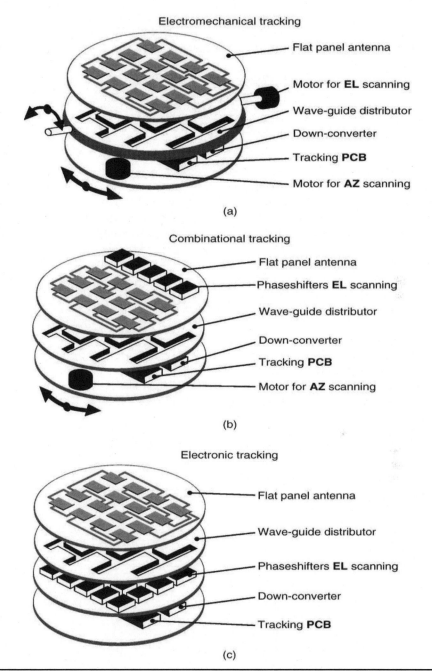

FIGURE 6.37 Assembly details of a planar array. (*a*) electromechanical tracking; (*b*) combined electromechanical and electronic tracking; (*c*) electronic tracking. (*Courtesy of Michael Parnes. Source: http://www.ascor. eltech.ru/ascor15.htm*)

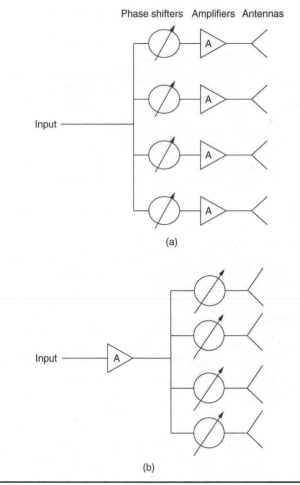

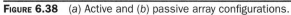

FIGURE 6.38 (*a*) Active and (*b*) passive array configurations.

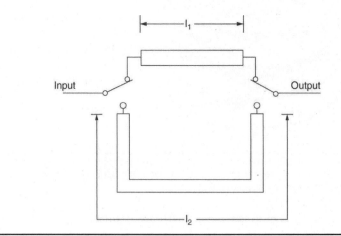

FIGURE 6.39 A transmission line phase shifter.

a short circuit, and with full reverse bias the diode ceases to conduct. In effect the diode behaves as a switch.

In practice PIN diode switches are usually wire-bonded into the phase changer, this being referred to as a *microwave integrated circuit* (MIC). The wire bond introduces a parasitic inductance, which sets an upper frequency limit, although they have been used at frequencies beyond 18 GHz. Two diodes are required for each DPST switch.

Metal semiconductor field effect transistors (MESFETs) are also widely used as microwave switches. In the MESFET, the charge in the channel between the drain and source electrodes is controlled by the bias voltage applied to the gate electrode. The channel can be switched between a highly conducting (ON) state and a highly resistive (OFF) state. MESFETs utilize gallium arsenide (GaAs) substrates, and they can be constructed along with the line elements as an integrated circuit, forming what is known as a *monolithic microwave integrated circuit* (MMIC). (MMICs may also contain other active circuits such as amplifiers and oscillators.) Figure 6.40 shows four MESFETs integrated into a switched line phase shifter.

The MEM switch is a small ON/OFF type switch that is actuated by electrostatic forces. In one form, a cantilever gold beam is suspended over a control electrode, these two elements forming an air-spaced capacitor. The dimensions of the beam are typically in the range of a few hundred microns (1 micron, abbreviated 1 μm is 10^{-6} m) with an air gap of a few microns.

The RF input is connected to one end of the beam, which contacts an output electrode when the beam is pulled down. The pull-down action occurs because of the electrostatic force arising when a direct voltage is applied between the control electrode and the beam. The voltage is in the order of 75 V, but little current is drawn. The power required to activate the switch depends on the number of cycles per second and the capacitance. In one example (see Reid, 2005), a voltage of 75 V, capacitance of 0.5 pF and a switching frequency of 10 kHz resulted in a power requirement of 14 μW.

A MEM switch can also be constructed where the beam, fixed at both ends, forms an air bridge across the control electrode (Brown, 1998). The top surface of the control electrode has a thin dielectric coating. The electrostatic force deflects the beam causing it to clamp down on the dielectric coating. The capacitance formed by the beam, dielectric coating, and control electrode provides the RF coupling between input and output. Thus, this is basically a contactless switch.

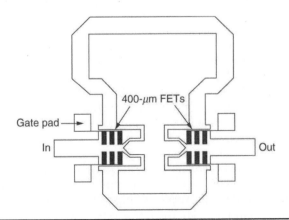

FIGURE 6.40 A MESFET switched line phase shifter. (*From http://parts.jpl.nasa.gov/mmic/3-IX.PDF*)

As mentioned earlier, it is also possible to alter the phase shift by altering the propagation coefficient. By definition, a sinusoidal electromagnetic wave experiences a phase change of 2π rad over distance of one wavelength λ, and therefore the phase change coefficient can be written simply as

$$\beta = \frac{2\pi}{\lambda} \qquad (6.47)$$

This is in radians per meter with λ in meters. As noted earlier, the connection between wavelength λ, frequency f, and phase velocity v_p is $\lambda f = v_p$. It is also known that the phase velocity on a transmission line having a dielectric of relative permittivity ε_r is $v_p = c/\sqrt{\varepsilon_r}$, where c is the free space velocity of light. Substituting these relationships in Eq. (6.45) gives:

$$\varphi = \frac{2\pi f \sqrt{\varepsilon_r}}{c} l \qquad (6.48)$$

This shows that, for a fixed length of line, a phase change can be obtained by changing the frequency f or by changing the relative permittivity (dielectric constant) ε_r. In one scheme (Nishio et al., 2004) a method is given for phasing a base station antenna array by means of frequency change. The modulated subcarrier is fed in parallel to many heterodyne frequency mixers. A common *local oscillator* (LO) signal is fed to each mixer to change the subcarrier up to the assigned carrier frequency, the output from each mixer feeding its own element in the antenna array. The phase change is introduced into the LO circuit by having a different, fixed length of line in each branch of the LO feed to the mixers. Thus the output from each mixer has its own fixed phase angle, determined by the phase shift in the oscillator branch.

Phase change can also be effected by changing the relative permittivity of a delay line. Efforts in this direction have concentrated on using *ferroelectric* material as a dielectric substrate for the delay line. Whereas the dielectric constant of a printed circuit board may range from about 2 to 10, ferroelectrics have dielectric constants measured in terms of several hundreds. The ferroelectric dielectric constant can be changed by application of an electric field, which may be in the order of 2000 kV/m. Thus to keep the applied voltage to reasonable levels, a thin dielectric is needed. For example, for a dielectric thickness of 0.15 mm and an electric field of 2000 kV/m the applied voltage is $2000\ e+03 \times 0.15e-03 = 300$ V. The characteristic impedance of a microstrip line is given by $Z_0/\sqrt{\varepsilon_r}$, where Z_0 is the impedance of the same line with an air dielectric. The characteristic impedance increases as a function of h/W, where W is the width of the line and h the thickness of the dielectric. Thus thinner dielectrics lead to lower characteristic impedance, and this combined with the high dielectric constant means that the line width W must be narrow. Values given in De Flaviis et al. (1997) for the ferroelectric material barium modified strontium titanium oxide ($Ba_{1-x}Sr_x TiO_3$) show a dielectric constant in the region of 600, dielectric thickness between 0.1 and 0.15 mm, and line width of 50 μm, for a characteristic impedance of 50 Ω. The bias voltage is 250 V.

The ferroelectric dielectric is used in several different ways. In the paper by De Flaviis et al., the material was used simply as the dielectric for a microstrip delay line. It has also been used as a lens to produce scanning by deflecting an antenna beam (see Ferroelectric Lens Phased Array at http://radar-www.nrl.navy.mil/Areas/Ferro). The lens is shown in Fig. 6.41. The ferroelectric dielectric in each column is biased to

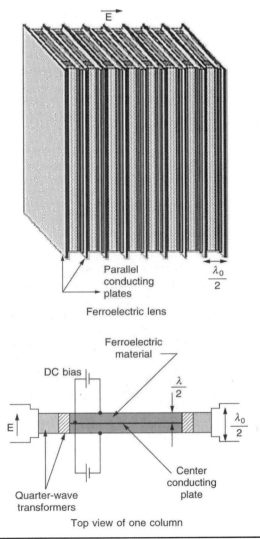

FIGURE 6.41   A ferroelectric lens. (*Courtesy of U. S. Naval Research Laboratory Radar Division, Washington, DC. Source: http://radar-www.nrl.navy.mil/Areas/Ferro*)

provide a progressive phase shift so that an incident plane wave normal to the edges of the dielectric columns emerge from the opposite edges in a direction determined by the phasing in the lens. A single lens produces one dimensional scanning. Ferroelectrics are also employed in reflectarrays described next.

6.20   Reflectarrays

A *reflectarray*, as the name suggests, is an array of antenna elements that acts as a reflector. A reflector array incorporates a planar array as a reflector, as shown in Fig. 6.42. The planar array basically replaces the parabolic reflector shown in Fig. 6.16.

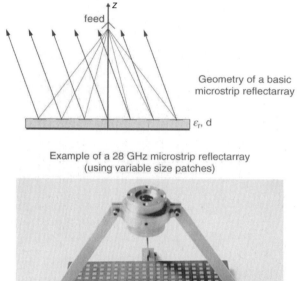

Geometry of a basic
microstrip reflectarray

Example of a 28 GHz microstrip reflectarray
(using variable size patches)

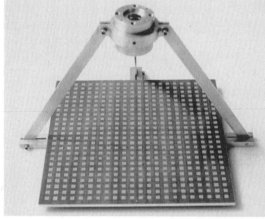

Patterns of 28 GHz reflectarray

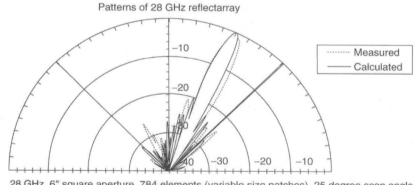

28 GHz, 6" square aperture, 784 elements (variable size patches), 25 degree scan angle,
corrugated conical horn feed, G = 31 dB, 51% aperture efficiency

FIGURE 6.42   A 784-element reflectarray. (*Courtesy of D.M. Pozar. Source: www.ecs.umass.edu/ ece/pozar/jina.ppt*)

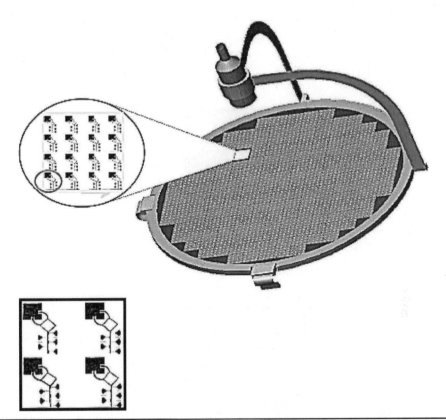

Figure 6.43 A 2832-element 19 GHz ferroelectric reflectarray concept. The callout shows a 16-element subarray patterned on a 3.1 × 3.1 cm, 0.25 mm thick MgO substrate. The array diameter is 48.5 cm. The unit cell area is 0.604 cm² and the estimated boresight gain is 39 dB. (*Courtesy of Robert R. Romanofsky, NASA Glenn Research Center, Cleveland Ohio. Source: gltrs.grc.nasa.gov/reports/2000/TM-2000-210063.pdf*)

Reflected waves from each of the elements in the array can be phased to produce beam scanning; and Fig. 6.42*b* shows the construction. The reflected wave is a combination of reflections from the antenna elements and the substrate. Figure 6.42*c* shows the polar diagram for a 784-element array. Further details of this array can be found in Pozar (2004).

Several methods of producing beam scanning have been proposed. Figure 6.43 shows a 2832 element, 19-GHz reflectarray that employs ferroelectric phase shifters for the elements. Further details can be found at www.ctsystemes.com/zeland/publi/TM-2000-210063.pdf. Varactor diodes have also been used to provide a phase shift that is controlled by an applied bias. A varactor is in effect a voltage-controlled capacitor, and changes in the capacitance introduce a corresponding phase shift. Figure 6.44 shows one arrangement for a five-element array. Further details are found at www.ansoft.com/news/articles/ICEAA2001.pdf.

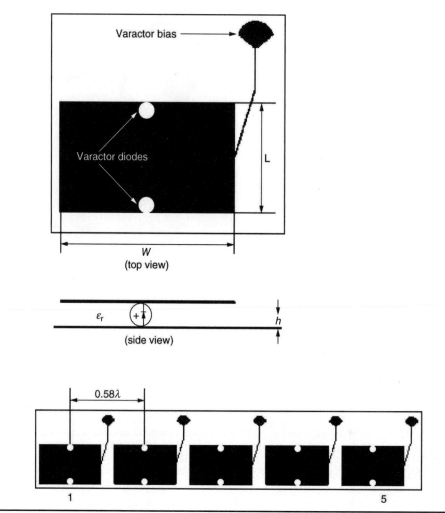

FIGURE 6.44   Phase shift using varactor diodes. (*Courtesy of L. Boccia. Source: www.ansoft.com/ news/articles/ICEAA2001.pdf*)

6.21   Array Switching

Switching of the phasing elements in the antenna arrays is usually carried out digitally. (There are analog phase shifters, some employing ferrite materials, which offer continuously variable phase change, but these are not considered here.) The digitally switched delay line type offers faster switching speed, which is an important consideration where beam scanning is employed. The phase shift increments are determined by successive division by 2 of 360°. These would follow the pattern 180°, 90°, 45°, 22.5°, 12.25°, and so on. Thus a 4-bit phase shifter would have $2^4 = 16$ states, providing increments of 180°, 90°, 45°, 22.5°. This is another limitation of digitally switched phase shifters, the resolution that can be achieved. One manufacturer (KDI Triangle Corp. states that 5.63 is the practical limit for digital, compared to 0.088 for analog types). The arrangement for a

4-bit phase shifter is shown in Fig. 6.45*a*. It is seen that four control lines are required, one for each logical bit. The *most significant bit* (MSB) switches in the 180 phase shift, and the *least significant bit* (LSB) the 22.5°, and if all four lines are activated the phase shift is the sum of the four delays, or 337.5°. With all four delay lines out of circuit the phase shift is back to zero, or 360°.

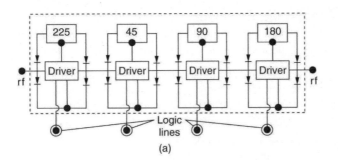

(a)

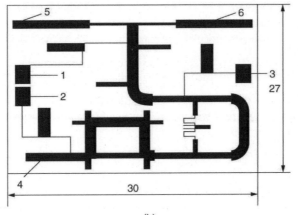

(b)

Nominal phase state	LEAD connections		
	cont. pl. 1	cont. pl. 2	cont. pl. 3
45	+	−	+
90	−	+	+
135	+	+	+
180	−	−	−
225	+	−	−
270	−	+	−
315	+	+	−

Terminology: "+" - control current "−" - reverse voltage.

(c)

FIGURE 6.45 Digital switching of phased arrays: (*a*) 4-bit (*Courtesy of T. J. Braviak. Source: www.kditriangle.com*), (*b*) Ku-band p-i-n diode (3-bit) phase shifter, (*c*) the switching logic for (*b*). (*Courtesy of Michael Parnes. Source: http://www.ascor.eltech.ru/ascor15.htm*)

Care must be taken how "bits" are interpreted. A "three bit" line has a MSB of 180 and a LSB of 45 and requires three logic control lines. However, in the 3-bit phase shifter shown in Fig. 6.45b positive and negative control voltages can be applied to the control lines, giving rise to the phase shifts shown in the table of Fig. 6.45c.

6.22   Problems and Exercises

6.1.   The power output from a transmitter amplifier is 600 W. The feed losses amount to –1 dB, and the voltage reflection coefficient at the antenna is 0.01. Calculate the radiated power.

6.2.   Explain what is meant by the *reciprocity theorem* as applied to antennas. A voltage of 100 V applied at the terminals of a transmitting dipole antenna results in an induced current of 3 mA in a receiving dipole antenna. Calculate the current induced in the first antenna when a voltage of 350 V is applied to the terminals of the second antenna.

6.3.   The position of a point in the coordinate system of Sec. 6.3 is given generally as $r(\theta, \phi)$. Determine the x, y, and z coordinates of a point $(3, 30°, 20°)$.

6.4.   What are the main characteristics of a radiated wave in the far-field region? The components of a wave in the far field region are $E_\theta = 3$ mV/m, $E_\phi = 4$ mV/m. Calculate the magnitude of the total electric field. Also calculate the magnitude of the magnetic field.

6.5.   The **k** vector for the wave specified in Prob. 6.4 is directed along the +x axis. Determine the direction of the resultant electric field in the yz plane.

6.6.   The **k** vector for the wave specified in Prob. 6.4 is directed along the +z axis. Is there sufficient information given to determine the direction of the resultant electric field in the xy plane? Give reason for your answer.

6.7.   The rms value of the electric field of a wave in the far-field region is 3 μV/m. Calculate the power flux density.

6.8.   Explain what is meant by the *isotropic power gain* of an antenna. The gain of a reflector antenna relative to a $(1/2)$ λ-dipole feed is 49 dB. What is the isotropic gain of the antenna?

6.9.   The directivity of an antenna is 52 dB, and the antenna efficiency is 0.95. What is the power gain of the antenna?

6.10.   The radiation pattern of an antenna is given by $g(\theta, \phi) = |\sin\theta\sin\phi|$. Plot the resulting patterns for (a) the xz plane and (b) the yz plane.

6.11.   For the antenna in Prob. 6.10, determine the half-power beamwidths, and hence determine the directivity.

6.12.   Explain what is meant by the *effective aperture* of an antenna. A paraboloidal reflector antenna has a diameter of 3 m and an illumination efficiency of 70 percent. Determine (a) its effective aperture and (b) its gain at a frequency of 4 GHz.

6.13.   What is the effective aperture of an isotropic antenna operating at a wavelength of 1 cm?

6.14.   Determine the half-power beamwidth of a half-wave dipole.

6.15.   A uniformly illuminated rectangular aperture has dimensions $a = 4\lambda$, $b = 3\lambda$. Plot the radiation patterns in the principal planes.

6.16. Determine the half-power beamwidths in the principal planes for the uniformly illuminated aperture of Prob. 6.15. Hence determine the gain. State any assumptions made.

6.17. Explain why the smooth-walled conical horn radiates copolar and cross-polar field components. Why is it desirable to reduce the cross-polar field as far as practical, and state what steps can be taken to achieve this.

6.18. When the rectangular aperture shown in Fig. 6.9 is fed from a waveguide operating in the TE_{10} mode, the far-field components (normalized to unity) are given by

$$E_\theta(\theta, \phi) = -\frac{\pi}{2}\sin\phi \frac{\cos X}{X^2 - \left(\frac{\pi}{2}\right)^2}\frac{\sin Y}{Y}$$

$$E_\phi(\theta, \phi) = E_\theta(\theta, \phi)\cos\theta\cot\phi$$

where X and Y are given by Eqs. (6.19) and (6.20). The aperture dimensions are $a = 3\lambda$, $b = 2\lambda$. Plot the radiation patterns in the principal planes.

6.19. Determine the half-power beamwidths in the principal planes for the aperture specified in Prob. 6.18, and hence determine the directivity.

6.20. A pyramidal horn antenna has dimensions $a = 4\lambda$, $b = 2.5\lambda$, and an illumination efficiency of 70 percent. Determine the gain.

6.21. What are the main characteristics of a parabolic reflector that make it highly suitable for use as an antenna reflector?

6.22. Explain what is meant by the *space attenuation function* in connection with the paraboloidal reflector antenna.

6.23. Figure 6.16b can be referred to xy rectangular coordinates with A at the origin and the x axis directed from A to S. The equation of the parabola is then $y^2 = 4fx$. Given that $y_{max} = \pm2.5$ m at $x_{max} = 0.9$ m, plot the space attenuation function.

6.24. What is the f/D ratio for the antenna of Prob. 6.23? Sketch the position of the focal point in relation to the reflector.

6.25. Determine the depth of the reflector specified in Prob. 6.23.

6.26. A 3-m paraboloidal dish has a depth of 1 m. Determine the focal length.

6.27. A 5-m paraboloidal reflector works with an illumination efficiency of 65 percent. Determine its effective aperture and peak gain at a frequency of 6 GHz.

6.28. Determine the half-power beamwidth for the reflector antenna of Prob. 6.27. What is the beamwidth between the first nulls?

6.29. Given a parabolic reflector antenna of diameter 20 wavelengths. Calculate the normalized main-beam pattern versus off-boresight angle in degrees to the −10 dB level using the Bessel approximation. What is the HPBW?

6.30. Repeat Prob. 6.29 for an antenna diameter of 100 wavelengths.

6.31. Repeat Prob. 6.29 using the Sinc² approximation.

6.32. Repeat Prob. 6.30 using the Sinc² approximation.

6.33. Compare the results of Probs. 6.29 and 6.30 using the normalized off-boresight angle ratio given in Eq. (6.33).

6.34. Describe briefly the *offset feed* used with paraboloidal reflector antennas, stating its main advantages and disadvantages.

6.35. Explain why double-reflector antennas are often used with large earth stations.

6.36. Briefly describe the main advantages to be gained in using an antenna array.

6.37. A basic dipole array consists of five equispaced dipole elements configured as shown in Fig. 6.30. The spacing between elements is 0.3λ. Determine the current phasing needed to produce an end-fire pattern. Provide a polar plot of the AF.

6.38. What current phasing is required for the array in Prob. 6.37 to produce a broadside pattern?

6.39. A four-element dipole array, configured as shown in Fig. 6.30, is required to produce maximum radiation in a direction $\phi_0 = 15°$. The elements are spaced by 0.2λ. Determine the current phasing required and provide a polar plot of the AF.

6.40. A rectangular patch antenna element has sides $a = 9$ mm, $b = 6$ mm. The operating frequency is 10 GHz. Plot the radiation patterns for the $\phi = 0$ and $\phi = 90$ planes.

6.41. For microstrip line, where the thickness t of the line is negligible compared to the dielectric thickness h, and the line width $W \geq h$ the effective dielectric constant is given by

$$\varepsilon_e \cong \frac{\varepsilon_r + 1}{2} + \frac{\varepsilon_r - 1}{2\sqrt{1 + 12\dfrac{h}{W}}}$$

ε_r is the dielectric constant of the dielectric material. The characteristic impedance is given by

$$Z_0 = \frac{120\pi}{\sqrt{\varepsilon_e}}\left[\frac{W}{h} + 1.393 + 0.667\ln\left(\frac{W}{h} + 1.444\right)\right]^{-1}$$

(See Chang, 1989.) Calculate the characteristic impedance for a microstrip line of width 0.7 mm, on an alumina dielectric of thickness 0.7 mm. The dielectric constant is 9.7.

6.42. For the microstripline of Prob. 6.41, calculate (*a*) the line wavelength (*b*) the phase shift coefficient in rad/m, and in degrees/cm. The frequency of operation is 10 GHz.

6.43. The dielectric constant of polyguide dielectric is 2.32. Calculate the characteristic impedance and phase shift coefficient for a microstrip line of width 2.45 mm, and dielectric thickness 1.58 mm.

6.44. The effective dielectric constant for a microstripline is 1.91. Design a switched-line phase shifter (see Fig. 6.39) to produce a phase shift of 22.5 at a frequency of 12 GHz. Show how switching might be achieved using PIN diodes.

6.45. Calculate the power required to drive a MEM switch, which must operate at a frequency of 8 kHz. The switch capacitance is 0.5 pF, and the drive voltage needed for switching is 75 V. (*Hint*: The energy stored in a capacitor is $1/2\,CV^2$ and power is J/s.)

References

Altair Engineering Inc., 2023, https://altair.com/electromagnetics-applications.

Balanis, C. 2005. *Antenna Theory: Analysis and Design*. Harper & Row, New York.

Brain, D. J., and A. W. Rudge. 1984. "Electronics and Power." *J. of the IEEE*, Vol. 30, No. 1, January, pp. 51–56.

Brown, R. E. 1998. "RF-MEMS Switches for Reconfigurable Integrated Circuits." *IEEE Trans. on Microwave Theory and Techniques*, Vol. 46, No. 11, November, pp. 1868–1880.

Chang, K. (ed.). 1989. *Handbook of Microwave and Optical Components*, Vol. 1. Wiley, New York.

De Flaviis, F., N. G. Alexopoulos, and O. M. Stafsudd. 1997. "Planar Microwave Integrated Phase Shifter Design with High Purity Ferroelectric Material." *IEEE Trans. on Microw. Theory and Tech.*, Vol. 45, No. 6, June, pp. 963–969 (see also www.ece.uci.edu/rfmems/publications/papers-pdf/J005.PDF).

Glazier, E. V. D., and H. R. L. Lamont. 1958. *The Services Textbook of Radio*, Vol. 5: *Transmission and Propagation*. Her Majesty's Stationery Office, London.

Hwang, Y. 1992. "Satellite Antennas." *Proc. IEEE*, Vol. 80, No. 1, January, pp. 183–193.

James, J. R., P. S. Hall, and C. Wood. 1981. *Microstrip Antenna Theory and Design*. Peter Peregrinus, UK.

Kraus, J. D. 1998. *Antennas*. McGraw-Hill, New York.

Kummer, W. H. 1992. "Basic Array Theory." *Proc. IEEE*, Vol. 80, No. 1, January, pp. 127–140.

Miya, K. (ed.). 1981. *Satellite Communications Technology*. KDD Engineering and Consulting, Japan.

Nishio, T., X. Hao, W. Yuanxun, and T. Itoh, 2004. "A Frequency Controlled Active Phased Array." *IEEE Microw. and Components Lett.*, Vol. 14, No. 3 March. pp. 115–117.

Olver, A. D. 1992. "Corrugated Horns." *Electron. Commun. Eng. J.*, Vol. 4, No. 10, February, pp. 4–10.

Pozar, D. M. 2004. "Microstrip Reflectarrays Myths and Realities." JINA Conference (see www.ecs.umass.edu/ece/pozar/jina.ppt).

Reid, J. R. 2005. "Microelectromechanical Phase Shifters for Lightweight Antennas," at www.afrlhorizons.com.

Rudge, A. W., K. Milne, A. D. Olver, and P. Knight (eds.). 1982. *The Handbook of Antenna Design*, Vol. 1. Peter Peregrinus, UK.

Rusch, W. V. T. 1992. "The Current State of the Reflector Antenna Art: Entering the 1990s." *Proc. IEEE*, Vol. 80, No. 1, January, pp. 113–126.

Vectors. 1993. Hughes In-House Magazine, Vol. XXXV, No. 3.

CHAPTER 7

The Space Segment

7.1 Introduction

A satellite communications system can be broadly divided into two segments—a ground segment and a space segment. The space segment obviously includes the satellites, but it also includes the ground facilities needed to keep the satellites operational—these are referred to as the *tracking, telemetry, and command* (TT&C) facilities. In many networks it is common practice to employ a ground station solely for the purpose of TT&C.

The equipment carried aboard the satellite also can be classified according to function. The *payload* refers to the equipment used to provide the service for which the satellite has been launched. The *bus* refers not only to the spacecraft, which carries the payload but also to the various subsystems that provide the power, attitude control, orbital control, thermal control, and command and telemetry functions required to service the payload.

In a communications satellite, the equipment that provides the connecting link between the satellite's transmit and receive antennas is referred to as the *transponder*. The transponder forms one of the main sections of the payload, the other being the antenna subsystems.

In this chapter, the emphasis is on large geostationary satellites, and the main characteristics of certain bus systems and payloads are described.

7.2 The Power Supply

The primary electrical power for operating the electronic equipment is obtained from solar cells. Individual cells can generate only small amounts of power, and therefore, arrays of cells in series-parallel connection are required. Figure 7.1 shows the cylindrical solar cell panels for the HS 376 satellite manufactured by Hughes Space and Communications Company. The spacecraft is 2.2 m in diameter and 6.6 m long, when fully deployed in orbit. During the launch sequence, the outer cylinder is collapsed over the inner one, to reduce the overall launch envelope, and only the outer panel generates electrical power during this phase. Once on the geostationary orbit, the spacecraft is fully extended so that both solar arrays are exposed to sunlight. At the beginning of life, the panels produce 940 W dc power, which are expected to degrade to 760 W at the end of 10 years because of radiation damage from solar exposure. During eclipse, power is provided by two nickel-cadmium (Ni-Cd) long-life batteries, which deliver 830 W. At the end of life, battery recharge time is less than 16 h, which satisfies the daily requirement of 24 h.

Figure 7.1 The HS 376 satellite. (*Courtesy of Hughes Aircraft Company Space and Communications Group.*)

The HS 376 spacecraft is a spin-stabilized spacecraft (the gyroscopic effect of the spin is used for mechanical orientational stability, as described in Sec. 7.3). Thus the arrays are only partially in sunshine at any given time (effective projected array area ~1/3), which places a limitation on the peak power.

Higher powers can be achieved with solar panels arranged in the form of rectangular *solar panels* or *solar sails*. Solar panels must be folded during the launch phase and extended when in geostationary orbit (see Fig. 7.2 and 7.26).

As shown, the solar panels are folded up on each side, and when fully extended in a north/south direction (normal to the orbit plane), they stretch to 20.4 m from tip to tip. The full complement of solar cells is exposed to the sunlight, and the solar panels are arranged to rotate (once per solar day) to track the Sun, so they are capable of ~3x the power output than cylindrical arrays having a comparable number of cells. In comparing the power capacity of cylindrical and solar-sail satellites, the cross-over point is estimated to be about 2 kW, where the solar-sail type is more economical than the cylindrical type (Hyndman, 1991).

FIGURE 7.2 Aussat B1 (renamed Optus B), Hughes first HS 601 communications satellite is prepared for environmental testing. (*Courtesy of Hughes Aircraft Company Space and Communications Group.*)

As discussed in Sec. 3.6, the Earth eclipses a geostationary satellite twice a year, during the spring and autumnal equinoxes, when the Sun passes through the equatorial plane. Daily eclipses start approximately 23 days before and end approximately 23 days after the equinox for both the spring and autumnal equinoxes and can last up to 72 min at the actual equinox days. Figure 7.3 shows the graph relating eclipse period to the day of year.

To maintain service during an eclipse, storage batteries must be provided, and based upon recent technology development for electric vehicles, lithium-ion (Li-ion) is considered the best-adapted battery technology because of its various advantages over the two other space technologies, namely: nickel-cadmium (Ni-Cd) and nickel-hydrogen (Ni-H$_2$).

The principle advantage of a space-based Li-ion battery is the mass reduction due to higher specific energy of a battery cell. For example, the specific energy of Li-ion is >125 Wh/kg that is more than double that of Ni-H$_2$. At the battery level, the corresponding

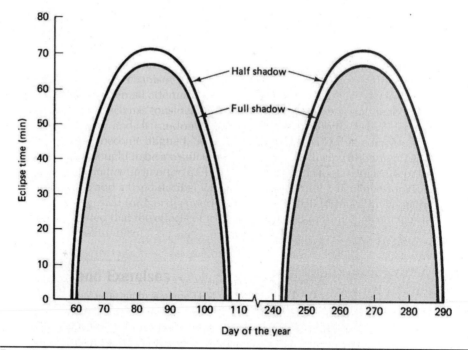

FIGURE 7.3   Satellite eclipse time as a function of the current day of the year. (*Courtesy of Spilker, 1977. Reprinted by permission of Prentice-Hall, Englewood Cliffs, NJ.*)

mass reduction is at least 40%, which results in >350 kg saving on a typical 20 kW satellite. Also, an additional 5 to 10% of the satellite mass savings can be linked to the lower thermal dissipation and higher Faradic efficiency of Li-ion that reduce solar panel and radiator sizes.

One final advantage, concerns reduced requirements for pre and post-launch operations of a Li-ion battery. Because the self-discharge of Li-ion is very low (0.03 percent of capacity loss per day compared to 10% for Ni-H$_2$), the management of the state of charge on a satellite during integration, launch pad operations and solar eclipse is easier. For example, the pre-launch requirement for charging the spacecraft battery on the launcher is not mandatory as it is with Ni-H$_2$ batteries. Because the memory effect observed on-orbit with Ni-Cd and Ni-H$_2$ does not exist with Li-ion batteries, the usual battery cycling reconditioning operation, associated with solar eclipse, may be eliminated.

7.3   Attitude Control

The *attitude* of a satellite refers to its orientation in space, and much of the equipment carried aboard a satellite is there for this purpose. Attitude control is necessary, for example, to ensure that directional antennas precisely point to the intended service areas. In the case of earth environmental satellites, the earth-sensing instruments must cover the required regions of the Earth, which also requires attitude control. Several forces, referred to as *disturbance torques*, can alter the attitude, some examples being the gravitational fields of the Earth and the Moon, solar radiation pressure, and meteorite impacts. Attitude control must not be confused with station keeping, which is the term used for maintaining a satellite in its correct orbital position, although the two are closely related.

To exercise attitude control, there must be available precise measurements in real-time (typically of order seconds) of a satellite's orientation in space. In one common method, infrared radiometric sensors, referred to as *horizon detectors*, are used to detect the rim of the "hot" Earth against the background of "cold" space. With the use of four such sensors, one for each quadrant, the center of the Earth can be readily established as a reference point to a relative accuracy of about 1 milliradian angular uncertainty. When more precision in attitude determination is required, *star trackers* may be used that provide one to two orders of magnitude improvement in attitude determination. These sensors are cameras that capture images of the current star field, and then use signal processing to derive the current spacecraft attitude from comparison with stored star maps. Star trackers are usually used for deep-space probes, but sometimes used for LEO and GEO satellites. Finally, for *safe-hold modes* for the satellite, there are *sun sensors* and *earth sensors* that provide very coarse attitude sensing (of order 0.1 radians uncertainty), but generally these modes are initiated in the case of emergencies to assure spacecraft survival.

Attitude control can be achieved using a Kalman filter that statistically combines previous attitude estimates with current sensor measurements to obtain an optimal estimate of the current attitude. Any change in attitude is detected by the sensors, and corresponding control signals are generated, which activates restoring torques. Usually, the attitude-control process takes place autonomously aboard the satellite, but it is also possible for control signals to be transmitted from the Earth based on attitude data obtained from the satellite. Where a shift in attitude is desired, an *attitude maneuver* is executed, and the control signals required are transmitted from an earth station.

The controlling torques may be generated in several ways. *Passive attitude control* refers to the use of mechanisms that stabilize the satellite without putting a drain on the satellite's energy supplies; at most, infrequent use is made of these supplies, for example, when thruster jets are fired to provide corrective torque. Examples of passive attitude control are *spin stabilization* and *gravity gradient stabilization*. The latter depends on the interaction of the satellite with the gravitational field of the central body and has been used, for example, with the Radio Astronomy Explorer-2 satellite, which was placed in orbit around the Moon (Wertz, 1984). For communications satellites in GEO, spin stabilization is often used, and this is described in detail in Sec. 7.3.1.

The other form of attitude control is *active control*. With active attitude control, there is no overall stabilizing torque present to resist the disturbance torques. Instead, corrective torques are applied as required in response to disturbance torques. Methods used to generate active control torques include momentum wheels, electromagnetic torquers, and mass expulsion devices, such as gas jets and ion thrusters. The electromagnetic torquer works on the principle that the Earth's magnetic field exerts a torque on a current-carrying coil and that this torque can be controlled through control of the current. However, the method is of use only for satellites relatively close to the Earth, i.e., for satellites in LEO or MEO, but not for GEOSATs. The use of momentum wheels is described in more detail in Sec. 7.3.2.

The three axes which define a satellite's attitude are its *roll, pitch,* and *yaw* (RPY) axes. These are shown relative to the Earth in Fig. 7.4. All three axes pass through the

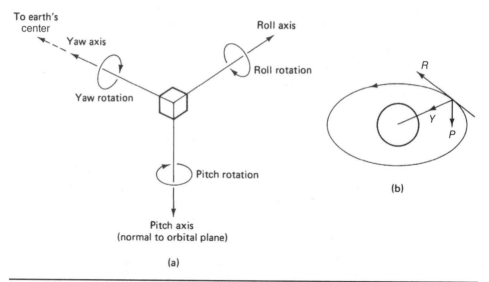

Figure 7.4 (a) Roll, pitch, and yaw axes. The yaw axis is directed toward the earth's center, the pitch axis is normal to the orbital plane, and the roll axis is perpendicular to the other two. (b) RPY axes for the geostationary orbit. Here, the roll axis is tangential to the orbit and lies along the satellite velocity vector.

center of gravity of the satellite. For an equatorial orbit, movement of the satellite about the roll axis moves the antenna footprint north and south; movement about the pitch axis moves the footprint east and west; and movement about the yaw axis rotates the alignment of the electric field vector (polarization rotation) for a linear polarized (vertical or horizontal) antenna. Thus, the yaw rotation requires a corresponding polarization alignment for the earth station antenna; however, when circular polarization is used, then the transmit and receiving antennas are automatically self-aligning.

7.3.1   Spinning Satellite Stabilization

Spin stabilization is commonly used with cylindrical satellites where the satellite is constructed so that it is mechanically balanced about one particular axis and is then set spinning around this axis. For geostationary satellites, the spin axis is adjusted to be parallel to the N-S axis of the Earth, as illustrated in Fig. 7.5. Spin rate is typically in the range of 50 to 100 rev/min. Spin is typically initiated during the launch phase by means of small gas jets.

In the absence of disturbance torques, the spinning satellite maintains its correct attitude relative to the Earth. Disturbance torques are generated in several ways, both external and internal to the satellite. Solar radiation pressure, gravitational gradients, and meteorite impacts are all examples of external forces that can give rise to disturbance torques. Motor-bearing friction and the movement of satellite elements such as the antennas also can give rise to disturbance torques. The overall effect is that the spin rate decreases, and the direction of the angular spin axis changes. Impulse-type thrusters, or jets, can be used to restore the spin rate again and to shift the axis back to its correct N-S orientation. *Nutation*, which is a form of wobbling, can occur because of the disturbance torques and/or from misalignment or unbalance of the control jets. This nutation must be damped out by means of energy absorbers known as *nutation dampers*.

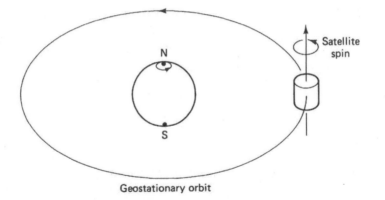

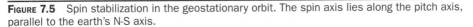

FIGURE 7.5 Spin stabilization in the geostationary orbit. The spin axis lies along the pitch axis, parallel to the earth's N-S axis.

Where an omnidirectional antenna is used (e.g., as shown for the INTELSAT I and II satellites in Fig. 1.1), the antenna, which points along the pitch axis, also rotates with the satellite. Where a directional antenna is used, which is more common for communications satellites, the antenna must be despun, giving rise to a dual-spin construction. An electric motor drive is used for despinning the antenna subsystem.

Figure 7.6 shows the Hughes HS 376 satellite in more detail. The antenna subsystem consists of a parabolic reflector and feed horns mounted on the despun shelf, which also carries the communications repeaters (transponders). The antenna feeds can therefore be connected directly to the transponders without the need for radio frequency (rf) rotary joints, while the complete platform is despun. Of course, control signals and power must be transferred to the despun section, and a mechanical bearing must be provided. The complete assembly for this is known as the *bearing and power transfer assembly* (BAPTA).

Figure 7.7 shows a photograph of the internal structure of the HS 376, with cylindrical solar panels removed. Certain dual-spin spacecraft obtain spin stabilization from a spinning flywheel rather than by spinning the satellite itself. These flywheels are termed *momentum wheels*, and their average momentum is referred to as *momentum bias*. Reaction wheels, described in the Sec. 7.3.2, operate at zero momentum bias.

7.3.2 Momentum Wheel Stabilization

Also, satellite attitude control can be achieved by utilizing the gyroscopic effect of a spinning flywheel. This approach is commonly used in satellites with cube-like bodies (such as shown in Fig. 7.1, Fig. 7.2, and Fig. 7.26), which are known as *body-stabilized* satellites. The complete unit, termed a momentum wheel, consists of a flywheel, the bearing assembly, the casing, and an electric drive motor with associated electronic control circuitry. The flywheel is attached to the rotor, which consists of a permanent magnet providing the magnetic field for motor action. The stator of the motor is attached to the body of the satellite. Thus the motor provides the coupling between the flywheel and the satellite structure. Speed and torque control of the motor is exercised through the currents fed to the stator. The housing for the momentum wheel is evacuated to protect the wheel from adverse environmental effects, and the bearings have controlled

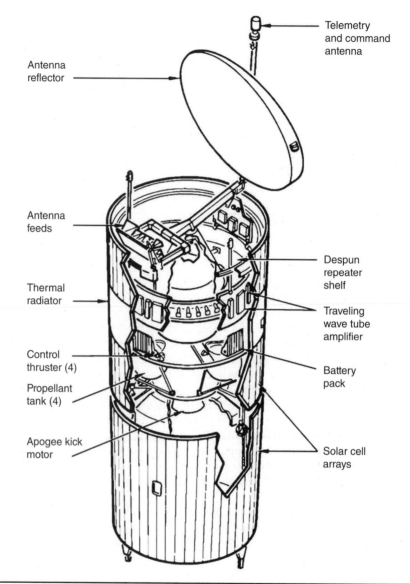

Telemetry
and command
antenna

Antenna
reflector

Antenna
feeds

Thermal
radiator

Control
thruster (4)

Propellant
tank (4)

Apogee kick
motor

Despun
repeater
shelf

Traveling
wave tube
amplifier

Battery
pack

Solar cell
arrays

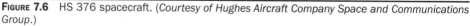

Figure 7.6 HS 376 spacecraft. (*Courtesy of Hughes Aircraft Company Space and Communications Group.*)

lubrication that lasts over the lifetime of the satellite. Typical momentum wheels range in diameters from 20 to 60 cm, and other details are found in Chetty (1991).

The term *momentum wheel* is usually reserved for wheels that operate at nonzero momentum, but, in Fig. 7.8a, this is termed a momentum bias condition. Here, a wheel provides passive stabilization for the yaw and roll axes, when the axis of rotation of the wheel lies along the pitch axis. Thus, control about the pitch axis is achieved by changing the speed of the wheel.

When a momentum wheel is operated with zero momentum bias, it is generally referred to as a *reaction wheel*. Reaction wheels are used in three-axis stabilized systems.

Figure 7.7 Technicians check the alignment of the Telestar 3 communications satellite, shown without its cylindrical panels. The satellite, built for the American Telephone and Telegraph Co., carries both traveling-wave tube and solid-state power amplifiers, as shown on the communications shelf surrounding the center of the spacecraft. The traveling-wave tubes are the cylindrical instruments. (*Courtesy of Hughes Aircraft Company Space and Communications Group.*)

Here, as the name suggests, each axis is stabilized by a reaction wheel, as shown in Fig. 7.8c. Reaction wheels can also be combined with a momentum wheel to provide the control needed (Chetty, 1991). Random and cyclic disturbance torques tend to produce zero momentum on average. However, there are always some disturbance torques that cause a cumulative increase in wheel momentum, and eventually at some point the wheel *saturates*. For example, geostationary communications satellites must rotate 360° about the pitch axis every day to maintain antenna pointing toward the Earth. Eventually, the momentum wheel reaches its maximum allowable angular velocity and can no longer take in any more momentum. Mass expulsion devices (thrusters) are then used to unload the wheel, that is, remove momentum from it (in the same way a brake removes energy from a moving vehicle). Of course, operation of the mass expulsion devices consumes part of the satellite's fuel supply.

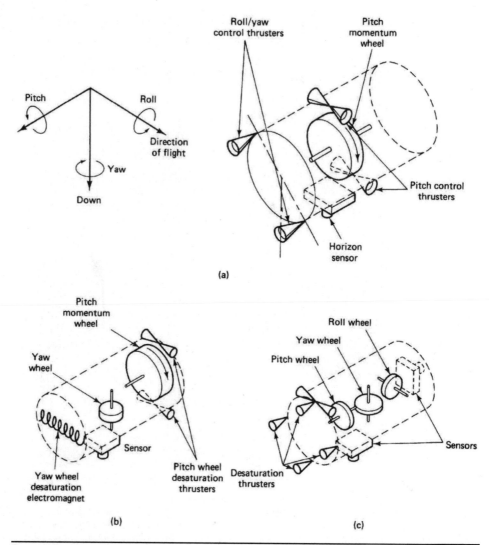

FIGURE 7.8   Alternative momentum wheel stabilization systems: (*a*) one-wheel, (*b*) two-wheel, (*c*) three-wheel. (*Reprinted with permission from Spacecraft Attitude Determination and Control, edited by James R. Wertz. Copyright, 1984 by D. Reidel Publishing Company, Dordrecht, Holland.*)

7.4   Station Keeping

In addition to having its attitude controlled, a geostationary satellite must be maintained at its assigned orbital longitude. This is especially important for the fixed satellite services, where the earth station antennas are either fixed pointed or have limited tracking capability to follow a satellite that has constant motion on orbit. Therefore, this topic is briefly described in this section; however, for a more rigorous treatment the reader is referred to Gordon and Morgan (1993).

As described in Sec. 2.8.1, the equatorial ellipticity of the Earth causes geostationary satellites to drift slowly along the orbit, to one of two stable points, at 75°E and 105°W. Consider a simple model presented in Fig. 7.9, which is used to explain this phenomenon. First, this diagram is distorted to exaggerate the location of the major axis, which connects these two stable orbit locations. These regions are represented as mass concentrations or *mascons* that result in increased gravity forces acting on a satellite. In addition, there are two *conditionally stable* orbit locations at 12°W and 162°E longitudes. So, these four orbit locations, approximately separated by 90° longitude, are where the gravity forces on the satellite are radially inward along the orbital radius. Given an analogy of gravity "hills and valleys," the locations 12°W and 162°E are represented by hilltops, and the locations 75°E and 105°W are the valleys between hills. If a ball were placed upon the hill-top, it remains there until it experienced some external horizontal force that causes the ball to roll downhill. On the other hand, a ball resting in a valley between hills remains there because any disturbing horizontal force are counteracted by the restoring force of gravity.

So, following this analogy, on average, a satellite located at longitude 75°E or 105°W remains there without the need for station keeping. Further, a satellite located at 12°W or 162°E requires considerably less station keeping adjustment to counteract the weak forces of solar and lunar gravitational attraction and other disturbing forces.

For satellite locations between these longitudes, there is a frequent need to make adjustments to keep the satellite from drifting outside of its assigned *orbital box*, which amount to corrections in the satellite orbital speed. For example, consider a satellite located at orbit position-B. At this location, because of the stronger gravitational

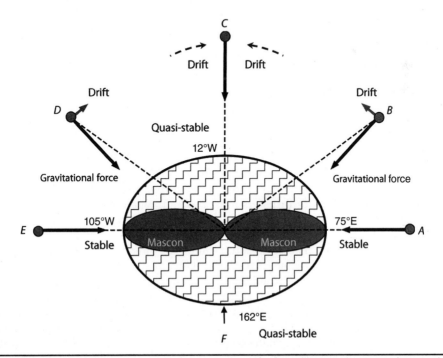

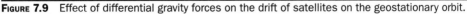

FIGURE 7.9   Effect of differential gravity forces on the drift of satellites on the geostationary orbit.

attraction force of the mascon at 75°E, there is a component of acceleration that reduces the satellite's orbital speed. When this occurs over time, there is a cumulative negative delta-V that causes the satellite position to change relative to the constant rotation of the Earth. However, counter to intuition, rather than move toward the stronger gravity, it drifts in the opposite direction, i.e., in a counter-clockwise direction. The reason for this is that the satellite obeys Kepler's laws, and external forces cause the orbit trajectory to change. So, when the orbital speed decreases below that for a circular orbit, then the satellite position becomes the apogee of a new slightly elliptical orbit, and the orbit radius at perigee decreases. Based upon Kepler's third law, the orbital period also decreases, and the satellite rotates in its orbit faster than the Earth's rotation, thereby causing it to move counter-clockwise relative to the surface of the Earth.

For another example, consider a satellite at the orbit position-D. Here the differential gravitational force, of the mascon at 105°W, is in the same direction as the satellite velocity; so over time the orbital speed increases. But, as before, the orbital drift is in the opposite direction (clockwise). Now, the reason is that the orbit speed is higher than the circular orbit, and the satellite's location becomes the perigee of a new elliptical orbit. Based upon Keppler's third law, the orbit period increases, and the satellite's rotation is now slower than that of the Earth. As a result, the satellite falls behind and has a drift in the clockwise direction as seen by an observer on the surface of the Earth.

So, to counter this drift, an oppositely directed velocity component is imparted to the satellite by means of small rockets called thrusters, and this occurs regularly every 2 or 3 weeks. So, after firing the thruster, this results in the satellite's drift changing direction, and the satellite moves past its nominal longitude position. However, since the differential gravity force continues to act on the satellite, it eventually overcomes the delta-V imparted by the thruster and the satellite stops drifting for a short period of time. Afterwards, the drift changes direction and the satellite moves past its nominal longitude position until the jets are pulsed once again. These periodic maneuvers are termed *east-west station-keeping maneuvers*, and the total excursion of the satellite along the orbit arc, remains within the bounds of the assigned box. For satellites in the 6/4-GHz band, the excursion must be kept within ±0.1° of the designated longitude, and for the 14/12-GHz band, within ±0.05°.

A satellite which is nominally geostationary also drifts in latitude due to perturbing forces of the gravitational pull of the Sun and the Moon. These forces cause the inclination to change at a rate of about 0.85°/year. If left uncorrected, the drift results in a cyclic change in the inclination, ranging from 0° to 14.7° over a period of 26.6 years (Spilker, 1977) and back to 0°, at which the cycle is repeated. To prevent the shift in inclination from exceeding specified limits, thrusters need to be fired at the appropriate time to return the inclination to zero. Counteracting thrusters must be fired, when the inclination is at 0° to halt the change in inclination. These maneuvers are termed *north-south station-keeping maneuvers*, and they are much more expensive in fuel usage than are east-west station-keeping maneuvers. The north-south station-keeping tolerances are the same as those for east-west station keeping, ±0.1° in the C-band and ±0.05° in the Ku-band.

Orbital correction is carried out by command from the TT&C earth station, which monitors the satellite position. East-west and north-south station-keeping maneuvers are usually carried out using the same thrusters as are used for attitude control. These periodic maneuvers results in a pendulum-like oscillation of the subsatellite point location, which occurs after station-keeping corrections are applied (see Fig. 7.10 for the Canadian Anik-C3 satellite).

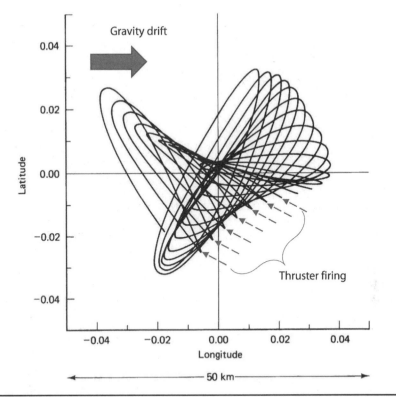

FIGURE 7.10 Typical satellite motion within the GEO-box. (*Courtesy of Telesat, Canada, 1983.*)

Satellite altitude also typically shows variations of about ±0.1 percent of the nominal geostationary height. So, given a typical satellite range of 37,000 km, the total variation in the height is 72 km. A C-band satellite therefore can be anywhere within a box bound by this height and the ±0.1° tolerances on latitude and longitude. Approximating the geostationary radius as 42,164.2 km (see Sec. 3.1), an earth central angle of 0.2° (0.0035 radians) subtends a GEO arc of approximately 147 km, which is the latitude and longitude sides of the box. This is given in Fig. 7.11, which also shows the relative beamwidths of a 30-m and a 5-m antenna. As shown by Eq. (6.29), the −3-dB beamwidth of a 30-m antenna is about 0.12° (0.0021 radians), and that of a 5-m antenna, about 0.7° (0.012 radians) at 6 GHz. For a typical slant range of 37,000 km, the diameter of the 30-m antenna footprint at the satellite is about 80 km, which does not encompass the entire box and therefore can miss the satellite. As a result, such narrow-beam Earth station antennas must track the satellite, and the impact of antenna mispointing loss is discussed in Sec. 12.4.1.

On the other hand, the 5-m antenna footprint diameter at the satellite is about 452 km, which does encompass the box, so tracking is not required. Also, the positional uncertainty of the satellite introduces an uncertainty in propagation time, which can be a significant factor in certain types of communications networks.

A potential solution may be achieved by placing the satellite in an inclined orbit, such that the north-south station-keeping maneuvers may be dispensed with. The savings in weight achieved by not having to carry fuel for these maneuvers allow the communications

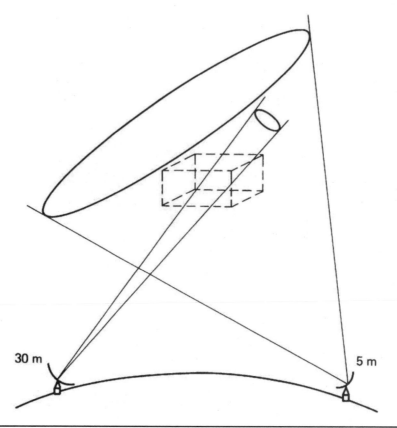

Figure 7.11 The rectangular box shows the positional limits for a satellite in geostationary orbit in relation to beams from a 30-m and a 5-m antenna.

payload to be increased. The satellite is placed in an inclined orbit of about 2.5° to 3°, in the opposite sense to that produced by drift. Over a period of about half the predicted lifetime of the mission, the orbit changes to equatorial and then continues to increase in inclination. However, this arrangement requires the use of tracking antennas at certain ground stations.

7.5 Thermal Control

Satellites are subject to large thermal gradients, receiving the Sun's radiation on one side while the other side faces into space. In addition, thermal radiation from the Earth and the Earth's *albedo*, which is the fraction of the radiation falling on the Earth, which is reflected, can be significant for low-altitude earth-orbiting satellites, although it is negligible for geostationary satellites. For geostationary COMSATs the communications subsystem is the major source of *waste heat*, which must be removed. For modern systems, the solar arrays produce between 10 and 20 kW of power, and the DC to RF radiated power efficiencies are <50%.

As a result, there is a major thermal design requirement to cool the electronic to allow an operating temperature typically between 0 and 50°C, and various methods are taken to achieve this. Thermal blankets, paints, and coatings are used to reflect incident solar energy, while capturing thermal infrared energy to balance the spacecraft temperature. Also, *heat pipes* are used to conduct the waste heat from the communications payload to *infrared radiators panels* that radiate IR energy to cold space and thereby to remove heat. Often, these radiator panels are covered by one-way mirrors that reflect solar energy, while allowing infrared energy to pass through. For example, the mirrored thermal radiator for the Hughes HS 376 satellite can be seen in Figs. 7.1 and 7.6. These mirrored drums surround the communications equipment shelves in each case and provide good radiation paths for the generated heat to escape into the surrounding space. One advantage of spinning satellites compared with body-stabilized is that the spinning body provides an averaging of the temperature extremes experienced from solar flux and the cold background of deep space.

Moreover, to maintain constant temperature conditions, heaters can be switched on to make up for the heat reduction, which occurs when transponders are switched off or when the satellite undergoes solar eclipse. Also, certain equipment requires active thermal control to keep the ambient temperature above a minimum value. For example, the INTELSAT VI satellite used heaters to maintain propulsion thruster and line temperatures (Pilcher, 1982).

7.6   TT&C Subsystem

The TT&C subsystem performs several routine functions aboard the spacecraft. The telemetry, or telemetering, function is a form of *measurement at a distance*, for engineering purposes of monitoring the health and safety of the satellite system. Specifically, it refers to the overall operation of making physical measurements and transmitting these data to selected earth stations. Telemetry data includes critical spacecraft attitude information and spacecraft house-keeping information such as temperatures, voltages and currents, momentum wheel spin rates, and other pertinent engineering data.

Certain RF channels have been designated by international agreement for satellite telemetry transmissions. During the transfer and drift orbital phases of the satellite launch, a special channel is used along with an omnidirectional antenna. Once the satellite is on station, one of the normal communications transponders may be used along with its directional antenna, unless some emergency arises, which makes it necessary to switch back to the special channel used during the transfer orbit.

Telemetry and command may be thought of as complementary functions. The telemetry subsystem transmits information about the satellite to the earth station, while the satellite command subsystem receives command signals from the earth station, often in response to telemetered information. The command subsystem demodulates and decodes the command signals and routes these to the appropriate equipment that execute the necessary action. Thus, attitude changes may be made, communication transponders switched in and out of circuits, antennas redirected, and station-keeping maneuvers carried out via external command. It is clearly important to prevent unauthorized commands from being received and decoded, and for this reason, the command signals are usually encrypted. This differs from the normal process of encoding which converts characters in the command signal into a code suitable for transmission.

Tracking of the satellite is accomplished by having the satellite transmit beacon signals that are received at the TT&C earth stations. Tracking is obviously important during the transfer and drift orbital phases of the satellite launch. Further, once the satellite is on station, the position of a geostationary satellite continually changes because of the various disturbing forces, described above. Therefore, it is necessary to continually track the satellite's movement and send correction signals as required. Tracking beacons may be transmitted in the telemetry channel, or by pilot carriers at frequencies in one of the main communications channels, or by special earth station tracking antennas.

Further, for satellites using *Time Division Multiple Access* (TDMA, discussed in Chap. 14), it is necessary to determine the satellites position in earth centric coordinates (typically once/s), based on simultaneous ranging from two or more TT&C the ground stations. Thus, the satellite location can be determined by measurement of the propagation delay of signals especially transmitted for ranging purposes.

In summary, telemetry, tracking, and command functions are complex operations, which require 24 h/7 days access to special ground facilities in addition to the TT&C subsystems aboard the satellite. Figure 7.12 shows in block diagram form the TT&C facilities used by Canadian Telesat for its satellites.

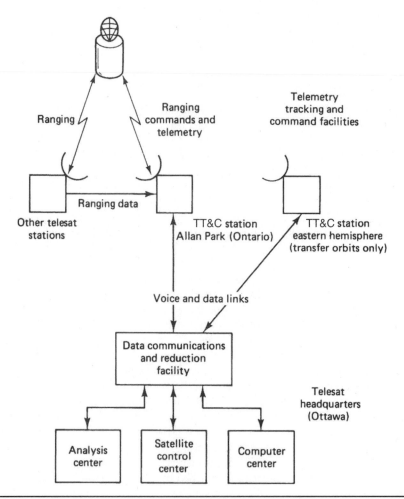

FIGURE 7.12   Satellite control system. (*Courtesy of Telesat Canada, 1983.*)

7.7   Transponders

Before satellites, in the 1950s, real-time video was only provided by local *over the air* television stations, and intracontinental "long-distance" telephone services were provided by RF terrestrial links that used simple line-of-sight repeaters. In addition, transoceanic telephone services were available by a few submarine cables that crossed the Atlantic Ocean between the United States and Europe.

So, shortly after the invention of Earth satellites, in the 1960s, the worldwide communications revolution began, and the justification for satellites was to provide real-time transoceanic communications for telephone and video services. In the beginning, these signals were analog, and the communications satellite were simple "bent pipe" repeaters in space (like terrestrial repeaters) that could be viewed by distant earth stations, separated by an ocean. Thus, the uplink signals were amplified and rebroadcast (without changing the signal content) using directional antennas to earth stations.

Shortly thereafter, in the 1970s to 1980s, came the transformation to digital communications technology, which was slowly adopted by the satellite communications community. In this section, we discuss the traditional communication satellite transponder, and in Sec. 7.12 we present a discussion of modern digital regenerative repeaters for satellite applications.

The traditional transponder is the series of interconnected RF electronics units that form multiple communications paths (channels) between the receive and transmit antennas of a communications satellite. As shown in the block diagram in Fig. 7.13*a*, a transponder normally comprises multi channels with most elements being common for all channels. For example, starting at the upper left, multiple signals (in the form of modulated RF carriers) are received by the satellite receiving antenna and are input to a bandpass filter to separate desired from unwanted signals. Next, the wanted carriers are linearly amplified in the *low-noise-amplifier* (LNA), heterodyne down-converted to a lower microwave frequency band, and then collectively linearly amplified again. At this point, the carriers are separated by the *input demutiplexer* (MUX) into parallel channels for final power amplification. The parallel approach is superior to a single power amplifier because of signal linearity and hardware redundancy considerations. Finally, the channels are recombined in the *output MUX*, and sent to the transmit antenna. This type of transponder is called a *bent pipe* because the same signal that enters the transponder, leaves the transponder.

Also shown in Fig. 7.13*b* is a typical transponder gain distribution that results in a transmitted carrier power in the 1 to 100 W range. Overall, the transponder provides power amplification of 100 to 130 dB gain.

Before going into more detail about the various units of a transponder, the overall frequency arrangement of a typical C-band communications satellite is examined briefly. The bandwidth allocated for C-band fixed satellite service is 500 MHz divided into *sub-bands*, one for each transponder channel. A typical channel bandwidth is 36 MHz for a 4-MHz *guard-band* between channels, which results in 12 transponder channels in the 500-MHz bandwidth. By making use of *polarization isolation*, provided by the satellite antennas, this number can be doubled. Polarization isolation refers to the fact that carriers, which may be on the same frequency but with opposite senses of polarization, can be isolated from one another by receiving antennas matched to the incoming polarization. With linear polarization, vertically and horizontally polarized carriers can be separated in this way, and with circular polarization, left-hand circular

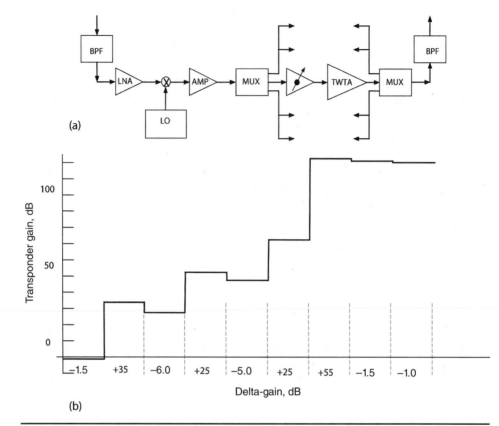

(a)

(b)

FIGURE 7.13 (a) Typical satellite transponder block diagram. (b) Gain distribution in decibels.

and right-hand circular polarizations can be separated. Because the carriers with opposite senses of polarization may overlap in frequency, this technique is referred to as *frequency reuse*. Figure 7.14 shows part of the frequency and polarization plan for a C-band communications satellite. Frequency reuse also may be achieved with spot-beam antennas, and these may be combined with polarization reuse to provide an effective bandwidth of 2000 MHz from the actual bandwidth of 500 MHz.

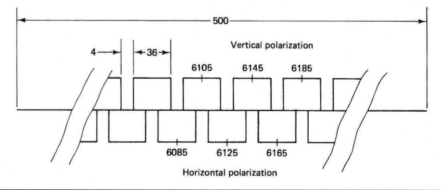

FIGURE 7.14 Section of an uplink frequency and polarization plan. Numbers refer to frequency in megahertz.

For one of the polarization groups, the incoming, or uplink, frequency range is 5.925 to 6.425 GHz. The carriers may be received on one or more antennas, all having the same polarization. The input filter passes the full 500-MHz band to the common receiver while rejecting out-of-band noise and interference such as might be caused by image signals. There are many modulated carriers within this 500-MHz passband, and all of these are amplified and frequency-converted in the common receiver. The frequency conversion shifts the carriers to the downlink frequency band, which is also 500 MHz wide, extending from 3.7 to 4.2 GHz. At this point the signals are channelized into frequency bands which represent the individual transponder bandwidths.

A transponder may handle one modulated carrier, such as a TV signal, or it may handle hundreds of separate carriers simultaneously, each modulated by its own telephony or other baseband channel.

7.7.1 The Wideband Receiver

In the GEOSAT design, single point failures that can end the mission are carefully avoided. For electronic boxes, this is accomplished by having multiple duplicate units that can be rapidly switched into the circuit to replace a degraded or failed unit. For increased reliability, RF switches are usually electro-mechanical latching switches rather than semiconductor variety. Given that the intended usage is a single event, the electromechanical switches have a higher reliability than the solid-state switch and switch driver circuitry combination.

The wideband receiver, shown in Fig. 7.15, is a critical single point failure that can disable 12 transponder channels. As a result, there are duplicate receivers that

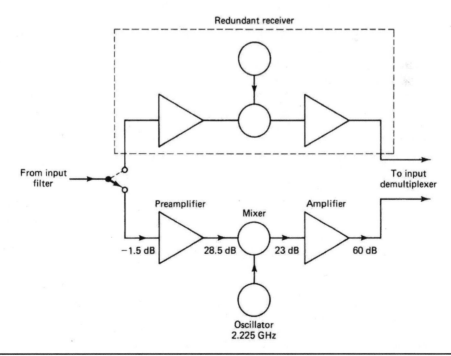

FIGURE 7.15 Satellite wideband receiver. (*Courtesy of CCIR, CCIR Fixed Satellite Services Handbook, final draft 1984.*)

are provided for redundancy, so that if one fails, another is automatically switched in. Further, it is common that one spare receiver is powered on to expedite the rapid restoration of transponder functionality.

The first stage in the receiver is a *low-noise amplifier* (LNA). This amplifier is designed to minimize the noise added to the carrier being amplified, while at the same time providing sufficient amplification for the carrier to override the higher noise level present in the following mixer stage. In calculations involving noise, it is usually more convenient to refer all noise levels to the LNA input, where the total receiver noise may be expressed in terms of an equivalent noise temperature (Ulaby and Long, 2014). In a well-designed receiver, the equivalent noise temperature referred to the LNA input is basically that of the LNA alone. Also, the overall noise temperature must consider the noise added from the antenna, and these calculations are presented in detail in Sec. 12.5. The equivalent noise temperature of a satellite receiver is on the order of a few hundred Kelvin.

The LNA output feeds into a mixer stage, which also requires a highly stable *local oscillator* (LO) signal with low-phase noise for the frequency-conversion process. A second amplifier follows the mixer stage to provide an overall receiver gain of about 60 dB, and an example of the signal levels in decibels referred to the input are shown in Fig. 7.13b. Splitting the gain between the preamplifier at 6 GHz and the second amplifier at 4 GHz prevents oscillation, which might occur if all the gain were to be provided at the same frequency.

7.7.2 The Input Demultiplexer

The input demultiplexer (MUX) separates the downconverted broadband input, covering the frequency range 3.7 to 4.2 GHz, into the transponder frequency channels. The output from the receiver is fed to a power splitter, which in turn feeds the two separate chains of circulators. The channelizing is achieved by means of channel filters connected to each circulator, where each filter has a bandwidth of 36 MHz and is tuned to the appropriate center frequency. For example, the separate channels labeled 1 through 12 are usually arranged in even-numbered (lower) and odd-numbered (upper) groups, as shown in Fig. 7.16, This provides greater frequency separation between adjacent channels in a group that reduces adjacent channel interference, while simplifying the filter design to reduce the roll-off requirements. Although there are considerable losses in the demultiplexer (typically –6 to –10 dB), these are easily made up in the overall gain for the transponder channels.

7.7.3 The Power Amplifier

Separate high-power amplifiers provide the output power for each transponder channel, as shown in Fig. 7.13a, and each is preceded by a variable gain driver amplifier, which comprises a fixed gain amplifier and a ground commandable attenuator. On orbit, it is necessary to adjust the input drive to each power amplifier to the desired level. For example, different input power *backoff* levels are required for various types of service (examples are discussed in Chaps. 12 and 14). Because this input power adjustment is an operational requirement, it is under the control of the ground TT&C network.

In GEOSAT transponders, a vacuum tube *traveling-wave tube amplifier* (TWTA) is widely used to provide the final high-power output stage to the output MUX and transmit antenna. Figure 7.17 shows the schematic of a *traveling wave tube* (TWT) and its power supplies.

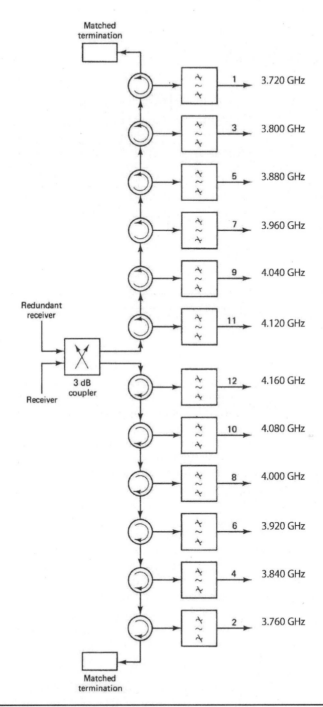

FIGURE 7.16   Input demultiplexer with satellite transponder channels. (*Courtesy of CCIR, CCIR Fixed Satellite Services Handbook, final draft 1984.*)

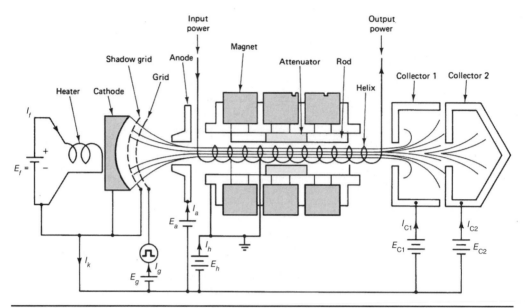

FIGURE 7.17   Schematic of a TWT and power supplies. (*Courtesy of Hughes TWT and TWTA Handbook; courtesy of Hughes Aircraft Company, Electron Dynamics Division, Torrance, CA.*)

In the TWT, an electron-beam gun assembly, consisting of a heater, a cathode, and focusing electrodes, is used to form a cylindrical electron beam, which is accelerated by a positive potential of the collectors at the other end of the vacuum enclosure. A series of permanent magnetics are used to contain the electron beam to travel inside of an RF wire helix transmission line.

The RF signal to be amplified is coupled into the helix at the end nearest the cathode and this results in a traveling wave along the helix. The RF wave travels around the helical path at close to the speed of light, but it is the axial component of wave velocity, which interacts with the flowing electron beam. This component is less than the speed of light approximately in the ratio of helix pitch to its circumference. Because of this effective reduction in RF wave phase velocity, the helix is referred to as a *slow wave structure*.

The RF wave is a sinusoidal waveform at the carrier frequency, and when the instantaneous RF electric field is positive, the electrons are accelerated, and conversely for a negative polarity electric field, they are decelerated. This causes a velocity modulated electron beam (in synchronism with the RF frequency) that results in the electron bunching periodically spaced along the beam. The average beam velocity, which is determined by the dc potential on the TWT collectors, is kept slightly greater than the phase velocity of the RF wave along the helix. Under these conditions, an energy transfer takes place, whereby the kinetic energy of the bunched electron beam induces a voltage in the helix that adds vectorially with the traveling RF wave, and amplification results.

A major advantage of the satellite TWT is its superior DC to RF radiated power efficiency over other types of vacuum tubes or semiconductor amplifiers, and its ability to support high power outputs on the order of 100s of Watts. Further, it can provide linear amplification over more than an octave bandwidth. However, input levels to the TWT must be carefully controlled, to minimize the effects of the nonlinear transfer characteristic of the TWT at high output powers, illustrated in Fig. 7.18a. At low-input powers, the output-input power relationship is linear; such that a given decibel change in input power produces the same decibel change in output power. However, at higher power inputs, the output power progressively saturates because of symmetrical clipping of the peaks the sinusoidal voltage waveform, which produces primarily odd harmonics of the carrier frequency ($3f_o$, $5f_o$, etc.). At the point of maximum power output (known as the *saturated power output*, P_{sat}), we define the backoff of the input power (BOi) and the backoff of the output power (BOo) to be 0 dB. Further increases of the input power (i.e., BOi = positive dB) cause voltage waveform to approach a square wave, which reduces the carrier power, while increasing the carrier harmonics. Of course these harmonic emissions are outside of the allowable communications frequency band, and they are removed by bandpass filtering (before the transmit antenna).

The linear region of the TWT is defined as the region bound by the thermal noise limit at the low end and by what is termed the *1-dB compression point* at the upper end. This is the point where the actual transfer curve drops −1 dB below the extrapolated straight line, as shown in Fig. 7.18a.

The selection of the operating point on the transfer characteristic is considered in more detail shortly, but first, the phase characteristics are described. The absolute time delay between input and output signals at a fixed input level is generally not significant. However, at higher input levels, where more of the beam energy is converted to output power, the average beam velocity is reduced, and the delay time is increased. Since phase delay is directly proportional to time delay, this results in a phase shift that varies with input level. Denoting the phase shift at saturation by θ_S and in general by θ, the phase difference relative to saturation is $\theta - \theta_S$. This is plotted in Fig. 7.18b as a function of input power. Thus, if the input signal power level changes, phase modulation results, which is termed *AM/PM conversion*. The slope of the phase shift characteristic gives the phase modulation coefficient, in degrees per decibel. The curve of the slope as a function of input power is also sketched in Fig. 7.18b.

For satellite communications, the signal modulation of the RF carriers is *angle modulation*, i.e., phase modulation (PM) or frequency modulation (FM); however, unwanted *amplitude modulation* (AM) can occur from ripples in the passband filter response, which occurs prior to the TWT input. When this AM signal is passed through the TWT nonlinear transfer function (Fig. 7.18a), then the process converts the unwanted AM to PM, which appears as noise on the output modulated carrier. Where only a single carrier is present in the transponder channel, it may be preconditioned by passing the carrier through a *hard limiter* before being amplified in the TWT. The hard limiter is a circuit that clips the carrier amplitude close to the zero baseline to remove any amplitude modulation, and the angle modulation is preserved in the zero crossover points and is not affected by the limiting.

On the other hand, often TWTs amplify two or more carriers simultaneously, this being referred to as *multicarrier operation*. The AM/PM conversion is then a complicated function of carrier amplitudes, but in addition, the nonlinear transfer characteristic introduces a more serious form of distortion known as *intermodulation distortion*. The nonlinear transfer characteristic may be expressed as a Taylor series expansion that relates input and output voltages:

$$e_o = ae_i + be_i^2 + ce_i^3 + \cdots \tag{7.1}$$

where, a, b, c, etc. are monotonic decreasing coefficients that depend on the TWT power transfer characteristic; e_i is the input voltage, and e_o is the output voltage, which consists of the vector sum of the individual sinusoidal carriers. The *third-order term* is ce_i^3, and this

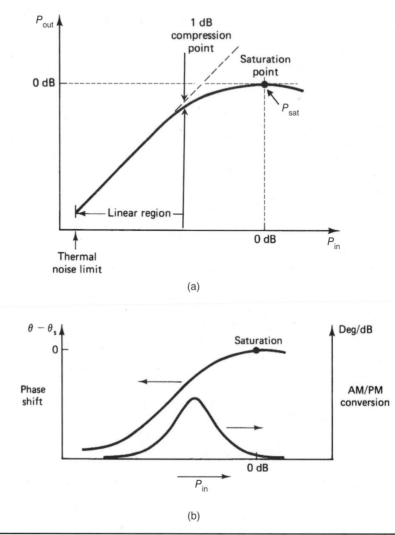

FIGURE 7.18 (a) *Power transfer* characteristics of a TWT, where the saturation point is used as 0-dB reference for both input and output. (b) Phase characteristics for a TWT, where θ is the input-to-output phase shift, and θ_s is the value at saturation. The AM/PM curve is derived from the slope of the phase shift curve.

and higher-order odd-power terms give rise to intermodulation products, known as *third-order intermodulation (3IM)*, but usually only the third-order contribution is significant.

For example, consider the case where multiple carriers (separated in frequency by Δf) are inputs to a TWTA. Now, pick an arbitrary pair of adjacent carriers f_1 and $f_2 = f_1 + \Delta f_1$, and calculate the resulting in-band intermodulation products in the TWTA output (see App. E). This calculation uses the third-order term of Eq. (7.1)

$$e_{3IM} = ce_i^3 = (e_1 + e_2)^3 = e_1^3 + 3e_1^2 e_2 + 3e_2^2 e_1 + e_2^3$$

where $e_1 = A\cos(2\pi f_1 t)$ and $e_2 = A\cos(2\pi f_2 t)$.

In the TWTA output, the in-band intermodulation signals are:

$$3e_1^2 e_2 \rightarrow @ \text{ frequency} = 2f_1 - f_2 = f_1 - \Delta f$$

and
$$3e_2^2 e_1 \rightarrow @ \text{ frequency} = 2f_2 - f_1 = f_2 + \Delta f,$$

and these 3IM products fall on the neighboring carrier frequencies as shown in Fig. 7.19. Similar intermodulation products arise from all other combinations of carrier pairs, and when the carriers are modulated, the intermodulation distortion appears as broadband noise across the transponder frequency band. This 3IM noise is discussed further in Sec. 12.11.

So, to reduce the 3IM noise generation, the operating point of the TWT is shifted closer to the linear portion of the curve, and the reduction in input power is referred to as *input backoff, BOi*. However, when multiple carriers are present, the TWT output power transfer function changes as shown in Fig. 7.20, where there are two distinct power transfer functions representing the single carrier (upper curve) and multiple carrier operation (lower curve). Note that for the multicarrier case, the output power curve is

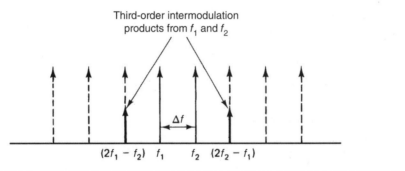

FIGURE 7.19   Third-order intermodulation products.

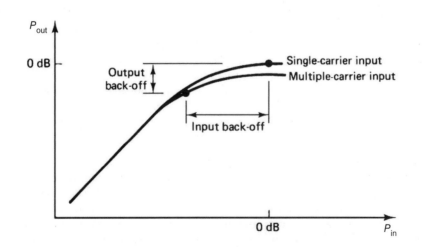

FIGURE 7.20   TWTA power transfer function for: (*a*) a single carrier input and (*b*) for a multiple-carrier input. Backoff for multiple-carrier operation is relative to saturation for single-carrier input.

the total for all carriers that have equal input amplitudes. At saturation, the TWT can only supply a total of P_{sat} Watts output power, but when there are "NC" carriers input, the maximum output power per carrier is $<P_{sat}/NC$, W, because of power going into intermodulation products (both in-band and out of band).

The analysis to calculate the output power associated with the 3IM product is difficult to perform, but fortunately, the input backoff BOi, required to achieve linear TWT operation, is based solely on the single-carrier operation case. Further, since the corresponding output backoff BOo is also based upon this same single-carrier operation case, the backoff values are always stated in decibels relative to the single-carrier saturation point. As a rule of thumb, to achieve linear TWT operation, the BOo $\geq$ –5 dB. Therefore, there is a significant penalty for multiple-carrier operation of a transponder channel. The need to incorporate backoff significantly reduces the carrier-to-noise ratio for the downlink, and the inclusion of backoff in the link budget calculations is discussed in detail in Secs. 12.7.2 and 12.8.1.

7.8 The Output Multiplexer

The output multiplexer is a crucial part of the transponder design. As discussed previously, losses of the input MUX are not considered an issue because they can be easily accommodated by slightly increasing the low-power transponder gain. On the other hand, the output MUX is the last component in the transponder chain before the transmit antenna; thus, any losses result in a reduction of the downlink carrier-to-noise ratio. Further, increasing the TWT output power to compensate for output MUX losses is not a good use of satellite resources because of the ripple effect that this has on increasing the size of the solar panel and the mass of the satellite, etc. So the solution is to build low-loss, high-power multiplexers to combine the multiple transponder channels into a single antenna input.

For satellite communications transponders, the important output MUX parameters are: in-band flatness of gain and group delay, out-of-band rejection, temperature stability, and low in-band insertion loss of typically <-0.5 dB (Wang and Yu, 2011). The generic approach incorporates a resonant design of a waveguide manifold with high-Q > 20,000 cavity channel filters. Also, because the MUX must handle 100s of Watts of RF power, even small ohmic dissipation losses can cause the unit to operate at high physical temperatures > 100°C. This means that special materials, such as gold-plated invar (a nickel-iron alloy), are used for filter cavities with low RF losses and low coefficients of thermal expansion. The latter requirement is important because the resonant frequency of the filter cavities depends upon their physical dimensions that change with thermal expansion. Fortunately, these technologies exist in a small community of industrial suppliers to service the needs of the aerospace suppliers of communications satellites.

7.9 The Antenna Subsystem

The antennas carried aboard a satellite provide the dual functions of receiving the uplink and transmitting the downlink signals. They range from dipole-type antennas where omnidirectional characteristics are required to the highly directional antennas required for telecommunications purposes and TV relay and broadcast. Parts of the antenna structures for the HS 376 and HS 601 satellites can be seen in Figs. 7.1, 7.2, and 7.6.

Directional beams are usually produced by means of reflector-type antennas, with diameters-to-wavelength ratios $10 < D/\lambda < 100$. The satellite antenna gain is directly proportional to $(D/\lambda)^2$ and the beamwidth inversely proportional to D/λ. Hence, the gain can be increased, and the beamwidth made narrower, by increasing the reflector size or decreasing the wavelength. In comparing C-band and Ku-band, the largest reflectors are those for the 6/4-GHz band. Comparable performance can be obtained with considerably smaller reflectors in the 14/12-GHz band. Satellites used for mobile services in the L-band employ much larger antennas (with reflector areas in the order of 100 m² to 200 m²) as described in Chap. 17.

Figure 7.21 shows the antenna subsystem of the INTELSAT VI satellite (Johnston and Thompson, 1982). This provides a good illustration of the level of complexity that

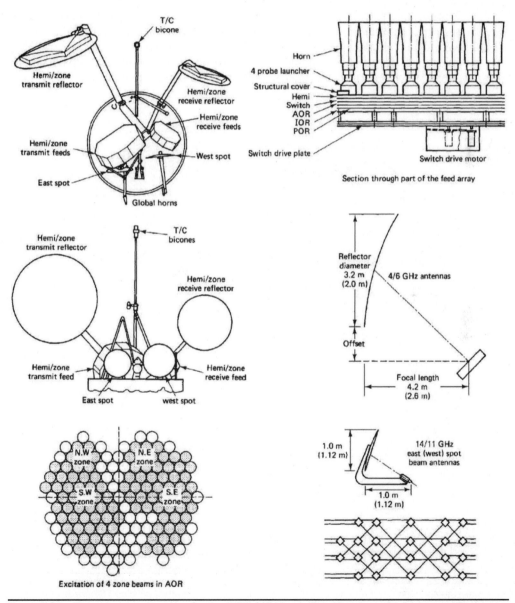

FIGURE 7.21 The antenna subsystem for the INTELSAT VI satellite. (*Courtesy of Johnston and Thompson, 1982, with permission.*)

has been reached in large communications satellites. The largest reflectors are for the 6/4-GHz hemisphere and zone coverages, as illustrated in Fig. 7.22, and these are off-set parabolic antennas that are fed from horn arrays and various groups of horns, which produce the beam shapes required. In this example, separate arrays are used for transmitting and receiving, and each array has 146 dual-polarization horns. In the 14/11-GHz band, off-set fed, 1 m diameter circular reflectors are used to provide spot beams, one for east and one for west, also shown in Fig. 7.22. These beams are mechanically steerable, with each beam fed by a single horn that is used for both transmitting and receiving.

Wide beams for global coverage are produced by simple horn antennas at 6/4 GHz. These horns beam the signal directly to the Earth without the use of reflectors. Also as shown in Fig. 7.21, a simple biconical dipole antenna is used for the tracking and controlling signals. The complete antenna platform and the communications payload are despun as described in Sec. 7.3 to keep the antennas pointing to their correct locations on the Earth.

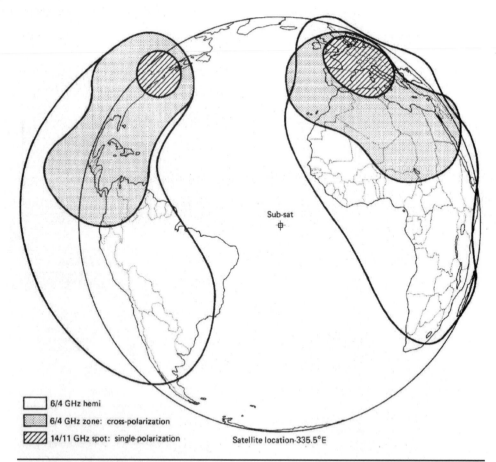

	6/4 GHz hemi
	6/4 GHz zone: cross-polarization
	14/11 GHz spot: single-polarization

Satellite location-335.5°E

FIGURE 7.22 INTELSAT V Atlantic satellite transmit capabilities. (Note: The 14/11-GHz spot beams are steerable and may be moved to meet traffic requirements as they develop.) (*Courtesy of Intelsat Document BG-28-72E M/6/77, with permission.*)

The same feed horn may be used to transmit and receive carriers with the same polarization. The transmit and receive signals are separated in a device known as a *diplexer*, and the separation is further aided by means of frequency filtering. Polarization discrimination also may be used to separate the transmit and receive signals using the same feed horn. For example, the horn may be used to transmit horizontally polarized waves in the downlink frequency band, while simultaneously receiving vertically polarized waves in the uplink frequency band. The polarization separation takes place in a device known as an *orthocoupler*, or *orthogonal mode transducer* (OMT). Separate horns also may be used for the transmit and receive functions, with both horns using the same reflector.

7.10   GEOSAT Spinner Example: Morelos I & II

Two GEOSATs, from the Hughes 376 spinner spacecraft series (Figs. 7.1 and 7.6), were launched by NASA in separate Space Shuttle missions in 1985. These were Mexico's first communications satellites; Morelos I launched in June at 113°W longitude and Morelos II in November at 116.8°W. A photograph of the Morelos I launch during the Space Shuttle Discovery/STS-51-G mission is shown in Fig. 7.23. Morelos-I was operational for 9 years, and Morelos-II was operational for 14 years.

Figure 7.23 Mexico's Morelos I satellite deploying from NASA Space Shuttle Discovery's payload bay STS 51G June 17, 1985. (*NASA Photo ID: STS51G-32-082.*)

The communications subsystem, shown in Fig. 7.24, is a *dual-band,* payload that operates C-band and Ku-band transponders. In the C-band, there are two solid-state receivers (connected to vertical and horizontal polarization ports of the uplink antenna OMT for frequency reuse). To provide additional isolation, the uplink channels are received on one polarization and the downlink channels transmitted on the opposite

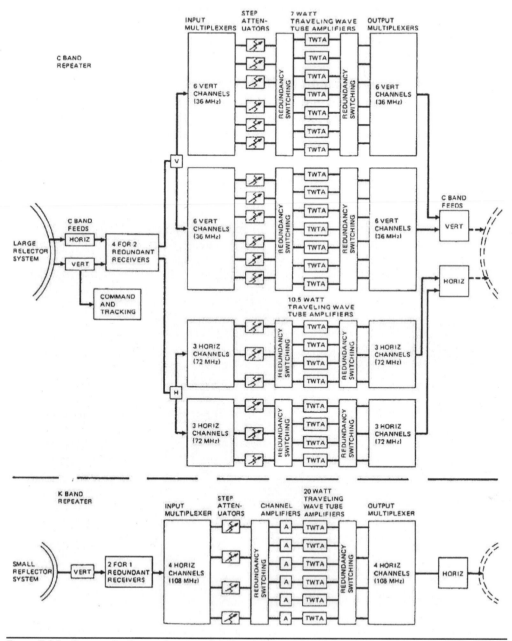

Figure 7.24 Communications subsystem functional diagram for Morelos. (*Courtesy of Hughes Aircraft Company Space and Communications Group.*)

polarization. To mitigate against a single point failure, a 4-for-2 redundancy is included, which means that for each of the 2 in-service units there is a redundant unit. The C-band frequency plan provides 12 narrowband channels, each 36-MHz wide, and 6 wideband channels, each 72-MHz wide. The 36-MHz channels use 7 W TWTAs with 14-for-12 redundancy, and the 72-MHz channels use 10.5 W TWTAs with 8-for-6 redundancy configuration.

At the Ku-band, the receivers are also solid-state designs, with a 2-for-1 redundancy for a single uplink (vertical) antenna polarization. There are four transponder channels, each 108-MHz wide that use six 20 W TWTAs with a 6-for-4 redundancy configuration.

The two antennas are mechanically joined and located on the satellite despun platform to view the Mexico service area. The C-band dual-linear polarized antenna is an offset-fed 1.8 m diameter parabolic dish that provides the shaped-beam EIRP pattern shown in Fig. 7.25*a*. The Ku-band reflector is elliptical in shape, with axes measuring 150 by 91 cm, and its own feed array, producing a footprint, which closely matches the C-band EIRP contours, as shown in Fig. 7.25*b*. Onboard tracking of a C-band beacon transmitted from the Tulancingo TT&C station ensures precise pointing of the antennas.

7.11   Anik Spinner and Body-Stabilized Satellites

The early Anik series (-A, 1972 to -D, 1995) were the world's first domestic satellites, designed to provide communications services (telephone and color TV) in Canada as well as cross-border services with the United States. Later versions (starting with the Anik-E series early 1990s) became much larger satellites of the three-axis body-stabilized design to provide international telecommunications coverage for other parts of the world. They were dual-band (C- and Ku-band) satellites, which had an equivalent capacity of 56 television channels, or more than 50,000 telephone circuits. Attitude control was of the momentum-bias, three-axis-stabilized type, and solar panels were used to provide power, the capacity being 3450 W at summer solstice and 3700 W at vernal equinox. Four NiH$_2$ batteries provided power during eclipse. Although the Anik-E has been superseded by the Anik-F series (first in November 2000), the Anik-E configuration as shown in Fig. 7.26 provides a good illustration of the body-stabilized type of satellite, and Fig. 7.27 shows a typical C-band transponder setup. Also, this configuration uses *solid-state power amplifiers* (SSPAs), which offer a significant improvement in reliability and weight saving over traveling-wave tube amplifiers, but they are not as efficient in the use of dc power. These antennas are fed through a *broadband feeder network* (BFN) to illuminate the large reflectors, which provides national, as distinct from regional, coverage.

The Ku-band transponder (not shown) uses TWTAs that may be switched to provide *4-for-2 redundancy*, as illustrated in Fig. 7.28. For example, table shown in Fig. 7.29 shows that channel 1A has amplifier 2 as its primary amplifier, and amplifiers 1 and 3 can be switched in as backup amplifiers by ground command.

7.12   Digital Regenerative Repeaters

As previously mentioned in the beginning of Sec. 7.7, the use of digital communications allows for a different transponder configuration known as a *digital receiver* or *regenerative repeater*. The advantage of this approach is presented in Sec. 12.12, and a summary is presented below.

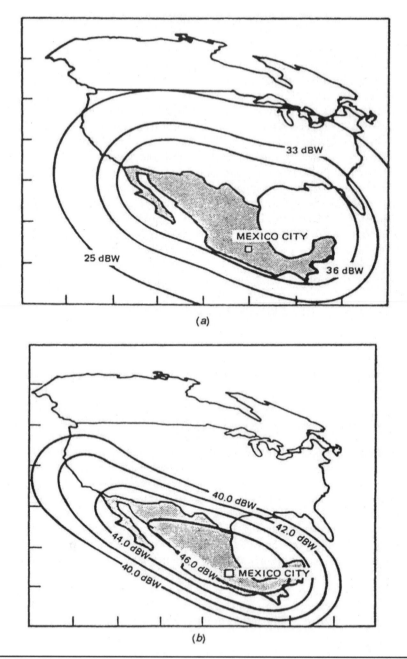

FIGURE 7.25 (a) C-band and (b) K-band transmit (EIRP) coverage for Morelos. (*Courtesy of Hughes Aircraft Company Space and Communications Group.*)

Figure 7.26 Anik-E spacecraft configuration. (*Courtesy of Telesat Canada.*)

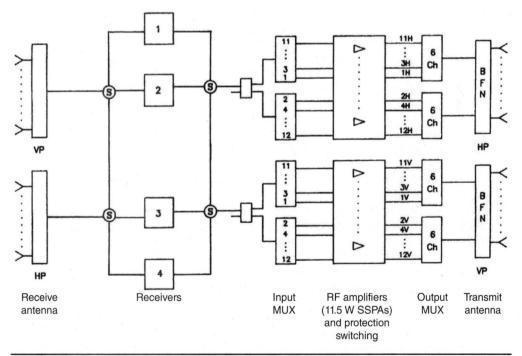

Receive antenna · Receivers · Input MUX · RF amplifiers (11.5 W SSPAs) and protection switching · Output MUX · Transmit antenna

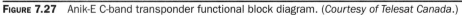

Figure 7.27 Anik-E C-band transponder functional block diagram. (*Courtesy of Telesat Canada.*)

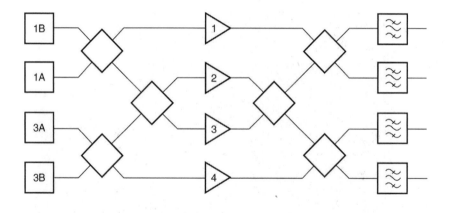

CHANNEL	1A	3A	1B	3B
TWTA				
Primary	2	3	1	4
Backup	1 or 3	2 or 4	2 or 3	2 or 3

FIGURE 7.28 A TWTA 4-for-2 redundancy switching arrangement for the Anik-E Ku-band transponder. (*Courtesy of Telesat Canada, 1983.*)

The communication system performance depends upon the ratio of signal power to noise power (SNR). This effect is especially significant for wireless communications, where the signal strength decreases dramatically with distance from the transmitter. To mitigate the reduction in signal power, comm repeaters (transponders) are used to amplify the weak signals and then retransmit them. The problem is that the signal-to-noise ratio continues to degrade after each repeater, and this is true whether the signal is analog or digital.

However, as discussed in Chap. 10, the nature of digital communications is more robust with respect to SNR, than is analog. This is because for digital, we need only make a simple *binary decision* as to whether the digital bit waveform is above or below a selected threshold, i.e., whether a *logic 1* or *logic 0* was sent. Poor SNR introduces *bit errors* in this digital decision process, but these can be mitigated or even eliminated by incorporating error correction encoding.

With most satellite communications becoming digital, there is an advantage for satellite transponders to abandon the simple bent-pipe repeater approach and to incorporate onboard digital signal processing. This can take several forms, but two common are: (a) active switching to route uplink signals to the appropriate transponder channel with the desired downlink antenna geographic coverage, and/or (b) synchronous detection of the uplink phase shift keying (PSK) modulated carriers to produce the baseband digital waveforms, optionally apply error correction, and then PSK modulate a carrier using the regenerated digital signal for downlink transmission.

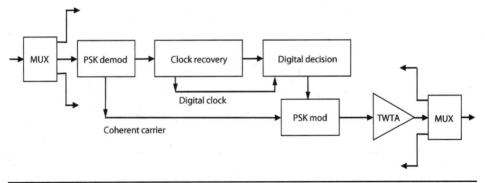

FIGURE 7.29 Functional block diagram for a satellite digital regenerative transponder.

A functional block diagram for such a satellite regenerative repeater is shown in Fig. 7.29. For simplicity, only the block diagram between the input and output MUXs is shown. Consider a frequency plan where there are 12 PSK modulated carriers at the downlink frequencies that are separated by the input MUX in a *single carrier per channel* (SCPC) manner.

Consider now, a single channel, where the PSK modulated carrier passes through a demodulation block to output a coherent RF carrier and the baseband digital signal plus additive gaussian noise. In the second block, the baseband signal and noise pass through the clock recovery circuit, which extracts a digital clock that is synchronized with the digital data. In the third block, the baseband digital signal and clock are input to the binary decision block, where a sample of the digital waveform is compared with a zero-voltage threshold to declare a logic-1 or logic-0 output. At this point, there is a restored "noise-free" digital signal that may or may not have bit errors introduced by the presence of noise.

If the uplink SNR > 23 dB, then the probability of bit error < 10^{-11}, which is negligible for most applications. If required, error correction algorithms are applied to remove bit errors, which results in an error-free digital signal that remodulates the RF carrier and is transmitted on the downlink. This example is overly simplified compared to the actual hardware implementation that may involve several heterodyne frequency translations, where various processes are performed. For example, the signal input to the first MUX may be at an IF frequency, which is processed and then upconverted to the microwave carrier frequency before amplification in the TWTA.

7.13 Problems and Exercises

7.1. Describe the TT&C facilities of a satellite communications system. Are these facilities part of the space segment or part of the ground segment of the system?

7.2. Explain why some satellites employ cylindrical solar arrays, whereas others employ solar-sail arrays to produce primary power. State the typical power output to be expected from each type. Why is it necessary for satellites to carry batteries in addition to solar-cell arrays?

7.3. Explain what is meant by satellite *attitude*, and briefly describe two forms of attitude control.

7.4. Define and explain the terms *roll, pitch*, and *yaw*.

7.5. Explain what is meant by the term *despun antenna*, and briefly describe one way in which the despinning is achieved.

7.6. Briefly describe the three-axis method of satellite stabilization.

7.7. Describe the east-west and north-south station-keeping maneuvers required in satellite station keeping. What are the angular tolerances in station keeping that must be achieved?

7.8. Referring to Fig. 7.11 and the accompanying text in Sec. 7.4, determine the minimum earth station antenna −3 dB beamwidth that accommodates the tolerances in satellite position without the need for tracking.

7.9. Explain what is meant by *thermal control* and why this is necessary in a satellite.

7.10. Explain why an omnidirectional antenna must be used aboard a satellite for telemetry and command during the launch phase. How is the satellite powered during this phase?

7.11. Briefly describe the equipment sections making up a transponder channel.

7.12. Draw to scale the uplink and downlink channeling schemes for a 500-MHz-bandwidth C-band satellite, accommodating the full complement of 36-MHz-bandwidth transponders. Assume the use of 4-MHz guard bands.

7.13. Explain what is meant by *frequency reuse* and briefly describe two methods by which this can be achieved.

7.14. Explain what is meant by *redundant receiver* in connection with communication satellites.

7.15. Describe the function of the input demultiplexer used aboard a communications satellite.

7.16. Describe briefly the most common type of high-power amplifying device used aboard a communications satellite and provide a sketch of the output versus input power transfer function.

7.17. What is the chief advantage of the TWTA used aboard satellites compared to other types of high-power amplifying devices? What are the main disadvantages of the TWTA?

7.18. Define and explain the term −1-dB *compression point*. What is the significance of this point in relation to the operating point of a TWTA?

7.19. Explain why operation near the saturation point of a TWTA is to be avoided when multiple carriers are being amplified simultaneously.

7.20. State the type of satellite antenna normally used to produce a wide beam radiation pattern, providing global coverage.

7.21. State the type of satellite antenna normally used to produce a spot beam radiation pattern. What determines the "spot diameter" on the Earth's surface?

7.22. Describe briefly how beam shaping of a satellite antenna radiation pattern may be achieved.

7.23. With reference to Fig. 7.28, explain what is meant by a *4-for-2 redundancy switching arrangement*.

References

CCIR. 1984. *Fixed Services Handbook*, final draft. Geneva.

Chetty, P. R. K. 1991. *Satellite Technology and Its Applications.* McGraw-Hill, New York.

Hyndman, J. E. 1991. *Hughes HS601 Communications Satellite Bus System Design Trades.* Hughes Aircraft Company, El Segundo, CA.

Gordon, Gary D., and Walter L. Morgan. 1993. *Principles of Communications Satellites.* John Wiley & Sons, Inc.

Johnston, E. C., and J. D. Thompson. 1982. "INTELSAT VI Communications Payload." *IEE Colloquium on the Global INTELSAT VI Satellite System*, Digest No. 1982/76. pp. 4/1–4/4.

Lilly, C. J. 1990. "INTELSAT's New Generation." *IEE Review*, vol. 36, no. 3, March. pp. 111–113.

Pilcher, L. S. 1982. "Overall Design of the INTELSAT VI Satellite." In: *3rd International Conference on Satellite Systems for Mobile Communications and Navigation*, IEE, London.

Schwalb, A. 1982a. "The TIROS-N/NOAA-G Satellite Series." *NOAA Technical Memorandum NESS 95*, Washington, DC.

Schwalb, A. 1982b. "Modified Version of the TIROS-N/NOAA A-G Satellite Series (NOAA E-J): Advanced TIROS N (ATN)." *NOAA Technical Memorandum NESS 116*, Washington, DC.

Spilker, J. J. 1977. *Digital Communications by Satellite.* Prentice-Hall, Englewood Cliffs, NJ.

Wang, Ying and Yu Ming, 2011. "Performance Analysis of the Enhanced Microwave Multiplexing Networks for Applications in Communications Satellites." *2011 IEEE MTT-S International Microwave Symposium.* Baltimore, MD, DOI: 10.1109/MWSTM.2011.5972740.

Wertz, J. R. (ed.). 1984. *Spacecraft Attitude Determination and Control.* D. Reidel, Holland.

CHAPTER 8

The Earth Segment

8.1 Introduction

The earth segment of a satellite communications system consists of the transmit and receive earth stations. The simplest of these are the home *TV receive-only* (TVRO) systems, and the most complex are the terminal stations used for international communications networks. Also included in the earth segment are those stations which are on ships at sea, and commercial and military land and aeronautical mobile stations.

As mentioned in Chap. 7, earth stations that are used for logistic support of satellites, such as providing the *telemetry, tracking, and command* (TT&C) functions, are considered as part of the space segment.

8.2 Receive-Only Home TV Systems

Planned broadcasting directly to home TV receivers takes place in the Ku (12-GHz) band. This service is known as *direct broadcast satellite* (DBS) service. There is some variation in the frequency bands assigned to different geographic regions. In the Americas, for example, the downlink band is 12.2 to 12.7 GHz, as described in Sec. 1.4.

The comparatively large satellite receiving dishes (ranging in diameter from about 1.83 m to about 3-m in some locations), which may be seen in some "backyards" are used to receive downlink TV signals at C-band (4 GHz). Originally, such downlink signals were never intended for home reception but for network relay to commercial TV outlets (VHF and UHF TV broadcast stations and cable TV "head-end" studios). Equipment is now marketed for home reception of C-band signals, and some manufacturers provide dual C-band/Ku-band equipment. A single mesh type reflector may be used that focuses the signals into a dual feedhorn, which has two separate outputs, one for the C-band signals and one for the Ku-band signals. Much of television programming originates as *first-generation signals*, also known as *master broadcast quality signals*. These are transmitted via satellite in the C-band to the network head-end stations, where they are retransmitted as compressed digital signals to cable and direct broadcast satellite providers. One of the advantages claimed by sellers of C-band equipment for home reception is that there is no loss of quality compared with the compressed digital signals.

To take full advantage of C-band reception the home antenna must be steerable to receive from different satellites, usually by means of a polar mount as described in Sec. 3.3. Another of the advantages, claimed for home C-band systems, is the larger number of satellites available for reception compared to what is available for direct broadcast satellite systems. Although many of the C-band transmissions are scrambled,

219

there are free channels that can be received, and what are termed "wild feeds." These are also free, but unannounced programs, of which details can be found in advance from various publications and Internet sources. C-band users can also subscribe to pay TV channels, and another advantage claimed is that subscription services are cheaper than DBS or cable because of the multiple-source programming available.

The most widely advertised receiving system for C-band system appears to be 4DTV manufactured by Motorola. This enables reception of:

1. Free, analog signals and "wild feeds"

2. VideoCipher ll plus subscription services

3. Free DigiCipher 2 services

4. Subscription DigiCipher 2 services

VideoCipher is the brand name for the equipment used to scramble analog TV signals. DigiCipher 2 is the name given to the digital compression standard used in digital transmissions. General information about C-band TV reception can be found at http://orbitmagazine.com/ (Orbit, 2005) and http://www.satellitetheater.com/ (Satellite Theater systems, 2005).

The major differences between the Ku-band and the C-band receive-only systems lies in the frequency of operation of the outdoor unit and the fact that satellites intended for DBS have much higher *equivalent isotropic radiated power* (EIRP), as shown in Table 1.4. As already mentioned C-band antennas are considerably larger than DBS antennas. For clarity, only the Ku-band system is described here.

Figure 8.1 shows the main units in a home terminal DBS TV receiving system. Although there are variations from system to system, the diagram covers the basic concept for analog (*frequency modulated* [FM]) TV. Direct-to-home digital TV, which is well on the way to replacing analog systems, is discussed in Chap. 16. However, the outdoor unit is similar for both systems.

8.2.1 The Outdoor Unit

This consists of a receiving antenna feeding directly into a low-noise amplifier/converter combination. A parabolic reflector is generally used, with the receiving horn mounted at the focus. A common design is to have the focus directly in front of the reflector, but for better interference rejection, an offset feed may be used as shown.

Huck and Day (1979) have shown that satisfactory reception can be achieved with reflector diameters in the range 0.6 to 1.6 m, and the two nominal sizes often quoted are 0.9 m and 1.2 m. By contrast, the reflector diameter for 4-GHz reception can range from 1.83 m to 3 m. As noted in Sec. 6.13, the gain of a parabolic dish is proportional to $(D/\lambda)^2$. Comparing the gain of a 3-m dish at 4 GHz with a 1-m dish at 12 GHz, the ratio D/λ equals 40 in each case, so the gains are about equal. Although the free-space losses are much higher at 12 GHz compared with 4 GHz, as described in Chap. 12, a higher-gain receiving antenna is not needed because the DBS operate at a much higher EIRP, as shown in Table 1.4.

The downlink frequency band of 12.2 to 12.7 GHz spans a range of 500 MHz, which accommodates 32 TV/FM channels, each of which is 24 MHz wide. Obviously, some overlap occurs between channels, but these are alternately polarized *left-hand circular* (LHC) and *right-hand circular* (RHC) or vertical/horizontal, to reduce interference to

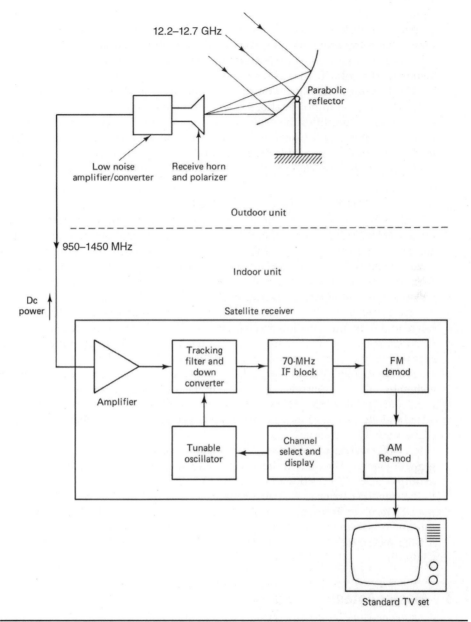

FIGURE 8.1   Block diagram showing a home terminal for DBS TV/FM reception.

acceptable levels. This is referred to as *polarization interleaving*. A polarizer that may be switched to the desired polarization from the indoor control unit is required at the receiving horn.

The receiving horn feeds into a *low-noise converter* (LNC) or possibly a combination unit consisting of a *low-noise amplifier* (LNA) followed by a converter. The combination is referred to as an LNB, for *low-noise block*. The LNB provides gain for the broadband

12-GHz signal and then converts the signal to a lower frequency range so that a low-cost coaxial cable can be used as feeder to the indoor unit. The standard frequency range of this downconverted signal is 950 to 1450 MHz, as shown in Fig. 8.1. The coaxial cable, or an auxiliary wire pair, is used to carry dc power to the outdoor unit. Polarization-switching control wires are also required.

The low-noise amplification must be provided at the cable input to maintain a satisfactory signal-to-noise ratio. An LNA at the indoor end of the cable is of little use because it also amplifies the cable thermal noise. Single-to-noise ratio is discussed in more detail in Sec. 12.5. Of course, having to mount the LNB outside means that it must be able to operate over a wide range of climatic conditions, and homeowners may have to contend with the added problems of vandalism and theft.

8.2.2   The Indoor Unit for Analog (FM) TV

The signal fed to the indoor unit is normally a wideband signal covering the range 950 to 1450 MHz. This is amplified and passed to a tracking filter that selects the desired channel, as shown in Fig. 8.1. As previously mentioned, polarization interleaving is used, and only half the 32 channels are present at the input of the indoor unit for anyone setting of the antenna polarizer. This eases the job of the tracking filter since alternate channels are well separated in frequency.

The selected channel is again downconverted, this time from the 950- to 1450-MHz range to a fixed intermediate frequency, usually 70 MHz although other values in the *very high frequency* (VHF) range are also used. The 70-MHz amplifier amplifies the signal up to the levels required for demodulation. A major difference between DBS TV and conventional TV is that with DBS, frequency modulation is used, whereas with conventional TV, amplitude modulation in the form of *vestigial single sideband* (VSSB) is used. The 70-MHz, FM *intermediate frequency* (IF) carrier therefore must be demodulated, and the baseband information used to generate a VSSB signal that is fed into one of the VHF/UHF channels of a standard TV set.

A DBS receiver provides several functions not shown on the simplified block diagram of Fig. 8.1. The demodulated video and audio signals are usually made available at output jacks. Also, as described in Sec. 13.3, an energy-dispersal waveform is applied to the satellite carrier to reduce interference, and this waveform must be removed in the DBS receiver. Terminals also may be provided for the insertion of IF filters to reduce interference from terrestrial TV networks, and a descrambler also may be necessary for the reception of some programs. The indoor unit for digital TV is described in Chap. 16.

8.3   Master Antenna TV System

A *master antenna TV* (MATV) system is used to provide reception of DBS TV/FM channels to a small group of users, for example, to the tenants in an apartment building. It consists of a single outdoor unit (antenna and LNA/C) feeding numerous indoor units, as shown in Fig. 8.2. It is basically like the home system already described, but with each user having access to all the channels independently of the other users. The advantage is that only one outdoor unit is required, but as shown, separate LNA/Cs and feeder cables are required for each sense of polarization. Compared with the single-user system, a larger antenna is also required (2- to 3-m diameter) to maintain a good signal-to-noise ratio at all the indoor units.

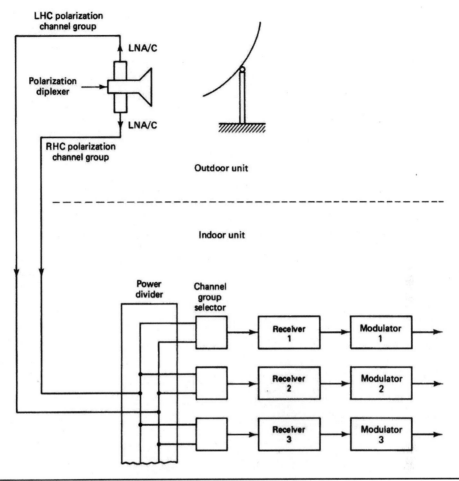

FIGURE 8.2   One possible arrangement for a master antenna TV (MATV) system.

Where more than a few subscribers are involved, the distribution system used is like the *community antenna* (CATV) system described in the following section.

8.4   Community Antenna TV System

The CATV system employs a single outdoor unit, with separate feeds available for each sense of polarization, like the MATV system, so that all channels are made available simultaneously at the indoor receiver. Instead of having a separate receiver for each user, all the carriers are demodulated in a common receiver-filter system, as shown in Fig. 8.3. The channels are then combined into a standard multiplexed signal for transmission over cable to the subscribers.

In remote areas where a cable distribution system may not be installed, the signal can be rebroadcast from a low-power VHF TV transmitter. Figure 8.4 shows a remote TV station that employs an 8-m antenna for reception of the satellite TV signal in the C-band.

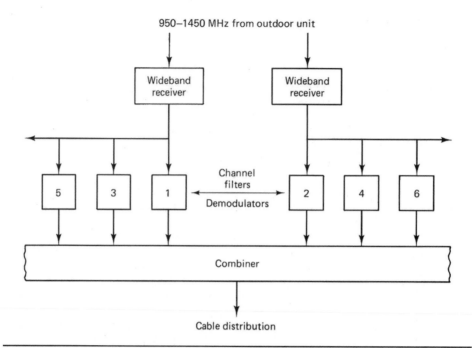

950–1450 MHz from outdoor unit

FIGURE 8.3   One possible arrangement for the indoor unit of a community antenna TV (CATV) system.

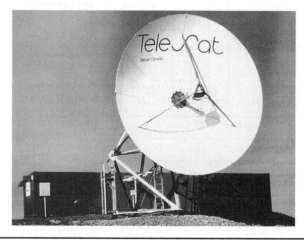

FIGURE 8.4   Remote television station. (*Courtesy of Telesat Canada, 1983.*)

With the CATV system, local programming material also may be distributed to subscribers, an option which is not permitted in the MATV system.

8.5   Transmit-Receive Earth Stations

In the previous sections, receive-only TV stations are described. Obviously, somewhere a transmit station must complete the uplink to the satellite. In some situations, a transmit-only station is required, for example, in relaying TV signals to the remote TVRO stations already described. Transmit-receive stations provide both functions and are required for telecommunications traffic generally, including network TV. The uplink facilities for digital TV are highly specialized and are covered in Chap. 16.

The basic elements for a redundant earth station are shown in Fig. 8.5. As mentioned in connection with transponders in Sec. 7.7.1, redundancy means that certain units are duplicated. A duplicate, or redundant, unit is automatically switched into a circuit to replace a corresponding unit that has failed. Redundant units are shown by dashed lines in Fig. 8.5.

The block diagram is shown in more detail in Fig. 8.6, where, for clarity, redundant units are not shown. Starting at the bottom of the diagram, the first block shows the interconnection equipment required between satellite station and the terrestrial network. For the purpose of explanation, telephone traffic is assumed. This may consist of a number of telephone channels in a multiplexed format. Multiplexing is a method of grouping telephone channels together, usually in basic groups of 12, without mutual interference. It is described in detail in Chaps. 9 and 10.

It may be that groupings different from those used in the terrestrial network are required for satellite transmission, and the next block shows the multiplexing equipment in which the reformatting is carried out. Following along the transmit chain, the multiplexed signal is modulated onto a carrier wave at an intermediate frequency, usually 70 MHz. Parallel IF stages are required, one for each microwave carrier to be transmitted. After amplification at the 70-MHz IF, the modulated signal is then upconverted to the required microwave carrier frequency. Several carriers may be transmitted simultaneously, and although these are at different frequencies they are generally specified by their nominal frequency, for example, as 6-GHz or 14-GHz carriers.

It should be noted that the individual carriers may be multidestination carriers. This means that they carry traffic destined for different stations. For example, as part of its load, a microwave carrier may have telephone traffic for Boston and New York. The same carrier is received at both places, and the designated traffic sorted out by filters at the receiving earth station.

Referring again to the block diagram of Fig. 8.6, after passing through the upconverters, the carriers are combined, and the resulting wideband signal is amplified. The wideband power signal is fed to the antenna through a diplexer, which allows the antenna to handle transmit and receive signals simultaneously.

The station's antenna functions in both, the transmit and receive modes, but at different frequencies. In the C-band, the nominal uplink, or transmit, frequency is 6 GHz and the downlink, or receive, frequency is nominally 4 GHz. In the Ku-band, the uplink frequency is nominally 14 GHz, and the downlink, 12 GHz. High-gain antennas are employed in both bands, which also means narrow antenna beams. A narrow beam is necessary to prevent interference between neighboring satellite links. In the case of C-band, interference to and from terrestrial microwave links also must be avoided. Terrestrial microwave links do not operate at Ku-band frequencies.

In the receive branch (the right-hand side of Fig. 8.6), the incoming wideband signal is amplified in an LNA and passed to a divider network, which separates out the individual microwave carriers. These are each downconverted to an IF band and passed on to the multiplex block, where the multiplexed signals are reformatted as required by the terrestrial network.

It should be noted that, in general, the signal traffic flow on the receive side differs from that on the transmit side. The incoming microwave carriers are different in number and in the amount of traffic carried, and the multiplexed output are telephone circuits not necessarily carried on the transmit side.

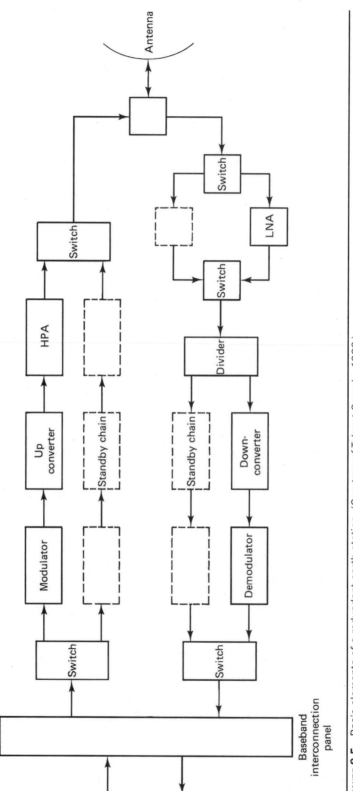

Figure 8.5 Basic elements of a redundant earth station. *(Courtesy of Telesat Canada, 1983.)*

226

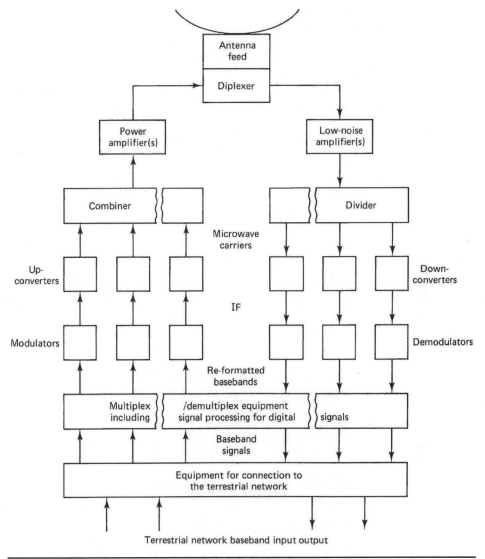

FIGURE 8.6 More detailed block diagram of a transmit-receive earth station.

Several different classes of earth stations are available, depending on the service requirements. Traffic can be broadly classified as heavy route, medium route, and thin route. In a thin-route circuit, a transponder channel (36 MHz) may be occupied by several single carriers, each associated with its own voice circuit. This mode of operation is known as *single carrier per channel* (SCPC), a multiple-access mode which is discussed further in Chap. 14. Antenna sizes range from 3.6 m for transportable stations up to 30 m for a main terminal.

A medium-route circuit also provides multiple access, either based on *frequency-division multiple access* (FDMA) or *time-division multiple access* (TDMA), multiplexed baseband signals being carried in either case. These access modes are also described in detail in Chap. 14. Antenna sizes range from 30 m for a main station to 10 m for a remote station.

In a 6/4-GHz heavy-route system, each satellite channel (bandwidth 36 MHz) can carry over 960 one-way voice circuits simultaneously or a single-color analog TV signal with associated audio (in some systems two analog TV signals can be accommodated). Thus the transponder channel for a heavy-route circuit carries one large-bandwidth signal, which may be TV or multiplexed telephony. The antenna diameter for a heavy-route circuit is at least 30 m. For international operation such antennas are designed to the INTELSAT specifications for a Standard A earth station (Intelsat, 1982). Figure 8.7 shows a photograph of a 32-m Standard A earth station antenna.

For large antennas, which may weigh in the order of 250 tons, the foundations must be very strong and stable. Such large diameters automatically mean very narrow beams, and therefore, any movement that deflect the beam unduly must be avoided. Where snow and ice conditions are likely to be encountered, built-in heaters are required. For the antenna shown in Fig. 8.7, deicing heaters provide reflector surface heat of 430 W/m² for the main reflectors and subreflectors, and 3000 W for the azimuth wheels.

Although these antennas are used with geostationary satellites, some drift in the satellite position does occur, as shown in Chap. 3. This, combined with the very narrow beams of the larger earth station antennas, means that some provision must be made for a limited degree of tracking. Step adjustments in azimuth and elevation may be made, under computer control, to maximize the received signal.

FIGURE 8.7   Standard-A (C-band 6/4 GHz) 32-m antenna. (*Courtesy of TIW Systems, Inc., Sunnydale, CA.*)

The continuity of the primary power supply is another important consideration in the design of transmit-receive earth stations. Apart from the smallest stations, power backup in the form of multiple feeds from the commercial power source and/or batteries and generators is provided. If the commercial power fails, batteries immediately take over with no interruption. At the same time, the standby generators start up, and once they are up to speed they automatically take over from the batteries.

8.6 Problems and Exercises

8.1. Explain what is meant by DBS service. How does this differ from the home reception of satellite TV signals in the C-band?

8.2. Explain what is meant by *polarization interleaving*. On a frequency axis, draw to scale the channel allocations for the 32 TV channels in the Ku-band, showing how polarization interleaving is used in this.

8.3. Why is it desirable to downconvert the satellite TV signal received at the antenna?

8.4. Explain why the LNA in a satellite receiving system is placed at the antenna end of the feeder cable.

8.5. With the aid of a block schematic, briefly describe the functioning of the indoor receiving unit of a satellite TV/FM receiving system intended for home reception.

8.6. In most satellite TV receivers the first IF band is converted to a second, fixed IF. Why is this second frequency conversion required?

8.7. For the standard home television set to function in a satellite TV/FM receiving system, a demodulator/remodulator unit is needed. Explain why.

8.8. Describe and compare the MATV and the CATV systems.

8.9. Explain what is meant by the term *redundant earth station*.

8.10. With the aid of a block schematic, describe the functioning of a transmit-receive earth station used for telephone traffic. Describe a multidestination carrier.

References

Huck, R. W., and J. W. B. Day. 1979. "Experience in Satellite Broadcasting Applications with CTS/HERMES." *XIth International TV Symposium*, Montreux, May 27–June 1.

INTELSAT. 1982. "Standard A Performance Characteristics of Earth Stations in the INTELSAT IV, IVA, and V Systems." BG-28-72E M/6/77.

Orbit, 2005, at http://orbitmagazine.com/.

Satellite Theater systems, 2005, at http://www.satellitetheater.com/.

CHAPTER 9

Analog Signals

9.1 Introduction

Analog signals are electrical replicas of the original signals such as audio and video. *Baseband signals* are those signals which occupy the lowest, or base, band of frequencies, in the frequency spectrum used by the telecommunications network. A baseband signal may consist of one or more information signals. For example, several analog telephony signals may be combined into one baseband signal by the process known as *frequency-division multiplexing* (FDM). Other common types of baseband signals are the multi-plexed video and audio signals that originate in the TV studio. In forming the multiplexed baseband signals, the information signals are *modulated* onto subcarriers. This modulation step must be distinguished from the modulation process, which places the multiplexed signal onto the microwave carrier for transmission to the satellite.

In this chapter, the characteristics of the more common types of analog baseband signals are described, along with representative methods of analog modulation.

9.2 The Telephone Channel

Natural speech, including that of female and male voices, covers a frequency range of about 80 to 8000 Hz. The somewhat unnatural quality associated with telephone speech results from the fact that a considerably smaller band of frequencies is used for normal telephone transmission. The range of 300 to 3400 Hz is accepted internationally as the standard for "telephone quality" speech, and this is termed the *speech baseband*. In practice, some variations occur in the basebands used by different telephone companies. The telephone channel is often referred to as a *voice frequency* (VF) channel, and in this book this is taken to mean the frequency range of 300 to 3400 Hz.

There are good reasons for limiting the frequency range. Noise, which covers a very wide frequency spectrum, is reduced by reducing the bandwidth. Also, reducing the bandwidth allows more telephone channels to be carried over a given type of circuit, as described in Sec. 9.4.

9.3 Single-Sideband Telephony

Figure 9.1*a* shows how the VF baseband may be represented in the frequency domain. In some cases, the triangular representation has the small end of the triangle at 0 Hz, even though frequency components below 300 Hz may not be present. Also, in some cases, the upper end is set at 4 kHz to indicate allowance for a guard band, the need for which is described later.

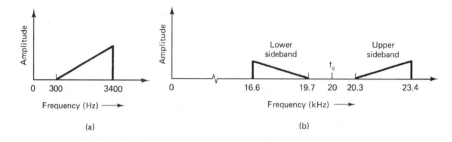

FIGURE 9.1 Frequency-domain representation of (a) a telephone baseband signal and (b) the double-sideband suppressed carrier (DSBSC) modulated version of (a).

When the telephone signal is multiplied in the time domain with a sinusoidal carrier of frequency f_c, a new spectrum results, in which the original baseband appears on either side of the carrier frequency. This is illustrated in Fig. 9.1b for a carrier of 20 kHz, where the band of frequencies below the carrier is referred to as the *lower sideband* and the band above the carrier as the *upper sideband*. To avoid distortion that would occur with sideband overlap, the carrier frequency must be greater than the highest frequency in the baseband.

The result of this multiplication process is referred to as *double-sideband suppressed-carrier* (DSBSC) modulation, since only the sidebands, and not the carrier, appear in the spectrum. Now, all the information in the original telephone signal is contained in either of the two sidebands. Since this information is redundant, it is necessary to transmit only one of these. A filter may be used to select either one or reject the other. The resulting output is termed a *single-sideband* (SSB) signal.

The SSB process utilizing the lower sideband is illustrated in Fig. 9.2, where a 20-kHz carrier is used as an example. It may be seen that for the lower sideband, the frequencies have been inverted, the highest baseband frequency being translated to the lowest transmission frequency at 16.6 kHz and the lowest baseband frequency to the highest transmission frequency at 19.7 kHz. This inversion does not affect the final baseband output since the demodulation process reinverts the spectrum. At the

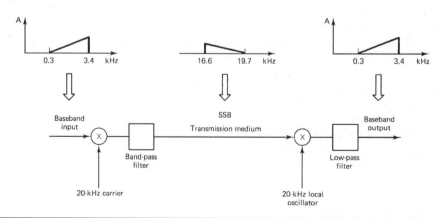

FIGURE 9.2 A basic SSB transmission scheme.

receiver, the SSB signal is demodulated (i.e., the baseband signal is recovered) by being passed through an identical multiplying process. In this case the multiplying sinusoid, termed the *local oscillator* (LO) signal, must have the same frequency as the original carrier. A low-pass filter is required at the output to select the baseband signal and reject other, unwanted frequency components generated in the demodulation process. This single-sideband modulation/demodulation process is illustrated in Fig. 9.2.

The way in which SSB signals are used for the simultaneous transmission of several telephone signals is described in the following section. It should be noted at this point that several different carriers are likely to be used in a satellite link. The *radiofrequency* (rf) carrier used in transmission to and from the satellite is much higher in frequency than those used for the generation of the set of SSB signals. These latter carriers are sometimes referred to as VF *carriers*. The term *subcarrier* is also used, and this practice is followed here. Thus the 20-kHz carrier shown in Fig. 9.2 is a subcarrier.

Companded single sideband (CSSB) refers to a technique in which the speech signal levels are compressed before transmission as a single sideband, and at the receiver they are expanded again back to their original levels. (The term *compander* is derived from *comp*ressor-exp*ander*.) In one companded system described by Campanella (1983), a 2:1 compression in decibels is used, followed by a 1:2 expansion at the receiver. It is shown in the reference that the expander decreases its attenuation when a speech signal is present and increases its attenuation when it is absent. In this way the "idle" noise on the channel is reduced, which allows the channel to operate at a reduced carrier-to-noise ratio. This in turn permits more channels to occupy a given satellite link, a topic that comes under the heading of *multiple access* and which is described more fully in Chap. 14.

9.4 FDM Telephony

FDM provides a way of keeping several individual telephone signals separate while transmitting them simultaneously over a common transmission link circuit. Each telephone baseband signal is modulated onto a separate subcarrier, and all the upper or all the lower sidebands are combined to form the frequency-multiplexed signal. Figure 9.3a shows how three voice channels may be frequency-division multiplexed. Each voice channel occupies the range 300 to 3400 Hz, and each is modulated onto its own subcarrier. The subcarrier frequency separation is 4 kHz, allowing for the basic voice bandwidth of 3.1 kHz plus an adequate guardband for filtering. The upper sidebands are selected by means of filters and then combined to form the final three-channel multiplexed signal. The three-channel FDM *pregroup* signal can be represented by a single triangle, as shown in Fig. 9.3b.

To facilitate interconnection among the different telecommunications systems in use worldwide, the *Comité Consultatif Internationale de Télégraphique et Téléphonique* (CCITT)[1] has recommended a standard modulation plan for FDM (CCITT G322 and G423, 1976). The standard group in the plan consists of 12 voice channels. One way to create such groups is to use an arrangement like that shown in Fig. 9.3a, except of course that 12 multipliers and 12 sideband filters are required. In the standard plan, the lower sidebands are selected by the filters, and the *group* bandwidth extends from 60 to 108 kHz.

---

[1]Since 1994, the CCITT has been reorganized by the International Telecommunications Union (ITU) into a new sector ITU-T.

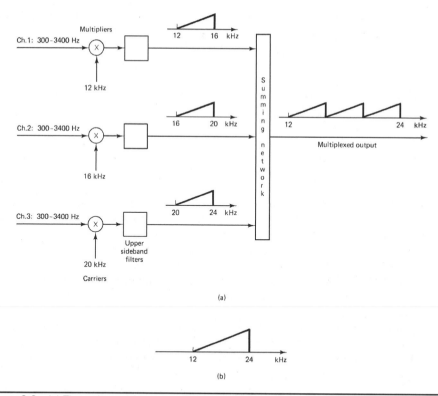

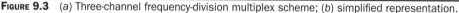

FIGURE 9.3 (a) Three-channel frequency-division multiplex scheme; (b) simplified representation.

As an alternative to forming a 12-channel group directly, the VF channels may be frequency-division multiplexed in threes by using the arrangement shown in Fig. 9.3a. The four 3-channel-multiplexed signals, termed pregroups, are then combined to form the 12-channel group. This approach eases the filtering requirements but does require an additional mixer stage, which adds noise to the process.

The main group designations in the CCITT modulation plan are:

Group. As already described, this consists of 12 VF channels, each occupying a 4-kHz bandwidth in the multiplexed output. The overall bandwidth of a group extends from 60 to 108 kHz.

Supergroup. A supergroup is formed by FDM five groups together. The lower sidebands are combined to form a 60-VF-channel supergroup extending from 312 to 552 kHz.

Basic mastergroup. A basic mastergroup is formed by FDM five supergroups together. The lower sidebands are combined to form a 300-VF-channel basic mastergroup. Allowing for 8-kHz guard bands between sidebands, the basic mastergroup extends from 812 to 2044 kHz.

Super mastergroup. A super mastergroup is formed by FDM three basic mastergroups together. The lower sidebands are combined to form the 900-VF-channel super mastergroup. Allowing for 8-kHz guardbands between sidebands, the super mastergroup extends from 8516 to 12,388 kHz.

In satellite communications, such multiplexed signals often form the baseband signal which is used to frequency modulate a microwave carrier (see Sec. 9.6). The smallest baseband unit is usually the 12-channel VF group, and larger groupings are multiples of this unit. Figure 9.4 shows how 24-, 60-, and 252-VF-channel baseband signals may be formed. These examples are taken from CCITT Recommendations G322 and G423. It can be observed that in each case a group occupies the range 12 to 60 kHz. Because of this, the 60-VF-channel baseband, which modulates the carrier to the satellite, differs somewhat from the standard 60-VF-channel supergroup signal used for terrestrial cable or microwave FDM links.

9.5   Color Television

The baseband signal for television is a composite of the visual information signals and synchronization signals. The visual information is transmitted as three signal components, denoted as the Y, I, and Q signals. The Y signal is a *luminance*, or *intensity*, component and is also the only visual information signal required by monochrome receivers. The I and Q signals are termed *chrominance components*, and together they convey information on the hue or tint and on the amount of saturation of the coloring that is present.

The synchronization signal consists of narrow pulses at the end of each line scan for horizontal synchronization and a sequence of narrower and wider pulses at the end of each field scan for vertical synchronization. Additional synchronization for the color information demodulation in the receiver is superimposed on the horizontal pulses, as described below.

The luminance signal and the synchronization pulses require a base bandwidth of 4.2 MHz for North American standards. The baseband extends down to and includes a dc component. The composite signal containing the luminance and synchronization information is designed to be fully compatible with the requirements of monochrome (black-and-white) receivers.

In transmitting the chrominance information, use is made of the fact that the eye cannot resolve colors to the extent that it can resolve intensity detail; this allows the chrominance signal bandwidth to be less than that of the luminance signal. The I and Q chrominance signals are transmitted within the luminance bandwidth by quadrature DSBSC (as seen later), modulating them onto a subcarrier which places them at the upper end of the luminance signal spectrum. Use is made of the fact that the eye cannot readily perceive the interference which results when the chrominance signals are transmitted within the luminance signal bandwidth. The baseband response is shown in Fig. 9.5.

Different methods of chrominance subcarrier modulation are employed in different countries. In France, a system known as *sequential couleur a mémoire* (SECAM) is used. In most other European countries, a system known as *phase alternation line* (PAL) is used. In North America, the NTSC system is used, where NTSC stands for *National Television System Committee*.

In the NTSC system, each chrominance signal is modulated onto its subcarrier using DSBSC modulation, as described in Sec. 9.3. A single oscillator source is used so that the I and Q signal subcarriers have the same frequency, but one of the subcarriers is shifted 90in phase to preserve the separate chrominance information in the I and Q baseband signals. This method is known as *quadrature modulation* (QM).

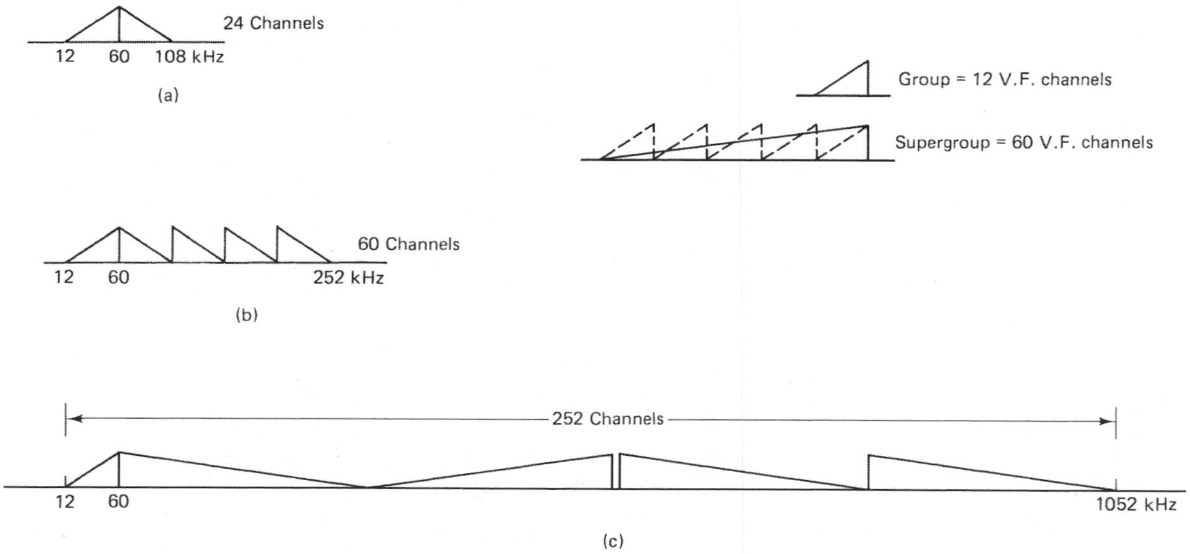

FIGURE 9.4 Examples of baseband signals for FDM telephony: (a) 24 channels; (b) 60 channels; (c) 252 channels.

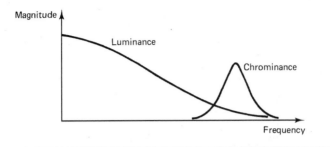

Figure 9.5   Frequency spectra for the luminance and chrominance signals.

The I signal is the chrominance signal that modulates the in-phase carrier. Its bandwidth in the NTSC system is restricted to 1.5 MHz, and after modulation onto the subcarrier, a single-sideband filter removes the upper sideband components more than 0.5 MHz above the carrier. This is referred to as a *vestigial sideband* (VSB). The modulated I signal therefore consists of the 1.5-MHz lower sideband plus the 0.5 MHz upper VSB.

The Q signal is the chrominance signal that modulates the quadrature carrier. Its bandwidth is restricted to 0.5 MHz, and after modulation, a DSBSC signal results. The spectrum magnitude of the combined I and Q signals is shown in Fig. 9.5.

The magnitude of the QM envelope contains the color saturation information, and the phase angle of the QM envelope contains the hue, or tint, information. The chrominance signal subcarrier frequency must be precisely controlled, and in the NTSC system it is held at 3.579545 MHz ± 10 Hz, which places the subcarrier frequency midway between the 227th and the 228th harmonics of the horizontal scanning rate (frequency). The luminance and chrominance signals are both characterized by spectra wherein the power spectral density occurs in groups which are centered about the harmonics of the horizontal scan frequency. Placing the chrominance subcarrier midway between the 227th and 228th horizontal-scan harmonics of the luminance-plus-synchronization signals causes the luminance and the chrominance signals to be interleaved in the spectrum of the composite NTSC signal. This interleaving is most apparent in the range from about 3.0 to 4.1 MHz. The presence of the chrominance signal causes high-frequency modulation of the luminance signal and produces a very fine stationary dot-matrix pattern in the picture areas of high color saturation. To prevent this, most of the cheaper TV receivers limit the luminance channel video bandwidth to about 2.8 to 3.1 MHz. More expensive "high resolution" receivers employ a *comb filter* to remove most of the chrominance signal from the luminance-channel signal while still maintaining about a 4-MHz luminance-channel video bandwidth.

Because the subcarrier is suppressed in the modulation process, a subcarrier frequency and phase reference carrier must be transmitted to allow the I and Q baseband chrominance signals to be demodulated at the receiver. This reference signal is transmitted in the form of bursts of 8 to 11 cycles of the phase-shifted subcarrier, transmitted on the "backporch" of the horizontal blanking pulse. These bursts are transmitted toward the end of each line sync period, part of the line sync pulse being suppressed to accommodate them. One line waveform including the synchronization signals is shown in Fig. 9.6.

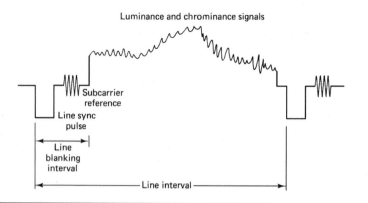

FIGURE 9.6 One line of waveform for a color TV signal.

Figure 9.7 shows in block schematic form the NTSC system. The TV camera contains three separate camera tubes, one for each of the colors red, blue, and green. It is known that colored light can be synthesized by *additive* mixing of red, blue, and green light beams, these being the three primary light beam colors. For example, yellow is obtained by adding red and green light. (This process must be distinguished from the subtractive process of paint pigments, in which the primary pigment colors are red, blue, and yellow.)

Color filters are used in front of each tube to sharpen its response. In principle, it would be possible to transmit the three-color signals and at the receiver reconstruct the color scene from them. However, this is not the best technical approach because such signals would not be compatible with monochrome television and would require extra bandwidth. Instead, three new signals are generated that do provide compatibility and do not require extra bandwidth. These are the luminance signal and the two chrominance signals which have been described already. The process of generating the new signals from the color signals is mathematically equivalent to having three equations in three variables and rearranging these in terms of three new variables which are linear combinations of the original three. The details are shown in the matrix M block of Fig. 9.7, and derivation of the equations from this is left as Prob. 9.9.

At the receiver, the three-color signals can be synthesized from the luminance and chrominance components. Again, this is mathematically equivalent to rearranging the three equations into their original form. The complete video signal is therefore a multiplexed baseband signal which extends from dc up to 4.2 MHz and which contains all the visual information plus synchronization signals.

In conventional TV broadcasting, the aural signal is transmitted by a separate transmitter, as shown in Fig. 9.8a. The aural information is received by stereo microphones, split into (L + R) and (L − R) signals, where L stands for left and R for right. The (L − R) signal is used to DSBSC modulate a subcarrier at $2f_h$ (31.468 kHz). This DSBSC signal is then added to the (L + R) signal and used to frequency modulate a separate transmitter whose rf carrier frequency is 4.5 MHz above the rf carrier frequency of the video transmitter. The outputs of these two transmitters may go to separate antennas or may be combined and fed into a single antenna, as is shown in Fig. 9.8a.

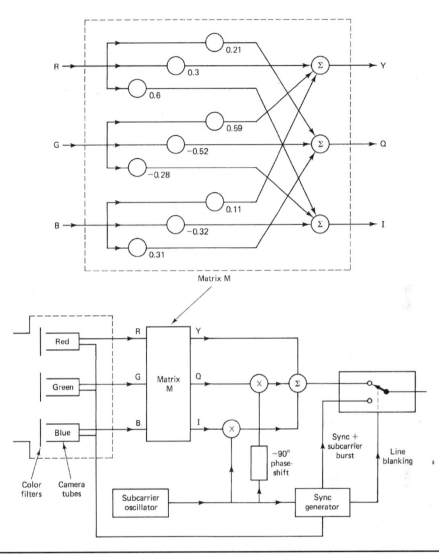

FIGURE 9.7   Generation of NTSC color TV signal. Matrix *M* converts the three-color signals into the luminance and chrominance signals.

The signal format for satellite analog TV differs from that of conventional TV, as shown in Fig. 9.8. To generate the uplink microwave TV signal to a communications satellite transponder channel, the composite video signal (going from 0 Hz to about 4.2 MHz for the North American NTSC standard) is added to two or three *frequency modulation* (FM) carriers at frequencies of 6.2, 6.8, and/or 7.4 MHz, which carry audio information. This composite FDM signal is then, in turn, used to frequency modulate the uplink microwave carrier signal, producing a signal with an rf bandwidth of about 36 MHz. The availability of three possible audio signal carriers permits the transmission of stereo and/or multilingual audio over the satellite link. Figure 9.8*b* shows a block diagram of this system.

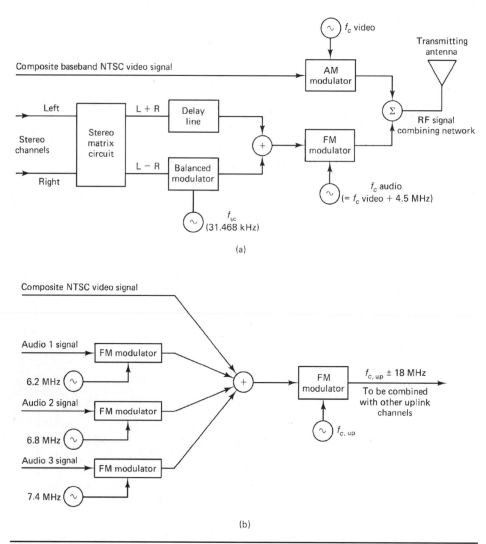

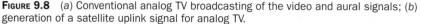

FIGURE 9.8 (a) Conventional analog TV broadcasting of the video and aural signals; (b) generation of a satellite uplink signal for analog TV.

As mentioned previously, three color TV systems—NTSC, PAL, and SECAM—are in widespread use. In addition, different countries use different line frequencies (determined by the frequency of the domestic power supply) and different numbers of lines per scan. Broadcasting between countries utilizing different standards requires the use of a converter. Transmission takes place using the standards of the country originating the broadcast, and conversion to the standards of the receiving country takes place at the receiving station. The conversion may take place through optical image processing or by conversion of the electronic signal format. The latter can be further subdivided into analog and digital techniques. The digital converter, referred to as *digital intercontinental conversion equipment* (DICE), is favored because of its good performance and lower cost (see Miya, 1981).

9.6 Frequency Modulation

The analog signals discussed in the previous sections are transferred to the microwave carrier by means of FM. Instead of being done in one step, as shown in Fig. 9.8*b*, this modulation usually takes place at an intermediate frequency, as shown in Fig. 8.6. This signal is then frequency multiplied up to the required uplink microwave frequency. In the receive branch of Fig. 8.6, the incoming (downlink) FM microwave signal is down-converted to an intermediate frequency, and the baseband signal is recovered from the *intermediate frequency* (IF) carrier in the demodulator. The actual baseband video signal is now available directly via a low-pass filter, but the audio channels must each undergo an additional step of FM demodulation to recover the baseband audio signals.

A major advantage associated with FM is the improvement in the postdetection signal-to-noise ratio at the receiver output compared with other analog modulation methods. This improvement can be attributed to three factors:

1. Amplitude limiting

2. A property of FM that allows an exchange between signal-to-noise ratio and bandwidth

3. A noise reduction inherent in the way noise phase modulates a carrier

These factors are discussed in more detail in the following sections.

Figure 9.9 shows the basic circuit blocks of an FM receiver. The receiver noise, including that from the antenna, can be lumped into one equivalent noise source at the receiver input, as described in Sec. 12.5. It is emphasized at this point that thermal-like noise only is being considered, the main characteristic of which is that the spectral density of the noise power is constant, as given by Eq. (12.15). This is referred to as a *flat spectrum*. (This type of noise is also referred to as *white noise* in analogy to white light, which contains a uniform spectrum of colors.) Both the signal spectrum and the noise spectrum are converted to the intermediate frequency bands, with the bandwidth of the IF stage determining the total noise power at the input to the demodulator. The IF bandwidth must be wide enough to accommodate the FM signal, as described in Sec. 9.6.2, but should be no wider.

9.6.1 Limiters

The total thermal noise referred to the receiver input modulates the incoming carrier in amplitude and in phase. The rf limiter circuit (often referred to as an instantaneous or "hard" limiter) following the IF amplifier removes the amplitude modulation, leaving only the phase-modulation component of the noise. The limiter is an amplifier designed to operate as a class A amplifier for small signals. With large signals, positive excursions are limited by the saturation characteristics of the transistor (which is operated at a low

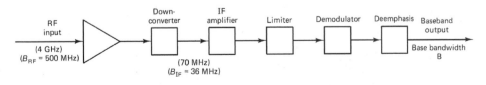

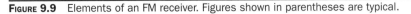

FIGURE 9.9   Elements of an FM receiver. Figures shown in parentheses are typical.

collector voltage), and negative excursions generate a self-bias that drives the transistor into cutoff. Although the signal is severely distorted by this action, a tuned circuit in the output selects the FM carrier and its sidebands from the distorted signal spectrum, and thus the constant amplitude characteristic of the FM signal is restored. This is the amplitude-limiting improvement referred to previously. Only the noise phase modulation contributes to the noise at the output of the demodulator.

Amplitude limiting is also effective in reducing the interference produced by impulse-type noise, such as that generated by certain types of electrical machinery. Noise of this nature may be picked up by the antenna and superimposed as large amplitude excursions on the carrier, which the limiter removes. Limiting also can greatly alleviate the interference caused by other, weaker signals that occur within the IF bandwidth. When the limiter is either saturated or cut off by the larger signal, the weaker signal has no effect. This is known as *limiter capture* (see Young, 1990).

9.6.2 Bandwidth

When considering bandwidth, it should be kept in mind that the word is used in a number of contexts. *Signal bandwidth* is a measure of the frequency spectrum occupied by the signal. *Filter bandwidth* is the frequency range passed by circuit filters. *Channel bandwidth* refers to the overall bandwidth of the transmission channel, which in general included several filters at different stages. In a well-designed system, the channel bandwidth matches the signal bandwidth.

Bandwidth requirements differ at different points in the system. For example, at the receiver inputs for C-band and Ku-band satellite systems, the bandwidth typically is 500 MHz, accommodating 12 transponders as described in Sec. 7.7. The individual transponder bandwidth is typically 36 MHz. In contrast, the baseband bandwidth for a telephony channel is typically 3.1 kHz.

In theory, the spectrum of a frequency-modulated carrier extends to infinity. In a practical satellite system, the bandwidth of the transmitted FM signal is limited by the intermediate-frequency amplifiers. The IF bandwidth, denoted by B_{IF}, must be wide enough to pass all the significant components in the FM signal spectrum that is generated. The required bandwidth is usually estimated by *Carson's rule* as

$$B_{IF} = 2(\Delta F + F_M) \tag{9.1}$$

where ΔF is the peak carrier deviation produced by the modulating baseband signal, and F_M is the highest frequency component in the baseband signal. These maximum values, ΔF and F_M, are specified in the regulations governing the type of service. For example, for commercial FM sound broadcasting in North America, $\Delta F = 75$ kHz and $F_M = 15$ kHz.

The deviation ratio D is defined as the ratio

$$D = \frac{\Delta F}{F_M} \tag{9.2}$$

Example 9.1 A video signal of bandwidth 4.2 MHz is used to frequency modulate a carrier, the deviation ratio being 2.56. Calculate the peak deviation and the signal bandwidth.

Solution

$$D \times F = 2.56 \times 4.2 = \underline{10.752 \text{ MHz}}$$
$$B_{IF} = 2(10.752 + 4.2) = \underline{29.9 \text{ MHz}}$$

A similar ratio, known as the *modulation index*, is defined for sinusoidal modulation. This is usually denoted by β in the literature. Letting Δf represent the peak deviation for sinusoidal modulation and f_m the sinusoidal modulating frequency gives

$$\beta = \frac{\Delta f}{f_m} \tag{9.3}$$

The difference between β and D is that D applies for an arbitrary modulating signal and is the ratio of the maximum permitted values of deviation and baseband frequency, whereas β applies only for sinusoidal modulation (or what is often termed *tone modulation*). Very often the analysis of an FM system is carried out for tone modulation rather than for an arbitrary signal because the mathematics is easier, and the results usually give a good indication of what to expect with an arbitrary signal.

Example 9.2   A test tone of frequency 800 Hz is used to frequency modulate a carrier, the peak deviation being 200 kHz. Calculate the modulation index and the bandwidth.

Solution

$$\beta = \frac{200}{0.8} = 250$$

$$B = 2(200 + 0.8) = 401.6 \text{ kHz}$$

Carson's rule is widely used in practice, even though it tends to give an underestimate of the bandwidth required for deviation ratios in the range $2 < D < 10$, which is the range most often encountered in practice. For this range, a better estimate of bandwidth is given by

$$B_{\text{IF}} = 2(\Delta F + 2F_M) \tag{9.4}$$

Example 9.3   Recalculate the bandwidths for Examples 9.1 and 9.2.

Solution   For the video signal,

$$B_{\text{IF}} = 2(10.75 + 8.4) = \underline{38.3 \text{ MHz}}$$

For the 800 Hz tone:

$$B_{\text{IF}} = 2(200 + 1.6) = \underline{403.2 \text{ kHz}}$$

In Examples 9.1 through 9.3 it can be seen that when the deviation ratio (or modulation index) is large, the bandwidth is determined mainly by the peak deviation and is given by either Eq. (9.1) or Eq. (9.4). However, for the video signal, for which the deviation ratio is relatively low, the two estimates of bandwidth are 29.9 and 38.3 MHz. In practice, the standard bandwidth of a satellite transponder required to handle this signal is 36 MHz.

The peak frequency deviation of an FM signal is proportional to the peak amplitude of the baseband signal. Increasing the peak amplitude results in increased signal power and hence a larger signal-to-noise ratio. At the same time, ΔF, and hence the FM signal bandwidth, increase as shown previously. Although the noise power at the demodulator input is proportional to the IF filter bandwidth, the noise power output after the demodulator is determined by the bandwidth of the baseband filters, and therefore, an increase in IF filter bandwidth does not increase output noise. Thus an improvement in signal-to-noise ratio is possible but at the expense of an increase in the IF bandwidth. This is the large-amplitude signal improvement referred to in Sec. 9.6 and considered further in Sec. 9.6.3.

9.6.3 FM Detector Noise and Processing Gain

At the input to the FM detector, the thermal noise is spread over the IF bandwidth, as shown in Fig. 9.10a. The noise is represented by the system noise temperature T_s, as described in Sec. 12.5. At the input to the detector, the quantity of interest is the carrier-to-noise ratio. Since both the carrier and the noise are amplified equally by the receiver gain following the antenna input, this gain may be ignored in the carrier-to-noise ratio calculation, and the input to the detector represented by the voltage source shown in Fig. 9.10b. The carrier *root-mean-square* (rms) voltage is shown as E_c.

The available carrier power at the input to the FM detector is $E_c^2/4R$, and the available noise power at the FM detector input is kT_sB_N (as explained in Sec. 12.5), so the input carrier-to-noise ratio, denoted by C/N, is

$$\frac{C}{N} = \frac{E_c^2}{4RkT_sB_N} \tag{9.5}$$

When a sinusoidal signal of frequency, f_m, frequency modulates a carrier of frequency, f_c: The instantaneous frequency is given by $f_i = f_c + \Delta f \sin 2\pi f_m t$, where Δf is peak frequency deviation. The output signal power following the FM detector is

$$P_s = A\Delta f^2 \tag{9.6}$$

where A is a constant of the detection process.

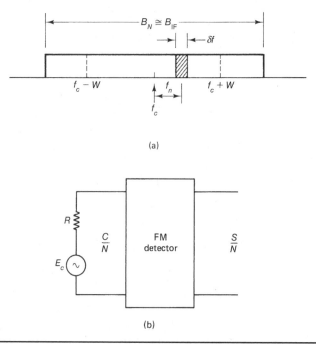

(a)

(b)

Figure 9.10 (a) The predetector noise bandwidth B_N is approximately equal to the IF bandwidth B_{IF}. The LF bandwidth W fixes the equivalent postdetector noise bandwidth at $2W$. δf is an infinitesimally small noise bandwidth. (b) Receiving system, including antenna represented as a voltage source up to the FM detector.

The thermal noise at the output of a bandpass filter, for which $f_c \gg B_N$ has a randomly varying amplitude component and a randomly varying phase component. (It cannot directly frequency modulate the carrier, the frequency of which is determined at the transmitter, which is at a great distance from the receiver and may be crystal controlled.) When the carrier amplitude is very much greater than the noise amplitude the noise amplitude component can be ignored for FM, and the carrier angle as a function of time is $\theta(t) = 2\pi f_c t + \phi_n(t)$, where $\phi_n(t)$ is the noise phase modulation. Now the instantaneous frequency of a phase modulated wave in general is given by $\omega_i = d\theta(t)/dt$ and since $\omega_i = 2\pi f_i$, the equivalent FM resulting from the noise phase modulation is

$$f_{eq.n} = f_c + \frac{1}{2\pi}\frac{d\varphi_n(t)}{dt} \tag{9.7}$$

What this shows is that the output of the FM detector, which responds to equivalent FM, is a function of the time rate of change of the phase change. Now as noted earlier, the available noise power at the input to the detector is $kT_s B_N$ and the noise spectral density, which is the noise power per unit bandwidth just kT_s. A result from Fourier analysis is that the power spectral density of the time derivative of a waveform is $(2\pi f)^2$ times the spectral density of the input. Thus the output spectral density as a function of frequency is $(2\pi f)^2 kT_s$. The variation of output spectral noise density as a function of frequency is sketched in Fig. 9.11a. Since voltage is proportional to the square root of power, the noise voltage spectral density is proportional to frequency as sketched in Fig. 9.11b.

Figure 9.11a shows that the output power spectrum is not a flat function of frequency. The available noise output power in a very small band δf would be given by $(2\pi f)^2 kT_s \delta f$. The total average noise output power would be the sum of all such increments, which is twice the area under the curve of Fig. 9.11a, twice because of the noise contributions from both sides of the carrier. The detailed integration required to evaluate the noise is not carried out here, but the result giving the signal power to noise ratio is

$$\frac{S}{N} = \frac{P_s}{P_n}$$

$$= 1.5\frac{C}{N}\frac{B_N \Delta f^2}{W^3} \tag{9.8}$$

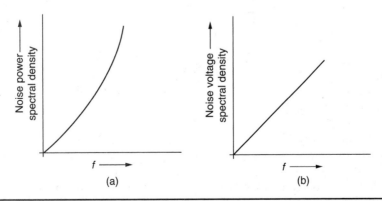

(a)   (b)

Figure 9.11 (a) Output noise power spectral density for FM. (b) The corresponding noise voltage spectral density.

The *processing gain* of the detector is the ratio of signal-to-noise ratio to carrier-to-noise ratio. Denoting this by G_p gives

$$G_P = \frac{S/N}{C/N}$$

$$= \frac{1.5\, B_N \Delta f^2}{W^3} \tag{9.9}$$

Using Carson's rule for the IF bandwidth, $B_{IF} = 2(\Delta f + W)$, and assuming $B_N \approx B_{IF}$, the processing gain for sinusoidal modulation becomes after some simplification

$$G_p = 3(\beta + 1\,)\beta^2 \tag{9.10}$$

Here, $\beta = \Delta f / W$ is the modulation index for a sinusoidal modulation frequency at the highest value of W. Equation (9.10) shows that a high modulation index results in a high processing gain, which means that the signal-to-noise ratio can be increased even though the carrier-to-noise ratio is constant.

9.6.4   Signal-to-Noise Ratio

The term *signal-to-noise ratio* introduced in Sec. 9.6.3 is used to refer to the ratio of signal power to noise power at the receiver output. This ratio is sometimes referred to as the *postdetector* or *destination* signal-to-noise ratio. In general, it differs from the carrier-to-noise ratio at the detector input (the words *detector* and *demodulator* may be used interchangeably), the two ratios being related through the receiver processing gain as shown by Eq. (9.9). Equation (9.9) may be written in decibel form as

$$10\,\log_{10}\frac{S}{N} = 10\,\log_{10}\frac{C}{N} + 10\,\log_{10}G_P \tag{9.11}$$

As indicated in App. G, it is useful to use brackets to denote decibel quantities where these occur frequently. Equation (9.11) therefore may be written as

$$\left[\frac{S}{N}\right] = \left[\frac{C}{N}\right] + [G_P] \tag{9.12}$$

This shows that the signal-to-noise in decibels is proportional to the carrier-to-noise in decibels. However, these equations were developed for the condition that the noise voltage should be much less than the carrier voltage. At low carrier-to-noise ratios this assumption no longer holds, and the detector exhibits a *threshold effect*. This is a threshold level in the carrier-to-noise ratio below which the signal-to-noise ratio degrades very rapidly. The threshold level is shown in Fig. 9.12 and is defined as the carrier-to-noise ratio at which the signal-to-noise ratio is −1 dB below the straight-line plot of Eq. (9.12). For conventional FM detectors (such as the Foster Seeley detector), the threshold level may be taken as 10 dB. *Threshold extension* detector circuits are available which can provide a reduction in the threshold level of between 3 and 7 dB (Fthenakis, 1984).

In normal operation, the operating point is always be above threshold, the difference between the operating carrier-to-noise ratio and the threshold level being referred to as the *threshold margin*. This is also illustrated in Fig. 9.12.

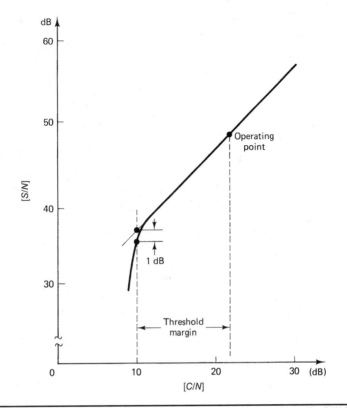

FIGURE 9.12 Output signal-to-noise ratio *S/N* versus input carrier-to-noise ratio *C/N* for a modulating index of 5. The straight-line section is a plot of Eq. (9.12).

Example 9.4 A 1-kHz test tone is used to produce a peak deviation of 5 kHz in an FM system. Given that the received $[C/N] = 30$ dB, calculate the receiver processing gain and the postdetector $[S/N]$.

Solution Since the $[C/N]$ is above threshold, Eq. (9.12) may be used. The modulation index is

$$\beta = 5 \text{ kHz}/1 \text{ kHz} = 5$$

Hence

$$G_p = 3 \times 5^2 \times (5+1) = 450$$

and

$$[G_p] = 26.5 \text{ dB}$$

From Eq. (9.12)

$$[S/N] = 30 + 26.5 = 56.5 \text{ dB}$$

9.6.5 Preemphasis and Deemphasis

As shown in Fig. 9.11*b*, the noise voltage spectral density increases in direct proportion to the demodulated noise frequency. As a result, the signal-to-noise ratio is worse at the high-frequency end of the baseband, a fact which is not apparent from the equation for signal-to-noise ratio, which uses average values of signal and noise power.

For example, if a test tone is used to measure the signal-to-noise ratio in a TV baseband channel, the result depends on the position of the test tone within the baseband, a better result being obtained at lower test tone frequencies. For FDM/FM telephony, the telephone channels at the low end of the FDM baseband would have better signal-to-noise ratios than those at the high end.

To equalize the performance over the baseband, a deemphasis network is introduced after the demodulator to attenuate the high-frequency components of noise. Over most of the baseband, the attenuation-frequency curve of the deemphasis network is the inverse of the rising noise-frequency characteristic shown in Fig. 9.11*b* (for practical reasons it is not feasible to have exact compensation over the complete frequency range). Thus, after deemphasis, the noise-frequency characteristic is flat, as shown in Fig. 9.13*d*. Of course, the deemphasis network also attenuates the signal, and to correct for this, a complementary preemphasis characteristic is introduced prior to the modulator at the transmitter. The overall effect is to leave the postdetection signal levels unchanged while the high-frequency noise is attenuated. The preemphasis, deemphasis sequence is illustrated in Fig. 9.13.

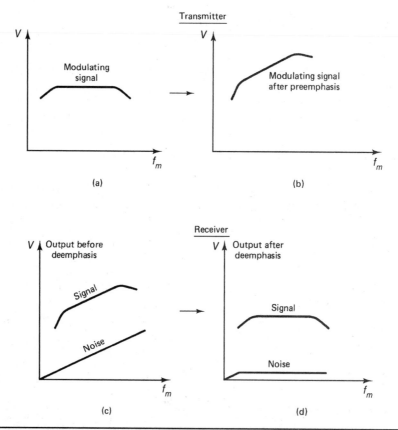

Figure 9.13 (*a* and *b*) Effect of preemphasis on the modulating signal frequency response at the transmitter. (*c* and *d*) Effect of deemphasis on the modulating signal and noise at the receiver output. The deemphasis cancels out the preemphasis for the signal while attenuating the noise at the receiver.

The resulting improvement in the signal-to-noise ratio is referred to variously as *preemphasis improvement*, *deemphasis improvement*, or simply as *emphasis improvement*. It is usually denoted by P, or [P] decibels, and gives the reduction in the total postdetection noise power. Preemphasis curves for FDM/FM telephony are given in CCIR Recommendation 275-2 (1978) and for TV/FM in CCIR Recommendation 405-1 (1982). CCIR values for [P] are 4 dB for the top channel in multichannel telephony, 13.1 dB for 525-line TV, and 13.0 dB for 625-line TV. Considering the emphasis improvement, Eq. (9.12) becomes

$$\left[\frac{S}{N}\right] = \left[\frac{C}{N}\right] + \left[G_P\right] + [P] \tag{9.13}$$

9.6.6 Noise Weighting

Another factor that generally improves the postdetection signal-to-noise ratio is referred to as *noise weighting*. This is the way in which the flat-noise spectrum must be modified to take into account the frequency response of the output device and the subjective effect of noise as perceived by the observer. For example, human hearing is less sensitive to a given noise power density at low and high audio frequencies than at the middle frequency range.

Weighting curves have been established for various telephone handsets in use by different telephone administrations. One of these, the CCIR curve, is referred to as the *psophometric weighting curve*. When this is applied to the flat-noise density spectrum, the noise power is reduced by 2.5 dB for a 3.1-kHz bandwidth (300–3400 Hz) compared with flat noise over the same bandwidth. The weighting improvement factor is denoted by [W], and hence for the CCIR curve [W] = 2.5 dB. (Do not confuse the symbol W used here with that used for bandwidth earlier.) For a bandwidth of b kHz, a simple adjustment gives

$$[W] = 2.5 + 10\log\frac{b}{3.1} = -2.41 + [b] \tag{9.14}$$

Here, b is the *numerical value of kHz* (a dimensionless number). A noise weighting factor also can be applied to TV viewing. The CCIR weighting factors are 11.7 dB for 525-line TV and 11.2 dB for 625-line TV. Taking weighting into account, Eq. (9.13) becomes

$$\left[\frac{S}{N}\right] = \left[\frac{C}{N}\right] + \left[G_P\right] + [P] + [W] \tag{9.15}$$

9.6.7 *S/N and Bandwidth for FDM/FM Telephony*

In the case of FDM/FM, the receiver processing gain, excluding emphasis and noise weighting, is given by Miya (1981) and Halliwell (1974)

$$G_P = \frac{B_{IF}}{b}\left(\frac{\Delta F_{rms}}{f_m}\right)^2 \tag{9.16}$$

Here, f_m is a specified baseband frequency in the channel of interest, at which G_P is to be evaluated. For example, f_m may be the center frequency of a given channel, or it may be the top frequency of the baseband signal. The channel bandwidth is b (usually 3.1 kHz), and ΔF_{rms} is the root-mean-square deviation per channel of the signal. The rms deviation is determined under specified test tone conditions, details of which are found in CCIR Recommendation 404-2 (1982). Some values are shown in Table 9.1.

Maximum Number of Channels	RMS Deviations per Channel (kHz)
12	35
24	35
60	50,100,200
120	50,100,200
300	200
600	200
960	200
1260	140,200
1800	140
2700	140

TABLE 9.1   FDM/FM RMS Deviations

Because ΔF_{rms} is determined for a test tone modulation, the peak deviation for the FDM waveform must consider the waveform shape through a factor g. This is a voltage ratio that is usually expressed in decibels. For a small number of channels, g may be as high as 18.6 dB (Fthenakis, 1984), and typical values range from 10 to 13 dB. For the number of channels n greater than 24, the value of 10 dB is often used. Denoting the decibel value as gdB, then the voltage ratio is obtained from

$$g = 10^{gdB/20} \tag{9.17}$$

The peak deviation also depends on the number of channels. This is taken into account through use of a *loading factor*, L. The relevant CCITT formulas are

$$\text{For } n < 240: 20 \log L = -15 + 10 \log n \tag{9.18}$$
$$\text{For } 12 \le n \le 240: 20 \log L = -1 + 4 \log n \tag{9.19}$$

Once L and g are found, the required peak deviation is obtained from the tabulated rms deviation as

$$\Delta F = g \cdot L \cdot \Delta F_{rms} \tag{9.20}$$

The required IF bandwidth can now be found using Carson's rule, Eq. (9.1), and the processing gain from Eq. (9.16). The following example illustrates the procedure.

Example 9.5   The carrier-to-noise ratio at the input to the demodulator of an FDM/FM receiver is 25 dB. Calculate the signal-to-noise ratio for the top channel in a 24-channel FDM baseband signal, evaluated under test conditions for which Table 9.1 applies. The emphasis improvement is 4 dB, noise weighting improvement is 2.5 dB, and the peak/rms factor is 13.57 dB. The audio channel bandwidth may be taken as 3.1 kHz.

Solution   Given data: $n = 24$; $gdB = 13.57$; $b = 3.1$ kHz; $[P] = 4$; $[W] = 2.5$; $[C/N] = 25$

From Eq. (9.17): $g = 10^{gdB/20} = 4.77$

From Eq. (9.19): $L = 10^{(-1+4\log n)/20} = 1.683$

From Table 9.1, for 24 channels $\Delta F_{\text{rms}} = 35$ kHz, and using Eq. (9.20),

$$\Delta F = g \cdot L \cdot \Delta F_{\text{rms}} \cong 281 \text{ kHz}$$

Assuming that the baseband spectrum is as shown in Fig. 9.4a, the top frequency is

$$f_m = 108 \text{ kHz}$$

and Carson's rule gives $\qquad B_{\text{IF}} = 2(\Delta F + f_m) = 778 \text{ kHz}$

From Eq. (9.16): $\qquad G_P = \dfrac{777.8}{3.1} \left(\dfrac{35}{108}\right)^2 \cong 26.36$

From Eq. (9.15): $\qquad \left[\dfrac{S}{N}\right] = \left[\dfrac{C}{N}\right] + 10 \log G_P + [P] + [W] = 25 + 14.21 + 4 + 2.5 = 45.7 \text{ dB}$

9.6.8 Signal-to-Noise Ratio for TV/FM

Television performance is measured in terms of the postdetector video signal-to-noise ratio, defined as (CCITT Recommendation 567-2, 1986)

$$\left[\frac{S}{N}\right]_V = \frac{\text{peak-to-peak video voltage}}{\text{rms noise voltage}} \qquad (9.21)$$

Because peak-to-peak video voltage is used, $2\Delta F$ replaces ΔF in Eq. (9.8). Also, since power is proportional to voltage squared,

$$\left(\frac{S}{N}\right)_V^2 = 1.5\left(\frac{C}{N}\right)\left(\left(B_N(2\Delta F)^2\right)/W^3\right) \qquad (9.22)$$

where W is the highest video frequency. With the deviation ratio $D = \Delta F/W$, and the processing gain for TV denoted as G_{PV},

$$G_{\text{PV}} = \frac{(S/N)_V^2}{C/N}$$

$$= 12D^2(D+1) \qquad (9.23)$$

Some workers include an implementation margin to allow for nonideal performance of filters and demodulators (Bischof et al., 1981).

With the implementation margin being a power ratio < 1, which is multiplicative that results in decibels denoted by a negative value [IMP], Eq. (9.15) becomes

$$\left[\left(\frac{S}{N}\right)_V^2\right] = \left[\frac{C}{N}\right] + \left[G_{\text{PV}}\right] + [P] + [W] + [\text{IMP}] \qquad (9.24)$$

Recall that the square brackets denote decibels, that is, $[X] = 10 \log_{10} X$. This is illustrated in the following example.

Example 9.6 A satellite TV link is designed to provide a video signal-to-noise ratio of 62 dB. The peak deviation is 9 MHz, and the highest video baseband frequency is 4.2 MHz. Calculate the carrier-to-noise ratio required at the input to the FM detector, given that the combined noise weighting, emphasis improvement, and implementation margin is 11.8 dB.

Solution $\qquad\qquad\qquad\qquad D = \dfrac{9}{4.2} = 2.143$

Equation (9.23) gives:

$$G_{PV} = 12 \times 2.143^2 \times (2.143 + 1) = 173.2$$

Therefore,

$$\left[G_{PV} \right] = 10 \log 173.2 = 22.4 \text{ dB}$$

Since the required signal-to-noise ratio is 62 dB, Eq. (9.24) can be written as

$$62 = \left[\frac{C}{N} \right] + 22.4 + 11.8$$

from which

$$\left[\frac{C}{N} \right] = 27.8 \text{ dB}$$

9.7 Problems and Exercises

9.1. State the frequency limits generally accepted for telephone transmission of speech and typical signal levels encountered in the telephone network.

9.2. Show that when two sinusoids of different frequencies are multiplied together, the resultant product contains sinusoids at the sum and difference frequencies only. Hence show how a multiplier circuit may be used to produce a DSBSC signal.

9.3. Explain how a DSBSC signal differs from a conventional amplitude modulated signal such as used in the medium-wave (broadcast) radio band. Describe one method by which an SSB signal may be obtained from a DSBSC signal.

9.4. Explain what is meant by FDM telephony. Sketch the frequency plans for the CCITT designations of group, supergroup, basic mastergroup, and super mastergroup.

9.5. With the aid of a block schematic, show how 12 VF channels could be frequency-division multiplexed.

9.6. Explain how a 252-VF-channel group is formed for satellite transmission.

9.7. Describe the essential features of the video signal used in the NTSC color TV scheme. How is the system made compatible with monochrome reception?

9.8. Explain how the sound information is added to the video information in a color TV transmission.

9.9. For the matrix network M shown in Fig. 9.7, derive the equations for the Y, Q, and I signals in terms of the input signals R, G, and B.

9.10. Explain what is meant by *frequency modulation*. A 70-MHz carrier is frequency modulated by a 1-kHz tone of 5-V peak amplitude. The frequency deviation constant is 15 kHz/V. Write down the expression for instantaneous frequency.

9.11. An angle-modulated wave may be written as $\sin \theta(t)$, where the argument $\theta(t)$ is a function of the modulating signal. Given that the instantaneous angular frequency is $\omega_i = d\theta(t)/dt$, derive the expression for the FM carrier in Prob. 9.10.

9.12. (*a*) Explain what is meant by *phase modulation*. (*b*) A 70-MHz carrier is phase modulated by a 1-kHz tone of 5-V peak amplitude. The phase modulation constant is 0.1 rad/V. Write down the expression for the argument $\theta(t)$ of the modulated wave.

9.13. Determine the equivalent peak frequency deviation for the phase-modulated signal of Prob. 9.12.

9.14. Show that when a carrier is phase modulated with a sinusoid, the equivalent peak frequency deviation is proportional to the modulating frequency. Explain the significance of this on the output of an FM receiver used to receive the PM wave.

9.15. In the early days of FM it was thought that the bandwidth could be limited to twice the peak deviation irrespective of the modulating frequency. Explain the fallacy behind this reasoning.

9.16. A 10-kHz tone is used to frequency modulate a carrier, the peak deviation being 75 kHz. Use Carson's rule to estimate the bandwidth required.

9.17. A 70-MHz carrier is frequency modulated by a 1-kHz tone of 5-V peak amplitude. The frequency deviation constant is 15 kHz/V. Use Carson's rule to estimate the bandwidth required.

9.18. A 70-MHz carrier is phase modulated by a 1-kHz tone of 5-V peak amplitude. The phase modulation constant is 0.1 rad/V. Find the equivalent peak deviation and, hence, use Carson's rule to estimate the bandwidth required for the PM signal.

9.19. Explain what is meant by *preemphasis* and *deemphasis* and why these are effective in improving signal-to-noise ratio in FM transmission. State typical improvement levels expected for both telephony and TV transmissions.

9.20. Explain what is meant by *noise weighting*. State typical improvement levels in signal-to-noise ratios which result from the introduction of noise weighting for both telephony and TV transmissions.

9.21. Calculate the loading factor L for (*a*) a 12-channel, (*b*) a 120-channel, and (*c*) an 1800-channel FDM/FM telephony signal.

9.22. Calculate the IF bandwidth required for (*a*) a 12-channel, (*b*) a 300-channel, and (*c*) a 960-channel FDM/FM telephony signal. Assume that the peak/rms factor is equal to 10 dB for parts (*a*) and (*b*) and equal to 18 dB for part (*c*).

9.23. Calculate the receiver processing gain for each of the signals given in Prob. 9.22.

9.24. A video signal has a peak deviation of 9 MHz and a video bandwidth of 4.2 MHz. Using Carson's rule, calculate the IF bandwidth required and the receiver processing gain.

9.25. For the video signal of Prob. 9.24, the emphasis improvement figure is 13 dB, and the noise weighting improvement figure is 11.2 dB. Calculate in decibels (*a*) the signal-to-noise power ratio and (*b*) the video signal-to-noise ratio as given by Eq. (9.21). The $[C/N]$ value is 22 dB. Assume a sinusoidal video signal.

References

Bischof, I. J., W. B. Day, R. W. Huck, W. T. Kerr, and N. G. Davies. 1981. "Anik-B Program Delivery Pilot Project. A 12-month Performance Assessment." CRC Report No. 1349, Dept. of Communications, Ottawa, December.

Campanella, S. J. 1983. *Companded Single Sideband (CSSB) AM/FDMA Performance.* Wiley, New York.

CCIR Recommendation 275-2. 1978. "Pre-emphasis Characteristic in Frequency Modulation Radio Relay Systems for Telephony Using Frequency-Division Multiplex." *14th Plenary Assembly*, vol. IX, Kyoto.

CCIR Recommendation 404-2. 1982. "Frequency Division for Analog Radio Relay Systems for Telephony Using Frequency Division Multiplex." *15th Plenary Assembly*, vol. IX, part 1, Geneva.

CCIR Recommendation 405-1. 1982. "Pre-emphasis Characteristics for Frequency Modulation Radio Relay Systems for Television." *15th Plenary Assembly*, vol. IX, part 1, Geneva.

CCITT G423. 1976. "Interconnection at the Baseband Frequencies of Frequency-Division Multiplex Radio-Relay Systems 1, 2." *International Carrier Analog Systems*, vol. III, part 2, Geneva.

CCITT Recommendation 567-2. 1986. "Transmission Performance of Television Circuits Designed for Use in International Circuits." *International Carrier Analog Systems*, vol. XII, Geneva.

CCITT Recommendation G322. 1976. "General Characteristics Recommended for Systems on Symmetric Pair Cables." *International Carrier Analog Systems*, vol. III, part 2, Geneva.

Freeman, Roger L. 1981. *Telecommunications Systems Engineering*. Wiley, New York.

Fthenakis, Emanuel. 1984. *Manual of Satellite Communications*. McGraw-Hill, New York.

Halliwell, B. J. (ed.). 1974. *Advanced Communication Systems*. Newnes-Butterworths, London.

Miya, K. (ed.). 1981. *Satellite Communications Technology*. KDD Engineering and Consulting, Tokyo, Japan.

Young, P. H. 1990. *Electronic Communication Techniques*. Merrill Publishing Company, New York.

CHAPTER 10

Digital Signals

10.1 Introduction

As already mentioned in connection with analog signals, baseband signals are those signals which occupy the lowest, or base, frequency band in the frequency spectrum used by the telecommunications network. A baseband signal may consist of one or more information signals. For example, many telephony signals in digital form may be combined into one baseband signal by the process known as *time-division multiplexing*. Further, analog signals may be converted into digital signals for transmission. Digital signals also originate in the form of computer and other data. In general, a digital signal is a coded version of the original data. In this chapter, the characteristics of the more common types of digital baseband signals are described, along with representative methods of digital modulation.

10.2 Digital Baseband Signals

Digital signals are coded representations of information. Keyboard characters, for example, are usually encoded in binary digital code. A *binary code* has two symbols, usually denoted as *logic-0* and *logic-1*, and these are combined to form binary words to represent the characters. For example, current keyboards use the American Standard Code for Information Interchange (ASCII) extended table. This is based on the Windows-1252 character set, which is the world standard for 8-bit character encoding with 256 characters and symbols. For example, an uppercase letter "A" is the ASCII Table entry # 65 that uses the combination **01000001** to represent this character.

Analog signals such as speech and video may be converted to a digital form through an *analog-to-digital* (A/D) converter. A particular form of A/D conversion, known as *pulse-code modulation*, is described in detail later. Some of digital data sources are illustrated in Fig. 10.1.

In *digital* terminology, a binary symbol is known as a *binit* from *bin*ary dig*it*. The *information* carried by a binit is, in most practical situations, equal to a unit of information known as a *bit*. Thus it has become common practice to refer to binary symbols as bits rather than binits.

The *digital* information is transmitted as a waveform. Some of the more common waveforms used for binary encoding are shown in Fig. 10.2. These are referred to as *digital waveforms*, although strictly speaking they are analog representations of the digital information being transmitted. As an illustration, the binary sequence shown in Fig. 10.2 is **1010111**. Detailed reasons for the use of different waveforms can be found in most books on digital communications (see Lathi and Ding, 2010).

Typical data rates

High speed wireless internet: 1 Gbps

Cell telephone (voice): 128 kbps

Wireless web cam: 10 Mbps

FIGURE 10.1 Examples of binary data sources.

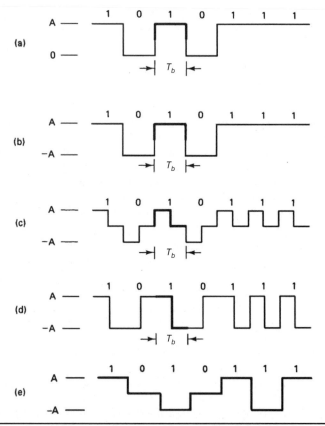

FIGURE 10.2 Examples of binary waveforms used for encoding digital data: (*a*) unipolar NRZ; (*b*) polar NRZ; (*c*) polar RZ; (*d*) split phase or Manchester; (*e*) alternate mark inversion (AMI).

The duration of a bit is referred to as the *bit period* and is shown as T_b. The bit rate is given by

$$R_b = \frac{1}{T_b} \tag{10.1}$$

with T_b in seconds, the bit rate is in bits per second, usually denoted by bps.

Figure 10.2*a* shows a *unipolar* waveform, meaning that the waveform excursions from zero are always in the same direction, either positive or negative, and they are shown as positive *A* in Fig. 10.2*a*. Because it has a dc component, the unipolar waveform is unsuitable for use on analog telephone lines and radio networks, including satellite links.

Figure 10.2*b* shows a *polar* waveform, which utilizes positive and negative polarities. (In Europe this is referred to as a *bipolar* waveform, but the term *bipolar* in North American usage is reserved for a specific waveform, described later.) For a long, random sequence of 1s and 0s, the dc component averages to zero. However, long sequences of the same or like symbols result in a gradual drift in the dc level, which creates problems at the receiver decoder. Also, the decoding process requires knowledge of the bit timing, which is derived from the zero crossovers in the waveform, and these are obviously absent in long strings of like symbols. Both the unipolar and polar waveforms shown in Fig. 10.2*a* and *b* are known as *non-return-to-zero* (NRZ) *waveforms*. This is so because the waveform does *not return* to the zero baseline at any point during the bit period.

Figure 10.2*c* shows an example of a polar *return-to-zero* (RZ) *waveform*. Here, the waveform returns to the zero baseline in the middle of the bit period, so transitions always occur even within a long string of like symbols, which enables bit timing to be extracted. However, even here, dc drift occurs with long strings of like symbols.

In the *split-phase* or *Manchester encoding* shown in Fig. 10.2*d*, a transition between positive and negative levels occurs in the middle of each bit. This ensures that transitions are always be present so that bit timing can be extracted, and because each bit is divided equally between positive and negative levels, there is no dc component.

A comparison of the frequency bandwidths required for digital waveforms can be obtained by considering the waveforms, which alternate at the highest rate between the two extreme levels. For the basic polar NRZ waveform of Fig. 10.2*b*, these appear as square waves, when the sequence is **. . . 101010 . . . .** The periodic time for this square wave is $2T_b$, and the fundamental frequency component is $1/2T_b$. For the split-phase encoding, the square wave with the highest repetition frequency occurs with a long sequence of like symbols such as . . . 1111111 . . . , as shown in Fig. 10.2*d*. The periodic time of this square wave is T_b, and hence the fundamental frequency component is twice that of the basic polar NRZ. Thus, the split-phase encoding requires twice the bandwidth compared with that for the basic polar NRZ, while the bit rate remains unchanged. Therefore, for this case, the utilization of bandwidth, measured in bits per second per Hertz, is less efficient.

An *alternate mark inversion* (AMI) *code* is shown in Fig. 10.2*e*. Here, the binary 0s are at the zero-baseline level, and the binary-1s alternate in polarity. In this way, the dc level is removed, while bit timing can be extracted easily, except when a long string of zeros occurs. Special techniques are available (such as digital scrambling) to counter this last problem. The highest pulse-repetition frequency occurs with a long string of **. . . 111111 . . .** the periodic time of which is $2T_b$, the same as the waveform of Fig. 10.2*b*. The AMI waveform is also referred to as a *bipolar* waveform in North America.

Bandwidth requirements may be reduced by utilizing multilevel digital waveforms, which enables sending multiple bits per symbol waveform. Figure 10.3*a* shows a polar NRZ signal for the sequence **11010010**. By arranging the bits in groups of two, four levels can be used. For example, these may be

11	3A
10	A
01	−A
00	−3A

This is referred to as *quaternary* encoding, and the waveform is shown in Fig. 10.3*b*. The encoding is symmetrical about the zero axis, the spacing between adjacent levels being 2A. Each level represents a *symbol*, the duration of which is the *symbol period*. For the quaternary waveform the symbol period is seen to be equal to twice the bit period, and the symbol rate is

$$R_{sym} = \frac{1}{T_{sym}} \tag{10.2}$$

The *symbol rate* is measured in units of *bauds* (Bd), where 1 Bd is one symbol per second (sps).

The periodic time of the square wave having the greatest symbol repetition frequency is $2T_{sym}$, which is equal to $4T_b$, and hence the bandwidth, compared with the basic binary waveform, is one-half. The bit rate (as distinct from the symbol rate) remains unchanged, and hence the bandwidth utilization is doubled, in terms of bits per second per Hertz.

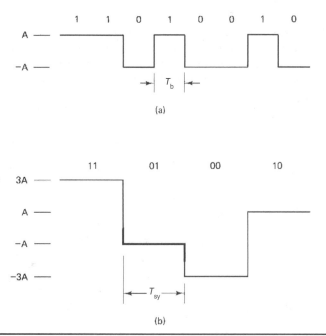

FIGURE 10.3   Encoding of **11010010** in (*a*) binary polar NRZ and (*b*) quaternary polar NRZ.

In general, a waveform may have M-levels (sometimes referred to as an *M-ary wave-form*), where each symbol represents m bits and

$$m = \log_2 M \tag{10.3}$$

The symbol period is therefore

$$T_{sym} = mT_b \tag{10.4}$$

and the symbol rate in terms of bit rate is

$$R_{sym} = \frac{R_b}{m} \tag{10.5}$$

For satellite transmission, the encoded message must be modulated onto the microwave carrier. Before examining the carrier modulation process, we describe the conversion of analog speech signals to a digital format using pulse code modulation.

10.3 Pulse Code Modulation

In the previous section describing baseband digital signals, the information is encoded in one of the digital waveforms shown in Figs. 10.2 and 10.3. However, speech and video begin as analog signals, and these must be converted to digital form prior to transmission over a digital link. In Fig. 10.1, the speech and video analog signals are shown converted to digital form using A/D converters. To promote ease of understanding, we begin with a signal processing technique known as *pulse amplitude modulation* (PAM), which is illustrated in Fig. 10.4, to covert the continuously-valued analog waveform into a sequence of discrete, or quantized, values.

The basis of digital communication of analog signals is the *sampling theorem*, which states that the information content in a continuous analog signal can be captured *without error* as a time series of samples of this signal, provided that two simple conditions are satisfied. First, the spectrum of the analog signal must be bandlimited, i.e., no energy exists beyond a frequency component of f_{max}. This condition can be easily accomplished

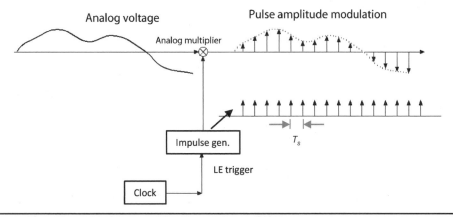

FIGURE 10.4 Example of pulse amplitude modulation (PAM).

by low-pass filtering of the signal before sampling. The second condition is that the *sampling frequency* must be at least twice f_{max}, which is known as the *Nyquist rate*

$$f_s \geq 2f_{max} \text{, Hz} \tag{10.6}$$

The resulting time interval between samples or the *sampling period* is

$$T_s = \frac{1}{f_s} \text{, sec} \tag{10.7}$$

As shown in Fig. 10.4, the samples are generated by the leading edge (rising voltage from negative to positive) of a square wave clock that triggers an impulse generator to issue a narrow pulse. The analog voltage and the impulse are analog multiplied, and the resulting signal is a time series narrow pulses, whose amplitudes follow the original analog signal envelope. The analog multiplication can conceptually be implemented by a simple single pole single throw electronic switch, i.e., the analog sample is created when the switch is closed. This is *pulse amplitude modulation* (PAM), which is the first of three-steps towards digital communication commonly known as *pulse code modulation* (PCM).

Two additional steps are required, namely; quantization and binary coding, which are performed using an analog to digital (A/D) converter (ADC) circuit. To understand the quantization process, a simplified model of a 2-bit linear quantizer is presented in Fig. 10.5, which is a building block of an A/D converter.

On the left size of the figure is a 1 V precision voltage source with a resistor voltage divider network comprising four equal valued resistors in series, which establishes reference voltages of: 0.25 V, 0.50 V and 0.75 V.

In the middle of the figure, there are three-voltage comparator digital circuits, and the inputs, to be compared, are the reference voltages and the input signal voltage (to be quantized). When the input signal voltage < reference voltage, then the module output is logic-0, and when the signal voltage > reference voltage, then the output is logic-1.

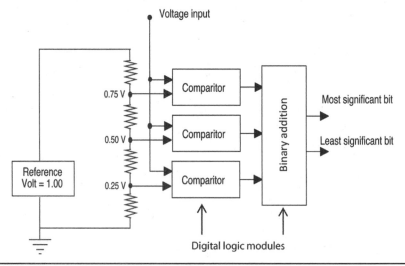

FIGURE 10.5 Example of a two-bit linear quantizer.

All comparator modules have identical transfer functions, except that the reference voltage is different to each.

Finally, on the right side is another digital circuit, which performs the binary addition of the comparator outputs to produce a coded digital word. The number of possible output symbols is a binary number $= 2^n$, where n is the number of bits of quantization. Also note that the number of bits equals the number of comparator modules less 1.

The following table gives the results for a variable input signal voltage between 0 and 1 volt.

Input Signal	Compare-1	Compare-2	Compare-3	Binary Sum
$0.25 \geq V(t) > 0$	0	0	0	00
$0.5 \geq V(t) > 0.25$	1	0	0	01
$0.75 \geq V(t) > 0.5$	1	1	0	10
$1.0 \geq V(t) > 0.75$	1	1	1	11

In the ADC, the signal voltage is quantized to become one of the four possible values, which by convention for a linear A/D equals the midpoint of the separation of the reference voltages or odd integer multiples of the A/D *voltage step* ΔV divided by 2. Thus, the binary word **00** $\Rightarrow 1 \times \Delta V/2 = 0.125$ V; the binary word **01** $\Rightarrow 3 \times \Delta V/2 = 0.125 + 0.25 = 0.375$ V; the binary word **10** $\Rightarrow 5 \times \Delta V/2 = 0.125 + 5.0 = 0.625$; and the binary word **11** $\Rightarrow 7 \times \Delta V/2 = 0.125 + 0.75 = 0.875$ V. Therefore, the maximum error in representing the value of the input signal voltage is ± 0.125 V, which is the A/D voltage step ΔV divided by 2

$$\epsilon_{max} = \pm \frac{\Delta V}{2} = \pm \frac{LSB}{2} \tag{10.8}$$

or $\pm$ (least significant bit value)/2, and this is known as the *quantization error*. Further, once the quantization occurs, the error cannot be removed by reversing the process. On the other hand, by increasing the # bits (n), the quantization error becomes progressively smaller because the binary number of quantization levels is

$$L = 2^n \tag{10.9}$$

Therefore, for digital communication, the number of bits of quantization is selected based upon the maximum allowable error of the analog signal, i.e., the number of bits is increased until the quantization error falls below the specified maximum analog error value.

Now, we return to our discussion of the PAM to PCM conversion using an A/D converter as shown in Fig. 10.6.

Each PAM sample is quantized and coded into a n-bit word, and this is *pulse code modulation*, PCM. Since, the PAM samples are a time series, each sample is converted to a n-bit coded word within the interval of T_s seconds. Afterwards, the n parallel bits are transmitted serially (either most-significant-bit first or least-significant-bit first) using one of the digital waveforms presented in Fig. 10.2. This digitization process is very fast and can occur in less than a few nanoseconds, but the average bit rate of the PCM has increased to n-times the average data sample rate of the PAM. Thus, the bandwidth of the communications channel increases linearly with the # bits of A/D conversion.

Sample data system

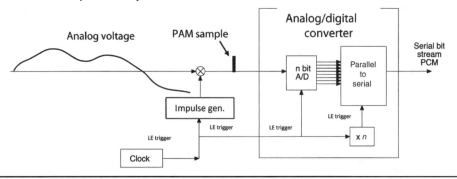

FIGURE 10.6 Example of pulse amplitude modulation (PAM) to pulse code modulation (PCM) conversion using an A/D converter.

In satellite communications, bandwidth is a limited and very important resource, so there is much effort devoted to increasing the digital transmission efficiency, which is measured in bits/Hz.

Commercially available integrated circuits known as PCM *codecs* (for coder-decoder) are used to implement PCM. Figure 10.7a shows a block schematic for the Motorola MC145500 series of codecs suitable for speech signals.

The analog signal enters at the Tx terminals and passes through a low-pass filter, followed by a high-pass filter to remove any 50/60-Hz interference which may appear on the line. The low-pass filter has a cutoff frequency of about 4 kHz, which allows for the filter roll-off above the audio limit of 3400 Hz. This provides a voice channel bandwidth extending from 300 to 3400 Hz that is considered satisfactory for speech quality signals. Also, band limiting the audio signal reduces noise. Finally, the analog signal is digitized by taking samples at the *Nyquist frequency* of 8 kHz, which is twice the upper cutoff frequency of the audio filter at 4 kHz.

Then, following the high-pass filter, the sampled voltage levels are encoded as binary digital numbers in the A/D converter. For practical reasons, speech quantization steps often follow a nonlinear law, with large signals being quantized into coarser steps than small signals. This is termed *compression*, and it is introduced to keep the signal-to-quantization noise ratio reasonably constant over the full dynamic range of the input signal, while maintaining the same number of bits per codeword. At the receiver end (the D/A block in Fig. 10.7a), the binary codewords are automatically decoded into the larger quantized steps for the larger signals, this being termed *expansion*. The expansion law is the inverse of the compression law, and the combined processing is termed *companding*.

Figure 10.7b shows how the MC145500 codec achieves compression by using a *chorded approximation*. The leading bit of the digital codeword is a *sign* bit, being 1 for positive and 0 for negative samples of the analog signal. The next three bits are used to encode the chord, in which the analog signal falls, with the three bits giving a total of eight chords. Each chord is made to cover the same number of input steps, but the step size increases from chord to chord. The chord bits are followed by four bits indicating the step in which the analog value lies. The normalized decision levels shown in Fig. 10.7b are the analog levels at which the comparator circuits change from one chord to the next

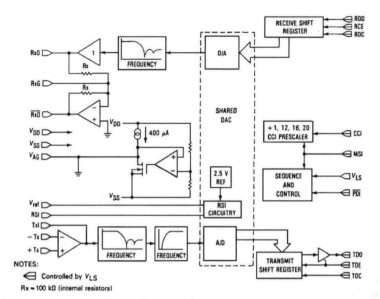

(a)

Chord number	Number of steps	Step size	Normalized encode decision levels	1 Sign	2 Chord	3 Chord	4 Chord	5 Step	6 Step	7 Step	8 Step	Normalized decode levels
			8159	1	0	0	0	0	0	0	0	8031
			7903									
8	16	256	:				:					:
			4319	1	0	0	0	1	1	1	1	4191
			4063									
7	16	128	:				:					:
			2143	1	0	0	1	1	1	1	1	2079
			2015									
6	16	64	:				:					:
			1055	1	0	1	0	1	1	1	1	1023
			991									
5	16	32	:				:					:
			511	1	0	1	1	1	1	1	1	495
			479									
4	16	16	:				:					:
			239	1	1	0	0	1	1	1	1	231
			223									
3	16	8	:				:					:
			103	1	1	0	1	1	1	1	1	99
			96									
2	16	4	:				:					:
			35	1	1	1	0	1	1	1	1	33
			31									
1	15	2	:				:					:
			3	1	1	1	1	1	1	1	0	2
			1									
	1	1	0	1	1	1	1	1	1	1	1	0

NOTES:
1. Characteristics are symmetrical about analog zero with sign bit = 0 for negative analog values.
2. Digital code includes inversion of all magnitude bits.

(b)

FIGURE 10.7 (a) MC145500/01/02/03/05 PCM CODEC/filter monocircuit block diagram. (b) μ-law encode-decode characteristics. (*Courtesy of Motorola, Inc.*)

and from one step to the next. These are normalized to a value 8159 for convenience in presentation. For example, the maximum value may be 8159 mV, and then the smallest step is 1 mV. The first step is shown as 1 (mV), but it should be kept in mind that ideally the first quantized level spans the analog zero so that 0^+ must be distinguished from 0^-. Thus the level representing zero has in fact a step size of ±1 mV.

As an example, suppose the sampled analog signal has a value +500 mV. This falls within the normalized range 479 to 511 mV, and therefore, the binary code is 10111111. It should be mentioned that normally the first step in a chord is encoded 0000, but the bits are inverted, as noted in Fig. 10.7b. This is so because low values are more likely than high values, and inversion increases the 1-bit density, which helps in maintaining synchronization.

The table in Fig. 10.7b shows mu-law encode-decode characteristics. The term *mu law*, usually written as μ-law, originated with older analog compressors, where μ was a parameter in the equation describing the compression characteristic. The μ-law characteristic is standard in North America and Japan, while in Europe and many other parts of the world a similar law known as the *A-law* is in use. Figure 10.8 shows the curves for $\mu = 255$ and $A = 87.6$, which are the standard values in use. These are shown as smooth curves, which are approached with the older analog compression circuits. The chorded approximation approaches these in straight-line segments, or *chords*, for each step.

Because of the similarity of the A-law and μ-law curves, the companding speech quality is similar in both systems, but otherwise the systems are incompatible, and conversion circuitry is required for interconnections such as might occur with international traffic. Of course the transmitting and receiving functions must be configured for the same law.

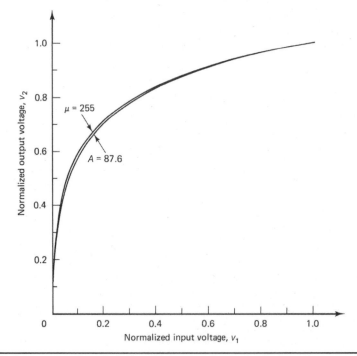

Figure 10.8 Compressor characteristics with input and output voltage scales normalized to the maximum values.

In the receiver, the output from the D/A converter is passed through a low-pass filter that selects the original analog spectrum from the quantized signal. Its characteristics are like those of the low-pass filter used in the transmitter. Apart from the quantization noise (which should be negligible), the final output is a replica of the filtered analog signal at the transmitter.

With a sampling rate of 8 kHz or 8000 samples per second and 8 bits for each sample codeword, the bit rate for a single-channel PCM audio signal is

$$R_b = 8000 \times 8 = 64 \text{ kbps} \tag{10.10}$$

The frequency spectrum occupied by a digital signal is proportional to the bit rate, and to conserve bandwidth, it may be necessary to reduce the bit rate. For example, if 7-bit codewords are used, the bit rate is 56 kbps. Various digital compression schemes are in use which give much greater reductions, and some of these can achieve bit rates as low as 2400 bps with some loss of quality (Hassanein et al., 1989 and 1992).

While the above discussion pertained to speech, this also applies to other forms of analog signals, for example, analog video that is transmitted as PCM.

Example 10.1 A security webcam produces a black & white TV image that comprises 525 scan lines (vertical resolution) by 700 (horizontal) resolution picture elements (pixels). This image is transmitted at the rate of 30 frames (images) per sec using PCM.

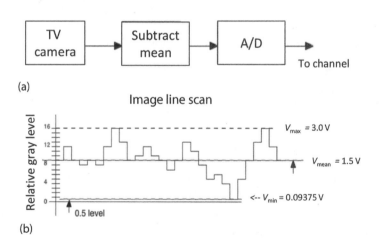

PCM comm system

(a)

Image line scan

(b)

A single scan line output voltage level of the TV camera corresponds to the gray scale intensities of the image. The pixel dynamic range (V_{max}/V_{min}) = 30 dB (brightest white to black) is used to set the number of bits for the linear A/D such that V_{min} equals the value of the LSB.

Before quantization, the signal is passed through differential amplifier to remove the dc level (i.e., the amplifier subtracts the mean value). The resulting output voltage range is ±1.5 V (a white pixel produces ~ +1.5 V and a black pixel produces ~ −1.5 V). Assume that the root mean square signal value is 0.90 V_{rms}.

a. What is the image size in pixels/frame?
b. Calculate # bits of A/D per pixel?
c. What is the PCM data rate in Mbps?

d. What is the maximum quantization error in volts?
e. What is the quantization noise power in Watts?
f. What is the signal-to-quantization noise power ratio (S/N_q)?
g. Using a low-pass filters with a roll-off factor $(\rho) = 0.50$, what is the minimum channel bandwidth required to transmit this TV signal?

Solution For part (a), we calculate the number of pixels in a single image (frame)

$$\# \text{ Pixels} = (\# \text{ horizontal pixels}) \times (\# \text{ vertical lines})$$
$$= 700 \times 525 = 367{,}500, \text{ pixels}$$

For part (b), we calculate the # bits A/D required, based upon the required dynamic range of the image gray scale, which is 30 dB. Since we are dealing with voltages, the conversion to ratio is

$$dB = 10 \log (\text{voltage ratio})^2$$

Therefore, $$\text{Voltage ratio} = \frac{V_{max}}{V_{min}} = \sqrt{10^{dB/10}} = \sqrt{1000} = 31.62$$

Using Eq. (10.10) find the number of levels of quantization that are $\geq$ the above voltage ratio

$$\# \text{Levels}, \ L = 2^n$$

and solving for the minimum # bits gives

$$L = 2^5 = 32 \text{ for 5 bits A/D}$$

For part (c), given the frame-rate = 30 frames/s, and 367,500 pixels/frame, and 5-bits/pixel, we calculate the bit rate as

$$R_b = 30 \times 367{,}500 \times 5 = 55{,}125{,}000 = 55.125 \text{ Mbps}$$

For part (d), using Eq. (10.8), the maximum quantization error value is

$$\epsilon_{max} = \pm \frac{\text{LSB}}{2} = \pm \frac{\Delta V}{2}$$

Assuming that the A/D converter is selected to accommodate instantaneous input signal of $\pm 1.5 \ V_{max}$, which is a peak-to-peak voltage: $V_{pk-pk} = 3$ V, then the value of the least significant bit (ΔV) equals

$$\Delta V = \frac{(V_{pk-pk})}{L} = \frac{3}{32} = 0.09375, \text{ V} \tag{10.11}$$
$$= 93.75 \text{ milliVolts}$$

For part (e), the quantization error is a uniformly distributed random variable x, with a probability distribution function (pdf) $p(x) = 1/\Delta V$, and the mean squared value (quantization noise power) is the second moment of the pdf given as

$$N_q = <x^2> = \int_{-\Delta V/2}^{+\Delta V/2} \frac{x^2}{\Delta V} dx = \frac{\Delta V^2}{12}$$

$$= \frac{(93.75e - 03)^2}{12} = 7.34e\text{-}04, \text{ W} \tag{10.12}$$

For part (f), the signal power = (signal rms value)^2 = $(0.90)^2$ = 0.81, W and the S/N_q is

$$\frac{S}{N_q} = \frac{0.81}{7.34e\text{-}04} = 1105.92 \Rightarrow 30.44 \text{ dB}$$

For part (g) using the low-pass filter given in Fig. 10.13 and Eq. (10.17), the bandwidth is

$$B = \frac{1+\rho}{2} R_b = (1.5)*55{,}125{,}000/2 = 41.344 \text{ MHz}$$

10.4 Time-Division Multiplexing

Many signals in binary digital form can be transmitted through a common channel by interleaving the pulses or groups of pulses in time, this being referred to as *time-division multiplexing* (TDM). To understand TDM consider a simple case (shown in Fig. 10.9) of two PAM signals that are multiplexed in time to create a PAM TDM transmitter. The two analog signals are low pass filtered and Nyquist sampled at a worst-case rate to create PAM signals. When the two LPFs have different cut-off frequencies, then the worst case is to double the highest cut-off frequency. Also, because there are two signals to be multiplexed, the square wave clock frequency (f_x) is two times the highest Nyquist frequency, and the sampling period is $T_x = 1/f_x$.

Conceptually, analog signal sampling is performed using a commutator (single-pole double-throw switch) that is controlled by the digital clock, and the PAM signals are created sequently by multiplying the analog signals with a narrow sampling pulse (triggered by the leading edge of the clock waveform) as shown in Figs. 10.9 and 10.10. The upper two graphs are the input analog signals; $X_1(t)$ and $X_2(t)$ respectively. The third graph is the time series of the sampling pulse, and the $X_1(t)$ and $X_2(t)$ PAM signals are interleaved in the bottom graph to produce the PAM TDM time series. At the receiver end, the signals are separated using a synchronized clock driving an inverse commutator.

Example 10.2 Three telemetry signals (of bandwidth 1 kHz, 2 kHz, and 5 kHz and of magnitude 1 V peak) (assume that the rms voltage is 0.707 V), are transmitted simultaneously using PCM-TDM. The max tolerable error in sample amplitudes is 0.40% of the peak signal amplitude (assume to be "full-scale" of the A/D converter = 1 volt), and the signals are Nyquist sampled. Determine: (a) What is the transmitter clock frequency in Hz? (b) The number of bits A/D?; (c) the data rate?; (d) (S/N_q), power ratio?; and (e) the minimum bandwidth required to transmit this PCM-TDM signal using low-pass filters with roll-off factor (ρ) = 1.0?

Solution First, determine the worse-case Nyquist sampling, which is 2×5 kHz for telemetry channel-3, and for part (a), the clock frequency = # signals × highest Nyquist frequency

$$f_{clock} = 3 * (2 * 5e{+}03) = 30.0 \text{ kHz}$$

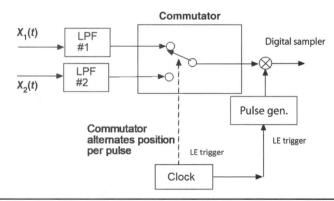

FIGURE 10.9 Example of a PAM TDM transmitter.

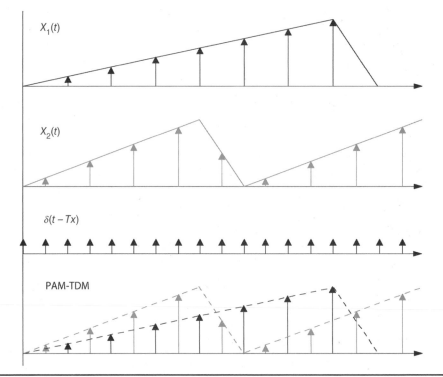

Figure 10.10 Example of PAM TDM multiplexed signals.

For part (b), the A/D converter is selected to accommodate a 2 V peak-to-peak input, and using Eq. (10.11), the voltage step is

$$\Delta V = \frac{(V_{\text{pk-pk}})}{L}$$

the max quantization error is given by Eq. (10.8)

$$\epsilon_{\text{max}} = \pm \frac{\Delta V}{2} \leq \text{Max analog error} = 0.40\% \text{ of } V \text{ peak}$$

$$= \frac{(2/L)}{2} \leq 0.004 * 1.0$$

and solving for L is

$$L \geq \frac{(2/0.004)}{2} = 250, \text{ but } L \text{ must be a binary value} = 256, \text{ so}$$

using Eq. (10.10), the # bits = 8 bits and the A/D $\Delta V = 2/256 = 7.813e - 03$, V
For part (c), the bit rate is

$$R_b = \text{clock freq} * \# \text{bits} = 30.0e{+}03 * 8 = 240.0 \text{ kbps}$$

For part (d), the signal power is

$$S = V_{\text{rms}}^2 = 0.707^2 = 0.50, \text{ W}$$

and from Eq. (10.12) the quantization noise power N_q is

$$N_q = \frac{\Delta V^2}{12} = \frac{(7.813e-03)^2}{12} = 5.09e-06, \text{ W}$$

Therefore the (S/N_q) is

$$(S/N_q) = \frac{0.50}{5.09e-06} = 9.83e+04 \Rightarrow 49.9 \text{ dB}$$

For part (e), the low-pass filter given in Fig. 10.13 and Eq. (10.17), with a roll-off factor $(\rho) = 1.0$, the bandwidth is

$$B = \frac{1+\rho}{2} R_b = (2)*240e+03/2 = 240.0 \text{ kHz}$$

To create a PCM TDM transmitter, an A/D converter produces a n-bit coded word for each PAM sample. Since the PAM signals are interleaved in time, the resulting PCM words are likewise interleaved. At the receiver end, with a synchronized clock, it is a trivial matter to separate the signals. Because the sampling time (T_x) is typically >100 microsec, there can be hundreds of analog signals multiplexed using PCM TDM, but the channel bandwidth required to transmit this signal is directly proportional to the number of signals multiplexed.

For example, for speech signals, a separate codec may be used for each voice channel, and the outputs from these being combined to form a TDM baseband signal, as shown in Fig. 10.11. At the baseband level in the receiver, the TDM signal is demultiplexed, and the PCM signals are routed to separate codecs for decoding. In satellite systems, the TDM waveform is used to modulate the carrier wave, as described later.

The time-division multiplexed signal format is best described with reference to the widely used Bell T1 system that is illustrated in Fig. 10.12a. Each PCM word contains 8 bits, and a T1 *frame* contains 24 PCM channels, also known as a *digital signal level-0* (DS0). In addition, a periodic *frame synchronization* signal is transmitted, and this is achieved by inserting a single bit from the frame synchronizing codeword at the beginning of every frame. At the receiver end, a special detector termed a *correlator* is used to detect the frame synchronizing codeword in the bit stream, which enables the frame timing to be established. The total number of bits in a frame is therefore $24 \times 8 + 1 = 193$. Now, as established earlier, the sampling frequency for voice is 8 kHz, and so the interval between PCM words for a given channel is $1/8000 = 125$ μs. For example, the leading bit in the PCM codewords for a given channel

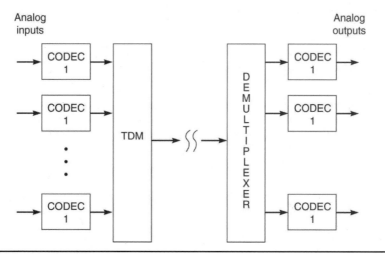

FIGURE 10.11 A basic TDM system.

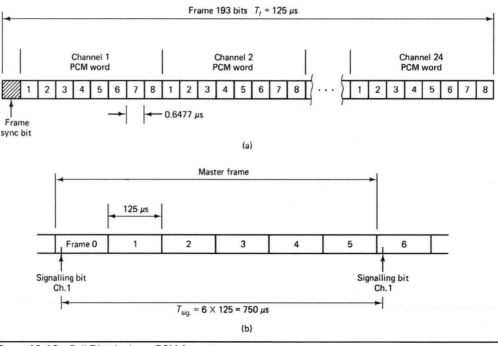

FIGURE 10.12 Bell T1 telephone PCM format.

must be separated in time by no more than 125 μs. As can be seen from Fig. 10.12*a*, this is also the frame period, and therefore, the bit rate for the T1 system is

$$R_b = \frac{193}{125 \times 10^{-6}} = 1.544 \text{ Mbps} \tag{10.13}$$

Signaling information is also carried as part of the digital stream. *Signaling* refers to such data as number dialed, busy signals, and other system information. Signaling can take place at a lower bit rate, and in the T1 system, the eighth bit (LSB) for every channel, in every sixth frame, is replaced by a signaling bit. This is illustrated in Fig. 10.12*b*. The time separation between signaling bits is $6 \times 125 \text{ μs} = 750 \text{ μs}$, and the signaling bit rate is therefore $1/(750 \text{ μs}) = 1.333 \text{ kbps}$.

10.5 Bandwidth Requirements

In a satellite transmission system, the baseband signal is modulated onto a carrier for transmission. Filtering of the signals takes place at several stages. The baseband signal itself is band-limited by filtering to prevent the generation of excessive sidebands in the modulation process. The modulated signal undergoes *bandpass filtering* (BPF) as part of the amplification process in the transmitter.

Where transmission lines form the channel, the frequency response of the lines also must be considered. With a satellite link, the main channel is the radio frequency path, which has little effect on the frequency spectrum, but it does introduce a propagation delay which must be considered.

At the receiver end, bandpass filtering of the incoming signal is necessary to limit the noise which is introduced at this stage. Thus the signal passes through several filtering stages, and the effect of these on the digital waveform must be considered.

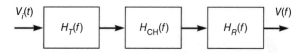

FIGURE 10.13 Frequency spectrum components of Eq. (10.14).

The spectrum of the output pulse at the receiver end is the product of the spectrum of the input pulse $V_i(f)$, the transmit filter response $H_T(f)$, the channel frequency response $H_{CH}(f)$, and the receiver filter response $H_R(f)$, as illustrated in Fig. 10.13, i.e.,

$$V(f) = V_i(f)H_T(f)H_{CH}(f)H_R(f) \qquad (10.14)$$

Inductive and capacitive elements are an inherent part of the filtering process, and the signal energy is periodically cycled between the signal magnetic and electric fields. The time required for this energy exchange results in a distortion of the output voltage waveform. Thus, an input rectangular voltage pulse (logical-1 bit) exhibits "ringing" at the voltage output of the receiver, which is illustrated in Fig. 10.14*a*. The output voltage

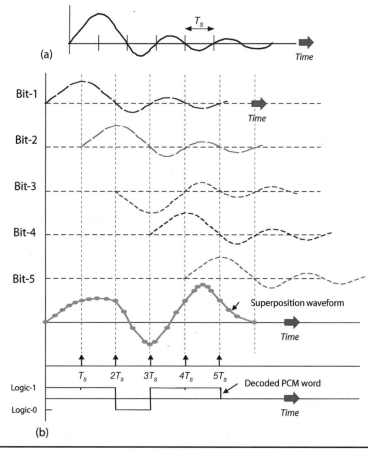

FIGURE 10.14 (*a*) Pulse ringing. (*b*) Sampling to avoid ISI.

waveform experiences a short-delay (not shown), and then it rises to a maximum at the end of the bit period (T_s), which is followed by a damped oscillation with zero crossings at integer multiples of this period.

Because the information is digitally encoded in the waveform, the distortion apparent in the pulse shape is not important provided that the receiver can distinguish the binary-1 pulse from the binary-0 pulse. This requires the waveform to be periodically sampled (period of T_s) at the end of the bit period to determine its polarity (binary-1 is positive and binary-0 is negative). With a continuous serial PCM bit stream, the "tails" which result from the "ringing" of all the preceding pulses can combine to interfere with the pulse being sampled. This is known as *intersymbol interference* (ISI), and it can be severe enough to produce an error in the detected signal polarity.

The ringing cannot be removed, but the pulses can be shaped such that the sampling of a given pulse occurs when all the waveforms are simultaneously at zero crossover points. This is illustrated in Fig. 10.14b, where five-bits in a PCM word are sampled. The waveform for each polar binary bit is identical, except the polarity of the binary-1 and binary-0 are reversed. Also, the next curve with "o" symbols is the composite summation of all the individual waveforms, which occurs at the filter output.

Of course there are unpredictable delays is the signal propagation over the satellite link, so time synchronization is required at the receiver. This is discussed in detail at the end of this chapter. Further, in practice, perfect pulse shaping cannot be achieved, and time synchronization errors do occur, so some ISI occurs; but through careful design, it can be reduced to negligible proportions.

The pulse shaping is carried out by controlling the spectrum of the received pulse as given by Eq. (10.14). One theoretical model for the spectrum is known as the *raised cosine response*, which is shown in Fig. 10.15. Although a theoretical model, it can be approached closely with practical designs. The raised cosine spectrum is described by

$$V(f) = \begin{cases} 1 & \text{for } f < f_1 \\ 0.5\left(1 + \cos\dfrac{\pi(f - f_1)}{B - f_1}\right) & \text{for } f_1 < f < B \\ 0 & \text{for } B < f \end{cases} \qquad (10.15)$$

The frequencies f_1 and B are determined by the symbol rate and a design parameter known as the *roll-off factor*, denoted here by the symbol ρ. The roll-off factor is a specified parameter in the range

$$0 \leq \rho \leq 1 \qquad (10.16)$$

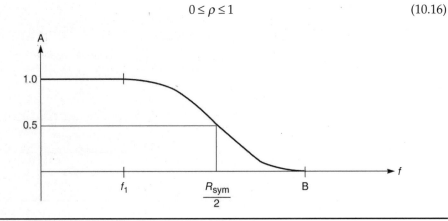

Figure 10.15   The raised cosine low-pass filter response.

In terms of ρ and the symbol rate, the bandwidth B is given by

$$B = \frac{1+\rho}{2} R_{\text{sym}} \tag{10.17}$$

and

$$f_1 = \frac{1-\rho}{2} R_{\text{sym}} \tag{10.18}$$

For binary transmission, the symbol rate simply becomes the bit rate. Thus, for the T1 signal, the required baseband bandwidth is

$$B = \frac{1+\rho}{2} \times 1.544 \times 10^6 = 0.772(1 + \rho) \text{ MHz} \tag{10.19}$$

For a roll-off factor of unity, the bandwidth for the T1 system becomes 1.544 MHz.

A satellite link also requires the same overall baseband frequency response for the avoidance of ISI. Fortunately, because the channel for a satellite link does not introduce frequency distortion, the pulse shaping can take place in the transmit and receive filters at baseband.

The modulation of the baseband signal onto a carrier is discussed in the following section.

10.6 Digital Carrier Systems

For transmission to and from a satellite, the baseband digital signal must be modulated onto a microwave carrier. In general, the digital baseband signals may be multilevel (*M*-ary), requiring multilevel modulation methods. The main binary modulation methods are illustrated in Fig. 10.16. They are defined as follows:

On-off keying (OOK), also known as amplitude-shift keying (ASK). The binary signal in this case is unipolar and is used to switch the carrier on and off.

Frequency-shift keying (FSK). The binary signal is used to frequency modulate the carrier, one frequency being used for a binary 1 and another for a binary 0. These are also referred to as the *mark-space frequencies*.

Binary phase-shift keying (BPSK). Polarity changes in the binary signal are used to produce 180° changes in the carrier phase. This may be achieved using double-side-band, suppressed-carrier modulation (DSBSC), with the binary signal as a polar NRZ waveform. In effect, the carrier amplitude is multiplied by a ±1 pulsed waveform. When the binary signal is +1, the carrier sinusoid is unchanged, and when it is −1, the carrier sinusoid is changed in phase by 180°. BPSK is also known as *phase-reversal keying* (PRK). The binary signal may be filtered at baseband before modulation, to limit the sidebands produced, and as part of the filtering needed for the reduction of ISI, as described in Sec. 10.5.

Differential phase-shift keying (DPSK). This is phase-shift keying in which the phase of the carrier is changed only if the current bit differs from the previous one. A reference bit must be sent at the start of message, but otherwise the method has the advantage of not requiring a reference carrier at the receiver for demodulation.

Quadrature phase-shift keying (QPSK). This is phase-shift keying for a four-symbol waveform, adjacent phase shifts being equispaced by 90°. The concept can be extended to more than four levels, when it is denoted as MPSK for *M-ary phase-shift keying*.

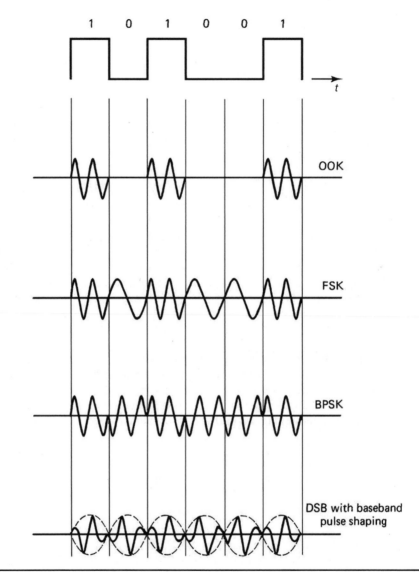

FIGURE 10.16   Some binary digital modulation formats: (*a*) On-off keying, (*b*) Freq-shift keying, (*c*) Binary phase-shift keying, and (*d*) Differential phase-shift keying.

Quadrature amplitude modulation (QAM).   This is also a multilevel (meaning higher than binary) modulation method in which the amplitude and the phase of the carrier are modulated.

Although all the methods mentioned find specific applications in practice, only BPSK and QPSK are described here, since they are widely used and many of the general properties can be illustrated through these methods.

The resulting modulated waveform is sketched in Fig. 10.16.

10.6.1 Binary Phase-Shift Keying

Binary phase-shift keying may be achieved by using the binary polar NRZ signal to multiply the RF carrier, as shown in Fig. 10.17a. It is important to note that this diagram is a simplified version of the transmitter block diagram, which omits the frequency upconversion to the designated microwave carrier frequency and the carrier power amplification before transmission. Since these details do not affect the signal modulation, they are omitted. Thus, for a binary signal $p(t)$, the modulated wave may be written as

$$e(t) = p(t)\cos\omega_0 t \tag{10.20}$$

Note that when $p(t) = +1$, $e(t) = \cos\omega_0 t$, and when $p(t) = -1$, $e(t) = -\cos\omega_0 t$, which is equivalent to $\cos(\omega_0 t \pm 180°)$.

The power spectrum of the modulated carrier has a sinc^2 shape, and bandpass filtering of this signal limits the radiated spectrum (see Sec. 10.6.3). Further, this bandpass filter also may incorporate the square root of the raised-cosine roll-off, described in Sec. 10.5, required to reduce ISI (for example, see Pratt, Bostian, and Allnutt, 2003). Thus, the modulator is basically the same as that used to produce the DSBSC signal described in Sec. 9.3. However, in this instance, the bandpass filter, following the modulator, is used to select the complete DSBSC signal rather than a single sideband.

Next, consider the BPSK receiver given in Fig. 10.17b, which is used to explain the demodulation process. As discussed above, the block diagram is simplified to remove nonessential components for understanding this process, namely; deleting the receiver low noise amplification and frequency down-conversion to an intermediate frequency (IF). At the input, the received signal undergoes additional bandpass filtering to complete the raised-cosine response and to limit input noise.

Next, the filtered modulated wave, $e'(t) = p'(t)\cos\omega_0 t$, is passed to the mixer stage, where it is analog multiplied by a phase-coherent replica of the carrier wave $\cos\omega_0 t$

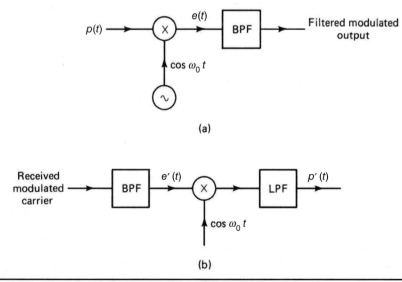

(a)

(b)

FIGURE 10.17 Simplified block diagram of: (a) BPSK transmitter (modulator) and (b) BPSK receiver with coherent detection.

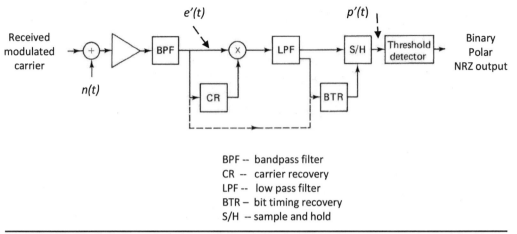

BPF -- bandpass filter
CR -- carrier recovery
LPF -- low pass filter
BTR -- bit timing recovery
S/H -- sample and hold

Figure 10.18 Block diagram of a coherent detector showing the carrier recovery section and the bit timing recovery.

(at the IF frequency). The output from the multiplier is equal to $p'(t)\cos^2 \omega_0 t$, which can be expanded as $p'(t)(0.5 + 0.5\cos 2\omega_0 t)$. Finally, the low-pass filter is used to remove the second harmonic component of the carrier, leaving the baseband output, which is $0.5p'(t)$, where $p'(t)$ is the filtered version of the input binary wave $p(t)$.

The receiver block diagram is expanded to give more details in Fig. 10.18. Note that the local oscillator (shown in Fig. 10.17b) is phased-locked to the input carrier wave using the "CR" block (see Sec. 10.7 for details). Thus, analog multiplication, of the BPSK modulated carrier with the phase-locked (unmodulated) carrier, in the mixer stage provides *coherent detection* (demodulation) of the BPSK signal.

As discussed in Sec. 10.5, to avoid ISI, sampling of the analog output waveform must be performed at the bit rate and precisely at the peaks of the output pulses. This requires that the sample-and-hold (S/H) circuit to be accurately synchronized to the bit rate, which is achieved using the (BTR) block (see Sec. 10.8).

Finally, thermal noise, $n(t)$ at the receiver input, results in unwanted phase modulation of the carrier, and so the demodulated waveform $p'(t)$ is accompanied by noise. The noisy $p'(t)$ signal is passed into the threshold detector that regenerates a noise-free output that may contains some bit errors because of the noise present on the waveform.

The QPSK signal has many features in common with BPSK, which is examined before describing in detail the carrier and bit timing recovery circuits and the effects of noise.

10.6.2   Quadrature Phase-Shift Keying

With QPSK, the binary data are converted into two-bit symbols that are then used to phase modulate the carrier. Since four combinations containing two bits are possible from a binary alphabet (logical 1s and 0s), the carrier phase can be shifted to one of four states.

A simplified block diagram given in Fig. 10.19a shows one-way that QPSK modulation is achieved. The entire digital circuit (within the dashed box) is controlled by the leading edges of a digital clock that is synchronized with the input bit stream. For illustration

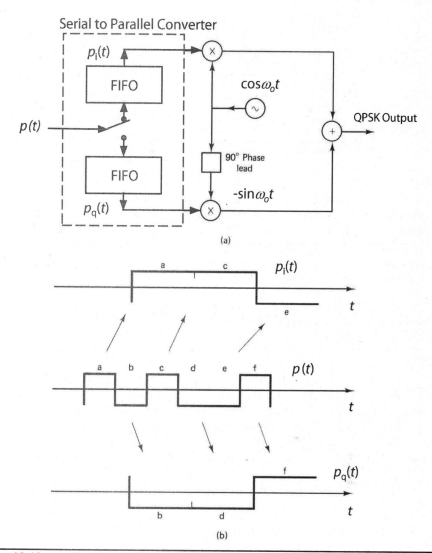

FIGURE 10.19   (a) QPSK modulator; (b) waveforms for (a).

purposes, the first six input bits in the $p(t)$ waveform are labeled a, b, c, d, e, and f as shown in Fig. 10.19b.

The process begins with the sorting of the incoming bit stream $p(t)$ into two binary streams using the serial-to-parallel converter. The first bit-a is connected to the I-FIFO ("first-in, first-out" two-bit storage register), and during the next clock pulse bit-b switches to the Q-FIFO. At the next clock pulse, both bits are moved to the FIFO output ports where the $p_i(t)$ bit stream is combined with a carrier $\cos(\omega_0 t)$ in a BPSK modulator, and the $p_q(t)$ bit stream is combined with a carrier $\sin(\omega_0 t)$, also in a BPSK modulator. At this point, these two BPSK analog waveforms are combined vectorially (using an RF hybrid) to produce the QPSK carrier, and the various combinations are

$p_i(t)$	$p_q(t)$	QPSK Vector Sum
1	1	$\cos \omega_0 t - \sin \omega_0 t = \sqrt{2} \cos(\omega_0 t + 45°)$
1	-1	$\cos \omega_0 t + \sin \omega_0 t = \sqrt{2} \cos(\omega_0 t - 45°)$
-1	1	$-\cos \omega_0 t - \sin \omega_0 t = \sqrt{2} \cos(\omega_0 t + 135°)$
-1	-1	$-\cos \omega_0 t + \sin \omega_0 t = \sqrt{2} \cos(\omega_0 t - 135°)$

TABLE 10.1 QPSK Modulator States (symbols)

shown in Table 10.1. Because it takes two bits to make a QPSF symbol (i.e., two bits are required to produce a given carrier phase), the symbol rate is one-half of the input bit rate.

During the third and fourth clock pulses, the FIFO outputs contain the bit-c and bit-d logic levels, which produce a new BPSK phase for each carrier, and this process repeats periodically thereafter. For the first six bits, the QPSK conversion is illustrated by the waveforms of Fig. 10.19b and Table 10.1.

The QPSK phase-modulation angles are shown in the phasor diagram of Fig. 10.20. Because the output from the I port modulates the carrier directly, it is termed the *in-phase component*, and hence the designation I. The output from the Q port modulates a quadrature carrier, one which is shifted by 90 from the reference carrier, and hence the designation Q. There are four symbols produced identified by the corresponding I,Q logic values, namely 00, 01, 10, and 11.

Because the modulation is carried out at half the bit rate of the incoming data, the bandwidth required by the QPSK signal is exactly half that required by a BPSK signal carrying the same input data rate. This is the advantage of QPSK compared with BPSK modulation; however, to achieve a desired digital bit error rate, QPSK requires twice the signal-to-noise power ratio than does BPSK, which are discussed in Sec. 10.6.4.

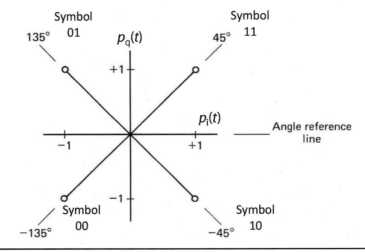

FIGURE 10.20 Phase diagram for QPSK modulation (four symbols).

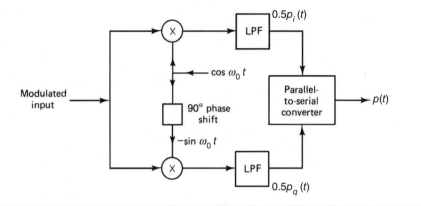

Figure 10.21   Simplified circuit for QPSK demodulation.

Demodulation of the downconverted IF QPSK carrier signal is illustrated using the simplified block diagram of Fig. 10.21. First, the QPSK signal is split using a power divider to simultaneously feed two analog mixers. Each mixer produces an analog multiplication of the QPSK signal with a local oscillator, which is phase coherent with the transmitter carrier. With the incoming carrier represented as $p_i(t)\cos\omega_0 t - p_q(t)\sin\omega_0 t$, after low-pass filtering, the baseband output of the upper (in-phase) BPSK demodulator is $0.5p_i(t)$, and the output of the lower (quadrature-phase) BPSK demodulator is $0.5p_q(t)$. These two baseband signals are binary summed in the parallel-to-serial converter to yield the desired output *2-bit coded symbol*. As with the BPSK signal, noise creates errors in the demodulated output of the QPSK signal.

10.6.3   Transmission Rate and Bandwidth for PSK Modulation

Equation (10.20), which shows the baseband signal $p(t)$ multiplied onto the carrier $\cos\omega_0 t$, is equivalent to double-sideband, suppressed-carrier modulation. The digital modulator circuit of Fig. 10.17a is like the single-sideband modulator circuit shown in Fig. 9.2, the difference being that after the multiplier, the digital modulator requires a bandpass filter, while the analog modulator requires a single-sideband filter. As shown in Fig. 9.1, the DSBSC spectrum extends to twice the highest frequency in the baseband spectrum. For BPSK modulation the latter is given by Eq. (10.17) with R_{sym} replaced with R_b:

$$B_{IF} = 2B = (1+\rho)R_b \tag{10.21}$$

Thus, for BPSK with a roll-off factor of unity, the IF bandwidth in Hertz is equal to twice the bit rate in bits per second.

As shown in the previous section, QPSK is equivalent to the sum of two orthogonal BPSK carriers, each modulated at a rate $R_b/2$, and therefore, the symbol rate is $R_{sym} = R_b/2$. The spectra of the two BPSK modulated waves overlap exactly, but interference is avoided at the receiver because of the coherent detection using quadrature carriers. Equation (10.15) is modified for QPSK to

$$B_{IF} = (1+\rho)R_{sym} = \frac{1+\rho}{2}R_b \tag{10.22}$$

An important characteristic of any digital modulation scheme is the ratio of data bit rate to transmission bandwidth. The units for this ratio are usually quoted as bits per second per Hertz (a dimensionless ratio in fact because it is equivalent to bits per cycle). Note that it is the data bit rate R_b and not the symbol rate R_{sym} which is used to calculate this ratio.

Another important bandwidth, for the receiver input filter, is the equivalent noise bandwidth B_N (see Sec. 12.5), which is equal to the R_{sym} rate for both BPSK and QPSK.

For BPSK, Eq. (10.21) gives an R_b/B_{IF} ratio of $1/(1 + \rho)$, and for QPSK, Eq. (10.22) gives an R_b/B_{IF} ratio of $2/(1 + \rho)$. Thus, QPSK is twice as efficient as BPSK in this respect. However, there is a hardware penalty in that more complex equipment is required to generate and detect the QSPK modulated signal.

10.6.4   Bit Error Rate for PSK Modulation

Referring to Fig. 10.18, the noise at the input to the receiver can cause logic errors in the detected signal. The noise voltage is a zero mean Gaussian random variable, which adds to the signal and causes this *signal plus noise voltage* to fluctuate randomly. If the signal-to-noise voltage ratio is < 4, then the noise excursions may often cause the polarity of the signal plus noise voltage to reverse, and thus the sampled value of signal plus noise may have the opposite polarity to that of the signal alone. This constitutes an error in the logic level of the received pulse. As the signal-to-noise ratio increases, logic errors become less common.

The noise is represented by a noise source at the input of the receiver, shown in Fig. 10.18 (this is discussed in detail in Sec. 12.5), and this noise is filtered by the input bandpass filter. Thus, this filter, in addition to contributing to minimizing the ISI, must maximize the received signal-to-noise ratio. In practice for satellite links (or wireless RF links), this usually can be achieved by making the transmit and receive filters identical, with each having a frequency response which is the square root of the raised-cosine response. Also, using identical filters is an advantage from the point of view of manufacturing.

The most encountered type of noise has a flat frequency spectrum, meaning that the noise power spectrum density, measured in joules (or W/Hz), is constant, and the noise spectrum density is denoted by N_0. When the filtering is designed to maximize the received signal-to-noise ratio, this ratio is found to be equal to $\sqrt{2E_b/N_0}$, where E_b is the average bit energy. The average bit energy can be calculated knowing the average received power P_R and the bit period T_b.

$$E_b = P_R T_b \tag{10.23}$$

The optimum bit decision and the probability of the detector making an error because of additive wideband Gaussian noise (AWGN) is derived in App. H, and for convenience, is given here

$$P_e = \frac{1}{2}\,\text{erfc}\left(\sqrt{\frac{E_b}{N_0}}\right) \tag{10.24}$$

where erfc stands for *complementary error function*, a library function in most engineering computational applications (e.g., Matlab, Mathcad, Python, and Excel, to name a few) or is available in books of mathematical tables. A related function, called the *error function*, denoted by erf(x) is sometimes used, where

$$\text{erfc}(x) = 1 - \text{erf}(x) \tag{10.25}$$

Equation (10.24) applies for polar NRZ baseband signals and for BPSK and QPSK modulation systems. The probability of bit error is also referred to as the *bit error rate* (BER) in units of errors/bit. Often, it is desirable to express bit errors in terms of average time between errors, which we define as *bit-error time* (BET) in units of errors/sec. Given this definition, then the BET is calculated as

$$\text{BET} = P_e * R_b, \text{ errors/s} \tag{10.26}$$

and the average time between errors becomes

$$\langle T_e \rangle = 1/\text{BET, seconds} \tag{10.27}$$

So, the reader is cautioned to determine the desired units associated with the bit errors and then to use the appropriate equation.

Finally, as discussed in App. H, Eq. (10.24) is sometimes expressed in the alternative form

$$P_e = Q\left(\sqrt{\frac{2E_b}{N_0}}\right) \tag{10.28}$$

Here, the $Q(x)$ function is simply another way of expressing the *complementary error function*, and in general

$$\text{erfc}(x) = 2Q(\sqrt{2x}) \tag{10.29}$$

These relationships are given for reference only and are not used further in this text.

The graph of P_e versus $[E_b/N_0]$ in decibels is shown in Fig. 10.22, but note that the energy ratio of E_b/N_0, <u>not the dB value</u>, must be used in Eq. (10.24). This is illustrated in the following example.

Example 10.3 The average power received in a binary polar transmission is 10 mW, and the bit period is 100 μs. If the noise power spectral density is 0.1 μJ, and optimum filtering is used, determine the bit error rate in errors/bit and the average time between bit errors.

Solution From Eq. (10.23): $E_b = 10 \times 10^{-3} \times 100 \times 10^{-6} = 10^{-6}$

and
$$\frac{E_b}{N_0} = \frac{10^{-6}}{10^{-7}} = 10$$

$$\text{erf}(\sqrt{10}) \cong 0.9999923$$

Combining Eqs. (10.24) and (10.25):

$$P_e = 0.5(1 - 0.9999923) = 3.9e - 06, \text{ errors/bit}$$

Next, to calculate the average time between bit errors, we must first calculate the bit rate

$$R_b = \frac{1}{T_b} = \frac{1}{100e\text{-}06} = 10.0 \text{ kbps}$$

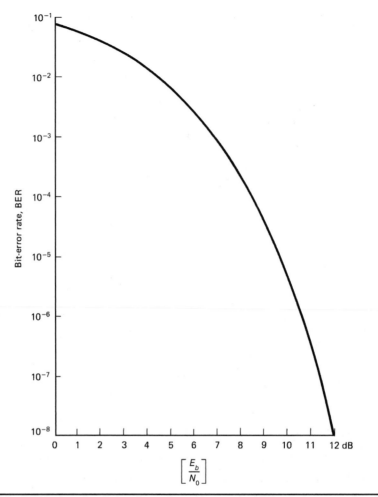

Figure 10.22   BER for baseband signaling using a binary polar NRZ waveform as a function of versus (E_b/N_0).

Using Eq. (10.26) the bit error time is

$$\text{BET} = R_b P_e = 10.0e + 03 * 3.9e - 06 = 3.9e - 02 \text{ errors/s}$$

and using Eq. (10.27) the average time between bit errors is

$$\langle T_e \rangle = \frac{1}{\text{BET}} = \frac{1}{3.9e - 02} = 25.64 \text{ sec}$$

An important parameter for carrier systems is the ratio of the average carrier power to the noise power density, usually denoted by $[C/N_0]$. The $[E_b/N_0]$ and $[C/N_0]$ ratios can be related as follows. The average carrier power at the receiver is P_R W. The energy per symbol is therefore P_R/R_{sym} J, with R_{sym} in symbols per second. Since each symbol contains m bits, the energy per bit is P_R/mR_{sym} J. But $mR_{sym} = R_b$, and therefore, the energy per bit, E_b, is

$$E_b = \frac{P_R}{R_b} \tag{10.30}$$

As before, let N_0 represent the noise power density. Then $E_b/N_0 = P_R/(R_b N_0)$, but P_R/N_0 is the carrier-to-noise density ratio, usually denoted by C/N_0, and therefore,

$$\frac{E_b}{N_0} = \frac{C/N_0}{R_b}$$ (10.31)

Rearranging this and putting it in decibel notation gives

$$\left[\frac{C}{N_0}\right] = \left[\frac{E_b}{N_0}\right] + [R_b]$$ (10.32)

Note that whereas $[E_b/N_0]$ has units of decibels, $[C/N_0]$ has units of dBHz, as explained in App. G.

Example 10.4 The downlink transmission rate in a satellite circuit is 61 Mbps, and to meet the desired P_e (errors/bit), the required $[E_b/N_0]$ at the ground station receiver is 9.5 dB. Calculate the required $[C/N_0]$.

Solution The transmission rate in decibels is $\left[R_b\right] = 10\log(61e + 06) = 77.85$ dB
Hence, from Eq. (10.32)

$$\left[\frac{C}{N_0}\right] = 9.5 + 77.85 = 87.35 \text{ dBHz}$$

The equations giving the probability of bit error are derived on the basis that the filtering provides maximum signal-to-noise ratio. In practice, there are several reasons why the optimal filtering may not be achieved. First, the raised-cosine response is a theoretical model that can only be approximated in practice. Second, for economic reasons, it is desirable to use production filters manufactured to the same specifications for the transmit and receive filter functions, and this may result in some deviation from the desired theoretical response.

In practice, the P_e that is acceptable for a given application is usually known. Thus, the corresponding ratio of bit energy to noise density can then be found from Eq. (10.24) or from Fig. 10.22. Once the theoretical value of E_b/N_0 is found, an *implementation margin*, amounting to a couple of decibels at most, is added to allow for imperfections in the filtering. This is illustrated in the following example.

Example 10.5 A BPSK satellite digital link is required to operate with a bit error rate of no more than 10^{-5} (errors/bit) and if the implementation margin = +2 dB, calculate the required $[E_b/N_0]$ in decibels.

Solution Using an iterative (trial and error) solution, for Eq. (10.24) find the $[E_b/N_0]$ that results in a $P_e \sim 10^{-6}$.
The corresponding value is $[E_b/N_0] = 10.53$ dB. This value is without an implementation margin. The required value, including a 2 dB implementation margin, is

$$[E_b/N_0] = 10.53 + 2 = 12.53 \text{ dB}$$

A table of probability of bit error versus $[E_b/N_0]$ in dB is given in Fig. 10.23.

To summarize, P_e and BER are specified requirements, which enable E_b/N_0 to be determined by using Eq. (10.24) or Fig. 10.22. The rate R_b also is specified, and hence the $[C/N_0]$ ratio can be found by using Eq. (10.32). The $[C/N_0]$ ratio is then used in the link budget calculations, as described in Sec. 12.7.

With purely digital systems, the P_e and BER are directly reflected in bit errors in the data being transmitted. With analog signals that have been converted to digital form through PCM, bit errors contribute to a degraded analog output signal-to-noise ratio, along with the quantization noise, as described in Sec. 10.3. Results showing the contributions of thermal noise and quantization noise to the signal-to-noise output for analog systems derived by (Taub and Schilling, 1986) are given in Fig. 10.24 and Eq. (10.33).

$$\frac{S}{N} = \frac{Q^2}{(1 + 4Q^2 P_e)}$$ (10.33)

where $Q = 2^n$ is the number of quantized steps, and n is the number of bits per sample.

Finally, the BER can be improved using error control coding. Error control coding is the topic of Chap. 11.

Eb/No, dB	Ratio	Pe	Eb/No, dB	Ratio	Pe	Eb/No, dB	Ratio	Pe
6	3.981	2.39E-03	8	6.310	1.91E-04	10	10.000	3.87E-06
6.05	4.027	2.27E-03	8.05	6.383	1.77E-04	10.05	10.116	3.43E-06
6.1	4.074	2.16E-03	8.1	6.457	1.63E-04	10.1	10.233	3.04E-06
6.15	4.121	2.05E-03	8.15	6.531	1.51E-04	10.15	10.351	2.68E-06
6.2	4.169	1.94E-03	8.2	6.607	1.39E-04	10.2	10.471	2.37E-06
6.25	4.217	1.84E-03	8.25	6.683	1.28E-04	10.25	10.593	2.08E-06
6.3	4.266	1.75E-03	8.3	6.761	1.18E-04	10.3	10.715	1.83E-06
6.35	4.315	1.65E-03	8.35	6.839	1.08E-04	10.35	10.839	1.61E-06
6.4	4.365	1.56E-03	8.4	6.918	9.97E-05	10.4	10.965	1.41E-06
6.45	4.416	1.48E-03	8.45	6.998	9.16E-05	10.45	11.092	1.24E-06
6.5	4.467	1.40E-03	8.5	7.079	8.40E-05	10.5	11.220	1.08E-06
6.55	4.519	1.32E-03	8.55	7.161	7.70E-05	10.55	11.350	9.47E-07
6.6	4.571	1.25E-03	8.6	7.244	7.05E-05	10.6	11.482	8.26E-07
6.65	4.624	1.18E-03	8.65	7.328	6.45E-05	10.65	11.614	7.19E-07
6.7	4.677	1.11E-03	8.7	7.413	5.89E-05	10.7	11.749	6.25E-07
6.75	4.732	1.05E-03	8.75	7.499	5.38E-05	10.75	11.885	5.43E-07
6.8	4.786	9.88E-04	8.8	7.586	4.91E-05	10.8	12.023	4.70E-07
6.85	4.842	9.30E-04	8.85	7.674	4.47E-05	10.85	12.162	4.07E-07
6.9	4.898	8.75E-04	8.9	7.762	4.07E-05	10.9	12.303	3.52E-07
6.95	4.955	8.22E-04	8.95	7.852	3.70E-05	10.95	12.445	3.03E-07
7	5.012	7.73E-04	9	7.943	3.36E-05	11	12.589	2.61E-07
7.05	5.070	7.26E-04	9.05	8.035	3.05E-05	11.05	12.735	2.25E-07
7.1	5.129	6.81E-04	9.1	8.128	2.77E-05	11.1	12.882	1.93E-07
7.15	5.188	6.38E-04	9.15	8.222	2.50E-05	11.15	13.032	1.65E-07
7.2	5.248	5.98E-04	9.2	8.318	2.27E-05	11.2	13.183	1.41E-07
7.25	5.309	5.60E-04	9.25	8.414	2.05E-05	11.25	13.335	1.21E-07
7.3	5.370	5.24E-04	9.3	8.511	1.85E-05	11.3	13.490	1.03E-07
7.35	5.433	4.90E-04	9.35	8.610	1.66E-05	11.35	13.646	8.75E-08
7.4	5.495	4.58E-04	9.4	8.710	1.50E-05	11.4	13.804	7.43E-08
7.45	5.559	4.27E-04	9.45	8.810	1.35E-05	11.45	13.964	6.30E-08
7.5	5.623	3.99E-04	9.5	8.913	1.21E-05	11.5	14.125	5.33E-08
7.55	5.689	3.72E-04	9.55	9.016	1.09E-05	11.55	14.289	4.50E-08
7.6	5.754	3.46E-04	9.6	9.120	9.74E-06	11.6	14.454	3.79E-08
7.65	5.821	3.22E-04	9.65	9.226	8.71E-06	11.65	14.622	3.19E-08
7.7	5.888	3.00E-04	9.7	9.333	7.79E-06	11.7	14.791	2.68E-08
7.75	5.957	2.79E-04	9.75	9.441	6.96E-06	11.75	14.962	2.25E-08
7.8	6.026	2.59E-04	9.8	9.550	6.20E-06	11.8	15.136	1.88E-08
7.85	6.095	2.40E-04	9.85	9.661	5.52E-06	11.85	15.311	1.57E-08
7.9	6.166	2.23E-04	9.9	9.772	4.91E-06	11.9	15.488	1.31E-08
7.95	6.237	2.06E-04	9.95	9.886	4.36E-06	11.95	15.668	1.09E-08

FIGURE 10.23   BER Table for selected $[E_b/N_0]$.

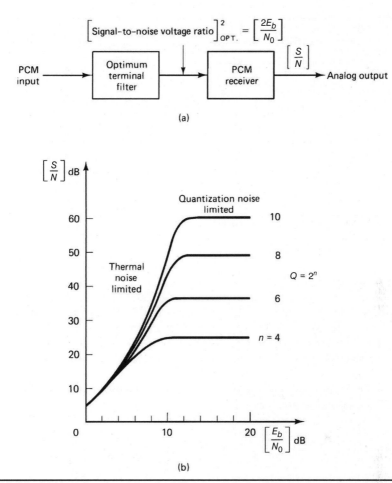

FIGURE 10.24 (a) Use of optimum terminal filter to maximize the signal-to-noise voltage ratio; (b) plot of Eq. (10.33).

10.7 Carrier Recovery Circuits

To implement coherent detection for PSK digital modulation, a *local oscillator* (LO), that is phase coherent with the transmitter carrier, must be provided at the receiver. As shown in Sec. 10.6.1, a BPSK signal is a *double sideband suppressed carrier* (DSBSC) type of signal, and therefore, the carrier is not directly available in the BPSK signal. However, the carrier can be recovered using a *phase-locked loop* (PLL), as shown in Fig. 10.25, provided that the digital PSK modulation can be removed beforehand.

For a BPSK signal, the phase modulation is removed by passing the signal through a 2× frequency multiplier. This frequency-doubler is a nonlinear circuit, which squares the input signal in the time domain. So squaring the BPSK modulated carrier given by Eq. (10.20) yields

$$e^2(t) = (p(t)\cos\omega_0 t)^2 = p^2(t)\cos^2\omega_0 t = p^2(t)\left(\frac{1}{2} + \frac{1}{2}\cos 2\omega_0 t\right) \qquad (10.34)$$

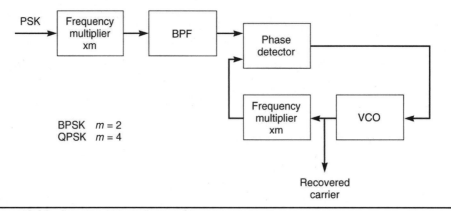

FIGURE **10.25** Functional block diagram for carrier recovery.

Note that in Eq. (10.20), the term $p(t)$ is equal to ± 1, which causes the BPSK modulation of $\pm 180°$. After squaring, $p^2(t) = +1$, and the carrier phase become modulo(360°) that removes the digital modulation phase changes. Since the squared signal is now the second harmonic frequency, the PLL incorporates a second frequency-doubler to produce the second harmonic frequency from the VCO. Thus, the VCO operating at the carrier frequency becomes the phase-coherent LO required to demodulate the BPSK carrier.

With QPSK, the signal can be represented by the formulas given in Table 10.1, which may be written generally as

$$e(t) = \sqrt{2} \cos\left(\omega_0 t \pm \frac{n\pi}{4}\right) \qquad (10.35)$$

Quadrupling the QPSK carrier removes the $\pm 90°$ and $\pm 135°$ phase modulation, and followed by some trigonometric simplification, results in

$$e^4(t) = \frac{3}{2} + 2\cos 2\left(\omega_0 t \pm \frac{n\pi}{4}\right) + \frac{1}{2}\cos 4\left(\omega_0 t \pm \frac{n\pi}{4}\right) \qquad (10.36)$$

The last term on the right-hand side is selected by the PLL bandpass filter and is

$$\frac{1}{2}\cos 4\left(\omega_0 t \pm \frac{n\pi}{4}\right) = \frac{1}{2}\cos(4\omega_0 t \pm n\pi) \qquad (10.37)$$

The fourth harmonic frequency of the carrier, including a constant-phase term that can be ignored. The fourth harmonic is selected by the bandpass filter, and the operation of the circuit proceeds in a similar manner to that for the BPSK signal.

Frequency multiplication can be avoided by use of a method known as the *Costas loop*. Details of this, along with an analysis of the effects of noise on the squaring loop and the Costas loop methods, can be found in Gagliardi (1991). Other methods are also described in detail in Franks (1980).

10.8   Bit Timing Recovery

Accurate bit timing is needed at the receiver to precisely sample the received waveform at the optimal points. In the most common arrangements, the clocking signal is recovered from the demodulated waveform, these being known as *self-clocking* or *self-synchronizing systems*. Where the waveform has a high density of zero crossings, a zero-crossing detector can be used to recover the clocking signal.

In practice, the received waveform is often badly distorted by the frequency response of the transmission link and by noise, and the design of the bit timing recovery circuit is quite complicated. In most instances, the spectrum of the received waveform does not contain a discrete component at the clock frequency. However, it can be shown that a periodic component at the clocking frequency is present in the squared waveform for digital signals (unless the received pulses are exactly rectangular, in which case squaring simply produces a dc level for a binary waveform).

A commonly used baseband scheme is shown in block schematic form in Fig. 10.26 (Franks, 1980). The filters A and B form part of the normal signal filtering (e.g., raised-cosine filtering). The signal for the bit timing recovery is tapped from the junction between A and B and passed along a separate branch which consists of a filter, a squaring circuit, and a bandpass filter which is sharply tuned to the clock frequency component present in the spectrum of the squared signal. This is then used to synchronize the clocking circuit, the output of which clocks the sampler in the detector branch.

The *early-late gate circuit* provides a method of recovering bit timing, which does not rely on a clocking component in the spectrum of the received waveform. The circuit utilizes a feedback loop in which the magnitude changes in the outputs from matched filters control the frequency of a local clocking circuit (for an elementary description see, for example, Roddy and Coolen, 1995). Detailed analyses of these and other methods can be found in Franks (1980) and Gagliardi (1991), as well as other standard digital communication textbooks.

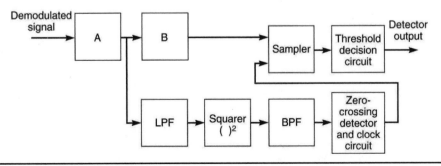

FIGURE 10.26   Functional block diagram for bit-timing recovery.

10.9   Problems and Exercises

10.1.   For a test pattern consisting of alternating binary-1s and -0s, determine the frequency spectra in terms of the bit period T_b for the following signal formats: (a) unipolar; (b) polar NRZ; (c) polar RZ; (d) Manchester.

10.2.   Plot the raised-cosine frequency response Eq. (10.15), for a bit rate of 1 bps and a roll-off factor of 1, for a symbol rate equal to the bit rate. Use the inverse Fourier transform to determine the shape of the pulse time waveform.

10.3.   Plot the compressor transfer characteristics for $\mu = 100$ and $A = 100$. The μ-law compression characteristic is given by

$$v_0 = \text{sign}(v_i)\frac{\ln(1+\mu|v_i|)}{\ln(1+\mu)}$$

where v_0 is the output voltage normalized to the maximum output voltage, and v_i is the input voltage normalized to the maximum input voltage. The A-law characteristic is given by

$$v_0 = \text{sign}(v_i)\frac{A|v_i|}{1+\ln A} \text{ for } 0 \le |v_i| \le \frac{1}{A}$$

and

$$v_0 = \text{sign}(v_i)\frac{1+\ln(A|v_i|)}{1+\ln A} \text{ for } \frac{1}{A} \le |v_i| \le 1$$

10.4.   Derive the expander transfer characteristics corresponding to the compressor characteristics given in Prob. 10.3.

10.5.   If the normalized levels shown in Fig. 10.7b represent millivolts, provide the digitally encoded words for input levels of: (a) ±90 mV; (b) ±100 mV; (c) ±190 mV; (d) ±3000 mV.

10.6.   Determine the decoded output voltage levels for the input levels given in Prob. 10.5. Also determine the quantization error in each case.

10.7.   (a) A test tone having the full peak-to-peak range is applied to a PCM system. If the number of bits per sample is 8, determine the quantization S/N_q. Assume uniform sampling of step size ΔV, for which the mean square noise voltage is $\Delta V^2/12$. (b) Given that a raised-cosine filter is used with $\rho = 1$, determine the bandwidth expansion factor B/W, where B is the PCM bandwidth and W is the upper cutoff frequency of the input.

10.8.   A PCM signal uses the polar NRZ format. Following optimal filtering, the $[E_b/N_0]$ at the input to the receiver decision detector is 10 dB. Determine the probability of bit error rate (BER) at the output of the decision detector.

10.9.   Using Eq. (10.24), calculate the probability of bit error for $[E_b/N_0]$ values of (a) 6 dB, (b) 12 dB, (c) 24 dB and (d) 48 dB.

10.10.   A PCM system uses 8 bits per sample and polar NRZ transmission. Determine the output $[S/N]$ for $[E_b/N_0]$ values of (a) 6 dB, (b) 12 dB, (c) 24 dB and (d) 48 dB. at the input to the decision detector.

10.11.   A binary periodic waveform of period $3T_b$ is low-pass filtered before being applied to a BPSK modulator. The low-pass filter cuts off at $B = 0.5/T_b$. Derive the trigonometric expansion for the modulated wave, showing that only sideband frequencies and no carrier are present. Given that the bit period is 100 ms and the carrier frequency is 100 kHz, sketch the spectrum, showing the frequencies to scale.

10.12. Explain what is meant by *coherent detection* as used for the demodulation of PSK bandpass signals. An envelope detector is an example of a *noncoherent detector*. Can such a detector be used for BPSK? Give reasons for your answer.

10.13. Explain how a QPSK signal can be represented by two BPSK signals. Show that the bandwidth required for QPSK signal is one-half that required for a BPSK signal operating at the same data rate.

10.14. The input data rate on a satellite circuit is 1.544 Mbps. Calculate the bandwidths required for BPSK modulation and for QPSK modulation, given that raised-cosine filtering is used with a roll-off factor of 0.2 in each case.

10.15. A QPSK system operates at a $[E_b/N_0]$ ratio of 8 dB. Determine the probability of bit error. If the bit rate is 1 kbps, find the average time between bit errors?

10.16. A BPSK system operates at a $[E_b/N_0]$ ratio of 16 dB. Determine the probability of bit error. If the bit rate is 10 Mbps, find the average time between bit errors?

10.17. The received power in a satellite digital communications link is 0.5 pW. The carrier is BPSK modulated at a bit rate of 1.544 Mbps. If the noise power density at the receiver is 0.5 e-19 J, determine the probability of bit error (BER) in units of errors/bit, the BET in units of errors/s, and the average time between errors in units of sec?

10.18. The received $[C/N_0]$ ratio in a digital satellite communications link is 86.5 dBHz, and the data bit rate is 50 Mbps. Calculate the $[E_b/N_0]$ ratio, the P_e (errors/bit) and the BET (errors/s) for the link.

10.19. For the link specified in Prob. 10.18, the $[C/N_0]$ ratio is improved to 87.5 dBHz. Determine the new P_e (errors/bit) and the BET (errors/s).

References

Franks, L. E. 1980. "Carrier and Bit Synchronization in Data Communication: A Tutorial Review." *IEEE Transactions On Communications*, vol. 28, no. 8, August, pp. 1107–1120.

Gagliardi, R. M. 1991. *Satellite Communications*, 2d ed. Van Nostrand Reinhold, New York.

Hassanein, H., A. B. Amour, and K. Bryden. 1992. "A Hybrid Multiband Excitation Coder for Low Bit Rates." Department of Communications, Communications Research Centre, Ottawa, Ontario, Canada.

Hassanein, H., A. B. Amour, K. Bryden, and R. Deguire. 1989. "Implementation of a 4800 bps Code-Excited Linear Predictive Coder on a Single TMS320C25 Chip." Department of Communications, Communications Research Centre, Ottawa, Ontario, Canada.

Lathi, B. P., and Z. Ding, 2010. *Modern Digital and Analog Communication Systems*, 4th ed. Oxford Univ. Press, New York.

Pratt, T., C. Bostian, and J Allnutt. 2003. *Satellite Communications*, 2nd ed. Wiley, New York.

Roddy, D., and J. Coolen. 1994. *Electronic Communications*, 4th ed. Prentice-Hall, Englewood Cliffs, NJ.

Taub, H., and D. L. Schilling. 1986. *Principles of Communications Systems*, 2nd ed. McGraw-Hill, New York.

CHAPTER 11

Error Control Coding

11.1   Introduction

As shown by Fig. 10.22, the probability of bit error (P_e) in a digital transmission can be reduced by increasing $[E_b/N_0]$, but there are practical limits to this approach. Equation (10.32) shows that for a given bit rate R_b, $[E_b/N_0]$ is directly proportional to $[C/N_0]$. An increase in $[C/N_0]$ can be achieved by increasing transmitted power and/or reducing the system noise temperature (to reduce N_0). Both these measures are limited by cost and, in the case of the onboard satellite equipment, size. In practical terms, a probability of bit error (P_e of Eq. 10.24) of about 10^{-4}, which is satisfactory for voice transmissions, can be achieved with off-the-shelf equipment. For lower P_e values such as required for some data, error control coding must be used. Error control performs two functions, error detection and error correction. Most codes can perform both functions, but not necessarily together. In general, a code is capable of detecting more errors than it can correct. Where error detection only is employed, the receiver can request a repeat transmission (a technique referred to as *automatic repeat request*, or ARQ). This is only of limited use in satellite communications because of the long transmission delay time associated with geostationary satellites, and of course radio and TV broadcast is essentially one-way so ARQ cannot be employed. What is termed *forward error correction* (FEC) allows errors to be corrected without the need for retransmission, but this is more difficult and costly to implement than ARQ.

A P_e value of 10^{-4} represents an average error rate of 1 bit in 10^4, and the error performance is sometimes specified as the *bit error rate* (BER). It should be recognized, however, that the probability of bit error P_e occurs as a result of noise at the input to the receiver, while the BER is the actual error rate at the output of the detector. When error control coding is employed, the distinction between P_e and BER becomes important. P_e is still determined by conditions at the input, but the error control will, if properly implemented, make the probability of bit error at the output (the BER) less than that at the input. Error control coding applies only to digital signals, and in most cases the signal is in binary form, where the message symbols are bits, or logic-1s and -0s.

Encoding refers to the process of adding coding bits to the uncoded bit stream, and *decoding* refers to the process of recovering the original (uncoded) bit stream from the coded bit stream. Both processes are usually combined in one unit termed a *codec*.

11.2   Linear Block Codes

A block code requires that the message stream be partitioned into blocks of bits (considering only binary messages at this stage). Let each block contain k bits, and let these k bits define a dataword. The number of datawords is 2^k. There is no redundancy in the

291

system, meaning that even a single bit error in transmission would convert one dataword into another, which of course would constitute an error.

The datawords can be encoded into codewords that consist of n bits, where the additional $n - k$ bits are derived from the message bits but are not part of the message. The number of possible codewords is 2^n, but only 2^k of these will contain datawords, and these are the ones that are transmitted. It follows that the rest of the codewords are redundant, but only in the sense that they do not contribute to the message. (The $n - k$ additional bits are referred to as *parity check bits*). If errors occur in transmission, there is high probability that they will convert the permissible codewords into one or another of the redundant words that the decoder at the receiver is designed to recognize as an error. It will be noted that the term *high probability* is used. There is always the possibility, however remote, that enough errors occur to transform a transmitted codeword into another legitimate codeword in error.

The code *rate r_c* is defined as the ratio of dataword bits to codeword bits (note that although it is called a *rate*, it is not a rate in bits per second)

$$r_c = \frac{k}{n} \tag{11.1}$$

The code is denoted by (n, k) for example a code that converts a 4-bit dataword into a 7-bit codeword would be a $(7, 4)$ code.

A repetition code illustrates some of the general properties of block codes. In a repetition code, each bit is considered to be a dataword, in effect, $k = 1$. For n-redundancy encoding, the output of the encoder is n bits, identical to the input bit. As an example, consider the situation when $n = 3$. A binary 1 at the input to the encoder results in a 111 codeword at the output, and a binary 0 at the input results in a 000 codeword at the output. At the receiver, the logic circuits in the decoder produce a 1 when 111 is present at the input and a 0 when 000 is present. It is assumed that synchronization is maintained between encoder and decoder. If a codeword other than 111 or 000 is received, the decoder detects an error and can request a retransmission (ARQ).

FEC can take place on the basis of a "majority vote." In this case, the logic circuits in the decoder are designed to produce a 1 at the output whenever two or three 1s occur in the received codeword (codewords 111, 101, 011, and 110) and a 0 whenever two or three 0s appear in the codeword (codewords 000, 001, 010, and 100). An odd number of "repeats" is used to avoid a tied vote.

Errors can still get through if the noise results in two or three successive errors in a codeword. For example, if the noise changes a 111 into a 000 or a 000 into a 111, the output will be in error whether error detection or FEC is used. If two errors occur in a codeword, then the "majority vote" method for FEC will result in an error. However, the probability of two or three consecutive errors occurring is very much less than the probability of a single error. This assumes that the bit energy stays the same, an aspect that is discussed in Sec. 11.7. Codes that are more efficient than repetitive encoding are generally used in practice.

As a matter of definition, a code is termed *linear* when any linear combination of two codewords is also a codeword. For binary codewords in particular, the linear operation encountered is modulo-2 addition. The rules for modulo-2 addition are:

$$0 + 0 = 0 \qquad 0 + 1 = 1 \qquad 1 + 0 = 1 \qquad 1 + 1 = 0$$

Modulo-2 addition is easily implemented in hardware using the exclusive-OR (XOR) digital circuit, which is one of its main advantages. All codes encountered in practice are linear, which has a bearing on the theoretical development (see Proakis and Salehi, 1994).

To illustrate this further consider the eight possible datawords formed from a 3-bit sequence. One parity bit will be attached to each dataword as shown in Table 11.1. The parity bits are selected to provide *even parity*, that is, the number of 1s in any codeword is even (including the all-zero codeword). The parity bits are found by performing a modulo-2 addition on the dataword.

With even parity it is seen that the bits in the codeword modulo-2 sum to zero. It would be possible to use *odd parity* where the parity bit is chosen so that the modulo-2 sum of the codeword is 1, however this would exclude the all-zero codeword. A linear code must include the all-zero codeword, and hence even parity is used.

The *Hamming distance* between two codewords is defined as the number of positions by which the two codewords differ. Thus the codewords 0000 and 1111 differ in four positions, so their Hamming distance is four. The *minimum Hamming distance*, usually just referred to as the *minimum distance* is the smallest Hamming distance between any two codewords. It can be shown that the minimum distance is given by the minimum number of binary 1s in any codeword, excluding the all-zero codeword. By inspection it will be seen that the minimum distance of the code in Table 11.1 is two. The greater the minimum distance the better the code, as this reduces the chances of one codeword being converted to another by noise.

The properties of linear block codes are best formulated in terms of matrices. Only a summary of some of these results are presented here, as background to aid in the understanding of coding methods used in satellite communications. A dataword (or message block) of size k is denoted by a row vector $\mathbf{d}$, for example the sixth dataword in Table 11.1 is $\mathbf{d}_6 = [101]$. Denoting the codeword by row vector $\mathbf{c}$, the corresponding codeword is $\mathbf{c}_6 = [1010]$. In general, the codeword is generated from the dataword by use of a *generator matrix* denoted by $\mathbf{G}$, where

$$\mathbf{c} = \mathbf{dG} \qquad (11.2)$$

Dataword	Modulo-2 Addition of Dataword	Codeword
000	0	0000
001	1	0011
010	1	0101
011	0	0110
100	1	1001
101	0	1010
110	0	1100
111	1	1111

TABLE **11.1** Even Parity Codewords

Design of the generator matrix forms part of coding practice and will not be gone into here. However an example will illustrate the properties. One example of a generator matrix for a (7, 4) code is

$$G = \begin{bmatrix} 1 & 0 & 0 & 0 & 1 & 1 & 1 \\ 0 & 1 & 0 & 0 & 1 & 1 & 0 \\ 0 & 0 & 1 & 0 & 1 & 0 & 1 \\ 0 & 0 & 0 & 1 & 0 & 1 & 1 \end{bmatrix} \qquad (11.3)$$

It will be noted that the matrix has 7 columns and 4 rows corresponding to the (7, 4) code, and furthermore, the first four columns form an *identity submatrix*. The identity submatrix results in the dataword appearing as the first four bits of the codeword, in this example. In general, a *systematic code* contains a sequence that is the dataword, and the most common arrangement is to have the dataword at the start of the codeword as shown in the example. It can be shown that any linear block code can be put into systematic form. The remaining bits in any row of **G** are responsible for generating the parity bits from the data bits. As an example, suppose it is required to generate a codeword for a dataword [1010]. This is done by multiplying **d** by **G**

$$C = \begin{bmatrix} 1 & 0 & 1 & 0 \end{bmatrix} \begin{bmatrix} 1 & 0 & 0 & 0 & 1 & 1 & 1 \\ 0 & 1 & 0 & 0 & 1 & 1 & 0 \\ 0 & 0 & 1 & 0 & 1 & 0 & 1 \\ 0 & 0 & 0 & 1 & 0 & 1 & 1 \end{bmatrix}$$

$$= \begin{bmatrix} 1 & 0 & 1 & 0 & 0 & 1 & 0 \end{bmatrix}$$

The dataword is seen to appear as the first four bits in the codeword, and the end three bits are the parity bits. The parity bits are generated from the data bits by means of the last three columns in the generator matrix. This submatrix is denoted by **P**:

$$P = \begin{bmatrix} 1 & 1 & 1 \\ 1 & 1 & 0 \\ 1 & 0 & 1 \\ 0 & 1 & 1 \end{bmatrix} \qquad (11.4)$$

The *transpose* of **P**, which enters into the decoding process is formed by interchanging rows with columns, that is, row 1 becomes column 1, and column 1 becomes row 1, row 2 becomes column 2 and column 2 becomes row 2, and so on. In full, the transpose of **P**, written as **P**^{T} is

$$P^T = \begin{bmatrix} 1 & 1 & 1 & 0 \\ 1 & 1 & 0 & 1 \\ 1 & 0 & 1 & 1 \end{bmatrix}$$

What is termed the *parity check matrix* (denoted by **H**) is now formed by appending an identity matrix to **P**^{T}:

$$H = \begin{bmatrix} 1 & 1 & 1 & 0 & 1 & 0 & 0 \\ 1 & 1 & 0 & 1 & 0 & 1 & 0 \\ 1 & 0 & 1 & 1 & 0 & 0 & 1 \end{bmatrix} \qquad (11.5)$$

The number of rows in **H** is equal to the number of parity bits, $n - k$, and the number of columns is n, that is the parity check matrix is a $(n - k, n)$ matrix. A fundamental property of these code matrices is that

$$\mathbf{GH}^\mathrm{T} = \mathbf{0} \tag{11.6}$$

When a codeword is received it can be verified as being correct on multiplying it by $\mathbf{H}^\mathrm{T}$. The product $\mathbf{cH}^\mathrm{T}$ should be equal to zero. This follows since $\mathbf{c} = \mathbf{dG}$ and therefore $\mathbf{cH}^\mathrm{T} = \mathbf{dGH}^\mathrm{T} = \mathbf{0}$. If a result other than zero is obtained, then an error has been detected. In general terms, the product $\mathbf{cH}^\mathrm{T}$ gives what is termed the *syndrome* and denoting this by **s**:

$$\mathbf{s} = \mathbf{cH}^\mathrm{T} \tag{11.7}$$

A received codeword can be represented by the transmitted codeword plus a possible error vector. For example if the transmitted codeword is [1010010] and the received codeword is [1010110] the error is in the fifth bit position from the left and this can be written as

$$[1010110] = [1010010] + [0000100]$$

More generally, if $\mathbf{c}_\mathrm{R}$ is the received codeword, $\mathbf{c}_\mathrm{T}$ the transmitted codeword and **e** the error vector then, with modulo-2 addition

$$\mathbf{c}_\mathrm{R} = \mathbf{c}_\mathrm{T} + \mathbf{e} \tag{11.8}$$

Substituting $\mathbf{c}_\mathrm{R}$ for **c** in Eq. (11.7) gives

$$\mathbf{s} = (\mathbf{c}_\mathrm{T} + \mathbf{e})\mathbf{H}^\mathrm{T}$$
$$= \mathbf{c}_\mathrm{T}\mathbf{H}^\mathrm{T} + \mathbf{eH}^\mathrm{T}$$

But as shown earlier, the product $\mathbf{cH}^\mathrm{T}$, which is $\mathbf{c}_\mathrm{T}\mathbf{H}^\mathrm{T}$ in this notation, is zero, hence,

$$\mathbf{s} = \mathbf{eH}^\mathrm{T} \tag{11.9}$$

This shows that the syndrome depends only on the error vector and is independent of the codeword transmitted. Since the error vector has the same number of bits n as the codeword there will be 2^n possible error vectors. Not all of these can be detected since the syndrome has only $n - k$ bits (determined by the number of rows in the **H** matrix), giving as 2^{n-k} the number of different syndromes. One of these will be the all zero syndrome, and hence the number of errors that can be detected is just $2^{n-k} - 1$. In practice the decoder is designed to correct the most likely errors, for example those with only 1-bit error. The received syndrome may be compared with values in a lookup table (tabulated against known error patterns), and the most likely match found. This is termed maximum likelihood decoding.

The Hamming codes described in the next section enable a single error to be corrected, and in fact the syndrome gives the bit position of the error. Suppose for example the codeword [1010010] as previously determined is transmitted but a bit error occurs in the fifth bit from the left, so that the received codeword is [1010110]. Applying Eq. (11.7)

$$s = [1\ 0\ 1\ 0\ 1\ 1\ 0] \begin{bmatrix} 1 & 1 & 1 \\ 1 & 1 & 0 \\ 1 & 0 & 1 \\ 1 & 1 & 1 \\ 1 & 0 & 0 \\ 0 & 1 & 0 \\ 0 & 0 & 1 \end{bmatrix}$$

$$= [1\ 0\ 0]$$

The fact that the syndrome is not zero indicates that an error has occurred. For the case of Hamming codes discussed in Sec. 11.3, and of which this is an example, the syndrome also shows which bit is in error. The syndrome [100] corresponds to the fifth column (from the left) of the parity check matrix **H** and this indicates that it is the fifth bit that is in error. In general, with the Hamming code, if the syndrome corresponds to column m then bit m is the one in error, and this can be "flipped" to the correct value.

11.3 Cyclic Codes

Cyclic codes are a subclass of linear block codes. They possess the property that a cyclic shift of a codeword is also a codeword. For example, if a codeword consists of the elements $\{c_1\ c_2\ c_3\ c_4\ c_5\ c_6\ c_7\}$, then $\{c_2\ c_3\ c_4\ c_5\ c_6\ c_7\ c_1\}$ is also a codeword. The advantage of cyclic codes is that they are easily implemented in practice through the use of shift registers and modulo-2 adders. Cyclic codes are widely used in satellite transmission, and the properties of the most important of these are summarized in the following sections. Only certain combinations of k and n are permitted in these codes. As pointed out in Taub and Schilling (1986), a code is an invention, and these codes are named after their inventors.

11.3.1 Hamming Codes

For an integer $m \geq 2$, the k and n values are related as $n = 2^m - 1$ and $k = n - m$. Thus some of the permissible combinations are shown in Table 11.2:

m	n	k
2	3	1
3	7	4
4	15	11
5	31	26
6	63	57
7	127	120

TABLE 11.2   *m, n, k* for some Hamming Codes

It will be seen that the code rate $r_c = k/n$ approaches unity as m increases, which leads to more efficient coding. However, only a single error can be corrected with Hamming codes.

11.3.2 BCH Codes

BCH stands for the names of the inventors, Bose, Chaudhuri, and Hocquenghen. These codes can correct up to t errors, and where m is any positive integer, the permissible values are $n = 2^m - 1$ and $k \geq n - mt$. The integers m and t are arbitrary, which gives the code designer considerable flexibility in choice. Proakis and Salehi (1994) give an extensive listing of the parameters for BCH codes, from which the values in Table 11.3 have been obtained. As usual, the code rate is $r_c = k/n$.

11.3.3 Reed-Solomon Codes

The codes described so far work well with errors that occur randomly rather than in bursts. However, there are situations where errors do occur in bursts; that is, a number of bits that are close together may experience errors as a result of impulse-type noise or impulse-type interference. *Reed-Solomon* (R-S) codes are designed to correct errors under these conditions. Instead of encoding directly in bits, the bits are grouped into symbols, and the datawords and codewords are encoded in these symbols. Errors affecting a group of bits are most likely to affect only one symbol that can be corrected by the R-S code.

Let the number of bits per symbol be k; then the number of possible symbols is $q = 2^k$ (referred to as a *q-ary alphabet*). Let K be the number of symbols in a dataword and N be the number of symbols in a codeword. Just as in the block code where k-bit datawords were mapped into n-bit codewords, with the R-S code, datawords of K symbols are mapped into codewords of N symbols. The additional $N - K$ redundant symbols are derived from the message symbols but are not part of the message. The number of possible codewords is 2^N, but only 2^K of these will contain datawords, and these are the ones that are transmitted. It follows that the rest of the codewords are redundant, but only in the sense that they do not contribute to the message. If errors occur

n	k	t
7	4	1
15	11	1
15	7	2
15	5	3
31	26	1
31	21	2
31	16	3
31	11	5
31	6	7

TABLE 11.3 Some Parameters for BCH Codes

in transmission, there is high probability that they will convert the permissible code-words into one or another of the redundant words that the decoder at the receiver is designed to recognize as an error. It will be noted that the term *high probability* is used. There is always the possibility, however remote, that enough errors occur to transform a transmitted codeword into another legitimate codeword even though this was not the one transmitted.

It will be observed that the wording of the preceding paragraph parallels that given in Sec. 11.2 on block codes, except that here the coding is carried out on symbols. Some of the design rules for the R-S codes are

$$q = 2^k$$

$$N = q - 1$$

$$2t = N - K$$

Here, t is the number of symbol errors that can be corrected. A simple example will be given to illustrate these. Let $k = 2$; then $q = 4$, and these four symbols may be labeled A, B, C, and D. In terms of the binary symbols (bits) for this simple case, we could have $A = 00$, $B = 01$, $C = 10$, and $D = 11$. One could imagine the binary numbers 00, 01, 10, and 11 being stored in memory locations labeled A, B, C, and D.

The number of symbols per codeword is $N = q - 1 = 3$. Suppose that $t = 1$; then the rule $2t = N - K$ gives $K = 1$; that is, there will be one symbol per dataword. Hence the number of datawords is $q^K = 4$ (i.e., A, B, C, or D), and the number of codewords is $q^N = 4^3 = 64$. These will include permissible words of the form AP_1P_2, BP_3P_4, CP_5P_6, and DP_7P_8, where P_1P_2, and so on are the parity symbols selected by the encoding rules from the symbol alphabet A, B, C, and D. This process is illustrated in Fig. 11.1.

At the decoder, these are the only words that are recognized as being legitimate and can be decoded. The other possible codewords not formed by the rules but which may be formed by transmission errors will be detected as errors and corrected. It will be observed that a codeword consists of 6 bits, and one or more of these in error will result in a symbol error. The R-S code is capable of correcting this symbol error, which in this simple illustration means that a burst of up to 6 bit errors can be corrected.

R-S codes do not provide efficient error correction where the errors are randomly distributed as distinct from occurring in bursts (Taub and Schilling, 1986). To deal with this situation, codes may be joined together or concatenated, one providing for random error correction and one for burst error correction. Concatenated codes are described in Sec. 11.6. It should be noted that although the encoder and decoder in R-S codes operate at the symbol level, the signal may be transmitted as a bit stream, but it is also suitable for transmission with multilevel modulation, the levels being determined by the symbols. The code rate is $r_c = K/N$, and the code is denoted by (N, K). In practice, it is often the case that the symbols are bytes consisting of 8 bits; then $q = 2^8 = 256$, and $N = q - 1 = 255$. With $t = 8$, a NASA-standard (255, 239) R-S code results.

Shortened R-S codes employ values $N' = N - l$ and $K' = K - l$ and are denoted as (N', K'). For example, DirecTV (see Chap. 16) utilizes a shortened R-S code for which $l = 109$, and *digital video broadcast* (DVB) utilizes one for which $l = 51$ (Mead, 2000). These codes are designed to correct up to $t = 8$ symbol errors.

data bits	01	00	11	00	01	11	01	01	11	01	10	10
datawords	B	A	D	A	B	D	B	B	D	B	C	C
codewords	BP_3P_4	AP_1P_2	DP_7P_8	AP_1P_2	BP_3P_4	DP_7P_8	BP_3P_4	BP_3P_4	DP_7P_8	BP_3P_4	CP_5P_6	CP_5P_6

Figure 11.1 Symbol (Reed-Solomon) encoding.

11.4 Convolution Codes

Convolution codes are also linear codes. A convolution encoder consists of a shift register which provides temporary storage and a shifting operation for the input bits and exclusive-OR logic circuits that generate the coded output from the bits currently held in the shift register.

In general, k data bits may be shifted into the register at once, and n code bits generated. In practice, it is often the case that $k = 1$ and $n = 2$, giving rise to a rate $1/2$ code. A rate $1/2$ encoder is illustrated in Fig. 11.2, and this will be used to explain the encoding operation.

Initially, the shift register holds all binary 0s. The input data bits are fed in continuously at a bit rate R_b, and the shift register is clocked at this rate. As the input moves through the register, the rightmost bit is shifted out so that there are always 3 bits held in the register. At the end of the message, three binary 0s are attached, to return the shift register to its initial condition. The commutator at the output is switched at twice the input bit rate so that two output bits are generated for each input bit shifted in. At any one time the register holds 3 bits that form the inputs to the exclusive-OR circuits.

Figure 11.3 is a tree diagram showing the changes in the shift register as input is moved in, with the corresponding output shown in parentheses. At the initial condition, the register stores 000, and the output is 00. If the first message bit in is a 1, the lower branch is followed, and the output is seen to be 11. Continuing with this example, suppose that the next three input bits are 001; then the corresponding output is 01 11 11. In other words, for an input 1001 (shown shaded in Fig. 11.3), the output, including the initial condition (enclosed here in brackets), is [00] 11 01 11 11. From this example it will be seen that any given input bit contributes, for as long as it remains in the shift register, to the encoded word. The number of stages in the register gives the constraint length of the encoder. Denoting the constraint length by m, the encoder is specified by (n, k, m). The example shows a $(2, 1, 3)$ encoder. Encoders are optimized through computer simulation.

At the receiver, the tree diagram for the encoder is known. Decoding proceeds in the reverse manner. If, for example, [00] 11 01 11 11 is received, the tree is searched for the matching branches, from which the input can be deduced. Suppose, however, that

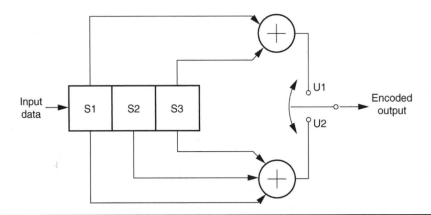

FIGURE 11.2 A rate 1/2 convolutional encoder.

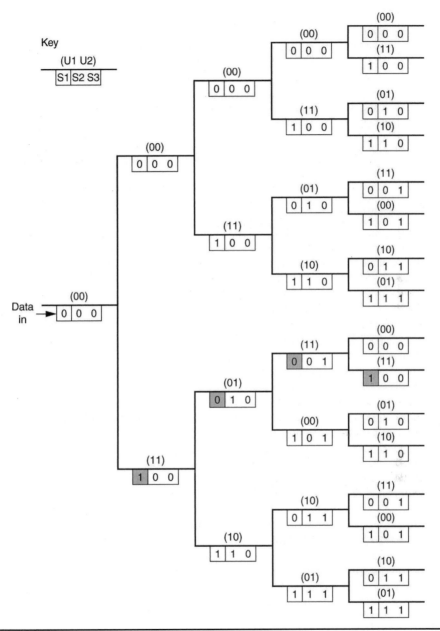

FIGURE 11.3 The tree diagram for the rate 1/2 convolutional encoder.

an error occurs in transmission, changing the received word to [00] 01 01 11 11; i.e., the error is in the first bit following the initial condition. The receiver decoder expects either a 00 or a 11 to follow the initial 00; therefore, it has to make the assumption that an error has occurred. If it assumes that 00 was intended, it will follow the upper branch, but now a further difficulty arises. The next possible pair is 00 or 11, neither of which matches the received code word. On the other hand, if it assumes that 11 was intended,

The CDV-10MIC is a single integrated circuit that implements all the functions required for a constraint length 7, rate 1/2, and punctured 2/3 or 3/4 rate, convolutional encoder, and Viterbi algorithm decoder. Important features of this chip are:

- Full decoder and encoder implementation for rates 1/2, 2/3, and 3/4
- Complies with INTELSAT IESS-308 and IESS-309 specifications
- Extremely low implementation margin
- No external components required for punctured code implementation
- Operates at all information rates up to 10 Mbits/s. Higher speed versions are under development
- All synchronization circuits are included on chip. External connection of ambiguity state counter and ambiguity resolution inputs allows maximum application flexibility
- Advanced synchronization detectors enable very rapid synchronization. Rate, 3/4 block and phase synchronization in less than 8200 information bits (5500 transmitted symbols)
- Soft decision decoder inputs (3 bits, 8 levels)
- Erasure inputs for implementing punctured codes at other rates
- Path memory length options to optimize performance when implementing high-rate punctured codes
- Error-monitoring facilities included on chip
- Synchronization detector outputs and control inputs to enable efficient synchronization in higher-speed multiplexed structures

Figure 11.4   Specifications for a single-chip Viterbi codec. (*Courtesy of Signal Processors, Ltd., Cambridge U.K.*)

it takes the lower branch, and then it can match all the following pairs with the branches in the decoding tree. On the basis of maximum likelihood, this would be the preferred path, and the correct input 1001 would be deduced.

Decoding is a more difficult problem than encoding, and as the example suggests, the search process could quickly become impracticable for long messages. The Viterbi algorithm is used widely in practice for decoding. An example of a commercially available codec is the CDV-10MIC single-chip codec made by Signal Processors Limited, Cambridge, U.K. The data sheet for this codec is shown in Fig. 11.4. The CDV-10MIC utilizes Viterbi decoding. It has a constraint length of 7 and can be adjusted for code rates of 1/2, 2/3, and 3/4 by means of what is termed *punctured coding*. With punctured coding, the basic code is generated at code rate 1/2, but by selectively discarding some of the output bits, other rates can be achieved (Mead, 2000). The advantage is that a single encoder can be used for different rates.

11.5   Interleaving

The idea behind interleaving is to change the order in which the bits are encoded so that a burst of errors gets dispersed across a number of codewords rather than being concentrated in one codeword. Interleaving as applied in block codes will be used here to illustrate the technique, but it also can be used with convolutional coding (Taub and Schilling, 1986).

Figure 11.5a shows part of the data bit stream where for definiteness the bits are labeled from b_1 to b_{24}. These are fed into shift registers as shown in Fig. 11.5b, where,

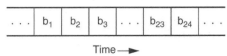

Time⟶

(a)

| Column No.⟍ | 1 | 2 | 3 | 4 | 5 | 6 |
Row No.⟍						
1	b_{24}	b_{23}	b_{22}	b_{21}	b_{20}	b_{19}
2	b_{18}	b_{17}	b_{16}	b_{15}	b_{14}	b_{13}
3	b_{12}	b_{11}	b_{10}	b_9	b_8	b_7
4	b_6	b_5	b_4	b_3	b_2	b_1
5	c_1	c_4	c_7	c_{10}	c_{13}	c_{16}
6	c_2	c_5	c_8	c_{11}	c_{14}	c_{17}
7	c_3	c_6	c_9	c_{12}	c_{15}	b_{18}

(b)

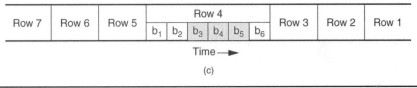

Time⟶

(c)

Figure 11.5   Illustrating interleaving (see Sec. 11.5).

again, for definiteness seven rows and six columns are shown. Rather than encoding the rows, the columns are encoded so that the parity bits fill up the last three rows. It will be seen, therefore, that the bits are not encoded in the order in which they appear in the data bit stream. The encoded bits are read out row by row as shown in Fig. 11.5c. Row 4 is shown in detail. If now an error burst occurs which changes bits b_5, b_4, and b_3, these will appear as separate errors in the encoded words formed by columns 2, 3, and 4. The words formed by the column bits are encoded to correct single errors (in this example), and therefore, the burst of errors has been corrected.

11.6 Concatenated Codes

Codes designed to correct for burst errors can be combined with codes designed to correct for random errors, a process known as *concatenation*. Figure 11.6 shows the general form for concatenated codes. The input data are fed into the encoder designed for burst error correction. This is the outer encoder. The output from the outer encoder is fed into the encoder designed for random error correction. This is the inner encoder. The signal is then modulated and passed on for transmission. At the receiver, the signal is demodulated.

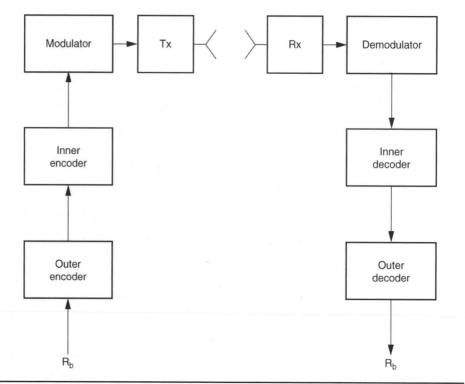

FIGURE 11.6   Concatenated coding (see Sec. 11.6).

The inner decoder matches the inner encoder and follows the demodulator. The output from the inner decoder is fed into the outer decoder, which matches the outer encoder. The term *outer* refers to the outermost encoding/decoding units in the equipment chain, and the term *inner* refers to the innermost encoding/decoding unit. In digital satellite television, the outer code is a R-S code, and the inner code is a convolutional code. The inner decoder utilizes Viterbi decoding.

11.7   Link Parameters Affected by Coding

Where no error control coding is employed, the message will be referred to as an *uncoded message*, and its parameters will be denoted by the subscript *U*. Figure 11.7*a* shows the arrangement for an uncoded message. Where error control coding is employed, the message will be referred to as a *coded message*, and its parameters will be denoted by the subscript *C*. Figure 11.7*b* shows the arrangement for a coded message. For comparison purposes, the $[C/N_0]$ value is assumed to be the same for both situations. The input bit rate to the modulator for the uncoded message is R_b, and for the coded message is R_c. Since n code bits must be transmitted for every k data bits, the bit rates are related as

$$\frac{R_b}{R_c} = r_c \qquad (11.10)$$

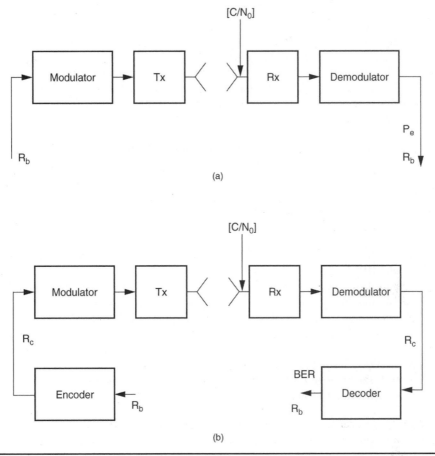

FIGURE 11.7 Comparing links with and without FEC.

Since r_c is always less than unity, then $R_c > R_b$ always. For constant carrier power, the bit energy is inversely proportional to bit rate (see Eq. 10.22), and therefore,

$$\frac{E_c}{E_b} = r_c \tag{11.11}$$

where E_b is the average bit energy in the uncoded bit stream (as introduced in Chap. 10), and E_c is the average bit energy in the coded bit stream.

Equation (10.24) gives the probability of bit error for *binary phase-shift keying* (BPSK) and *quadrature phase-shift keying* (QPSK) modulation. With no coding applied, E_b is just the E_b of Eq. (10.24), and the probability of bit error in the uncoded bit stream is

$$P_{eU} = \frac{1}{2} \text{erfc}\left(\sqrt{\frac{E_b}{N_0}}\right) \tag{11.12}$$

For the coded bit stream, the bit energy is $E_c = r_c E_b$, and therefore, Eq. (10.24) becomes

$$P_{eC} = \frac{1}{2} \text{erfc}\left(\sqrt{\frac{r_c E_b}{N_0}}\right) \tag{11.13}$$

This means that $P_{eC} > P_{eU}$, or the probability of bit error with coding is worse than that without coding. It is important to note, however, that the probability of bit error applies at the input to the decoder. For the error control coding to be effective, the output BER should be better than that obtained without coding. More will be discussed about this later.

The limitation imposed by bandwidth also must be considered. If the time for transmission is to be the same for the coded message as for the uncoded message, the bandwidth has to be increased to accommodate the higher bit rate. The required bandwidth is directly proportional to bit rate (see Eq. 10.22), and hence it has to be increased by a factor $1/r_c$. If, however, the bandwidth is fixed (the system is band limited), then the only recourse is to increase the transmission time by the factor $1/r_c$. For a fixed number of bits in the original message, the bit rate R_b entering into the encoder is reduced by a factor r_c compared with what it could have been without coding.

As an example, it is shown in Sec. 10.4 that the TI message rate is 1.544 Mb/s. When 7/8 FEC is applied, the transmission rate becomes $1.544 \times 8/7 = 1.765$ Mb/s. From Eq. (10.22), the required bandwidth is $B_{IF} = 1.765 \times (1.2)/2 = 1.06$ MHz.

11.8   Coding Gain

As shown by Eqs. (11.12) and (11.13), the probability of bit error for a coded message is higher (therefore, worse) than that for an uncoded message, and therefore, to be of advantage, the coding itself must more than offset this reduction in performance. In order to illustrate this, the messages will be assumed to be BPSK (or QPSK) so that the expressions for error probabilities as given by Eqs. (11.12) and (11.13) can be used. Denoting by BER_U the bit error rate after demodulation for the uncoded message and by BER_C the bit error rate for the coded message after demodulation and decoding, then for the uncoded message

$$BER_U = P_{eU} \tag{11.14}$$

Certain codes known as *perfect codes* can correct errors up to some number t. The BER for such codes is given by (see Roddy and Coolen, 1995)

$$BER_C = \frac{(n-1)!}{t!(n-1-t)!} P_{eC}^{t+1} \tag{11.15}$$

where $x! = x(x-1)(x-2) \ldots 3.2.1$ (and n is the number of bits in a codeword). The Hamming codes are perfect codes that can correct one error. For this class of codes and with $t = 1$, Eq. (11.15) simplifies to

$$BER_C = (n-1) P_{eC}^2 \tag{11.16}$$

A plot of BER_C and BER_U against $[E_b/N_0]$ is shown in Fig. 11.8 for the Hamming (7, 4) code. The crossover point occurs at about 4 dB, so for the coding to be effective, $[E_b/N_0]$ must be higher than this. Also, from the graph, for a BER of 10^{-5}, the $[E_b/N_0]$ is 9.6 dB for the uncoded message and 9 dB for the coded message. Therefore, at this BER value the Hamming code is said to provide a coding gain of 0.6 dB.

Some values for coding gains given in Taub and Schilling (1986) are block codes, 3 to 5 dB; convolutional coding with Viterbi decoding, 4 to 5.5 dB; concatenated codes using R-S block codes and convolutional decoding with Viterbi decoding, 6.5 to 7.5 dB. These values are for a P_e value of 10^{-5} and using hard decision decoding as described in the following section.

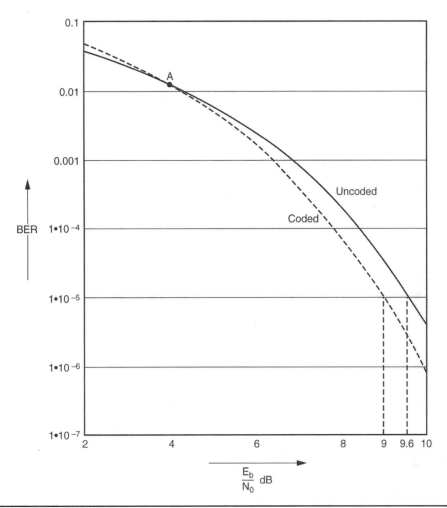

FIGURE 11.8 Plot of BER versus [E_b/N_0] for coded and uncoded signals.

11.9   Hard Decision and Soft Decision Decoding

With hard decision decoding, the output from the optimum demodulator is passed to a threshold detector that generates a "clean" signal, as shown in Fig. 11.9a. Using triple redundancy again as an example, the two codewords would be 111 and 000. For binary polar signals, these might be represented by voltage levels 1 V 1 V 1 V and –1 V –1 V –1 V. The threshold level for the threshold detector would be set at 0 V. If now the sampled signal from the optimum demodulator is 0.5 V 0.7 V –2 V, the output from the threshold detector would be 1 V 1 V –1 V, and the decoder would decide that this was a binary 1 1 0 codeword and produce a binary 1 as output. In other words, a firm or hard decision is made on each bit at the threshold detector.

With soft decision decoding (Fig. 11.9b), the received codeword is compared in total with the known codewords in the set, 111 and 000 in this example. The comparison is

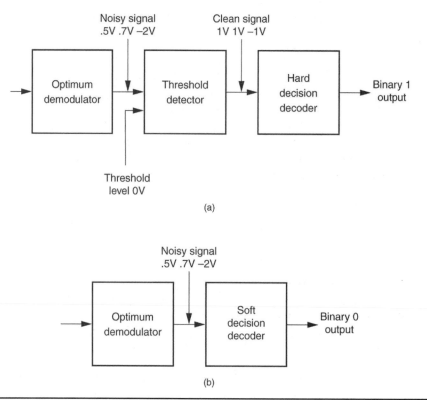

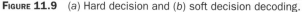

FIGURE 11.9 (a) Hard decision and (b) soft decision decoding.

made on the basis of minimum distance (the minimum distance referred to here is a *Euclidean distance* as described shortly. This is not the same as the minimum distance introduced in Sec. 11.2). To illustrate this, consider the first two points in an x, y, z coordinate system. Let point P_1 have coordinates x_1, y_1, z_1 and point P_2 have coordinates x_2, y_2, z_2. From the geometry of the situation, the distance d between the points is obtained from

$$d^2 = (x_1 - x_2)^2 + (y_1 - y_2)^2 + (z_1 - z_2)^2$$

Treating the codewords as vectors and comparing the received codeword on this basis with the stored version of 111 results in

$$(0.5 - 1)^2 + (0.7 - 1)^2 + (-2 - 1)^2 = 9.34$$

Comparing it with the stored version of 000 results in

$$[0.5 - (-1)]^2 + [0.7 - (-1)]^2 + [-2 - (-1)]^2 = 6.14$$

The distance determined in this manner is often referred to as the *Euclidean distance* in acknowledgment of its geometric origins, and the distance squared is known as the *Euclidean distance metric*. On this basis, the received codeword is closest to the 000 codeword, and the decoder would produce a binary 0 output.

Soft decision decoding results in about a 2-dB reduction in the $[E_b/N_0]$ required for a given BER (Taub and Schilling, 1986). This reference also gives a table of comparative values for soft and hard decision coding for various block and convolutional codes. Clearly, soft decision decoding is more complex to implement than hard decision decoding and is only used where the improvement it provides must be had.

11.10 Shannon Capacity

In a paper on the mathematical theory of communication (Shannon, 1948) Shannon showed that the probability of bit error could be made arbitrarily small by limiting the bit rate R_b to less than (and at most equal to) the *channel capacity*, denoted by C. Thus

$$R_b \leq C \tag{11.17}$$

For random noise where the spectrum density is flat (this is, the N_0 spectral density previously introduced) the channel capacity is given by

$$C = W \log_2 \left(1 + \frac{S}{N}\right) \tag{11.18}$$

Here, W is the baseband bandwidth, and S/N is the baseband signal to noise power ratio (not decibels). Shannon's theorem can be written as

$$R_b \leq W \log_2 \left(1 + \frac{S}{N}\right) \tag{11.19}$$

Letting P_R represent the average signal power, and T_b the bit period then as shown by Eq. (10.23) the bit energy is $E_b = P_R T_b$. The noise power is $N = WN_0$ and the signal to noise ratio is

$$\frac{S}{N} = \frac{P_R}{N} = \frac{E_b}{T_b W N_0} \tag{11.20}$$

The bit rate is $R_b = 1/T_b$. Substituting this in Eq. (11.20) then in the inequality (11.19) gives

$$\frac{R_b}{W} \leq \log_2 \left(1 + \frac{R_b E_b}{W N_0}\right) \tag{11.21}$$

As noted in Sec. 10.6.3, the ratio of bit rate to bandwidth (R_b/W in the inequality (11.21) is an important characteristic of any digital system. The greater this ratio, the more efficient the system. The limiting case is when the inequality sign is replaced by the equal sign.

$$\frac{R_b}{W} = \log_2 \left(1 + \frac{R_b E_b}{W N_0}\right) \tag{11.22}$$

Keep in mind that this relationship is for the condition of arbitrarily small probability of bit error. A plot of R_b/W as a function of E_b/N_0 is shown in Fig. 11.10. Note that although the graph shows E_b/N_0 in decibels ($[E_b/N_0]$ in our previous notation) the power ratio must be used in evaluating Eq. (11.22).

In any practical system there will be a finite probability of bit error, and to see how this fits in with the Shannon limit, consider the BER graph of Fig. 10.22, which applies for BPSK and QPSK. From Fig. 10.22 [or from calculation using Eq. (10.24)], for a probability

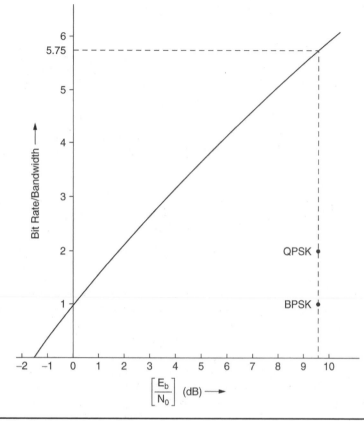

FIGURE 11.10 Graph showing the Shannon limit, Eq. (11.22). The units for the y-axis are (bits/s)/Hz (a dimensionless ratio in fact). The points for BPSK and QPSK are evaluated for a $P_e = 10^{-5}$.

of bit error (or BER in this case) of 10^{-5} the $[E_b/N_0]$ is about 9.6 dB. (See also the uncoded curve of Fig. 11.8.) As shown in Sec. 10.6.3 the bit rate to bandwidth ratio is $1/(1 + \rho)$ for BPSK and $2/(1 + \rho)$ for QPSK. For purposes of comparison ideal filtering will be assumed, for which $\rho = 0$. Thus on Fig. 11.10, the points (1, [9.6]) for BPSK and (2, [9.6]) can be shown. At an $[E_b/N_0]$ of 9.6 dB, the Shannon limit indicates that a bit rate to bandwidth ratio of about 5.75 : 1 should be achievable, and it is seen that BPSK and QPSK are well below this. Alternatively, for a bit-rate/bandwidth ratio of 1, the Shannon limit is 0 dB (or an E_b/N_0 ratio of unity), compared to 9.6 dB for BPSK.

11.11 Turbo Codes and LDPC Codes

Till 1993 all codes used in practice fell well below the Shannon limit. In 1993, a paper (Berrou et al., 1993) presented at the IEEE International Conference on Communications made the claim for a digital coding method that closely approached the Shannon limit (a pdf file for the paper will be found at www-elec.enst-bretagne.fr/equipe/berrou/Near%20Shannon%20Limit%20Error.pdf). Subsequent testing confirmed the claim to be true. This revitalized research into coding, resulting in a number of "turbo-like" codes, and a renewed interest in codes known as *low-density parity check* (LDPC) codes (see Summers, 2004).

Turbo codes and LDPC codes use the principle of *iterative decoding* in which "soft decisions" (i.e., a probabilistic measure of the binary 1 or 0 level) obtained from different encoding streams for the same data, are compared and reassessed, the process being repeated a number of times (iterative processing). This is sometimes referred to as *soft input soft output* to describe the fact that during the iterative process no hard decisions (binary 1 or 0) are made regarding a bit. Each reassessment generally provides a better estimate of the actual bit level, and after a certain number of iterations (fixed either by convergence to a final value, or by a time limit placed on the process) a hard decision output is generated.

Turbo codes are so named because the iterative or feedback process was likened to the feedback process in a turbo-charged engine (see Berrou et al., 1993). The turbo principle can be applied with concatenated block codes (known as *Turbo Product codes* or TPCs, see Comtech, 2002). However, the more common arrangement is to use parallel concatenation using convolution encoders. Because of the continuous nature of convolution coding, data and code *sequences* rather than words are involved. The convolution encoders shown in Fig. 11.11 (Burr, 2001) are *recursive convolution encoders*. They differ from the convolution encoder of Fig. 11.2 in that feedback is employed. (This *recursive* feedback is part of the encoding process and is quite separate from the iterative feedback to be described shortly, which gives turbo codes their name.) It can be shown that the recursive feedback (see Burr, 2001) assists in maintaining a large minimum Hamming distance (see Sec. 11.2) for code sequences. Another difference between the encoders of Fig. 11.11 and that of Fig. 11.2 is that the data sequence is fed directly to the multiplexer of Fig. 11.11, making the output systematic (see Sec. 11.2).

Parity-1 bits are generated directly from the data bits and parity-2 bits from the interleaved data bits, so that two independent streams of parity bits are generated. Interleaving, as described in Sec. 11.5, was used there as a means of combating bursty errors. With turbo encoding the purpose of interleaving is different, it is used to provide independent parity bits for the same input. A number of different methods of interleaving are available, and the design of the interleaver is a crucial aspect of turbo code design.

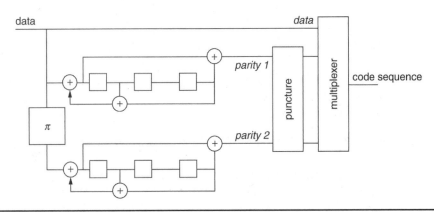

FIGURE 11.11 A turbo encoder with puncturing. The π symbol indicates an interleaver, and the $\oplus$ symbol indicates modulo-2 addition. (*Courtesy of A. Burr and the IEEE.*)

The output coded sequence is *data, parity-1, parity-2*. Since each encoder generates a parity bit for every data bit the code rate is 1/3. This is relatively low but can be increased by puncturing, as described in Sec. 11.4 and shown in Fig. 11.11. For example, one parity bit might be discarded in turn from each of the encoders, resulting in a 1/2 code rate. Other rates are possible with puncturing. If puncturing is used with the encoder, dummy bits are inserted at the decoder to replace those discarded. The dummy bit level is set midway between the binary 1 and 0 levels, so they do not affect the decoding process.

The block schematic for the decoder is shown in Fig. 11.12 (Burr, 2001). The demultiplexer provides outputs for the data sequence and parity-1 and parity-2 sequences. These outputs are "soft," meaning that some measure of the bit level is used rather than a hard decision output of binary 1 or 0. For example, assuming a threshold decision level of 0.5V, the demultiplexer output might be 0.9V, 0.7V 0.1V, 0.2V, 0.9V, 0.65V, 0.3V, suggesting a hard decision binary sequence of 1 1 0 0 1 1 0. However, the hard decision output does not make use of the *likelihood* of the hard decision being correct. The 0.9V level is obviously more likely to be a binary1 than the 0.65V level. A statistical measure termed the *log-likelihood ratio* (LLR) is most commonly used. For a given received value r, let $p(1/r)$ represent the probability that a 1 was transmitted, and $p(0/r)$ the probability that a 0 was transmitted. The log likelihood ratio is defined as

$$\text{LLR} = \log_e\left(\frac{p(1/r)}{p(0/r)}\right) \tag{11.23}$$

Where the transmission of 1s and 0s are equiprobable (the probability of either a 1 or 0 occurring being 1/2 , rather like the probability of heads or tails of a the toss of a fair coin) the LLR becomes:

$$\text{LLR} = \log_e\left(\frac{p(r/1)}{p(r/0)}\right) = \log_e p(r/1) - \log_e p(r/0) \tag{11.24}$$

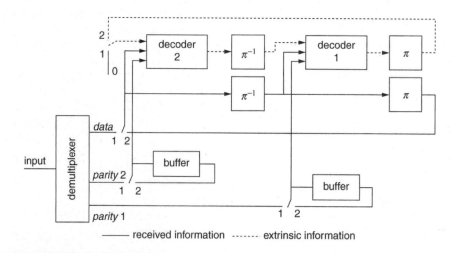

Figure 11.12   The turbo decoder for the encoder of Figure 11.11. The symbol π^{-1} represents a deinterleaver. (*Courtesy of A Burr and the IEEE.*)

where $p(r/1)$ is the probability of receiving value r, given that a 1 was transmitted and $p(r/0)$ the probability of receiving value r, given that a 0 was transmitted. If the voltage levels are normalized so that 1V represents a probability of 1, a certainty and 0V, zero probability, then for $r = 0.9$V for example, $p(r/1) = 0.9$ and $p(r/0) = 1 - .9 = 0.1$, so that LLR = 2.197. With $r = 0.3$, LLR = -0.847. In general, LLR yields a positive number for r closer to 1 and a negative number for r closer to 0. The magnitude of LLR is a measure of "how close." These two pieces of information are included in the soft sequences that form a part of the output of the multiplexer and that is the input to the decoders. The outputs from the decoders are also "soft" and the system is referred to as *soft-input soft-output* (SISO).

As shown in Fig. 11.12, the switches are in position 1 for the first iteration of the decoding step. Following the first iteration the switches are switched to position 2, and each decoder makes use of the soft information obtained from the other decoder to obtain a better estimate of bit values. Recall that two independent parity sequences are available for a given data sequence. The decoded data is adjusted to take into account the new estimates, and the process is repeated a number of times, typically for 4 to 10, before a final hard decision is made. The information that is obtained from the received data bits is termed *intrinsic information*, the intrinsic information flow paths being shown by the solid line in Fig. 11.12. The information that is passed from one decoder to the other is termed *extrinsic information*, the paths for the extrinsic information flow being shown by the dotted line. After the final iteration the output of the second decoder is switched to the output line (not shown in Fig. 11.12). It will be a 1 for a positive LLR and a 0 for a negative LLR. A more detailed description of the encoder of Fig. 11.11 and the decoder of Fig. 11.12 will be found in Burr (2001).

11.11.1   Low-Density Parity Check (LDPC) Codes

LDPC refers to the fact that the parity check matrix (Sec. 11.2) is sparse, that is, it has few binary 1s compared to binary 0s. The LDPC codes were first introduced by Gallagher (1962) who showed that a low density parity check matrix resulted in excellent minimum distance properties (as defined in Sec. 11.2), and they are comparatively easy to implement. As mentioned above a feature common to LDPC codes and turbo codes is that SISO decoding is employed, and a series of iterations performed to (hopefully) improve the probability estimate of a bit being a 1 or 0. Only after a predetermined number of iterations is a "hard decision" arrived at.

An example of a parity check matrix for a LDPC code (Summers, 2004) is

$$\mathbf{H} = \begin{bmatrix} 1 & 1 & 1 & 0 & 0 & 0 & 0 & 0 & 0 & 1 & 0 & 0 & 0 & 0 & 0 & 0 \\ 0 & 0 & 0 & 1 & 1 & 1 & 0 & 0 & 0 & 0 & 1 & 0 & 0 & 0 & 0 & 0 \\ 0 & 0 & 0 & 0 & 0 & 0 & 1 & 1 & 1 & 0 & 0 & 1 & 0 & 0 & 0 & 0 \\ 1 & 0 & 0 & 1 & 0 & 0 & 1 & 0 & 0 & 0 & 0 & 0 & 1 & 0 & 0 & 0 \\ 0 & 1 & 0 & 0 & 1 & 0 & 0 & 1 & 0 & 0 & 0 & 0 & 0 & 1 & 0 & 0 \\ 0 & 0 & 1 & 0 & 0 & 1 & 0 & 0 & 1 & 0 & 0 & 0 & 0 & 0 & 1 & 0 \\ 0 & 0 & 0 & 0 & 0 & 0 & 0 & 0 & 0 & 0 & 0 & 0 & 1 & 1 & 1 & 1 \end{bmatrix}. \tag{11.25}$$

As shown in connection with Eq. (11.5) the number of rows in $\mathbf{H}$ is equal to the number of parity bits $n - k$, and the number of columns is equal to the length n of the codeword. In this case $n - k = 7$ and $n = 16$, hence $k = 9$, and the $\mathbf{H}$ matrix represents a (16, 9) code. From Eq. (11.7) the syndrome is obtained on multiplying the received codeword by $\mathbf{H}^T$, the transpose of $\mathbf{H}$, and ideally, an error-free codeword is indicated by an all-zero syndrome.

Standard practice is to index bit positions starting from zero, thus a 16-bit codeword would have the bits labeled $c_0, c_1, c_2, \dots c_{15}$. Likewise, elements in the **H** matrix are labeled h_{pq} where the first element (top left-hand corner) is h_{00}.

In general the row number (indexed from zero) gives the number of the syndrome element, and the 1s in the columns indicate which codeword bits are used. The seven parity check equations obtained from the **H** matrix, are, on setting the syndrome equal to 0.

$$C_0 \oplus C_1 \oplus C_2 \oplus C_9 = 0$$
$$C_3 \oplus C_4 \oplus C_5 \oplus C_{10} = 0$$
$$C_6 \oplus C_7 \oplus C_8 \oplus C_{11} = 0$$
$$C_0 \oplus C_3 \oplus C_6 \oplus C_{12} = 0 \qquad (11.26)$$
$$C_1 \oplus C_4 \oplus C_7 \oplus C_{13} = 0$$
$$C_2 \oplus C_5 \oplus C_8 \oplus C_{14} = 0$$
$$C_{12} \oplus C_{13} \oplus C_{14} \oplus C_{15} = 0$$

As noted in connection with Eq. (11.3) a systematic code has the dataword at the beginning of the codeword, thus it follows that the columns 0 to 8 of the **H** matrix operate on the datawords. The fact that each column has two 1s means that two of the dataword bits appear in each parity check equation determined by these columns. A standard way of showing the parity check equations and the codeword bits is by means of a *Tanner graph* (Tanner, 1981) in which circles represent the bit nodes and squares represent the parity check equations, Fig. 11.13. The lines (technically referred to as *edges*) join the bit nodes to their respective parity check nodes. Edges occur wherever a 1 appears in the **H** matrix. Thus for row 0, 1s appear in the $c_0, c_1, c_2,$ and c_9 positions [(and as shown by the first parity line in Eq. (11.26)]. In Fig. 11.13, only the parity equations for rows 0, 1, and 3 are shown for clarity, but the complete Tanner graph would show all the edges. Messages pass along the edges. Initially, the output from the channel demodulator provides the first "soft" estimate of a bit. If p^1 is the probability that, the bit is a 1, the probability that it is a zero is $1 - p^1$. These estimates are sent to their respective parity check equations where the equation is used to derive probability estimates for a bit. Considering the first equation for example, an estimate for the probability that c_0 is a 1 can be obtained from the probabilities for $c_1, c_2,$ and c_9 being 1. For the

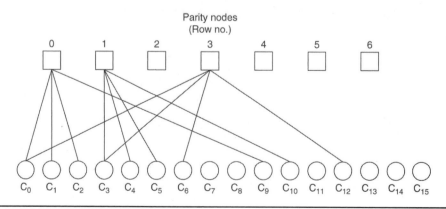

FIGURE 11.13 Illustrating a Tanner graph.

group $c_1c_2c_9$ the combinations that would result in c_0 being 1 are 100, 010, 100, and 111. The sum of the corresponding probabilities gives an estimate, p'_0 for the probability of c_0 being 1:

$$p'_0 = p_1^1\left(1-p_2^1\right)\left(1-p_9^1\right)+p_2^1\left(1-p_1^1\right)\left(1-p_9^1\right)$$

$$+p_9^1\left(1-p_1^1\right)\left(1-p_2^1\right)+p_1^1p_2^1p_9^1 \qquad (11.27)$$

The estimates from the parity nodes are returned to the respective bit nodes. The bit node now has estimates from the parity check nodes and from the channel, which enables a new estimate for probability to be calculated. For example if bit node 0 receives estimates from parity check nodes A, B, and C, denoted by p^1_{A}, p^1_{B}, p^1_{C} respectively, and p^1_{CH} from the channel, the new estimates sent to these parity check nodes are the products $K(p^1_{CH}\, p^1_{B}p^1_{C})$ to parity node A, $K(p^1_{CH}\, p^1_{A}p^1_{C})$ to parity node B, and $K(p^1_{CH}\, p_A p_B)$ to parity node C, where K is a normalizing constant. It will be noticed that the estimate from any given parity node is not included in the new estimate sent to that parity node. This process is repeated a number of times (iterated) before a final "hard decision" is made.

Turbo codes and LDPC codes are being employed in a number of satellite systems with a resulting improvement in channel performance, for example the *Digital Video Broadcast S2 standard* (DVB-S2) employs LDPC as the inner code in its FEC arrangement (Breynaert, 2005, Yoshida, 2003), and the DVB-RCS plans to use turbo codes (talk Satellite, 2004). The performance of a number of codes is shown in Fig. 11.14.

11.12   Automatic Repeat Request (ARQ)

Error detection without correction is more efficient than FEC in terms of code utilization. It is less complicated to implement, and more errors can be detected than corrected. Of course, it then becomes necessary for the receiver to request a retransmission when

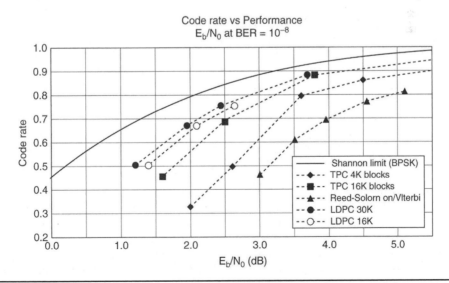

FIGURE 11.14   FEC performance of Comtech AHA 4701 LDPC coder. (*Courtesy of A. Summers Comtech AHA.*)

an error has been detected. This is an automatic procedure, termed *automatic repeat request* (ARQ). The request for retransmission must be made over a low-bit-rate channel where the probability of bit error can be kept negligibly small. Because of the long round-trip delay time, on the order of half a second or more, encountered with geostationary satellites, ARQ is only suitable for transmission that is not sensitive to long delays. ARQ is normally used with block encoding.

An estimate of the probability of an error remaining undetected can be made. With an (n, k) block code, the number of datawords is 2^k, and the total number of codewords is 2^n. When a given dataword is transmitted, an undetected error results when the transmission errors convert the received codeword into one that contains a permissible dataword but not the one that was transmitted. The number of such datawords is $2^k - 1$. An upper bound on the probability of an error getting through can be made by assuming that all codewords are equiprobable. The ratio of number of possible error words to total number of codewords then gives the average probability of error. In practice, of course, all the codewords will not be equiprobable, those containing datawords being more probable than those that do not. The ratio, therefore, gives an upper bound on the probability of error:

$$\text{BER} \leq \frac{2^k - 1}{2^n}$$

(11.28)

$$< 2^{-(n-k)}$$

where $n - k$ is the number of redundant bits in a codeword. For example, a (15, 11) code has an upper bound of approximately 0.06, while a (64, 32) code has an upper bound of approximately 2.3×10^{-10}.

It will be assumed, therefore, that the data are sent in coded blocks (referred to simply as *blocks* in the following). The receiver acknowledges receipt of each block by sending back a positive acknowledgment or ACK signal if no errors are detected in the block and a negative acknowledgment or NAK signal if errors are detected. In what is termed *stop and wait* ARQ, the transmitter stores a copy of the block just transmitted and waits for the acknowledgment signal. If a NAK signal is received, it retransmits the block, and if an ACK signals is received, it transmits the next block. In either case, the delay between transmissions is about half a second, the round-trip time to and from a geostationary satellite. This would be unacceptable in many applications.

What is required is continuous transmission of blocks incorporating the retransmission of corrupted blocks when these are detected. *Go back N ARQ* achieves this by having the blocks and the acknowledgment signals numbered. The transmitter must now be capable of storing the number of blocks N transmitted over the round-trip time and updating the storage as each ACK signal is received. If the receiver detects an error in block i, say, it transmits an NAK_i signal and refuses to accept any further blocks until it has received the correct version of block i. The transmitter goes back to block i and restarts the transmission from there. This means that block i and all subsequent blocks are retransmitted. It is clear that a delay will only be encountered when an NAK signal is received, but there is the additional time loss resulting from the retransmission of the good blocks following the corrupted block. The method can be further improved by using what is termed *selective repeat* ARQ, where only a correct version of the corrupted block is retransmitted, and not the subsequent blocks. This creates a storage requirement at the receiver because it must be able to store the subsequent blocks while waiting for the corrected version of the corrupted block.

It should be noted that in addition to the ACK and NAK signals, the transmitter operates a timeout clock. If the acknowledgment signal (ACK or NAK) for a given block is not received within the timeout period, the transmitter puts the ARQ mechanism into operation. This requires the receiver to be able to identify the blocks so that it recognizes which ones are repeats.

The *throughput* of an ARQ system is defined at the ratio of the average number of data bits accepted at the receiver in a given time to the number of data bits that could have been accepted if ARQ had not been used. Let P_A be the probability that a block is accepted; then, as shown in Taub and Schilling (1986), the throughputs are

$$\text{Go back } N = \frac{k}{n} \frac{P_A}{P_A + N(1 - P_A)} \tag{11.29}$$

$$\text{Selective repeat} = \frac{k}{n} P_A \tag{11.30}$$

Typically, for a BCH (1023, 973) code and $N = 4$, the throughput for go back N is 0.915, and for selective repeat, 0.942.

It is also possible to combine ARQ methods with FEC in what are termed *hybrid ARQ systems*. Some details will be found in Pratt, Bostian, and Allnutt (2003).

11.13   Problems and Exercises

11.1.   Explain in your own words how *error detection* and *error correction* differ. Why would FEC normally be used on satellite circuits?

11.2.   A transmission takes place where the average probability of bit error is 10^{-6}. Given that a message containing 10^8 bits is transmitted, what is the average number of bit errors to be expected?

11.3.   A transmission utilizes a code for which $k = 7$ and $n = 8$. How many codewords and how many datawords are possible?

11.4.   A triple redundancy coding scheme is used on a transmission where the probability of bit error is 10^{-5}. Calculate the probability of a bit error occurring in the output when (*a*) error detection only is used and (*b*) when error correction with majority voting is implemented.

11.5.   Using the generator matrix given in Eq. (11.3) find the codewords for the datawords (*a*) [0000]; (*b*) [1111]; and (*c*) [0010].

11.6.   Using the parity check matrix of Eq. (11.5) find the syndrome for the codeword (1110100). Comment on this.

11.7.   Using the parity check matrix of Eq. (11.5) find the syndrome for the codeword (1000110). Comment on this.

11.8.   A received codeword is 1011000. Determine, using the parity check matrix of Eq. (11.5) if this is a valid codeword, and if not, write out the error vector on the assumption that only one error is present.

11.9.   Calculate the code rate for a Hamming (31, 26) code.

11.10.   Calculate the code rate and the number of errors that can be corrected with a BCH (63, 36) code.

11.11.   An R-S code is byte oriented with $k = 8$. Given that there are eight redundant symbols, calculate the number of symbol errors that can be corrected.

11.12.   Determine the values for N' and K' for the shortened R-S codes used in (a) DirecTV and (b) DVB.

11.13.   Describe how convolution coding is achieved. State some of the main advantages and disadvantages of this type of code compared with block codes.

11.14.   Explain what is meant by *interleaving* when applied to error control coding and why this might be used.

11.15.   Explain what is meant by *concatenated codes* and why these might be used.

11.16.   Explain what is meant by a FEC code. FEC coding at a code rate of 3/4 is used in a digital system. Given that the message bit rate is 1.544 Mb/s, calculate the transmission rate.

11.17.   The bit rate for a baseband signal is 1.544 Mb/s, and FEC at a code rate of 7/8 is applied before the signal is used to modulate the carrier. Given that the system uses raised-cosine filtering with a rolloff factor of 0.2, determine the bandwidth required for (a) BPSK, and (b) QPSK.

11.18.   A BPSK signal provides an $[E_b/N_0]$ of 9 dB at the receiver. Calculate the probability of bit error.

11.19.   For the signal in Prob. 11.18, calculate the new value of bit error probability if FEC is applied at a code rate of 3/4, given that the carrier power remains unchanged.

11.20.   Derive Eq. (11.11).

11.21.   Explain what is meant by *coding gain* as applied to error correcting coding. When FEC coding is used on a digital link, a coding gain of 3 dB is achieved for the same BER as the uncoded case. What decibel reduction in transmitted carrier power does this imply?

11.22.   A certain (15, 11) block code is capable of correcting one error at most. Given that this is a perfect code (see Sec. 11.8), plot, on the same set of axes, the BER for the coded and uncoded cases for an $[E_b/N_0]$ range of 2 to 12 dB. Calculate the coding gain at a BER of 10^{-6}.

11.23.   State briefly the difference between hard and soft decision decoding. Following the description given in Sec. 11.9, determine the output produced by (a) hard decision and (b) soft decision decoding when the sampled signal from the demodulator is −0.4 V, 0.85 V, and −0.4 V for triple redundancy coding.

11.24.   From Fig. 11.10 find the minimum $[E_b/N_0]$ as determined by the Shannon limit curve. Explain the significance of this.

11.25.   For equiprobable bit transmission a received bit level is 0.55V. Assuming this is normalized where 1V represents a certainty of the bit being a 1, calculate the LLR.

11.26.   Referring to Eq. (11.26), identify the parity equations which contain code bit c_7.

11.27.   Complete the Tanner graph of Fig. 11.13 for all the parity equations given in Eq. 11.26.

11.28.   For the parity equation for row 5, (Eq. 11.26), the probabilities are: $p_5^1 = 0.7$, $p_8^1 = 0.6$, $p_{14}^1 = 0.3$. Calculate the estimated probability p_2.

11.29.   Research the literature and write brief comparative notes on the use of turbo codes and LDPC codes in satellite communications.

11.30.   A (31, 6) block code is used in an ARQ scheme. Determine the upper bound on the probability of bit error.

References

Berrou, C., A. Glavieux, and P. Thitmajshima. 1993. "Near Shannon Limit error-correcting coding: Turbo codes." *Proc. of the IEEE Int. Conf. Commun.*, Switzerland, at www-elec .enst-bretagne.fr/equipe/berrou/Near%20Shannon%20Limit%20Error.pdf.

Breynaert, D. 2005. Analysis of the Bandwidth Efficiency of DVB-S2 in a Typical Data Distribution Network. CCBN, Beijing, March, at www.newtecamerica.com/news/ articles/DVB-S2%20whitepaper.pdf.

Burr, A. 2001. "Turbo Codes: The Ultimate Error Control Codes?" *Electron. Commun. Eng. J.*, August, pp. 155–165.

Comtech, 2002. The Case for Turbo Product Coding in Satellite Communications, at www.linksite/manuals/whitepapers/Comtech%20EF%20Data%20Case%20for%20 Turbo%20Product%20Coding.pdf.

Gallagher, R. G. 1962. "Low Density Parity Check Codes." *IRE Trans. Info. Theory 8*, pp. 21–28.

Mead, D. C. 2000. *Direct Broadcast Satellite Communications*. Addison-Wesley, Reading, MA.

Pratt, Timothy, C. W. Bostian, and J. E. Allnutt. 2003. *Satellite Communications*. Wiley, New York.

Proakis, J. G., and M. Salehi. 1994. *Communication Systems Engineering*. Prentice-Hall, Englewood Cliffs, NJ.

Roddy, D., and J. Coolen. 1994. *Electronic Communications*. 4th ed. Prentice-Hall, Englewood Cliffs, NJ.

Shannon, C. E. 1948. "A Mathematical Theory of Communications." *BSTJ*, vol. 27, pp. 379–423, 623–656, July, October.

Summers, T. 2004. "LDPC: Another Key Step toward Shannon." *CommsDesign*, October, at www.commsdesign.com/design_corner/showArticle.jhtml?articleID=49901136-50k.

Talk Satellite 2004. Spectra Licensing Group Announces Gilat's Entrance into the Turbo Code Licensing Program for DVB-RCS Satellite Applications. December 15, at www.talksatellite.com/EMEAdoc105.

Tanner, R. M. 1981. "A Recursive Approach to Low Complexity Codes." *IEEE Trans. Inf. Theory*, vol. 27, no. 5, September, pp. 533–547.

Taub, H., and D. L. Schilling. 1986. *Principles of Communications Systems*. 2nd ed. McGraw-Hill, New York.

Yoshida, J. 2003. "Hughes Goes Retro in Digital Satellite TV Coding." *EETimesUK*, November, at www.eetuk.com/tech/news/OEG20031110S0081-42k-.

CHAPTER **12**

The Space Link

12.1 Introduction

Analog communications suffer from the effects of additive wideband Gaussian noise (AWGN), and the communications system performance depends upon the ratio of signal-power to noise-power ratio (SNR) at the receiver output. This is especially significant for SATCOM, where the signal strength decreases dramatically with the distance from the transmitter. To mitigate this effect, satellite communications repeaters (transponders) are used to receive the weak signals, amplify them, and then retransmit. The issue is that the SNR continues to degrade after each repeater such that the end-to-end SNR is smaller than the weakest link.

Digital communications also suffer from the effects of AWGN, and the digital quality (measured in bit error rate) is also determined by the output SNR. However, digital signal processing provides additional noise immunity compared to analog, but more importantly, regenerative digital repeaters are viable that "restore the SNR." Thus, for digital SATCOM, a high SNR may be maintained regardless of the distance between the transmitter and the receiver (and the number of repeaters).

This chapter describes the RF link connectivity of two or more earth stations through a satellite relay consisting of an earth-to-space (uplink) and a space-to-earth (downlink). For most GEOSAT communication systems, the satellite transponder serves as a "bent pipe" repeater, where the input and the output are the same signal content, but for practical reasons (discussed in Sec. 7.7), the uplink and downlink operate at different RF carrier frequencies. This chapter describes how the link-budget calculations, of SNR, are made for these systems, and, for satellite communications links with digital regenerative satellite receivers.

We begin with the calculations for signals and later include the effects of noise. For a single RF carrier, the ideal satellite communications system given in Fig. 12.1 is applicable to both the uplink and the downlink scenarios. For a free-space, direct line-of-sight (LOS) propagation path, our objective is to determine the required transmitter output power (P_t) to yield the desired signal power at the receiver input (P_r). For most systems, the receiver is linear, and the output SNR is the same as the input; however, some receivers improve the output SNR with signal processing gain. Both types of systems are described in this text.

In the derivations presented in Chap. 12, both linear and decibel (dB) forms of the link equations are presented, where the notation of [square] brackets is used for dB quantities. Occasionally formulas are given in both formats, and for these cases, a single

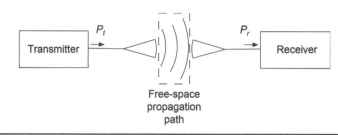

FIGURE 12.1   Simplified SATCOM Link block diagram.

equation number is assigned with linear having the subscript "a" and with decibel having the subscript "b".

12.2   Equivalent Isotropic Radiated Power

A key parameter in link budget calculation is the *"equivalent isotropic radiated power"* (EIRP). From Eqs. (6.4) and (6.5), the maximum power flux density at a distance d from a transmitting antenna of gain G_t is

$$\psi_M = \frac{G_t P_t}{4\pi d^2}, \, \text{W/m}^2 \tag{12.1}$$

Given that an isotropic radiator (with an input power equal to $G_t P_t$) produces the same flux density in all directions, this is the definition of EIRP

$$\text{EIRP} = G_t P_t, \, \text{W} \tag{12.2a}$$

Usually, EIRP is expressed in decibels relative to 1 W, or dBW as

$$[\text{EIRP}] = [G_t] + [P_t], \, \text{dBW} \tag{12.2b}$$

where $[P_t]$ is also in dBW and $[G_t]$ is in dBi, which is relative to isotropic antenna gain.

Example 12.1   A satellite link at 12 GHz operates with a transmit power of 6 W and an antenna gain of 48.93 dB, calculate the EIRP in dBW.

Solution   From Eq. (12.2b):

$$[\text{EIRP}] = [48.93] + 10 \log\left(\frac{6 \, \text{W}}{1 \, \text{W}}\right) = 56.71, \, \text{dBW}$$

For a parabolic dish antenna, the peak (maximum) isotropic power gain is given by Eq. (6.28) and repeated here as

$$G_{pk} = \eta_i \left(\frac{\pi D}{\lambda}\right)^2, \, \text{power ratio}$$

where $\lambda = c/f$ meters and c = speed of light (~3.0e + 08 m/s) and f is the carrier frequency in Hz, D is the reflector diameter in meters, and η_i is the aperture illumination efficiency (ratio). A typical value for aperture efficiency is between 0.55 to 0.73 (Andrew Antenna, 1985).

Example 12.2   Calculate the gain in decibels of a 3-m parabolic dish antenna operating at a frequency of 12 GHz, with an aperture efficiency of 0.55.

Solution $\lambda = \dfrac{3.0e+08}{12.0e+09} = 0.025$, m

From Eq. (6.28): $G_0 = 0.55 \left(\dfrac{\pi \times 3}{0.025} \right)^2 = (7.817e+04) \Rightarrow 48.93$ dB

Note: the symbol "$\Rightarrow$" is used to denote the conversion from (ratio) to [dB] or vice versa

12.3   Ideal Free-Space Propagation Path Losses

The next parameter in the ideal link-budget calculation is the free-space propagation *path loss*, which is defined as the reduction of the power flux density of the EM wave as it propagates from the transmitting to the receiving antenna. This path-loss (also known as *spreading-loss*) is the result of the spherical propagation nature of EM waves in free-space (both in a vacuum and in a physical media such as the atmosphere).

Because engineers understand the concept of gain of a linear system, in this text edition, we choose to extend this concept to express losses as gains that have <1.0 power ratio. In this way, the SATCOM link transfer function of cascaded blocks is simplified as the product of block gains (some of that may be losses). We feel that this is more intuitive than the alternative approach of treating losses as power ratios >1.0 and then dividing them in the transfer function. This topic is discussed further in this chapter and in App. G.

To determine the power loss associated with the spreading of the EM wave in space, we first derive the end-to-end transfer function for the cascade of three blocks (Fig. 12.1), namely the transmitter, the path loss, and the receiver. The output of the transmitter block (that includes the transmit antenna) is the EIRP, and this produces a power flux density at the output of the second block equal to

$$\psi_M = \frac{\text{EIRP}}{4\pi d^2}, \ \text{W/m}^2 \tag{12.3}$$

where d the distance (range) in meters between the transmitting and receiving antennas.

Finally, the output of the third block is the power delivered to the input of the matched receiver (P_r). This is equal to the product of the power flux density and the effective aperture of the receiving antenna (A_{eff}), given by Eqs. (12.3) and (6.15), respectively. Thus, the received power is

$$P_r = \psi_M A_{\text{eff}}, \ \text{W} = \left(\frac{\text{EIRP}}{4\pi d^2} \right) \left(\frac{\lambda^2 G_r}{4\pi} \right) = \text{EIRP} \ G_r \left(\frac{1}{4\pi \left(\dfrac{d}{\lambda} \right)} \right)^2 = \text{EIRP} \ G_r L_p \tag{12.4a}$$

where G_r is the power gain of the receiving antenna. Note that the right-hand side of Eq. (12.4a) is separated into the three terms associated with, respectively, the transmitter, the receiver, and the free space propagation.

The received power (expressed in dBW) is the sum of the transmitted EIRP in dBW plus the receiver antenna gain in dBi (above isotropic) and the path loss in dB,

$$[P_r] = [\text{EIPR}] + [G_r] + [L_p], \ \text{dBW} \tag{12.4b}$$

The free-space propagation path loss is written as

$$L_p = \left(\frac{1}{4\pi\left(\frac{d}{\lambda}\right)}\right)^2, \text{ ratio} \qquad (12.5a)$$

Note that the ratio (d/λ) is used to remind the reader that the path loss depends upon the distance in wavelengths (rather than in meters). Normally, the frequency (rather than wavelength) is known, so it is necessary to calculate the wavelength as $\lambda = c/f$, using the approximation of the speed of light as $c = 3.0e + 08$ m/s and the frequency in Hertz. Then, the decibel free-space loss is

$$[L_p] = 10 \log((1/4\pi)^2) + 10 \log((\lambda/d)^2)$$

$$= [-21.98] + 10 \log((\lambda/d)^2), \text{ dB} \qquad (12.5b)$$

Note that we use λ in all equations rather than having alternate formulas that use frequency as an input.

Equations (12.4) and (12.5) are applicable to both the uplink and the downlink of a SATCOM link.

Throughout this text, the reader should be careful to include the sign of the dB value and to note that gains <1 are negative dB values (such as path loss in Eqs. 12.4b and 12.5b). Further, since the equations for power ratios are defined as products of terms, when expressed as dB, the terms are always added (see App. G for further discussion).

Example 12.3 The LOS distance (range), between a ground station at 70° latitude and a GEO-satellite, $d = 40,400$ km ($4.040e + 07$ m). Calculate the free-space loss at a frequency $f = 6.0$ GHz.

Solution From Eq. (12.5):

$$\lambda = c/f = 3.0e + 08/6.0e + 09 = 0.050, \text{ m}$$

$$[L_p] = -21.98 + 10 \log((0.05/4.04e + 07)^2) = -21.98 - 178.16 = -200.14 \text{ dB}$$

For a mid-latitude (40°) earth station, the LOS distance $d = 37,500$ km and the corresponding path loss is

$$[L_p] = -21.98 + 10 \log((0.05/3.75e + 07)^2) = -199.49 \text{ dB}$$

Note that the L_p for a geosynchronous satellite is ~ $-200 < \pm10$ dB (accounting for typical slant ranges and wavelengths); so, this is an excellent "quality control test" (i.e., a necessary but not sufficient test) to check link calculations. Also, note that if the distance is incorrectly used in km, then the L_p is ~ $-140 < \pm10$ dB; or if the wavelength is incorrectly used in cm, then the L_p is ~ $-220 < \pm10$ dB.

This L_p is a very large loss that must be overcome by the positive gains of the transmitter and receiver. For example, if the [EIRP] is +55.98 dBW (as in Example 12.1) and the receiving antenna gain is +50 dB, then the received power is

$$[P_r] = [55.98] + [50] + [-200.14] = -94.16 \text{ dBW} \Rightarrow 3.84e\text{-}10, \text{ W}$$

Note that the received power increases by the receiving antenna gain, which is proportional to $(D/\lambda)^2$ (for both transmit and receive). Hence, if both the transmitting and receiving antenna diameters are fixed, then increasing the frequency of operation (and thereby decreasing wavelength) increases the received power. However, because the free-space loss (L_p) is proportional to $(\lambda/d)^2$, so these two effects partially cancel, and the P_r increases as the square of the frequency.

12.4   Link Budget with Realistic Losses

Next, consider the expanded SATCOM link block diagram with losses given in Fig. 12.2. In this text, the combined "feeder losses" that occur in the connection between the transmitter and the transmitting antenna are gains <1 that are denoted as L_{ta}. In a similar manner, losses between the receiver and the receiving antenna are also gains <1 that are denoted as L_{ra}. These losses are associated with both ohmic power absorption (also known as dissipation) and reflective power division losses that occur in the connecting transmission lines, filters, and couplers. For convenience, we combine all practical losses into a single system loss, so that the resulting formula (Link Equation) for P_r becomes

$$P_r = P_t G_t L_p G_r L_{sys}, \text{ W} \tag{12.6a}$$

where P_t is the transmitter output in Watts, G_t and G_r are LOS antenna gains expressed as power ratios, L_p is the spreading or free-space path loss expressed as a ratio, and L_{sys} is the product of all applicable losses in ratio. The corresponding equation in dB is

$$[P_r] = [P_t] + [G_t] + [L_p] + [G_r] + [L_{sys}], \text{ dB} \tag{12.6b}$$

12.4.1   Antenna Misalignment Losses

When a satellite link is established, the ideal situation is to have the earth station and satellite antennas aligned for maximum gain, as shown in Fig. 12.3a. Unfortunately, because of the geographically diverse locations of earth station users, often this is not possible; thus, for the general case, there are two possible sources of off-axis antenna gain loss as shown in Fig. 12.3b. So, the link-budget must be modified to use the LOS antenna gains, when calculating the received power (for both uplinks and downlinks).

12.4.1.1   Earth-Station Antenna Mispointing Loss

Consider first an earth station antenna that is pointed to the center of the "satellite box" on the geostationary arc. As previously discussed (Sec. 7.4), the satellite location is constantly changing about the assigned orbital position, but it is bounded by a box that is

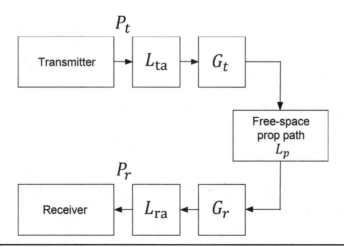

Figure 12.2   SATCOM Link with realistic losses.

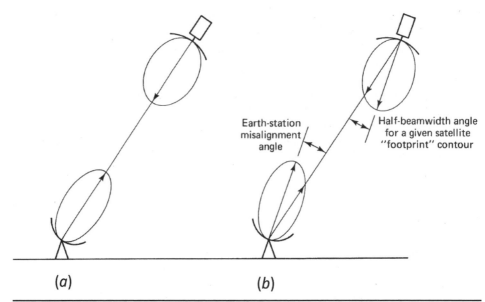

Earth-station
misalignment
angle

Half-beamwidth angle
for a given satellite
"footprint" contour

(a)   (b)

Figure 12.3 Alignment of earth station and satellite antenna gain patterns for a geostationary satellite geometry, (a) ideal, with peak gains coaligned, and (b) practical, with both gain patterns misaligned.

±0.1° North/South and ±0.1° East/West for C-band SATCOM (±0.05° for Ku-band). At any instant, there may be an earth station antenna mispointing loss associated with the stationary transmit antenna for the uplink (or with the earth station (es) receiving antenna for the downlink).

In this text, the earth station (transmitting or receiving) antenna mispointing loss (L_m) is a gain <1 that is multiplied by the peak antenna gain, to yield the LOS gain

$$G_{\text{LOS,es}} = G_o L_m, \text{ power ratio} \tag{12.7}$$

By convention, the earth station peak antenna gain is used in the uplink link equation as G_t (G_r for downlink) and the L_m is an additional multiplicative term in the link budget (additive dB term) included in L_{sys} in Eq. (12.6). For the worst-case analysis, the L_m is calculated assuming that the satellite is located at a corner of the box that corresponds to an off-boresight angle $\theta = 0.14°$ for C-band (= 0.07° for Ku-band). The resulting antenna mispointing loss is typically less than a few tenths of a dB depending upon the earth station antenna (λ/D); however, for large earth station antennas, limited satellite tracking capability is frequently used to adjust the antenna pointing to mitigate unacceptably large, time-varying, mispointing losses.

Example 12.4   Given an earth station antenna of diameter $D = 3.0$ m and efficiency $\eta = 55.0\%$, calculate the peak antenna gain and the corresponding antenna mispointing loss (L_m) for a GEOSAT link at Ku-band frequency $f = 12.0$ GHz.

Solution $\lambda = c/f = 3.0e + 08/12.0e + 09 = 0.025$, m

Reflector diameter $D = 3.0$ m and $D/\lambda = 120$

From Eq. (6.28):

$$G_o = \eta \left(\frac{\pi D}{\lambda} \right)^2 = 0.55 \, (120 \, \pi)^2 = 7.817e + 04 \Rightarrow 48.93 \text{ dB}$$

For Ku-band mispointing loss, the worst-case off-boresight angle, $\theta = 0.0707°$
  Using the **Bessel normalized pattern approximation** (Eq. (6.31), Sec. 6.13.1),

$$G(\theta)_{\text{Bessel}} = 64 \left[\frac{J_2(u)}{u^2} \right]^2$$

where $u = \dfrac{\pi}{\lambda/D} \sin \theta = (120 \, \pi) \sin(0.071°) = 0.465$

$L_{m\_b} = G(\theta)_{\text{Bessel}} = 0.9645 \Rightarrow -0.16 \text{ dB}$

Alternatively, using the **Sinc² normalized pattern approximation** (Eq. (6.35), Sec. 6.13.1), the argument is

$$x = 0.690 \, u = 0.690 * 0.465 = 0.3210, \text{ radians}$$

$$L_{m\_s} = G(\theta)_{\text{sinc}} = \left(\frac{\sin(x)}{x} \right)^2 = 0.966 \Rightarrow -0.15 \text{ dB}$$

Example 12.5   Given an earth station antenna of diameter $D = 10.0$ m and efficiency $\eta = 70.0\%$, calculate the peak antenna gain and the antenna mispointing loss (L_m) at a C-band transmit frequency $f = 6.0$ GHz. This analysis applies to both uplink and downlink.

Solution $\lambda = c/f = 3.0 \text{ E} + 08/6.0 \text{ E} + 09 = 0.05$, m
Reflector diameter $D = 10.0$ m and $D/\lambda = 200$
  From Eq. (6.28):

$$G_o = \eta \left(\frac{\pi D}{\lambda} \right)^2, \text{ power ratio}$$

$$= 0.70 * (200\pi)^2 = 2.763 \text{ E} + 05 \Rightarrow 54.41 \text{ dB}$$

For C-band, the worst-case off-boresight angle, $\theta = 0.1414°$
  Using the Bessel normalized pattern approximation

$$G(\theta)_{\text{Bessel}} = 64 \left[\frac{J_2(u)}{u^2} \right]^2$$

where $u = \left(\pi \dfrac{D}{\lambda} \right) \sin(\theta) = (200\pi) \sin(0.14°) = 1.5506$

$L_{m\_b} = G(\theta)_{\text{Bessel}} = 0.663 \Rightarrow -1.79 \text{ dB}$

Similarly, using the Sinc² normalized pattern approximation, the argument is

$$x = 0.690 \, u = 0.690 * 1.5506 = 1.070, \text{ radians}$$

$$L_{m\_s} = G(\theta)_{\text{Sinc}} = \left(\frac{\sin(x)}{x} \right)^2 = 0.672 \Rightarrow -1.73 \text{ dB}$$

Note: A large earth station antenna such as this usually employs limited pointing capability to track the GEOSAT position to mitigate this unacceptably large mispointing loss.

12.4.1.2   GEOSAT Antenna LOS Gain

For the satellite antenna (receiving on the uplink or transmitting on the downlink), the LOS gain is more involved. In practice, the antenna radiation pattern is often a shaped beam, and the simple boresight gain given by Eq. (6.28) is not applicable, and the measured satellite antenna patterns are normally used for the link analysis. However, for this text, to expedite the analysis procedure, the satellite antenna is defined by two parameters, namely: the boresight gain G_o and the effective (λ/D). Using these, the absolute LOS gain is

$$G_{\text{LOS,sat}} = G_o * G(\theta), \text{ power ratio} \tag{12.8}$$

where $G(\theta)$ is the normalized antenna gain pattern (see Sec. 6.13.1) evaluated at the off-boresight angle (θ) between the antenna boresight and the earth station geographic locations.

Example 12.6   Given a GEOSAT at longitude $\phi_S = -95.0°$ (west) with an uplink earth station located at latitude $\lambda_E = +35.0°$ (north) by longitude $\phi_E = -90.0°$ (west). Assume that the satellite antenna has a peak gain = 34.77 dB and an effective $(\lambda/D) = 0.0314$ radians, which is boresighted at the earth surface latitude $\lambda_o = +28.0°$ by longitude $\phi_o = -92.5°$. Calculate the satellite antenna LOS gain at a frequency of 6.0 GHz.

As a note of caution for the calculations that follow, (λ/D) is a dimensionless number, and when used to calculate the normalized off-boresight angle ratio $= \left(\dfrac{\theta}{\lambda/D}\right)$, the off-boresight angle (θ) must be radians.

Solution   To calculate the satellite antenna LOS gain, we must determine the off-boresight angle between the antenna boresight vector and the LOS direction to the desired earth station. Consider the GEOSAT/earth-station geometry previously presented in Sec. 3.3 and Fig. 3.2 that is reproduced here as Fig. 12.4 (a).

The first step is to calculate the distance (slant range, d_1) from the GEOSAT to the earth-station. From Eq. (3.8): the earth central angle between the subsatellite point and the earth station is

$$b_1 = \arccos\{\cos\lambda_E * \cos(\phi_E - \phi_S)\}$$
$$= \arccos\{\cos(35°) * \cos(-90° - (-95°))\} = 35.31°$$

From the law of cosines Eq. (3.10): the slant range (d_1) from the earth station to the satellite is

$$d_1 = 37{,}142.82 \text{ km}$$

Next, we use Fig. 12.4a to repeat this calculation of slant range (d_2) to the antenna boresight point on the earth's surface.

$$b_2 = \arccos\{\cos\lambda_o * \cos(\phi_o - \phi_S)\}$$
$$= \arccos\{\cos(28°) * \cos(-92.5° - (-95°))\} = 28.10°$$
$$d_2 = 36{,}661.30 \text{ km}$$

Now, consider Fig. 12.4b, which gives the geometry for calculating the off-boresight angle θ. Using the latitudes and longitudes for the earth station and the boresight point, we can use

$$\cos\alpha = \cos\lambda_E \cos\lambda_o \cos(\phi_E - \phi_o) + \sin\lambda_E \sin\lambda_o \tag{12.9}$$

to solve for the corresponding earth central angle: $\alpha = 7.32°$.

Next, we solve for the antenna off-boresight pointing angle:

$$\cos\theta = \frac{(d_1^2 + d_2^2 - 2R^2(1 - \cos\alpha))}{2d_1 d_2} \tag{12.10}$$

$$\theta = 0.01778 \text{ radians, } (1.02°)$$

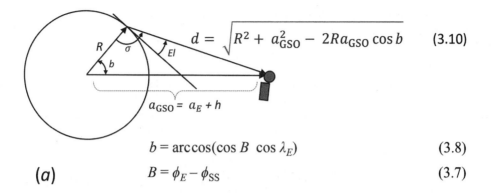

$$d = \sqrt{R^2 + a_{GSO}^2 - 2Ra_{GSO}\cos b} \qquad (3.10)$$

$$a_{GSO} = a_E + h$$

$$b = \arccos(\cos B \ \cos \lambda_E) \qquad (3.8)$$

(a) $\qquad B = \phi_E - \phi_{SS} \qquad (3.7)$

$$\cos \alpha = \cos \lambda_E \cos \lambda_o \cos(\phi_E - \phi_o) + \sin \lambda_E \sin \lambda_o \qquad (12.9)$$

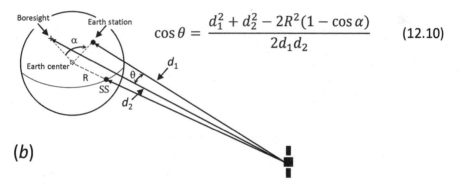

$$\cos \theta = \frac{d_1^2 + d_2^2 - 2R^2(1 - \cos \alpha)}{2d_1 d_2} \qquad (12.10)$$

(b)

Figure 12.4 *(a)* GEOSAT to earth-station geometry in the incident plane and *(b)* satellite geometry for calculating the antenna off-boresight angle.

The final step is to calculate the $G(\theta)$ using the Bessel or Sinc² antenna pattern approximations, given the satellite antenna effective

$$(\lambda/D) = 0.0314 \text{ radians or } \frac{1}{\lambda/D} = 31.85, \text{ radians}$$

Bessel approximation (Sec. 6.13.1)

The argument of the Bessel function is

$$u = \frac{\pi}{\lambda/D}\sin \theta = (31.85 \ \pi) \sin(1.02°) = 1.779$$

$$G(\theta)_{\text{Bessel}} = 64\left[\frac{J_2(u)}{u^2}\right]^2 = 0.579, \text{ ratio} \Rightarrow -2.37 \text{ dB}$$

The LOS gain in dB is

$$[G_r]_{\text{Bessel}} = [G_o] + [G(\theta)] = 34.77 + (-2.37) = 32.40 \text{ dB}$$

Sinc² approximation (Sec. 6.13.1)

From above, the argument $u = 1.779$

$$x = 0.69\, u = 0.69 * 1.779 = 1.228,\ \text{radians}$$

$$G(\theta)_{\text{sinc}} = \left(\frac{\sin(x)}{x}\right)^2 = 0.588,\ \text{power ratio} \Rightarrow -2.30\ \text{dB}$$

The LOS receiving gain in dB is

$$[G_r]_{\text{sinc}} = 34.77 + (-2.30) = 32.47\ \text{dB}$$

This value is used as the G_r in the "Link Equation" Eq. (12.6b) (or as G_t for the downlink). In practice, the satellite antenna pattern is measured and used to calculate the LOS gain. However, for this text, we use these normalized patterns for the purposes of working examples. It should be noted these two approximations are nearly equal over the antenna main beam pattern, but only the Bessel approximation should be used for estimating the extended antenna pattern including the first few sidelobes.

12.4.2 Antenna Polarization Losses

In addition to the above, losses may also result at the antenna from misalignment of the polarization direction, but these losses are usually small, and treated as if they are included in the above L_m. Also, most SATCOM links use circular polarization, which are self-aligning and do not suffer loss due to polarization misalignment. Further, it is noted that the antenna polarization misalignment losses have a random component that must be estimated, from a statistical analysis of errors observed, for many earth stations, and of course, the separate antenna misalignment losses for the uplink and the downlink must be considered.

12.4.3 Atmospheric and Ionospheric Losses

Clear-sky atmospheric gases (oxygen and water vapor) result in small losses by absorption (L_{atm}), as described in Sec. 4.2 and by Eq. (4.1). These losses are frequency-dependent over the SATCOM range of 1 to 40 GHz and decrease with higher latitudes and are generally < -1 dB. On the other hand, rain effects, especially in the tropics, can cause severe attenuation (both scattering and absorption loses), which can reduce SNRs by more than -10 dB, and this is the major source of link failures. Also, as discussed in Sec. 5.5, the ionosphere introduces a depolarization loss given by Eq. (5.19), and in subsequent calculations, the decibel value for this is denoted by $[L_{\text{ion}}]$.

12.4.4 The Link Power Budget Summary

For convenience of the reader, the generic SATCOM link equation for a single carrier is repeated here. First, for the uplink case (see Sec. 12.7)

$$P_r = P_t G_t L_p G_r L_{\text{sys}},\ \text{W}$$

where P_r *is the received power at the input to the satellite receiver,* P_t *is the earth station high-power amplifier (HPA) output in Watts,* G_t *is the peak earth station transmit antenna gain,* G_r *is the satellite receiving antenna LOS gain,* L_p *is the spreading or free-space path loss, and* L_{sys} *is the combined applicable losses.* The corresponding equation expressed in dB is

$$[P_r] = [P_t] + [G_t] + [L_p] + [G_r] + [L_{\text{sys}}],\ \text{dB}$$

In contrast to previous editions of this text, for this version, the losses are treated as gains <1, and the dB quantities are negative in Eq. (12.6b).

In a similar manner, for the downlink case (see Sec. 12.8), the generic SATCOM link equation Eq. (12.6a) is applicable, but now, P_r is the received power at the input to the earth station receiver, P_t is the satellite HPA output in Watts, G_t is the satellite LOS transmit antenna gain, G_r is the peak earth station receiving antenna gain, L_p is the spreading or free-space path loss, and L_{sys} is the combined applicable losses.

Example 12.7 A satellite uplink operating at 14.0 GHz has a free-space (path) loss $L_p = -207$ dB, transmitter feeder losses, $L_{ta} = -2.0$ dB, transmit antenna mispointing loss, $L_{m,t} = -0.5$ dB, atmospheric absorption loss $L_{atm} = -0.2$ dB, losses between the receive antenna and the receiver, $L_{ra} = -1.5$ dB, and other losses may be neglected. Calculate the total link loss for clear-sky conditions.

Solution The total link loss is the sum of all the losses in dB:

$$[\text{LOSSES}] = [L_p] + [L_{sys}] = [L_p] + [L_{ta}] + [L_{m,t}] + [L_{atm}] + [L_{ra}] + [L_{other}]$$

$$= (-207) + (-2.0) + (-0.5) + (-0.2) + (-1.5) = -211.2 \text{ dB}$$

12.5 System Noise

It is shown in Sec. 12.3 that the receiver power in a satellite link is very small, on the order of picowatts (1.0e-12 W). By itself, this is not a problem because amplification can be used to bring the signal strength up to an acceptable level (~1 milliwatt) for detection. However, electrical noise is always present at the input, and unless the signal is significantly greater than the noise, amplification is of no help because it increases the signal and noise by the same gain factor. Moreover, the situation is worsened by additional noise created during the amplification process.

The major source of electrical noise in RF electronics arises from the random thermal motion of electrons in various resistive and active devices (transistors, diodes, etc.) in the receiver. Also, thermal noise is generated in the passive lossy components of antennas and transmission lines, and blackbody radiation noise is received by the antennas (Ulaby and Long, 2014).

The available noise power from a thermal noise source is given by

$$P_N = kT_N B_N \tag{12.11}$$

Here, T_N is known as the equivalent noise temperature in Kelvin, B_N is the equivalent noise bandwidth in Hz, and $k = 1.380\text{e-}23$, J/K is Boltzmann's constant, and the noise power is in Watts.

The equivalent noise bandwidth is equal to that of an ideal rectangular bandpass filter centered on the carrier frequency given by

$$B_N = \frac{1}{G_o} \int_0^\infty G(f)\, df \tag{12.12}$$

where $G(f)$ is the receiver power gain frequency response and G_o is the maximum power gain at the carrier frequency. Note that the noise power bandwidth is always wider than the −3-dB signal bandwidth of the receiver bandpass filter, and a useful rule of thumb is that the noise bandwidth is equal to 1.12 times the −3-dB bandwidth, or $B_N \approx 1.12 \times B_{-3\text{dB}}$.

Thus, the noise power (P_N) and the noise temperature (T_N) are linearly proportional with the constant of proportionality being (kB_N). In dealing with noise, the conventional approach is to work with noise temperature in Kelvin (rather than power in Watts); however, the transformation to noise power is made by multiplying the noise temperature by kB_N.

A key characteristic of thermal noise assumed in the RF and MW frequency bands is that the thermal noise can be treated as having a *flat power spectrum with respect to frequency*. The noise power per unit bandwidth (W/Hz) is a constant $= kT_N$, which is known as the *noise power spectral density, N_o,* expressed as

$$N_o = P_N/B_N = k\,T_N,\ \text{J} \tag{12.13}$$

Moreover, the noise temperature is directly related to the physical temperature of the noise source, but it is not always equal, as described in the following sections. Because the various sources of noise voltages are independent Gaussian random variables with uniformly distributed random phase, when they are combined, their powers (variances) are additive to yield the total noise power.

12.5.1 Antenna Noise

Antennas operating in the receiving mode introduce noise into the satellite circuit, which is undesirable but totally unavoidable. Thus, noise is introduced by both the satellite receive and the earth station receive antennas, and although the physical origins of the noise in either case are similar, the magnitudes of the effects differ significantly. For example, satellite antennas are generally pointed toward the earth. They receive the thermal radiation from the earth's surface, and equivalent noise temperature at the antenna aperture (T_{ap}) is approximately 290 K. However, it should be noted that this is the typical noise temperature of land surfaces, whereas the typical value for fresh water or ocean is ~100 K for horizontal polarized antennas and ~200 K for vertical polarization or ~150 K for circular polarization antennas. Therefore, for convenience, throughout this discussion, we assume a worst-case noise temperature for the Earth to be 290 K.

On the other hand, for an upward-looking (zenith) earth station receiving antenna, the sky noise is a random, nonpolarized radiation flux that originates from all matter at finite temperatures throughout the universe that is described by Planck's blackbody radiation law (Ulaby and Long, 2014). At the receiving antenna aperture, the equivalent noise temperature of the sky, is given by the lower curve in Fig. 12.5. Over the SATCOM frequency range of 1 GHz to 10 GHz, the atmosphere is very transparent, and the noise temperature is approximately 3 K that corresponds to the background noise temperature of deep space. Above 10 GHz, the atmosphere is partially opaque because of molecular absorption by water vapor and oxygen (as shown in Fig. 4.2), and the resulting noise temperature increases with frequency. Also, as antenna elevation angle decreases, the T_{ap} increases because of the longer atmospheric path such that

$$T_{ap}(\theta) = (T_{ap})_{zenith} * \sec(\theta),\ \text{K} \tag{12.14}$$

where θ is the incident angle of the receiving antenna LOS relative to the zenith direction. Moreover, for satellite communications links, the dominate source of atmospheric noise (not shown) is rainfall that introduces additional noise (over clear sky) and signal attenuation, which are discussed in Secs. 4.4 and 12.9.

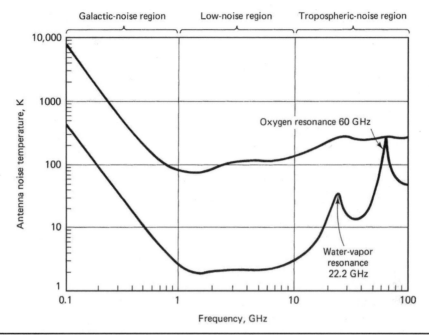

FIGURE 12.5 Typical noise temperature (T_{ap}) of an ideal, ground-based antenna (very narrow beam without sidelobes or RF losses). The maximum (upper curve) has two regions of beam pointing, namely: (a) frequencies 0.1 to 1 GHz, pointing to the galactic center and (b) for frequencies >1 GHz, pointing just above the horizon (~5° at C band and 10° at Ku-band) that includes some earth radiation contributions. The minimum (lower curve) also has two regions of beam pointing, namely: (a) frequencies 0.1 to 1 GHz, pointing to the galactic pole and (b) for frequencies >1 GHz, zenith-pointing. The low-noise region between 1 and 10 GHz is most amenable to application of special, low-noise antennas. (*From Philip F. Panter, "Communications Systems Design," McGraw-Hill Book Company, New York, 1972. With permission.*)

12.5.2   Noise Model for Linear Circuits

Before developing the noise model for a communications receiver, it is instructive to consider a noisy linear component (device) given in Fig. 12.6 with parameters: power gain (G), noise bandwidth (B_N), and the circuit noise figure (NF), which represents the additional noise added by the device. This is a two-port network (input and output ports), but for illustrative purposes, we separate the signal and noise components of power; however, they physically occur at the same port.

The input signal power, S_{in}, and output signal power, S_{out}, are related by the component gain,

$$S_{out} = G S_{in} \tag{12.15}$$

but the noise output power is

$$N_{out} = G N_{in} + \Delta N_{out} \tag{12.16}$$

where the first term is the transmission (amplification) of the input noise power and the second term is the additional noise power contributed by the noisy linear system.

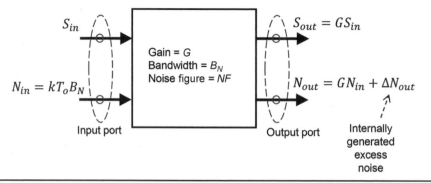

Figure 12.6 Noise Model for two-port linear circuit.

The noise factor (F) is a measured parameter that relates the input and output SNRs given as

$$F = \text{SNR}_{in}/\text{SNR}_{out}, \text{ power ratio} \qquad (12.17a)$$

which can be written as

$$F = \frac{S_{in}}{N_{in}} * \frac{N_{out}}{S_{out}} = \frac{1}{G} * \frac{GN_{in} + \Delta N_{out}}{N_{in}} = 1 + \frac{\Delta N_{out}}{GN_{in}} \qquad (12.17b)$$

Substituting $N_{in} = kT_oB$ and solving for ΔN_{out} yields

$$\Delta N_{out} = GkB_N * (F - 1) * T_o = GkB_N T_{e,in}, \text{ W} \qquad (12.18)$$

where

$$T_{e,in} = (F - 1)T_o, \text{ K} \qquad (12.19)$$

is the equivalent excess noise temperature referred to the device input port. For noise factor measurements, $T_o = 290$ K by IEEE standard. Also, the noise figure (NF) is the name given for the noise factor expressed as dB, i.e., $NF = 10 \text{ Log}(F)$.

The equivalent noise diagram is presented in Fig. 12.7, where the two-port network is now "noise-free." It should be noted, that when using this diagram, both signal and noise are treated identically, and the respective outputs are the input signal or noise powers times the device gain. Further, the total input noise (known as *system noise*) is the sum of $(N_{in} + N_{e,in})$. Moreover, it is standard practice to work in *noise temperature* (rather than *noise power*), so the input system noise temperature $T_{sys} = (T_o + T_{e,in})$.

Finally, the noise output can be given as

$$N_{out} = GkT_{sys}B_N \qquad (12.20)$$

where

$$T_{sys} = T_{in} + T_{e,in}, \text{ K} \qquad (12.21)$$

and the noise power density at the output is

$$N_{o,out} = N_{out}/B_N, \text{ J} \qquad (12.22)$$

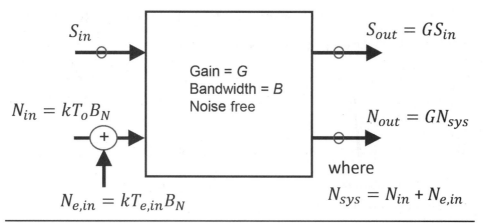

FIGURE 12.7   Noise equivalent model of a noisy device.

It is important to note that using this model, the resulting SNR is constant such that: $(\text{SNR}_{in}) = (\text{SNR}_{out})$. On the other hand, the noise power density is not constant such that at the output is

$$N_{o,out} = GN_{o,in}, \text{ J} \tag{12.23}$$

12.5.3   Low Noise Amplifier Noise Temperature

Consider now, the noise representation of the first amplifier component of a typical receiver, which is the *low noise amplifier* (LNA) shown in Fig. 12.8a. Since the noise power spectrum is flat (i.e., constant over the frequency range of interest), it is convenient to work with the noise energy in Joules. The input noise energy to the LNA is the sum from the antenna $(N_{o,ant} = kT_{ant})$ and the internal noise of the LNA referred to the input $(N_{o,lna} = kT_{lna})$. From Eq. (12.23), the noise energy output is

$$N_{o,out} = Gk(T_{ant} + T_{lna}) = GkT_{sys}, \text{ J} \tag{12.24}$$

where $T_{sys} = T_{ant} + T_{lna}$, K.

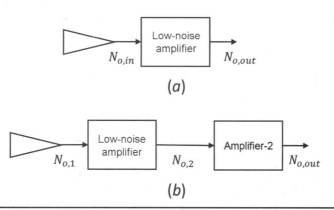

FIGURE 12.8   Circuit used in finding equivalent noise temperature of (a) an amplifier and (b) two amplifiers in cascade.

Example 12.8 An earth station antenna viewing a satellite has an antenna noise temperature of 35 K, and it is matched into a receiver that has a noise temperature of 100 K and a noise bandwidth of 3.0 MHz. Calculate the LNA input (*a*) noise power density and (*b*) noise power.

Solution

(*a*) $T_{ant} = 35$ K and $T_{lna} = 100$ K, thus

$$T_{sys} = T_{ant} + T_{lna} = 135 \text{ K}$$
$$N_o = kT_{sys} = (135) \times 1.380\text{e-}23 = 1.86\text{e-}21, \text{ J}$$

(*b*) $N_{in} = N_o B_N$

$$= 1.86\text{e-}21 \times 3.0\text{e+}06 = 0.0056, \text{ pW} \Rightarrow -142.5, \text{ dBW}$$

12.5.4   Noise Temperature for Amplifiers in Cascade

The cascade connection for the LNA and the second amplifier is shown in Fig. 12.8*b*. For this arrangement, the overall power gain is

$$G_{sys} = G_{lna} G_2 \tag{12.25}$$

The noise energy of amplifier-2 referred to its own input is kT_{e2}

The noise input to amplifier-2 from the LNA output is:

$$G_{lna} k (T_{ant} + T_{e1}),$$

therefore, the total noise energy referred to amplifier-2 input is

$$N_{o,2} = G_{lna} k (T_{ant} + T_{e1}) + kT_{e2} \tag{12.26}$$

This noise energy may be referred to the input of amplifier-1 (LNA) by dividing by the power gain

$$N_{o,1} = N_{o,2}/G_{lna} = k\left(T_{ant} + T_{e1} + \frac{T_{e2}}{G_{lna}} \right) \tag{12.27}$$

The system noise energy is kT_{sys} ; therefore, for the cascaded amplifiers

$$T_{sys} = T_{ant} + T_{e1} + \frac{T_{e2}}{G_1} \tag{12.28}$$

This is a very important result because it shows that, when referred to the receiver input, the effect of noise temperature of the second stage is divided by the power gain of the first stage. Because we desire to keep the overall system noise as low as possible, the first stage should have high power gain as well as a low noise temperature, i.e., be an LNA.

This result may be generalized to "*n*" number of stages in cascade, giving

$$T_{sys} = T_{ant} + T_{e1} + \frac{T_{e2}}{G_1} + \frac{T_{e3}}{G_1 G_2} + \cdots + \frac{T_{en}}{G_1 G_2 \dots G_{n-1}} \tag{12.29}$$

Example 12.9 An LNA is connected to a receiver which has a noise figure of 12 dB. The gain of the LNA is 40 dB, and its noise temperature is 120 K. Calculate the overall receiver noise temperature (T_{rec}) referred to the LNA input.

Solution   Case-1:

$$NF = [12 \text{ dB}] \Rightarrow 15.849, \text{ power ratio; therefore,}$$
$$T_{e2} = (15.849 - 1) \times 290 = 4306.2 \text{ K}$$

A gain of 40 dB is a power ratio of 10^4, and therefore,

$$T_{sys} = 120 + \frac{4306.2}{10^4}$$

$$= 120.43 \text{ K}$$

Case-2: Suppose that an LNA is used with a lower gain = 30 dB, what is the impact on the T_{rec}?

$$G = 30 \text{ dB} \Rightarrow 10^3, \text{ power ratio}$$

$$T_{sys} = 120 + \frac{4306.2}{10^3} = 124.3 \text{ K}$$

The change in the system noise density is

$$\Delta N_{o,sys} = \frac{124.3}{120.43} = 1.032 \Rightarrow 0.14 \text{ dB}$$

Case-3: Suppose that an LNA with a lower gain = 20 dB is used, what is the impact on the T_{rec}?

$$G = 20 \text{ dB} \Rightarrow 100, \text{ power ratio}$$

$$T_{sys} = 120 + \frac{4306.2}{100} = 163.1 \text{ K}$$

The change in the system noise density is

$$\Delta N_{o,sys} = \frac{163.1}{120.43} = 1.354 \Rightarrow 1.32 \text{ dB}$$

Also, even though the main receiver has a very high noise temperature, its effect is made negligible by the high gain of the LNA; however, reducing the LNA gain to less than 30 dB has a significant degradation of the system noise threshold.

12.5.5 Noise Temperature of Absorptive Passive Networks

An *absorptive passive network* is one that contains only resistive elements (i.e., no active elements that provide amplification). These introduce losses by absorbing energy from the signal (and noise) and converting this energy into heat. Resistive attenuators, transmission lines, and waveguides are all examples of absorptive networks. Even rainfall, which absorbs energy from radio signals passing through it, can be considered a form of absorptive network. Because an absorptive network absorbs energy, it also generates broadband thermal RF/MW noise emissions as well as introducing signal losses.

Since we choose to treat the losses as a transmission gain <1, we can calculate the cascaded gain by simply multiplying ratios, which we believe is a more intuitive approach than dividing by losses. Using losses as gains <1 is especially useful for spreadsheet calculations, where values of dB are often used. For our approach, all cells in ratio are multiplied and the equivalent cells in dB are added; rather than having to multiply and divide cells in ratio and add the cells for positive dB's and subtract the losses (in dB). For absorptive passive networks, the power loss ratio is simply the ratio of output signal power to input signal power, and this is always less than unity.

Consider an absorptive network shown in Fig. 12.9, which has a power loss $L < 1$. Let the network be matched at both ends with terminating resistors (matched loads of the transmission line). Since both ports are matched, there are no reflections, and the entire network is at uniform physical temperature T_{phy}.

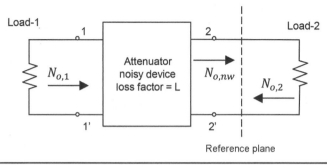

FIGURE 12.9a   Noise Model for an absorptive passive network in thermal equilibrium.

The noise energy transferred from the Load-1 into the network is

$$N_{o,1} = kT_{phy}, \text{ J} \tag{12.30}$$

and the network output noise energy is

$$N_{o,nw} = LkT_{phy} + kT_{e,out} \tag{12.31}$$

where $T_{e,out}$ is the equivalent network noise temperature at the output. Because the network is in thermal equilibrium, the energy transfer from the left-hand side of the reference plane must equal the noise energy transferred from the Load-2, into the network, which equals $N_{o,2} = kT_{phy}$. Solving for $T_{e,out}$ yields

$$T_{e,out} = (1 - L)\, T_{phy} \tag{12.32}$$

Next, referring this output noise energy to the network input (by dividing by the gain) is

$$T_{e,in} = (1 - L)\, T_{phy} / L \tag{12.33}$$

The noise equivalent model for signal and noise flow is shown in Fig. 12.9b. If the ambient physical temperature were $T_{phy} = T_o = 290$ K, then from Eq. (12.17a), the noise factor is

$$F = \left(\frac{1}{L}\right)\left(\frac{T_{e,in}}{T_{e,out}}\right) = \left(\frac{1}{L}\right), \text{ power ratio}$$

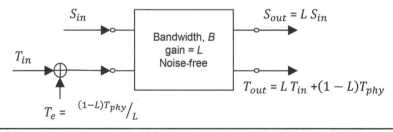

FIGURE 12.9b   Noise equivalent model for an absorptive passive circuit.

Now, the output noise temperature is

$$T_{out} = LT_{in} + (1-L)T_{phy} \qquad (12.34)$$

where the first term is the transmitted input noise, and the second term is the self-emission noise.

Note that the term $(1 - L)$ is the decimal fraction of the input noise temperature absorbed, and in blackbody radiation theory, this is the noise emissivity coefficient. The magnitude of the noise self-emission temperature is the product of the emissivity coefficient times the physical temperature of the loss in Kelvin.

Example 12.10 Consider an antenna with ohmic losses at a physical temperature T_{phy}, represented by the block diagram below. Calculate the input noise temperature (T_{ant}) delivered to the receiver as a function of the ohmic loss (L_{ant}) and the apparent noise temperature (T_{ap}).

Solution This problem can be modeled using Eq. (12.34), where T_{in} is the apparent noise temperature collected by the lossless antenna (Fig. 12.10), and $L = L_{ant}$. The results are displayed in tabular form given below.

Calculated T_{ant} for selected T_{ap} and L_{ant}, and a physical temperature = 290 K.

T_{ap}, K	L_{ant}, dB	L_{ant}, Ratio	T_{ant}, K	T_{ap}, K	L_{ant}, dB	L_{ant}, Ratio	T_{ant}, K
10	−0.1	0.9772	16.4	20	−0.1	0.9772	26.1
	−0.2	0.9550	22.6		−0.2	0.9550	32.2
	−0.4	0.9120	34.6		−0.4	0.9120	43.8
	−0.6	0.8710	46.1		−0.6	0.8710	54.8
	−0.8	0.8318	57.1		−0.8	0.8318	65.4
	−1	0.7943	67.6		−1	0.7943	75.5
	−1.5	0.7079	91.8		−1.5	0.7079	98.9
	−2	0.6310	113.3		−2	0.6310	119.6
	−2.5	0.5623	132.5		−2.5	0.5623	138.2
	−3	0.5012	149.7		−3	0.5012	154.7

T_{ap}, K	L_{ant}, dB	L_{ant}, Ratio	T_{ant}, K	T_{ap}, K	L_{ant}, dB	L_{ant}, Ratio	T_{ant}, K
100	−0.1	0.9772	104.3	290	−0.1	0.9772	290.0
	−0.2	0.9550	108.6		−0.2	0.9550	290.0
	−0.4	0.9120	116.7		−0.4	0.9120	290.0
	−0.6	0.8710	124.5		−0.6	0.8710	290.0
	−0.8	0.8318	132.0		−0.8	0.8318	290.0
	−1	0.7943	139.1		−1	0.7943	290.0
	−1.5	0.7079	155.5		−1.5	0.7079	290.0
	−2	0.6310	170.1		−2	0.6310	290.0
	−2.5	0.5623	183.2		−2.5	0.5623	290.0
	−3	0.5012	194.8		−3	0.5012	290.0

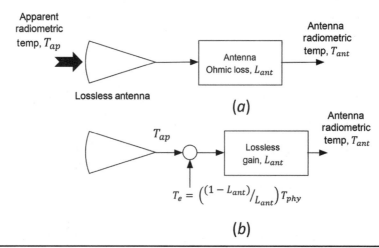

Figure 12.10   (a) Model for an antenna with ohmic losses and (b) Noise equivalent model.

12.5.6   Overall System Noise Temperature

A typical earth station SATCOM receiving system is shown in Fig. 12.11a. By applying the results of the previous sections, the system noise temperature is

$$T_{sys} = T_{ant} + T_{e1} + T_{e2}/G_{lna} + T_{e3}/(G_{lna}L_{coax}) \qquad (12.35)$$

from Eq. (12.33), $T_{e2} = (1 - L_{coax})T_{phy}/L_{coax}$ and

from Eq. (12.19), $T_{e3} = (F_{recv} - 1)290$

that becomes

$$T_{sys} = T_{ant} + T_{lna} + ((1 - L_{coax})T_{phy}/L_{coax})/G_{lna} + (F_{recv} - 1)290/(G_{lna}L_{coax}) \qquad (12.36)$$

The significance of the individual terms is illustrated in the following examples.

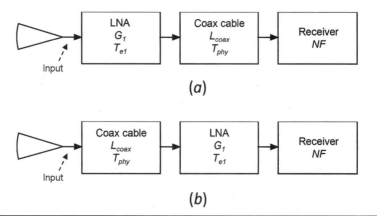

Figure 12.11   Typical earth station satellite TV receiver (a) low-noise amplifier located at the antenna feed and (b) low-noise amplifier located at the in-doors receiver.

Example 12.11 Consider a satellite TV receiver shown in Fig. 12.11a, with a small dish antenna with preamplifier (LNA) mounted on the home roof and a 10 m coax cable connecting the LNA to the TV receiver inside the home. For this configuration, the equipment rental is $15/month plus the TV subscription.

An alternate configuration of the similar equipment in Fig. 12.11b has the antenna connected to the preamplifier by the coax rather than directly to the antenna. Because the indoor preamplifier need not be weatherproof, the equipment is cheaper to build so rental is reduced to $10/month.

For both configurations the component RF specifications are identical, namely: LNA noise temperature $T_{LNA} = 150$ K and RF power gain $G_{LNA} = 40$ dB; cable loss $L_{coax} = -3.0$ dB; receiver noise figure $NF = 12$ dB; $T_{ant} = 35$ K; and the ambient temperature $T_{phy} = 290$ K.

Is the increased cost justified for configuration-(a) and explain why or why not? (Hint) calculate the noise temperature referred to the antenna output for both configurations and discuss relative advantages/disadvantages.

Solution   Part-a

From Eq. (12.33), $T_{e2} = ((1 - L)T_{phy}/L)$, and $[L_{coax}] = -3.0$ dB $\Rightarrow 0.5012$;

therefore, $T_{e2} = 288.6$ K

From Eq. (12.19), $T_{e3} = (F - 1)\ 290$, where $F = 10^{(12/10)} = 15.849$;

therefore, $T_{e3} = 4,306.2$ K

From Eq. (12.36), we calculate the T_{sys}, where $T_{ant} = 35$; $T_{e2} = 150$; $G_{LNA} = 1.0e + 04$;

$$T_{sys,a} = 35 + 150 + \frac{288.6}{10,000} + \frac{4306.2}{0.5 * 10,000}$$

$$= 185 + 0.03 + 0.86 = 185.9 \text{ K}$$

Part-b   In this case the cable precedes the LNA, and therefore, the equivalent noise temperature referred to the antenna output is:

$$T_{receiver,b} = T_{e1} + \frac{T_{e2}}{L_{coax}} + \frac{T_{e3}}{(L_{coax} * G_{LNA})}$$

where $T_{ant} = 35$; $T_{e1} = (1 - 0.5012)\ 290/0.5012 = 288.6$ K, $T_{e2} = 150$ and $T_{e3} = 4,306.5$

From Eq. (12.36), we can calculate the systems temperature $= T_{ant} + T_{receiver,b}$

$$T_{sys,b} = 35 + 288.6 + 150/0.5012 + \frac{4306.5}{0.5012 * 10,000}$$

$$= 35 + 288.6 + 299.3 + 0.86 = 623.8 \text{ K}$$

What is the increase in the noise power density $N_{o,sys}$ for the second case compared to the first?

$$\Delta N_{o,sys} = \frac{T_{sys,b}}{T_{sys,a}} = \frac{623.8}{185.9} = 3.36 \Rightarrow 5.26 \text{ dB}$$

So, the $5/month higher equipment rental provides significant noise reduction in clear sky; but this becomes even more important during rain events. Therefore, if the subscriber lives where it rains, then the extra cost ensures a much more reliable service without rain outages.

This example illustrates why the LNA must be placed ahead of the cable, which is consistent with preamplifiers mounted directly to the feed of the dish in most home satellite TV systems. This problem is revisited in Sec. 12.6.

12.6 Carrier-to-Noise Ratio

In communications theory, the quality of the link is a function of the signal power to noise power ratio. Moreover, for satellite communications, we frequently use nonlinear HPAs on the satellite because of their wide-bandwidth and improved DC to RF power efficiency. As a result, only carrier angle-modulation (phase PM or frequency FM) can pass through the HPA without distortion. The satellite transponder bandwidth required for a single modulated carrier depends upon the modulation scheme used, but according to Parseval's theorem, the total power in the modulated carrier spectrum is equal to the unmodulated continuous wave carrier power. Assuming that the transponder/earth station receiver bandwidth is sufficient to capture the entire modulated carrier spectrum, then, the signal power is known, independent of the modulation used. Thus, we use the unmodulated carrier power as the signal power and calculate the equivalent carrier-to-noise ratio (C/N or CNR). However, it is important to recognize that occasionally for digital modulation, it is advantageous for the transponder bandwidth to be less than the modulated carrier spectrum width. For these cases, there is an additional loss that must be included in the L_{ta} loss to account for the reduction of the signal power.

A measure of the performance of a satellite link is the C/N, which is calculated at the receiver input. Link-budget calculations are usually concerned with determining this ratio given as

$$\frac{C}{N} = P_r(1/P_N), \text{ power ratio} \tag{12.37a}$$

where P_R is the signal power and the P_N is the noise power at the receiver input, or in terms of decibels,

$$\left[\frac{C}{N}\right] = [P_r] + [1/P_N] \tag{12.37b}$$

Note: that by expressing Eq. (12.37a) as a product of power ratios, Eq. (12.37b) becomes the sum of the terms, expressed as dB; and when the terms, expressed as a power ratios, are <1, then the dB is negative, as is the case for $[1/P_N]$. This approach is convenient for using spreadsheets, where the cells are multiplicative for ratio and additive for dB.

Equations (12.6) and (12.11) are used for $[P_R]$ and $[P_N]$, respectively, resulting in

$$\left[\frac{C}{N}\right] = [\text{EIRP}] + [G_R] + [\text{LOSSES}] + [1/k] + [1/T_{\text{sys}}] + [1/B_N] \tag{12.38}$$

Since the G/T ratio is a key parameter in specifying the receiving system performance, the antenna gain G_R and the system noise temperature T_{sys} are combined as

$$[G/T] = [G_R] + [1/T_{\text{sys}}], \text{ dBK}^{-1} \tag{12.39}$$

Therefore, the link equation becomes

$$\left[\frac{C}{N}\right] = [\text{EIRP}] + [G/T] + [\text{LOSSES}] + [1/k] + [1/B_N] \tag{12.40}$$

However, for digital SATCOM links, the ratio of carrier power to noise power density (P_r/N_o) is the desired reference quantity. Since $P_N = kT_N B_N = N_o B_N$, then

$$\frac{C}{N} = \frac{C}{N_o B_N} = \frac{1}{B_N}\frac{C}{N_o}, \text{ power ratio} = \text{Hz}^{-1} * \text{W/J (dimensionless)}$$

Solving for (C/N_o) yields

$$\frac{C}{N_o} = \left(\frac{C}{N}\right)B_N$$

and therefore

$$\left[\frac{C}{N_o}\right] = \left[\frac{C}{N}\right] + [B_N] \tag{12.41}$$

where $[C/N]$ is the carrier-to-noise power ratio in dB, and $[B_N]$ is the noise bandwidth expressed in dB relative to 1 Hz, or dBHz. Thus, the units for $[C/N_o]$ are dBHz.

Combining Eqs. (12.40) and (12.41) cancels the $[B_N]$ terms and yields

$$\left[\frac{C}{N_o}\right] = [\text{EIRP}] + [G/T] + [\Sigma \text{ LOSSES}] + [1/k]$$

$$= [\text{EIRP}]_i + [G/T]_j + [L_p]_j + [L_{ta}]_j + [L_m]_j + [L_{atm}]_j + [L_{ra}]_j + [1/k] \tag{12.42}$$

This equation is applicable to both the uplink and downlink, and the subscript j = "up" or "dn" are used, respectively.

Example 12.12 In a link-budget calculation at 12.0 GHz, the EIRP = 48.0 dBW, $[G/T]$ = 19.5 dB/K, free-space loss is –206.0 dB, transmit antenna pointing loss is –0.4 dB, atmospheric absorption is –2.0 dB, and receiver feeder losses are –1.0 dB. Calculate the carrier-to-noise spectral density ratio.

Solution The data are best presented in tabular form and readily lend themselves to spreadsheet-type computations, which are made for both linear (power ratio) and dB units. While this may seem redundant, it is good practice (especially for novices) because it provides the ability to detect errors that may have occurred during the conversion from ratio to dB or vice versa. These conversions are always required because the link parameters are customarily given in both units (e.g., transmitter powers and distances are normally given in linear units, while gains and losses are normally given in dB). By constructing spreadsheets in this manner, there are two ways of calculating the same parameter, and by converting the linear result to dB, the two answers should match. This is illustrated in the following table—see the bottom row and cell to the right of the "check ⇒," which is 10*log(linear product).

Link Parameter	Linear	dB
EIRP	6.310E+04	48.0
G_r/T_{sys}	8.913E+01	19.5
Free-space path loss, L_p	2.512E-21	−206.0
Atmos absorption loss, L_{atm}	0.63	−2.0
Xmt *Ant* mispointing loss, L_{mT}	0.912	−0.40
Recv *Ant* feeder loss, L_{ra}	0.794	−1.0
(1/k)	7.246E+22	228.6
C/N_o	4.679E+08	86.7
	check ⇒	86.7

If the noise bandwidth is 2.0 MHz, what is the C/N?

Link Parameter	Linear	dB
EIRP	6.310E+04	48.0
G_r/T_{sys}	8.913E+01	19.5
Free-space path loss, L_p	2.512E-21	−206.0
Atmos absorption loss, L_{atm}	0.63	−2.0
Xmt *Ant* mispointing loss, L_{mT}	0.912	−0.40
Recv *Ant* feeder loss, L_{ra}	0.794	−1.0
$(1/k)$	7.246E+22	228.6
$(1/B_N)$	5.00E-07	−63.01
C/N	233.93	23.7
	check $\Rightarrow$	23.7

Example 12.13 In the Sec. 12.5, we discussed the system noise for a home satellite TV receiving system, so now consider the C/N for this system. Given that the carrier power at the receiving antenna output $C = -120.0$ dBW in clear sky and that the equivalent noise bandwidth $B_N = 5.0$ MHz, find the C/N for the two system configurations (Fig. 12.11).

Solution Part-a
From Example 12.11 and Fig. 12.11*a*, a satellite TV receiver has a small dish antenna and the preamplifier (LNA) mounted on the home roof, which resulted in $T_{sys} = 185.9$ K.
The noise power is

$$N = kT_{sys}B_N$$
$$= 1.38\text{e-}23 * 185.9 * 5.0\text{e}+06 = 1.28\text{e-}14, \text{W} \Rightarrow -138.9 \text{ dBW}$$

The carrier power $= -120.0$ dBW $\Rightarrow 1.0\text{e-}12$, W

The $C/N = 1.0\text{e-}12/1.28\text{e-}14 = 78.0 \Rightarrow 18.92$, dB

The minimum acceptable TV quality requires a threshold $[C/N]_{Th} = 11.0$ dB, which gives a margin of 7.9 dB for rain interference. This is a very acceptable value to prevent frequent rain degradation of service.

Part-b
From Example 12.11 and Fig. 12.11*b*, a satellite TV receiver has the same antenna with a 10 m coax cable connecting the LNA/receiver inside the home, which resulted in a $T_{sys} = 623.8$ K.
The noise power is

$$N = kT_{sys}B_N$$
$$= 1.38\text{e-}23 * 623.8 * 5.0\text{e}+06 = 4.30\text{e-}14, \text{W} \Rightarrow -133.7 \text{ dBW}$$

The carrier power $= -120.0$ dB $\Rightarrow 1.0\text{e-}12$, W

The $C/N = 1.0\text{e-}12/4.30\text{e-}14 = 23.2 \Rightarrow 13.7$, dB

Compared to the minimum acceptable threshold $[C/N]_{Th} = 11.0$ dB, this gives a margin of only 2.7 dB for rain interference. Except for desert regions, where rain is infrequent, this low margin is likely to result in severe TV quality degradation during random rain events with durations of 10s of seconds to minutes for a single event. Given this large difference in the probability of rain outages, the small cost differential for system-a is considered a reasonable solution.

12.7   The Uplink

The uplink of a satellite circuit is the one in which the earth station is transmitting the signal and the satellite is receiving it. Equation (12.42) can be applied to the uplink with the subscript "*up*" used to denote specifically that the uplink is being considered. Thus:

$$\left[\frac{C}{N_o}\right]_{up} = [EIRP]_{up} + [G/T]_{up} + [L_p]_{up} + [L_{ta}]_{up} + [L_m]_{up} + [L_{atm}]_{up} + [L_{ra}]_{up} + [1/k]$$

where $[EIRP]_{up}$ is the earth station single carrier EIRP, $[G/T]_{up}$ is the satellite receiving antenna LOS gain divided by the uplink system noise temperature, $[L_p]_{up}$ is uplink path loss, $[L_{ta}]_{up}$ is the earth station transmitter feeder loss, $[L_m]_{up}$ is the earth station transmit antenna mispointing loss, $[L_{atm}]_{up}$ is the atmospheric absorption loss at the uplink frequency, and $[L_{ra}]_{up}$ is the satellite receiver feeder loss. The resulting carrier-to-noise density ratio is at the input to the satellite receiver. Further, in some instances, the uplink saturation flux density at the satellite receive antenna is specified (rather than the earth-station EIRP), and for this case, Eq. (12.43) is modified as presented next.

12.7.1   Saturation Flux Density

As explained in Sec. 7.7.3, the *traveling-wave tube amplifier* (TWTA) in the satellite transponder exhibits power output saturation, as shown in Fig. 7.20. The flux density at the satellite receiving antenna aperture required to produce saturation of the TWTA is termed the *saturation flux density* (Ψ_S), which is a specified quantity in link budget calculations. If Ψ_S is known, then it is possible to calculate the corresponding saturation $EIRP_S$ at the transmit earth station. To show this, consider again Eq. (12.3) which gives the saturation flux density in terms of $EIRP_S$, repeated here for convenience.

$$\Psi_S = \frac{EIRP_S}{4\pi d^2}$$

Solving for the saturated $EIRP_S$ and using decibel notation is

$$[EIRP_S]_{up} = [\Psi_S] + [4\pi d^2]_{up} \tag{12.43}$$

where $[4\pi d^2]_{up} = 10\log(4\pi d^2)$.

This equation is derived on the basis that the only loss present is the spherical spreading loss; but as shown in the previous sections, the other applicable losses: are the transmitter feeder loss $[L_{ta}]$, the atmospheric absorption loss $[L_{atm}]$ at the uplink frequency, and the transmitting antenna misalignment loss (including polarization mismatch) $[L_m]$. When allowance is made for these, the required EIRP becomes

$$\Psi_S = \frac{EIRP_S}{4\pi d^2} * L_{ta} * L_{atm} * L_m = \frac{EIRP_S}{4\pi d^2} * L_{sys} \tag{12.44}$$

Solving for $EIRP_S$ is

$$EIRP_S = \Psi_S(4\pi d^2)\left(\frac{1}{L_{ta}}\right)\left(\frac{1}{L_{atm}}\right)\left(\frac{1}{L_m}\right) \tag{12.45a}$$

Converting to dB yields

$$[EIRP_S]_{up} = [\Psi_S] + [4\pi d^2]_{up} + [1/L_{ta}]_{up} + [1/L_{atm}]_{up} + [1/L_m] \tag{12.45b}$$

This is for clear-sky conditions that gives the *minimum* value of [EIRP] that produces the required saturation flux density at the satellite, with saturation values denoted by the subscript "$_s$."

Example 12.14   For an uplink, the saturation flux density is $[\Psi_S] = -120$ dB(W/m²). The uplink slant range $d = 38{,}200$ km, and the system losses $[L_{sys}] = -2.0$ dB or ($[1/L_{sys}] = +2.0$ dB). Calculate the earth-station $[\text{EIRP}_S]$ required for saturation, assuming clear-sky conditions.

Solution   From Eq. (12.45b)

$$[\text{EIRP}_S]_{up} = [\Psi_S] + [4\pi d^2]_{up} + [1/L_{sys}]_{up}$$

$$4\pi d^2 = 4\pi (3.82e + 07)^2 = 1.83e + 16 \Rightarrow 162.63, \text{ dB}$$

$$[\text{EIRP}_S]_{up} = -120 + 162.63 + 2.0 = 44.63, \text{dBW}$$

Example 12.15   An uplink operates at 14.0 GHz, and the flux density required to saturate the transponder $[\Psi_S] = -120$ dB (W/m²). The free-space loss $[L_p] = -207.0$ dB, and the other propagation losses $[L_{sys}] = -2.0$ dB. Calculate the earth-station $[\text{EIRP}_S]$ required for saturation, assuming clear-sky conditions.

Solution   In this case, because the slant range (d) to the GEOSAT is not given, we cannot use the same approach as above. However, since the path loss is given, we can solve for the slant range using Eq. (12.5a).

First calculate the wavelength: $\lambda = c/f = 3.0e + 08/14.0e + 09 = 0.0214$ m

Then, solve for the slant range: $d = \dfrac{\lambda}{4\pi\sqrt{L_p}}$

$$[L_p] = -207.0, \text{dB} \Rightarrow 2.00e\text{-}21, \text{ratio}$$

$$\sqrt{L_p} = 4.467e - 11, \text{ratio}$$

$$d = 0.0214/(4\pi*4.467e\text{-}11) = 3.82e + 07, \text{m}$$

Next, find: $[4\pi*d^2] = 10 \log (4\pi*d^2) = 162.63, \text{dB}$

From Eq. (12.45b); $[\text{EIRP}_S]_{up} = [\Psi_S] + [4\pi d^2]_{up} + [1/L_{sys}]_{up}$

$$= -120.0 + 162.63 + 2.0 = 44.63, \text{dBW}$$

12.7.1.1 Alternate Calculation for $[\text{EIRP}_S]_{up}$

This approach may be used when the saturation flux density is given but the range to the GEOSAT is unknown. To understand this derivation, consider again Eq. (12.3), which gives the saturation flux density in terms of EIRP_S:

$$\Psi_S = \frac{\text{EIRP}_S}{4\pi d^2}$$

But, from Eq. (12.5a) the free-space path loss is

$$L_p = \left(\frac{1}{4\pi \left(\dfrac{d}{\lambda} \right)} \right)^2 = \left(\frac{\lambda^2}{4\pi} \right) \left(\frac{1}{4\pi d^2} \right)$$

Rearranging terms becomes

$$\left(\frac{1}{4\pi d^2}\right) = \left(\frac{4\pi}{\lambda^2}\right) L_p$$

and substituting into Eq. (12.5a) becomes

$$\Psi_S = \text{EIRP}_S L_p\left(\frac{4\pi}{\lambda^2}\right) \tag{12.46}$$

The $\lambda^2/4\pi$ term has dimensions of area, and in fact, from Eq. (6.15) it is the effective area of an isotropic antenna. Denoting this by A_o gives

$$A_o = \frac{\lambda^2}{4\pi} \tag{12.47}$$

Combining this with Eq. (12.46) and solving for the EIRP_S is

$$\text{EIRP}_S = \Psi_S A_o\left(\frac{1}{L_p}\right)_{up} \tag{12.48a}$$

$$[\text{EIRP}_S] = [\Psi_S] + [A_o] + \left[\frac{1}{L_p}\right]_{up} \tag{12.48b}$$

This equation is derived on the basis that the only loss present is the spreading path loss, denoted by $[L_p]$. But, as shown in Sec. 12.7.1, the other propagation losses are present as system losses $[L_{sys}]$. When allowance is made for these, Eq. (12.48a) becomes

$$\text{EIRP}_{Sup} = \Psi_S A_o\left(\frac{1}{L_p}\right)_{up}\left(\frac{1}{L_{sys}}\right)_{up} \tag{12.49a}$$

$$[\text{EIRP}_S] = [\Psi_S] + [A_o] + [1/L_p]_{up} + [1/L_{ta}]_{up} + [1/L_{atm}]_{up} + [1/L_m] \tag{12.49b}$$

Example 12.16 An uplink operates at 14.0 GHz, and the flux density required to saturate the transponder $[\Psi_S] = -120$ dB(W/m²). The free-space loss $[L_p] = -207.0$ dB, and the other propagation losses amount to $[L_{sys}] = -2.0$ dB. Calculate the earth-station $[\text{EIRP}_S]$ required for saturation, assuming clear-sky conditions.

Solution Using the alternate approach, to solve for the saturation EIRP_S:

At 14.0 GHz, $\qquad\qquad \lambda = c/f = 3.0e + 08/14.0e + 09 = 0.0214$ m

Next, from Eq. (12.48), $\qquad [A_o] = 10\log\left(\frac{\lambda^2}{4\pi}\right) = -44.37$ dB

Hence, from Eq. (12.49b),

$$[\text{EIRP}_S]_{up} = -120 - 44.37 + 207 + 2 = 44.63, \text{dBW}$$

12.7.2 Input Backoff

As described in Sec. 12.7.3, when multiple carriers are present simultaneously in a TWTA, the operating point must be backed off to a linear portion of the transfer characteristic to reduce the effects of intermodulation distortion. Such multiple carrier operation occurs with *frequency-division multiple access* (FDMA), which is described in Chap. 14. The point to be made here is that *backoff* (BO$_i$) of the input power to the TWTA must be included in the uplink-budget calculations.

Given the saturation flux density (Ψ_S) for single-carrier operation, the input backoff BO$_i$ (relative to the single-carrier saturation level) can be specified for multiple-carrier operation. It is important to note that the EIRP/*carrier* for the multiple carriers (usually from different earth stations in the transponder channel) must be carefully matched to yield

$$[\text{EIRP}]_{\text{up}} = [\text{EIRP}_S]_{\text{up}} + [\text{BO}_i] \tag{12.50}$$

where $[\text{BO}_i]$ is a negative dB value, which often becomes part of the uplink system loss.

Although some control of the input to the transponder power amplifier is possible through the ground TT&C station (as described in Sec. 12.7.3 and shown in Fig. 7.13), usually the input backoff is achieved through reduction of the EIRP of the multiple earth stations that are accessing the transponder. Note that the (C/N_0) for all carriers are equal, and Eq. (12.42) for each is

$$\left[\frac{C}{N_0}\right]_{\text{up}} = [\Psi_S] + [A_0] + [\text{BO}_i] + [G/T]_{\text{up}} + [1/L_{\text{ra}}]_{\text{up}} + [1/k] \tag{12.51}$$

Example 12.17 An uplink at 14.0 GHz requires a saturation flux density $[\Psi_S] = -91.4$ dBW/m^2 and an input $[\text{BO}_i] = -11.0$ dB. The satellite $[G/T] = -6.7$ dBK^{-1}, and receiver feeder losses $[L_{\text{ra}}] = -0.6$ dB. Calculate the carrier-to-noise density ratio.

Solution As in Example 12.11, the calculations are best carried out in tabular form. $[A_0] = -44.37$ dBm^2 is calculated for a frequency of 14.0 GHz by using Eq. (12.47).

Link Parameter	Linear	dB
Sat flux density	7.24436E-10	-91.4
A_0 @ 14 GHz	3.66E-05	-44.37
Input BO$_i$	7.94E-02	-11.00
Satellite (G_r/T_{sys})	2.14E-01	-6.70
(1/k) Boltzmann const	7.25E+22	228.60
L_{ra}	8.71E-01	-0.60
C/N_0	2.839E+07	74.53
	check $\Rightarrow$	74.53

Note that $[k] = -228.6$ dB, so $[1/k] = +228.6$ dB and input $[\text{BO}_i]$ and $[L_{\text{ra}}]$ are negative dB. The total gives the uplink carrier-to-noise density ratio at the satellite receiver as 74.5 dBHz. Since fade margins have not been included at this stage, Eq. (12.51) applies for *clear-sky* conditions. Usually, the most serious fading is caused by rainfall, as described in Secs. 4.4 and 12.9.

12.7.3 The Earth Station HPA

The earth station HPA carries multiple carriers, and as such, it is a linear amplifier to prevent the production of intermodulation noise products. Because of the satellite transponder class-C power amplifier (TWTA), it is important that all received carriers at the satellite have equal flux densities at the input (see Sec. 12.8). In addition, the earth station HPA power output per carrier must be compensated for the reduction of the earth station transmit feeder losses (denoted here by $[L_{ta}]$ dB). These include waveguide, filter, and coupler losses between the HPA output and the transmit antenna. So, in the link equation, the P_t is the HPA output-power/carrier, and the feeder losses are included in the system losses.

Further, the HPA output power must be adjustable to provide the input BO_i to set the proper transponder level and to compensate for rain fades on the uplink (as discussed in Sec. 12.9.1). Also, to ensure linear operation, an earth station HPA with a comparatively high saturation level is used. It is noted that the large physical size and power consumption associated with earth station high power vacuum tubes (e.g., Klystrons) do not carry the same penalties they do when used aboard the satellite.

12.8   Downlink

The downlink of a satellite circuit is the one in which the satellite is transmitting the signal and the earth station is receiving it. Equation (12.42) can be applied to the downlink, but now the subscript *"dn"* is used to denote specifically that the downlink is being considered. Thus, the carrier to noise power density becomes

$$\left[\frac{C}{N_o}\right]_{dn} = [EIRP]_{dn} + [G/T]_{dn} + [LOSSES]_{dn} + [1/k] \qquad (12.52)$$

Since the transponder may have several carriers, from different user earth stations using FDMA, we consider the general case where the number of carriers is ≥ 1. In this formula, the first link parameter is the single carrier satellite EIRP that may differ by several dB for the group of users. By design, the transmit power $[P_t]$ for each carrier should be nearly identical because all input powers $[P_r]$s are equal, and the TWTA input backoff makes the transponder linear. However, the transmit antenna LOS gains $[G_t]$ are different because the variable off-boresight angle of the individual earth station LOS directions. The other parameters should be similar, but not identical, such that there may be overall several dB of variability in the downlink $[C/N]$ for individual users.

The resulting carrier-to-noise density ratio is calculated at the earth station receiver input, but because the receiver is linear, the C/N_o is identical at the earth station receiver output (i.e., the receiver amplifies both signal and noise by the same gain factor, so the ratio is constant at any point in the cascaded receiver circuit).

When the carrier-to-noise ratio is the specified quantity (rather than carrier-to-noise density ratio), the downlink C/N is

$$\left[\frac{C}{N}\right]_{dn} = [EIRP]_{dn} + [G/T]_{dn} + [\Sigma \ LOSSES]_{dn} + [1/k] + [1/B_N] \qquad (12.53)$$

Example 12.18   A satellite TV signal occupies the full transponder bandwidth of 36 MHz, and it must provide a $[C/N] = +22$ dB at the destination earth station. Given that the total transmission losses are -200.0 dB and the destination earth-station $[G/T] = +31.0$ dB/K, calculate the satellite $[EIRP]_{dn}$ required.

Solution   Equation (12.53) expressed as power ratio can be solved for $EIRP_{dn}$ as

$$EIRP_{dn} = \left(\frac{C}{N}\right)_{dn} \left(\frac{1}{G/T}\right)_{dn} \left(\frac{1}{L_p * L_{sys}}\right)_{dn} k B_N$$

Thus, expressing as dB yields

$$[EIRP]_{dn} = \left[\frac{C}{N}\right]_{dn} + \left[\frac{1}{G/T}\right]_{dn} + \left[\frac{1}{(L_p L_{sys})}\right]_{dn} + [k] + [B_N]$$

Setting this spreadsheet format,

Link Parameter	Linear	dB
C/N	1.585E+02	22.0
$1/(G_r/T_{sys})$	7.943E-04	−31.0
$1/(L_p * L_{sys})$	1.000E+20	200.0
(k)	1.380E-23	−228.6
(B_N)	3.60E+07	75.56
EIRP, dBW	6254.34	38.0
	check ⇒	38.0

The required EIRP is 38 dBW or, equivalently, 6.3 kW.

12.8.1 TWTA Output Backoff

As described in Sec. 12.7.2, when the input BO_i is employed, there is a corresponding output backoff BO_o that must be included in the satellite EIRP calculation. In practice, this relationship between BO_o and BO_i is measured for the selected TWTA, but for now consider a plot of a typical TWTA transfer function given in Fig. 12.12, where the y axis is the power output expressed in dBW and the x axis is the input backoff in dBm, where $[BO_i] = 0$ dB corresponds to the maximum output power $[P_{sat}]$, and note

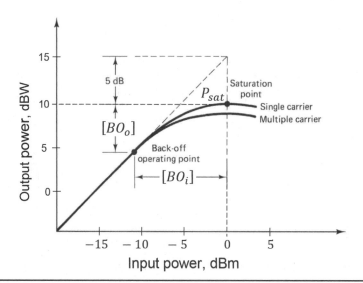

FIGURE 12.12 Input and output back-off relationship: $[BO]_i = [BO]_o + 5$ dB for a satellite TWTA.

that the $[BO_i]$ values are negative dB, which corresponds to reduced signal input power. For low level signal input, the TWTA has a constant power gain (typically 45–60 dB), and the output power increases linearly with increasing input power. Eventually, the TWTA output signal begins "to saturate," and the corresponding TWTA gain $G(BO_i)_{twta}$ progressively decreases until the maximum output power (defined as P_{sat}) is reached.

So, we define the transmitter output power P_t as a function of BO_o as

$$P_t = P_{sat} * BO_o, \text{Watts} \qquad (12.54a)$$

where BO_o is a monotonic decreasing, nonlinear function of BO_i that is derived from the measured TWTA transfer function during pre-launch acceptance testing of flight hardware. Thus, the transmitted power, when expressed as dBW, is

$$[P_t] = [P_{sat}] + [BO_o], \text{dBW} \qquad (12.54b)$$

Note that both backoffs (input and output) are negative dBs.

For our purposes, we choose to provide a simple relationship between BO_o and BO_i for the purposes of performing link calculations; but it is important to recognize that in practice actual measured values should be used. So, with this caveat, we give the following "rule of thumb" that is frequently used. From Fig. 12.12, we select the $[P_{sat}]$ point, which is $[BO_i] = 0$ dB and $[BO_o] = 0$ dB, which is –5 dB below the extrapolated linear portion of the TWTA transfer function. Next, we select the backoff operating point to be the point where the TWTA transfer curve becomes nonlinear, which for this plot, corresponds to an input backoff of $[BO_i] = -11$ dB below the saturation input power. Since the linear portion gives a 1:1 change in decibels, the relationship between input and output BO is

$$[BO_o] = [BO_i] + 5 \text{ dB} \qquad (12.55)$$

For example, the corresponding output backoff is $[BO]_o = -11 + 5 = -6$ dB, but this simple relationship only applies for $BO_i < -10$ dB.

Since the satellite EIRP for saturation conditions is specified as $[EIRP_s]_{dn} = [P_{sat}] + [G_t]$, then

$$[EIRP]_{dn} = [EIRP_s]_{dn} + [BO_o] \qquad (12.56)$$

and Eq. (12.53) becomes

$$\left[\frac{C}{N_0}\right]_{dn} = [EIRP_s]_{dn} + [BO_o] + [G/T]_{dn} + [\Sigma \text{ LOSSES}]_{dn} + [1/k] \qquad (12.57)$$

Example 12.19 The specified parameters for a downlink are satellite saturation value of EIRP, = 25 dBW; output $[BO_o] = -6$ dB; free-space loss $[L_p] = -196$ dB; allowance for other downlink losses $[L_{sys}] = -1.5$ dB; and earth-station $[G/T] = +41$ dBK^{-1}. Calculate the carrier-to-noise density ratio at the earth station.
Solution As with the uplink budget calculations, the calculation is best using the spreadsheet format for Eq. (12.57) with the specified downlink parameters.

Link Parameter	Linear	dB
EIRP$_{sat}$	316.23	25.0
BO$_o$	0.25	−6.0
earth station (G_r/T_{sys})	1.26E+04	41.0
Path loss, L_p	2.51E-20	−196.0
System loss	0.708	−1.5
(1/k) Boltzmann const	7.25E+22	228.60
C/N_o	1.29E+09	91.10
	check ⇒	91.10

Note, for the uplink, the saturation flux density at the satellite receiver is a specified quantity, but for the downlink, there is no reason to know the saturation flux density at the earth-station receiver. The reason being that because this is a terminal point, the signal is not used to saturate a power amplifier.

12.8.2 Satellite TWTA Output

For a given transponder channel, the satellite power amplifier (TWTA) supplies the required radiated power (P_t/carrier) plus makeup power for the transmit feeder losses L_{ta}, which include the waveguide, output mux, and harmonic filter losses. For a single carrier, the maximum transmitter power is P_{sat}, but for multicarrier operation (with number of carriers equal to NC), the P_t/carrier must be $< P_{sat}/NC$. On the other hand, if the BO$_o$ is sufficient to achieve linear TWTA operation, then the

$$\frac{P_t}{\text{carrier}} = P_{sat}\text{BO}_o L_{ta}\left(\frac{1}{NC}\right) \qquad (12.58a)$$

or expressed as dB as

$$\left[\frac{P_t}{\text{carrier}}\right] = [P_{sat}] + [\text{BO}_o] + [L_{ta}] + \left[\frac{1}{NC}\right] \qquad (12.58b)$$

Or expressed in terms of the downlink EIRP/carrier = P_t/carrier G_t.
Solving for P_{sat} becomes

$$P_{sat} = (\text{EIRP/carrier})\left(\frac{1}{G_t}\right)\left(\frac{1}{\text{BO}_0}\right)\left(\frac{1}{L_{ta}}\right)(NC) \qquad (12.59a)$$

Or expressed as dB

$$[P_{sat}] = [\text{EIRP/carrier}] + \left[\frac{1}{G_t}\right] + \left[\frac{1}{\text{BO}_o}\right] + \left[\frac{1}{L_{ta}}\right] + [NC] \qquad (12.59b)$$

Example 12.20   Using FDMA, NC = 4 carriers that share a single transponder channel, and for a single carrier the required [EIRP/carrier] = +44 dBW. Assuming that the TWTA is linear for a [BO$_o$] = −6 dB, that the transmitter feeder losses [L_{ta}] = −2 dB, and that the antenna gain [G_t] = +40 dB, calculate the required TWTA saturated power P_{sat}.

Solution   From Eq. (12.59b)

$$[P_{sat}] = [\text{EIRP/carrier}] + [1/G_t] + [1/\text{BO}_o] + [1/L_{ta}] + [NC]$$

$$= 44 + (-40) + 6 + 2 + 6 = 18 \text{ dBW} \Rightarrow 63.1 \text{ W}$$

Using the linear approximation TWTA relationship between [BO$_o$] and [BO$_i$] given by Eq. (12.56), what is the P_t/carrier value for the four carriers/channel FDMA operation?

Solution Since the [BO$_o$] = –6 dB, then the [BO$_i$] = –11 dB, and assuming that the TWTA is linear, the transmitted power/carrier is

$$P_t = \frac{\text{EIRP}_{\text{carrier}}}{G_t} \text{ Watts or as dB}$$

$$[P_t] = \left[\left(\frac{\text{EIRP}}{\text{carrier}}\right)\right] + [1/G_t]$$

$$= +44 + (-40) = +4.0 \text{ dBW} \Rightarrow 2.51 \text{ W}$$

What is the total TWTA output power for all carriers?

Solution $P_{t,\text{total}} = 4 * 2.51 \text{ W} = 10.04 \text{ W}$, which is $\leq P_{\text{sat}} = 63.1 \text{ W}$ (a necessary condition).

12.9 Effects of Rain

Up to this point, calculations have been made for clear-sky conditions, meaning the absence of weather-related phenomena that might adversely affect the signal strength. In the Ku-band, and to a much less extent the C-band, rainfall is the most significant cause of signal fading (Hogg and Chu, 1975). As discussed in Sec. 4.4, rainfall attenuates the propagation of microwaves by both absorption of energy and scattering from raindrops. The combined effect is called extinction (Ulaby and Long, 2014). Also, the atmospheric noise emission significantly increases with the absorption associated with increasing rain rate and carrier frequency, and an example is shown in Fig. 4.5.

Further, studies have shown (CCIR Report 338-3, 1978) that the rain extinction for horizontal polarization is considerably greater than for vertical polarization. Raindrops when falling through the atmosphere, are flattened in shape by air drag, thereby becoming elliptical rather than spherical. When a radio wave with some arbitrary polarization passes through raindrops, the component of electric field in the direction of the major axes of the raindrops is affected differently from the component along the minor axes. This produces a depolarization of the wave that causes the wave to become elliptically polarized (see Sec. 5.6). This is true for both linear and circular polarizations, and the effect seems to be much worse for circular polarization (Freeman, 1981). Where only a single polarization is involved in the satellite link, the effect is not serious. On the other hand, where frequency reuse is achieved, using orthogonal polarizations (as described in Chap. 5), this is an issue. The mixing of channels in the antenna reception results in crosstalk noise that degrades (C/N) for both channels. Thus, in regions with frequent high rain rates, depolarizing devices that compensate for the rain depolarization may be installed to mitigate this form of noise interference.

Finally, when the earth-station antenna is operated under cover of a radome, the effect of the rain on the radome are more serious. Rain falling on a hemispherical radome forms a thin water layer that introduces losses, both by absorption and by reflection. Results presented by Hogg and Chu (1975) show an attenuation of about –14 dB for a 1-mm-thick water layer at frequencies >10 GHz. Therefore, it is desirable that earth station antennas be operated without radomes, where possible. Without a radome, water may coat the antenna reflector, but the net attenuation produced by this is much less serious than that produced by the wet radome.

As noted, rain produces significant performance impacts on the satellite circuit. At any given location, the occurrence of rainfall can be treated as a random variable. The statistics of rain extinction (denoted as $[A_p]$ in Eq. (4.10) and in this chapter as $[L_{rain}]$) are usually available in the form of a cumulative probability distribution showing the fraction of time that a given attenuation is exceeded or, equivalently, the probability that a given attenuation is exceeded (see Hogg et al., 1975; Lin et al., 1980; Webber et al., 1986, and Pratt et al., 2003). An example of typical exceedance values, for the ITU climate zone-K over Southern Canada and Northern United States, is given for C- and Ku-band in Fig. 4.8 and is tabulated in Table 4.6.

The left panel of Fig. 4.8 is the R_p exceedance curve, where the y axis is the cumulative probability in %, and the x axis is the R_p in mm/h. The cumulative probability is the % of one-year that the abscissa value is exceeded. For example, a y value of 10^{-2} (0.01%) has a corresponding x value of 42 mm/h, which means that the R_p randomly exceeds 42 mm/h a total of only 0.01% of the time (~52.6 min/yr). Alternatively, one could interpret this as 99.99% of the time, the $R_p < 42$ mm/h.

12.9.1 Uplink Rain-Fade Margin

Rainfall produces an attenuation of the propagating carrier (L_{rain}) and an increase in the atmospheric noise density (ΔN_o) such that the uplink $[C/N_o]_{rain}$ becomes

$$\left(\frac{C}{N_o}\right)_{rain} = \left(\frac{C}{N_o}\right)_{CS} * \left(\frac{L_{rain}}{\Delta N_o}\right) \tag{12.60a}$$

$$\left[\frac{C}{N_o}\right]_{rain} = \left[\frac{C}{N_o}\right]_{CS} + [L_{rain}] + \left[\frac{1}{\Delta N_o}\right] \tag{12.60b}$$

where the "subscript cs" denotes clear sky. However, since the satellite receiving antenna views a "radiometrically hot" earth, the rain has very little effect on the apparent noise temperature at the antenna aperture, i.e., $\Delta N_o \sim 1$. Thus, the change in the uplink system noise temperature is negligible and may be ignored.

On the other hand, for certain modes of operation, it is important that the uplink carrier power at the satellite must be held within close limits, and therefore, some form of *uplink power control* may be necessary to compensate for rain fades. The power output from the satellite may be monitored by a central control station or in some cases by each earth station, and the power output for the affected uplink Earth station may be increased, to compensate for rain fading. Thus, the Earth-station HPA must have sufficient capability to increase transmit power to meet the fade margin requirement. However, heavy rain-fades are random and very transient and may only last for 10s of seconds to a few minutes. Given this, it is very difficult to completely cancel the fade by increasing the Earth station transmit power, so the link (C/N_o) may be degraded on limited occasions.

Typical rain-fade values given in Table 4.6 show that, for the Ku-band uplink frequency of 14 GHz, the $[A_{01}]$ exceeds −2 dB for 0.1 percent of the time. Thus, for 99.9 percent of the time, the earth station must be capable of providing a +2 dB margin over the clear-sky conditions.

Many links require a 99.99 percent availability, and for this case, we must consider the R_p exceedance for 0.01 percent, which is a $R_{01} = 42$ mm/h and an associated $[A_{01}]$ that exceeds −8.6 dB at 14 GHz. For a C-band uplink frequency of 6 GHz, the corresponding $[A_{01}]$ is −1.5 dB.

Example 12.21 For a satellite uplink at $f = 14.0$ GHz, under clear-sky conditions, the following apply: uplink $[C/N] = 26$ dB, receiver noise temperature $T_{rec} = 200$ K, antenna noise temperature $T_{ap} = 290$ K (measured at the receiver input), receiver feed loss $[L_{ra}] = 0$ dB, and receiver noise bandwidth $B_N = 4.0$ MHz. For 0.1 percent of the time, the uplink experiences $R_{01} = 12$ mm/h and a corresponding rain extinction $[L_{rain}] = -2.04$ dB $\Rightarrow 0.617$ power ratio. Calculate the resulting $[C/N_o]_{rain}$ during this rain event.

Solution To solve this problem, we use Eq. (12.60b) given above.

Because the uplink satellite receiving antenna views the radiometrically "hot" earth, we assume that the increased system noise $[\Delta N_o]$ is zero dB.

Therefore, calculate the clear-sky uplink $[C/N_o]_{CS}$ using Eq. (12.42)

$$\left[\frac{C}{N_o}\right]_{CS} = \left[\frac{C}{N}\right]_{CS} + [B_N] = 26.0 + 10\log(4.0e+06) = 26.0 + 66.02 = 92.02 \text{ dBHz}$$

and for rain

$$\left[\frac{C}{N_o}\right]_{rain} = \left[\frac{C}{N_o}\right]_{CS} + [L_{rain}] = 92.02 + (-2.04) = 89.98 \text{ dBHz}$$

Now, we justify the assumption that $[\Delta N_o] = $ zero dB. First, we are given that $T_{ant} = 290$ K for clear sky, and the system temperature

$$T_{sys,cs} = 290 + 200 = 490 \text{ K}.$$

In the presence of rain, the atmosphere partially absorbs the scene brightness and re-emits noise at its effective noise temperature. As a result

$$T_{ant,rain} = 290 * L_{rain} + (1 - L_{rain}) * T_{eff} = 290 * 0.625 + (1 - 0.625) * 273 = 283.7 \text{ K}$$

$$T_{sys,rain} = T_{ant,rain} + T_{rec} = 283.5 + 200 = 483.5 \text{ K}$$

so that the delta noise power density is

$$\Delta N_o = \frac{kT_{sys,rain}}{kT_{sys,CS}} = \frac{483.5}{490} = 0.987 \Rightarrow -0.06 \text{ dB}$$

which is negligible.

12.9.2 Downlink Rain-Fade Margin

For the downlink, Eq. (12.40) $[C/N]$ and Eq. (12.42) $[C/N_o]$ are applicable for clear-sky conditions; however, during rainfall, there is an additional atmospheric extinction by absorption and scattering by raindrops. Based upon the results presented in Fig. 4.6, raindrop scattering effects are significant (>1%) for only $R_p > 10$ mm/h and at frequencies >10 GHz, therefore, we neglect them for the present discussion concerning C-band (4–6 GHz), but this assumption for the Ku-band (12–14 GHz) is revisited later in this section.

Given these constraints, we define $[L_{rain}] = [A]$ Eq. (4.10) to be the *rain attenuation caused by absorption alone*. Now consider Sec. 12.5.5, where the effective noise temperature of absorptive passive networks is presented, which is applicable to rain on the downlink, for frequencies <10 GHz. Replacing L in Eq. (12.32) with the total atmospheric absorption ($L_{rain} * L_{cs}$), the effective noise temperature of the rain at the receiving antenna aperture is

$$T_{rain} = T_a(1 - L_{rain} * L_{CS}) \qquad (12.61)$$

where T_a is known as the *effective atmospheric (absorber) temperature* in units Kelvin, which is a function of several factors including the physical temperature of the rain, the raindrop diameter (proportional to rain rate), and the scattering effect of the rain cell on the thermal noise incident upon it (Hogg and Chu, 1975; Ulaby and Long, 2014). The range of T_a lies between 270 and 290 K, with measured values for North America being close to or just below freezing (273 K). For example, the measured value given by Webber et al. (1986) is 272 K, but for this text, we use 273 K.

So, the system noise power density becomes

$$N_{o\,\text{rain}} = kT_{\text{sys rain}} = k(T_{\text{rain}} + T_{\text{rec}}) \tag{12.62}$$

Thus, rainfall degrades the downlink $[C/N_o]$ in two ways, namely: by attenuating the received carrier power

$$C_{\text{rain}} = C_{\text{CS}} L_{\text{rain}} \tag{12.63}$$

and by increasing the system noise power density by a ratio (ΔN_o) such that

$$N_{o\,\text{rain}} = \Delta N_o * N_{o\text{CS}} \tag{12.64}$$

solving for ΔN_o becomes

$$\Delta N_o = \frac{N_{o\,\text{rain}}}{N_{o\text{CS}}} = \frac{T_{\text{sys rain}}}{T_{\text{sys CS}}} \tag{12.65a}$$

or expressed as dB

$$[\Delta N_o] = [T_{\text{sys rain}}] + [1/T_{\text{sys CS}}] \tag{12.65b}$$

So, the resulting carrier power to noise power density becomes

$$\left(\frac{C}{N_o}\right)_{\text{rain}} = \left(\frac{C}{N_o}\right)_{\text{CS}} L_{\text{rain}} \left(\frac{1}{\Delta N_o}\right) \tag{12.66a}$$

$$\left[\frac{C}{N_0}\right]_{\text{rain}} = \left[\frac{C}{N_0}\right]_{\text{CS}} + [L_{\text{rain}}] + \left[\frac{1}{\Delta N_o}\right] \tag{12.66b}$$

Example 12.22 **Part-A** For a C-band satellite downlink at $f = 4.0$ GHz, under clear-sky conditions, the following apply: downlink $[C/N] = 20$ dB, receiver noise temperature $T_{\text{rec}} = 200$ K, antenna noise temperature $T_{\text{ap}} = 5.0$ K (measured at the receiver input), receiver feed loss $[L_{\text{ra}}] = 0$ dB, and receiver noise bandwidth $B_N = 4.0$ MHz. Using the rain exceedance Table 4.6 for 0.01% of the time, the downlink experiences a rain rate $R_1 \geq 42$ mm/h and a corresponding absorption $[L_{\text{rain}}] = -0.64$ dB $\Rightarrow 0.863$, power ratio.

Calculate the resulting $[C/N_o]_{\text{rain}}$ during this rain event.

Solution First, convert the clear sky $[C/N]_{\text{CS}}$ to $[C/N_o]_{\text{CS}}$ by Eq. (12.41)

$$\left[\frac{C}{N_o}\right]_{\text{cs}} = \left[\frac{C}{N}\right]_{\text{cs}} + [B_N] = 20.0 + 10\log(4.0e+06) = 86.02, \text{dBHz}$$

Next using Eq. (12.66b) $\qquad \left[\frac{C}{N_o}\right]_{\text{rain}} = \left[\frac{C}{N_o}\right]_{\text{cs}} + [L_{\text{rain}}] + \left[\frac{1}{\Delta N_o}\right]$

calculate the increased noise density ratio ΔN_o due to rain

$$\Delta N_o = \frac{N_{o\,\text{rain}}}{N_{o\text{cs}}} = \left(\frac{T_{\text{sys rain}}}{T_{\text{sys cs}}}\right) \tag{12.67}$$

Numerical calculation: First, consider the clear sky case, for Eq. (12.61) set $L_{rain} = 1.0$ (i.e., no rain), thus, $T_{cs} = T_a(1 - L_{cs})$

Solving for L_{cs} becomes

$$L_{cs} = \left(1 - \frac{T_{cs}}{T_a}\right) = \left(1 - \frac{5}{273}\right) = 0.982 \Rightarrow -0.08 \text{ dB}$$

and the system temperature for clear sky is

$$(T_{sys})_{cs} = (T_{cs} + T_{rec}) = 5 + 200 = 205 \text{ K}$$

Next, calculate the noise temperature during rain using Eq. (12.61)

$$T_{rain} = T_a(1 - L_{rain} * L_{CS}) = 273\,(1 - 0.863*0.982) = 42 \text{ K}$$

and

$$T_{sysrain} = (T_{rain} + T_{rec}) = 42 + 200 = 242 \text{ K}$$

Now, using Eq. (12.67), calculate the increased noise density ratio ΔN_o due to rain

$$\Delta N_o = \left(\frac{T_{sysrain}}{T_{syscs}}\right) = \frac{242}{205} = 1.18 \Rightarrow 0.72 \text{ dB}$$

Finally, using Eq. (12.66b)

$$\left[\frac{C}{N_o}\right]_{rain} = \left[\frac{C}{N_o}\right]_{cs} + [L_{rain}] + \left[\frac{1}{\Delta N_o}\right] = 86.02 + (-0.64) + (-0.72) = 84.66 \text{ dBHz}$$

and from Eq. (12.41)

$$\left[\frac{C}{N}\right]_{rain} = \left[\frac{C}{N_o}\right]_{rain} + \left[\frac{1}{B_N}\right] = 84.66 + (-66.02) = 18.64 \text{ dB}$$

For low microwave frequencies and low rainfall rates, the rain attenuation is almost entirely absorptive; however, at higher rainfall rates, scattering becomes significant, especially at higher frequencies. When both rain scattering and absorption are present, we must use the results of Sec. 4.4 and Table 4.4 to solve for the absorptive attenuation, which is then used to calculate the increase in noise temperature. It is important to note that the rain effects on the system noise are negligible for the uplink; so this must be considered for only for the downlink. Further, the downlink operates at the lower frequencies of the band, which provides additional immunity against rain thermal emission.

Example 12.22   Part-b For a Ku-band satellite downlink at $f = 12.0$ GHz, under clear-sky conditions, the following apply: downlink $[C/N] = 20$ dB, receiver noise temperature $T_{rec} = 200$ K, antenna noise temperature $T_{ap} = 6.2$ K (measured at the receiver input), receiver feed loss $[L_{ra}] = 0$ dB, and receiver noise bandwidth $B_N = 4.0$ MHz. For 0.01% of the time, the downlink experiences a rain rate $R_1 \geq 42$ mm/h and a corresponding absorption $[L_{rain}] = -6.22$ dB $\Rightarrow 0.239$, power ratio.

Calculate the resulting $[C/N_o]_{rain}$ during this rain event.

Solution   First, convert the clear sky $[C/N]_{cs}$ to $[C/N_o]_{cs}$ by Eq. (12.41)

$$\left[\frac{C}{N_o}\right]_{cs} = \left[\frac{C}{N}\right]_{cs} + [B_N] = 20.0 + 10 \log(4.0e + 06) = 86.02, \text{ dBHz}$$

Next using Eq. (12.66b)

$$\left[\frac{C}{N_o}\right]_{rain} = \left[\frac{C}{N_o}\right]_{cs} + [L_{rain}] + \left(\frac{1}{\Delta N_o}\right)$$

But before, we must determine the absorptive component of the rain attenuation, which is used to calculate the increased noise density ratio ΔN_o due to rain. Based upon the results in Sec. 4.4 and Table 4.4 that $k_{rain} = k_a + k_s$, dB/km, it follows that

$$[L_{rain}] = [L_{a,rain}] + [L_{s,rain}]$$

Solving for the absorptive component $[L_{a,rain}]$ in terms of the rain albedo, AB, yields

$$[L_{a,rain}] = (1 - AB) * [L_{rain}] \tag{12.68}$$

Numerical calculation: First, consider the clear-sky case, for Eq. (12.61) set $L_{rain} = 1.0$ (i.e., no rain), thus, $T_{cs} = T_a(1 - L_{cs})$

Solving for L_{cs} becomes

$$L_{cs} = \left(1 - \frac{T_{cs}}{T_a}\right) = \left(1 - \frac{6.2}{273}\right) = 0.977 \Rightarrow -0.10 \text{ dB}$$

and the system temperature for clear sky is

$$T_{syscs} = (T_{cs} + T_{rec}) = 6.2 + 200 = 206 \text{ K}$$

Since rain scattering cannot be neglected, we must use the results of Table 4.4 to solve for the absorptive component of attenuation, which is used to calculate the increase in noise temperature. Using linear interpolation of Table 4.4 for a $R_{01} = 42$ mm/h yields a scattering albedo $AB = 0.235$, so the absorptive component

$$[L_{abs,rain}] = (1 - AB) * [L_{rain}]$$
$$= (1 - 0.235)(-6.22) = -4.76 \text{ dB} \Rightarrow 0.334$$

Next, calculate the noise temperature during rain using Eq. (12.61)

$$T_{rain} = T_a(1 - L_{rain} * L_{CS}) = 273 \ (1 - 0.334*0.977) = 184 \text{ K}$$

and $T_{sysrain} = (T_{rain} + T_{rec}) = 184 + 200 = 384$ K

Now, using Eq. (12.65a), calculate the increased noise density ratio ΔN_o due to rain

$$\Delta N_o = \left(\frac{T_{sysrain}}{T_{syscs}}\right) = \frac{384}{206} = 1.864 \Rightarrow 2.70 \text{ dB}$$

Finally, using Eq. (12.66b)

$$\left[\frac{C}{N_o}\right]_{rain} = \left[\frac{C}{N_o}\right]_{cs} + [L_{rain}] + \left[\frac{1}{\Delta N_o}\right] = 86.02 + (-6.22) + (-2.70) = 77.10 \text{ dBHz}$$

and from Eq. (12.41)

$$\left[\frac{C}{N}\right]_{rain} = \left[\frac{C}{N_o}\right]_{rain} + \left[\frac{1}{B_N}\right] = 77.10 + (-66.02) = 11.08 \text{ dB}$$

Example 12.23   Under clear-sky conditions, the downlink $[C/N]_{cs} = 20$ dB, and the effective noise temperature $T_{sys,cs} = 400$ K.

(a) If rain attenuation exceeds -1.9 dB for 0.1 percent of the time, and assuming that the rain attenuation is totally absorptive and that the effective noise temperature of rain, $T_a = 273$ K. Calculate the corresponding value of $[C/N]_{rain}$.

Solution   An attenuation of -1.9 dB $\Rightarrow 0.646$ power ratio; and since the rain is totally absorptive, the equivalent noise temperature of the rain is

$$T_{rain} = 273 \ (1 - 0.646) = 96.6 \text{ K}$$

The $T_{\text{sys,rain}} = 400 + 96.6 = 496.6$ K, and the

$$\Delta N_o = \left(\frac{T_{\text{sysrain}}}{T_{\text{syscs}}}\right) = \frac{496.6}{400} = 1.242 \Rightarrow +0.94 \text{ dB}$$

At the same time, the carrier is reduced by −1.9 dB, and therefore, the

$$[C/N]_{\text{rain}} = [C/N]_{\text{cs}} + [L_{\text{rain}}] + [1/\Delta N_o] = 20 - 1.9 + (-0.94) = 17.16 \text{ dB}$$

This is the value below which $[C/N]_{\text{dn}}$ occurs for 0.1 percent of the time.

(b) If rain attenuation exceeds −1.9 dB for 0.1 percent of the time, and assuming, that the rain albedo = 0.15, calculate the corresponding value of $[C/N]_{\text{rain}}$.

To calculate the equivalent noise temperature of the rain, we must determine the absorptive component of the rain attenuation. From Ex. 12.22,

$$[L_{\text{abs,rain}}] = (1 - AB) * [L_{\text{rain}}] = (1 - 0.15) * (-1.9) = -1.62 \text{ dB} \Rightarrow 0.689$$

So, the equivalent noise temperature of the rain is

$$T_{\text{rain}} = 273 \, (1 - 0.689) = 84.8 \text{ K}$$

The $T_{\text{sys,rain}} = 400 + 84.8 = 484.8$ K, and the

$$\Delta N_o = \left(\frac{T_{\text{sysrain}}}{T_{\text{syscs}}}\right) = \frac{484.8}{400} = 1.212 \Rightarrow +0.83 \text{ dB}$$

At the same time, the carrier is reduced by −1.9 dB, and therefore, the

$$[C/N]_{\text{rain}} = [C/N]_{\text{cs}} + [L_{\text{rain}}] + [1/\Delta N_o] = 20 - 1.9 + (-0.83) = 17.27 \text{ dB}$$

Thus, only a change of: $17.27 - 17.16 = +0.11$ dB difference, which is not very significant. However, as the rain attenuation increases, so does the scattering proportion, and the changes become significant as the attenuation approaches −10 dB.

It is left as an exercise for the student to show that where the rain power attenuation A (not dB) is entirely absorptive, then the downlink C/N power ratios (not dBs) are related to the clear-sky value by approximately

$$\left(\frac{N}{C}\right)_{\text{rain}} = \left(\frac{N}{C}\right)_{\text{CS}} \left(\frac{1}{L_{\text{rain}}}\right) \left(1 + (1 - L_{\text{rain}}) \frac{T_a}{T_{\text{Sys,CS}}}\right) \qquad (12.69)$$

Note that noise-to-carrier ratios, rather than carrier-to-noise ratios are required by Eq. (12.69).

As discussed in Chap. 9, a minimum value of $[C/N]$ is required for satisfactory reception. In the case of frequency modulation (FM), the minimum value is set by the threshold level of the FM detector, and a *threshold margin* is normally allowed, as shown in Fig. 9.12. Sufficient margin must be allowed so that rain-induced fades do not take the $[C/N]$ below threshold more than a specified percentage of the time, as presented below. Note, for FM modulation, there is processing gain that occurs during the demodulation process that improves the output SNR over the receiver input (C/N).

Example 12.24   In an FM satellite system, the clear-sky downlink $[C/N]_{\text{cs}} = 17.4$ dB and the FM detector threshold $[C/N]$ is 10 dB, as shown in Fig. 9.12.

(a) Calculate the threshold margin at the FM detector, assuming the threshold $[C/N]$ is determined solely by the downlink value, i.e., the $[C/N]_{\text{up}} \gg [C/N]_{\text{dn}}$.

(b) Given that $T_a = 273$ K and that $T_{sys,cs} = 544$ K, calculate the percentage of time the system stays above threshold. The curve of Fig. 12.13 may be used for the downlink, and it may be assumed that the rain attenuation is entirely absorptive.

Solution

(a) Since it is assumed that the overall $[C/N]$ ratio is equal to the downlink value, the clear-sky input $[C/N]$ to the FM detector = 17.4 dB. The threshold level for the detector is 10 dB, and therefore, the rain-fade margin is $17.4 - 10 = 7.4$ dB.

(b) The rain attenuation can reduce the $[C/N]$ to the threshold level of 10 dB (i.e., it reduces the margin to zero), which is a (C/N) power ratio of 10.0 or a downlink N/C power ratio of 0.10.

For clear-sky conditions, $[C/N]_{CS} = 17.4$ dB $\Rightarrow 54.95$, which gives an $N/C = 0.0182$. Substituting these values in Eq. (12.69) gives

$$\left(\frac{N}{C}\right)_{rain} = \left(\frac{N}{C}\right)_{CS}\left(\frac{1}{L_{rain}}\right)\left(1 + (1 - L_{rain})\frac{T_a}{T_{Sys,CS}}\right)$$

$$0.1 = 0.0182\left(\frac{1}{L_{rain}}\right)\left(1 + \frac{(1 - L_{rain}) \times 273}{544}\right)$$

Solving this equation for $L_{rain} = 0.251 \Rightarrow -6.01$ dB. From the curve of Fig. 12.13, the probability of exceeding the -6 dB value is $P_e = 2.5 \times 10^{-4}$, and therefore, the availability is $1 - 2.5 \times 10^{-4} = 0.99975$, or 99.975 percent.

For digital signals, the required $[C/N_o]$ ratio is determined by the acceptable BER, which must not be exceeded for more than a specified percentage of the time. Figure 10.22 relates the BER to the $[E_b/N_o]$ ratio, and this in turn is related to the $[C/N_o]$ by Eq. (10.32), as discussed in Sec. 10.6.4.

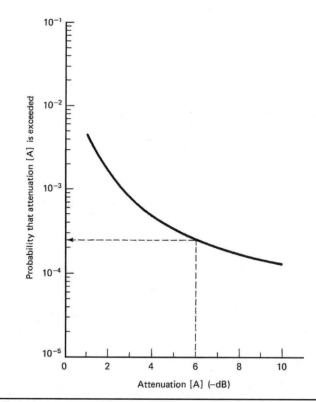

Figure 12.13 Typical exceedance curve used in Example 12.22.

For the downlink, the user does not have control of the satellite [EIRP], and thus, the downlink equivalent of uplink power control, described in Sec. 12.9.1, cannot be used. To provide the rain-fade margin needed, the gain of the receiving antenna may be increased by using a larger dish and/or a receiver front end having a lower noise temperature. Both measures increase the receiver $[G/T]$ ratio and thus increase $[C/N_o]$ as shown by Eq. (12.52).

12.10   Combined Uplink and Downlink *C/N* Ratio

As previously discussed, for both analog and digital signals, the link performance depends upon the end-to-end C/N power ratio at the receiving earth station output. The following analysis applies to a satellite transponder or a "bent pipe" repeater system connecting an uplink and a downlink earth station, as illustrated in Fig. 12.14a. Note that the analysis is performed on a single carrier basis.

Noise is introduced on the uplink, at the satellite receiver input, as the noise power density (per unit bandwidth) P_{NU}, and the corresponding average carrier power is P_{RU}. Thus, the carrier-to-noise density ratio on the uplink is $(C/N_o)_u = (P_{RU}/P_{NU})$. It is important to note that powers in Watts, and not decibels, are used here.

At the downlink end of the space link, the average received carrier output power $P_R = \gamma * P_{RU}$ where P_{RU} is the carrier power input at the satellite, and γ is the system power gain from satellite input to earth-station input, as shown in Fig. 12.14a. Note that the system gain includes: the uplink satellite feeder losses, the satellite transponder gain, the satellite transmit antenna gain, the downlink losses (including path loss), and the earth-station receive antenna gain and feeder losses.

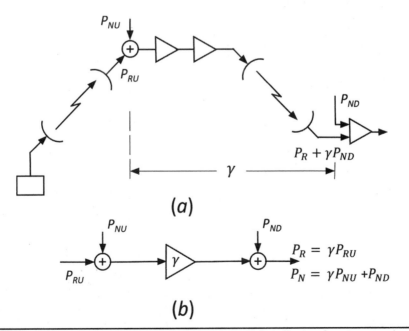

Figure 12.14 (a) Combined uplink and downlink; (b) power flow diagram for a transponder satellite communication system.

The noise at the satellite input (P_{NU}) also appears at the downlink earth station input multiplied by γ, and in addition, the earth station introduces its own noise, with the total being denoted by P_{ND}. Thus, the end-to-end link noise = $\gamma P_{NU} + P_{ND}$.

The power flow diagram is shown in Fig. 12.14b, and the C/N_o ratio for the downlink alone, not counting the γP_{NU} contribution, is P_R/P_{ND}, and the combined C/N_o ratio at the input to the ground receiver is

$$\frac{C}{N_o} = \frac{P_R}{(\gamma P_{NU} + P_{ND})}$$

The combined carrier-to-noise ratio can be determined in terms of the individual link values. To show this, it is more convenient to work with the noise-density to carrier ratios rather than the carrier-to-noise density ratios, and again, these must be expressed as power ratios, not decibels. Denoting the combined noise-to-carrier ratio value by $(N_o/C)_{tot}$, the uplink value by $(N_o/C)_{up}$, and the downlink value by $(N_o/C)_{dn}$, then,

$$\left(\frac{N_o}{C}\right)_{tot} = \frac{P_N}{P_R} = \frac{\gamma P_{NU} + P_{ND}}{P_R} = \frac{\gamma P_{NU}}{P_R} + \frac{P_{ND}}{P_R} = \frac{\gamma P_{NU}}{\gamma P_{RU}} + \frac{P_{NU}}{P_R}$$

$$= \left(\frac{N_o}{C}\right)_{up} + \left(\frac{N_o}{C}\right)_{dn} \tag{12.70}$$

This shows that to obtain the combined (end-to-end) value of C/N_o, first the reciprocals of the individual values must be added to obtain the combined N_o/C ratio. Next, the reciprocal of this is taken to get C/N_o. To understand the rational for this approach of summing the reciprocals method is that a single signal power is being transferred through the system, while the various noise powers, introduced at the uplink and downlink, are additive. Thus, (N_o/C)s add to yield the end-to-end value, which is equivalent to combining resistors in parallel by finding the equivalent conductance and then reciprocating. Also note that a necessary (but not sufficient condition) quality control test is that *the end-to-end $(C/N_o)_{tot}$ ratio is smaller than the smallest C/N_o ratio being combined.* Similar reasoning applies to the carrier-to-noise ratio, (C/N).

Example 12.25 For a satellite circuit the individual link carrier-to-noise spectral density ratios are: uplink 100 dBHz; downlink 87 dBHz. Calculate the combined end-to-end C/N_o ratio.

Solution

$$\frac{N_o}{C} = 10^{-10} + 10^{-8.7} = 2.095e - 09$$

Therefore,

$$\left[\frac{C}{N_o}\right] = \left[\frac{N_o}{C}\right] = 10\log(1/2.095e - 09) = 86.79 \text{ dBHz}$$

This example illustrates the point that when one of the link C/N_o ratios is much less than the others, the combined C/N_o ratio is approximately equal to the lower (worst) one. The downlink C/N_o is usually (but not always) less than the uplink C/N_o, and in many cases it is much less. This is true primarily because of the limited EIRP available from the satellite.

The next example illustrates how input BO_i is considered in the link-budget calculations and how it affects the C/N_o ratio.

Example 12.26 A multiple carrier satellite circuit operates in the 6/4-GHz band with the following characteristics.

Uplink:
Saturation flux density $[\psi_s] = -67.5$ dBW/m²; $[BO_i] = -11$ dB; satellite $G/T = 11.6$ dBK⁻¹.

Downlink:
Satellite saturation $[EIRP_{sat}] = 26.6$ dBW; $[BO_o] = -6$ dB; free-space loss $[L_p] = -196.7$ dB; earth station $G/T = 40.7$ dBK⁻¹. For this example, the other losses may be ignored. Calculate the carrier-to-noise density ratios for both links and the combined value.

Solution   As in the previous examples, the data are best presented in tabular form, and values are shown in both power ratios and dB.

Uplink	
Saturation flux density	−67.5
$[A_o]$ at 6 GHz	−37
Input BO	−11
Satellite saturation $[G/T]$	−11.6
$-[k]$	228.6
$[C/N_o]$ from Eq. (12.51)	101.5
Downlink	
Satellite [EIRP]	26.6
Output BO	−6
Free-space loss	−196.7
Earth station $[G/T]$	40.7
$-[k]$	228.6
$[C/N_o]$ from Eq. (12.57)	93.2

Application of Eq. (12.70) provides the combined $[C/N_o]$:

$$\frac{N_o}{C} = 10^{-10.15} + 10^{-9.32} = 5.49e - 10$$

$$\left[\frac{C}{N_o}\right] = 10\log(1/5.49e - 10) = 92.6 \text{ dBHz}$$

Again, it is seen from this example that the combined C/N_o value is close to the lowest value, which is the downlink value.

So far, only thermal and antenna noise has been considered in calculating the combined value of C/N_o ratio. Another source of noise to be considered is intermodulation noise, which is discussed in Sec. 12.11.

12.11   Intermodulation Noise

Intermodulation occurs where multiple carriers pass through a device that has nonlinear characteristics. In satellite communications systems, this most commonly occurs in the traveling-wave tube HPA aboard the satellite, as described in Sec. 7.7.3. Both amplitude and phase nonlinearities give rise to intermodulation products.

As shown in Fig. 7.19, third-order intermodulation products fall "in-band" on neighboring carrier frequencies, where they result in interference. Where many modulated carriers are present, the intermodulation products are not distinguishable separately but instead appear as a type of noise that is termed *intermodulation noise* or *3IM noise*.

The carrier-to-intermodulation-noise ratio is usually found experimentally, or in some cases it may be determined by computer methods (Ha, 1990). Once this ratio is known, it can be combined with the carrier-to-thermal-noise ratio by the addition of the reciprocals in the manner described in Sec. 12.10. Note in this section, we perform the link analysis using carrier-to-noise ratios (rather than carrier to noise density ratios, which is equivalent and provides the same results). Denoting the intermodulation term by $(C/N)_{IM}$ and bearing in mind that the reciprocals of the C/N power ratios (and not the corresponding dB values) must be added, Eq. (12.70) is extended to

$$\left(\frac{N}{C}\right)_{tot} = \left(\frac{N}{C}\right)_{U} + \left(\frac{N}{C}\right)_{D} + \left(\frac{N}{C}\right)_{IM} \tag{12.71}$$

A similar expression applies for noise-density-to-carrier (N_o/C) ratios.

Example 12.27 For a satellite circuit the carrier-to-noise ratios are uplink 23 dB, downlink 20 dB, intermodulation 24 dB. Calculate the overall carrier-to-noise ratio in decibels.

Solution From Eq. (12.71),

$$\left(\frac{N}{C}\right)_{tot} = 10^{-2.4} + 10^{-2.3} + 10^{-2} = 0.0019$$

$$\left[\frac{C}{N}\right]_{tot} = 10 \log(1/0.0019) = 17.2 \text{ dBHz}$$

To reduce intermodulation noise, the TWTA must be operated in a back-off condition as described previously. As shown in Fig. 12.15, the $[C/N]_{IM}$ ratio improves (for a typical TWTA) as the input $[BO_i]$ is increased (larger negative dB, which reduces the input signal power). Also, these curves are a function of the number of carriers input, with the total 3IM noise increasing with the number of input carriers. At the same time, increasing the $[BO_i]$ decreases both $[C/N]_{up}$ and $[C/N]_{dn}$, as shown by Eqs. (12.51) and (12.57). The result is that there is an optimal point where the overall carrier-to-noise ratio is a maximum. The component $[C/N]$ ratios as functions of the TWTA input are sketched in Fig. 12.16. The TWT input in dB is ($[\Psi]_S + [BO_i]$), (where BO_i is a negative dB), and therefore, Eq. (12.51) plots as a straight line. Equation (12.57) reflects the curvature in the TWTA characteristic through the output $[BO_o]$, which is not linearly related to the input $[BO_i]$, as shown in Fig. 12.12. The intermodulation curve is not easily predictable, and only the general trend is shown. The overall $[C/N]_{tot}$, which is calculated from Eq. (12.71), is also sketched, and the optimal operating point is defined by the peak of this curve.

12.12   Regenerative Digital Satellite Links

As previously discussed, it was noted that regenerative digital repeaters can restore the C/N_o for certain digital satellite systems (see Sec. 7.12); however, this does not apply to "bent pipe" transponders. For regenerative systems, the digital data are decoded onboard the satellite with error correction coding applied that produces very low probability of bit errors. Afterwards, these data are reformatted and transmitted on the satellite downlink to the receiving earth stations. This process of regenerative decoding effectively removes the uplink noise, and as a result, the end-to-end C/N_o, is determined by only the downlink. Thus, for digital SATCOM, a high SNR may be maintained regardless of the distance between the transmitter and the receiver (and the number of intersatellite cross-links). In other words, each satellite acts as a regenerative repeater that restores the effective C/N_o to a high value (negligible bit errors).

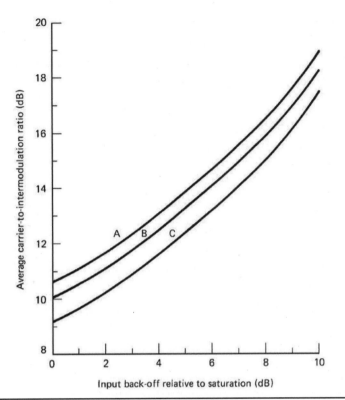

Figure 12.15   Intermodulation in a typical TWT. Curve A, 6 carriers; curve B, 12 carriers; curve C, 500 carriers. (*From CCIR, 1982b. With permission from the International Telecommunications Union.*)

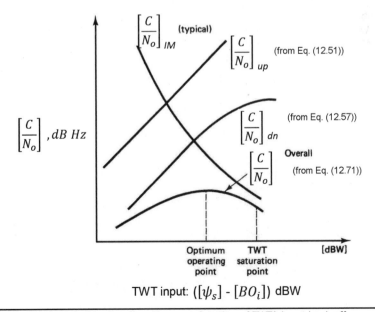

Figure 12.16   Carrier-to-noise density ratios as a function of TWTA input back-off.

An example of this type of the digital communications system is the current LEO Starlink system (see Chap. 15) for broadband satellite internet and future telephone services, which currently comprises over 5000 satellites in low Earth orbit (350–400 km) in various orbit planes to provide global connectivity. Future plans are to expand the network to about 42,000 satellites in the constellation.

12.13 Inter-Satellite Links

Inter-satellite links (ISLs) are radio frequency or optical links that provide a connection between satellites without the need for intermediate ground stations. Although many different links are possible, the most useful ones in operation are:

- *low earth orbiting* (LEO) satellites—LEO $\leftrightarrow$ LEO
- *geostationary earth orbiting* (GEO) satellites—GEO $\leftrightarrow$ GEO
- LEO $\leftrightarrow$ GEO

Consider first some of the applications to GEOs. As shown in Chap. 3, the antenna angle of elevation is limited to a minimum of about 5 degrees because of noise induced from the earth. The limit of visibility as set by the minimum angle of elevation is a function of the satellite longitude and earth-station latitude and longitude as shown in Sec. 3.2. Figure 12.17 shows the situation where earth station A is beyond the range of satellite $S2$, a problem that can be overcome using two satellites connected by an ISL. Thus, a long-distance link between earth stations A and B can be achieved by this means. A more extreme example is where an intercontinental service may require several "hops." For example, a Europe-Asia circuit requires three hops (Morgan, 1999): Europe to eastern U.S.A.; eastern U.S.A. to western U.S.A.; western U.S.A. to Asia; and of course, each hop requires an uplink and a downlink. By using an ISL only one uplink and one downlink is required. Also, as discussed shortly, the ISL frequencies are well outside the standard uplink and downlink bands so that spectrum use is conserved. The cost of the ISL is more than offset by not having to provide the additional earth stations required by the three-hop system.

The distance d for the ISL is easily calculated. From Figs. 12.17 and 12.18:

$$d = 2a_{\text{GSO}} \sin \frac{\Delta\phi}{2} \qquad (12.72)$$

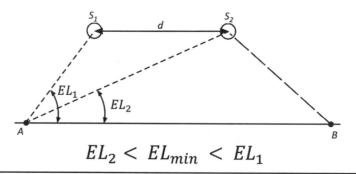

$$EL_2 < EL_{min} < EL_1$$

FIGURE 12.17 Angle of elevation for an earth station antenna as determined by an ISL.

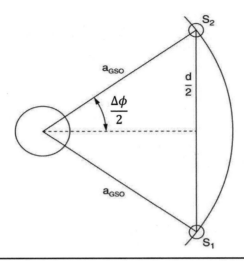

Figure 12.18 Satellite/earth station geometry for finding the cross-link LOS distance *d* between two GEO satellites.

where $\Delta\phi$ is the longitudinal separation between satellites $S1$ and $S2$, and α_{GSO} is the radius of the geostationary orbit [see Eq. (3.2)], equal to 42,164 km. For example, the western limits for the continental United States (CONUS) arc are at 55° and 136° W (see Prob. 3.11). Although there are no satellites positioned exactly at these longitudes they can be used to get an estimate of distance *d* for an ISL spanning CONUS.

$$d = 2 \times 42,164 \times \sin\frac{(136° - 55°)}{2} = 54,767 \text{ km}$$

Although this may seem large, the range from earth station to satellite is in the order of 41,000 km (Prob. 3.11), so distance involved with three uplinks and three downlinks is 246,000 km! We note that the large difference in the time delays (latencies) between the two options.

GEO satellites are often arranged in clusters at some nominal longitude. For example, there are several EchoStar satellites at longitude 119° W. The separation between satellites is typically about 100 km, the corresponding longitudinal separation for this distance being, from Eq. (12.72), approximately 0.136°. Because the satellites are relatively close together they are subject to the same perturbing and drift forces, which simplifies positional control. Also, all satellites in the cluster are within the main lobe of the earth-station antenna.

Because LEO satellites are not continuously visible from a given earth location, an intricate network of satellites is required to provide continuous coverage of any region. A typical LEO satellite network utilizes several orbit planes, with equi-spaced satellites in each orbit. For example, the Iridium system uses six orbital planes with 11 equi-spaced satellites in each plane, for a total of 66 satellites. Communication between two earth stations via the network appear seamless as message handover occurs between satellites via intersatellite links.

Radio frequency ISLs make use of frequencies that are highly attenuated by the atmosphere, so that interference to and from terrestrial systems using the same

Frequency Band, GHz	Available Bandwidth, MHz	Designation
22.55–23.55	1000	ISL-23
24.45–24.75	300	ISL-24
25.25–27.5	2250	ISL-25
32–33	1000	ISL-32
54.25–58.2	3950	ISL-56
59–64	5000	ISL-60
65–71	6000	ISL-67
116–134	18000	ISL-125
170–182	12000	
185–190	5000	

Source: Morgan (1999).

TABLE 12.1   ISL Frequency Bands

frequencies is avoided. Figure 4.2 shows the atmospheric absorption peaks at 22.3 GHz and 60 GHz. Table 12.1 shows the frequency bands in use.

Antennas for the ISL are steerable and the beamwidths are sufficiently broad to enable a tracking signal to be acquired to maintain alignment. Table 12.2 gives some values for an ISL used in the iridium system. The details differ for Starlink, but it is generally similar.

	East-West ISL (Without Sun)
Frequency, GHz	23.28
Range, km	4400.0
Transmitter	
Power, dBW	5.3
Antenna gain, dB	36.7
Circuit loss, $[L_{ta}]$, dB	–1.8
Pointing loss, $[L_{m,t}]$, dB	–1.8
Receiver	
Antenna gain, dB	36.7
Pointing loss, $[L_{m,r}]$, dB	–1.8
Link Margin, $[L_{margin}]$, dB	–1.8
Noise bandwidth, dBHz	71

Source: Motorola (1992).

TABLE 12.2   Iridium ISL

Example 12.28 Calculate the received power for the ISL parameters tabulated in Table 12.2.

Solution The [EIRP] is

$$[\text{EIRP}] = [P_T] + [G_T] = 5.3 + 36.7 = 42.0 \text{ dBW}$$

From Eq. (12.5b), the path loss is $[L_p] = 10 \log((1/4\pi)^2) + 10 \log((\lambda/r)^2)$

$$\lambda = \frac{c}{f} = \frac{3.0e + 08}{23.28e + 09} = 0.0129 \text{ m}$$

$$[L_p] = -21.98 + 10 \log(0.0129/4400)^2 = -21.98 + (-170.67) = -192.65 \text{ dB}$$

The total system losses are

$$[L_{\text{sys}}] = [L_{\text{ta}}] + [L_{m,t}] + [L_{\text{margin}}] + [L_{m,r}]$$
$$= (-1.8) + (-1.8) + (-1.8) + (-1.8) = -7.2$$

from Eq. (12.6b) the received power is

$$[P_r] = [\text{EIRP}] + [G_r] + [L_p] + [L_{\text{sys}}]$$
$$= 42.0 + 36.7 + (-192.65) + (-7.2) = -121.15 \text{ dBW}$$

Radio ISLs have the advantage that the technology is mature, so the risk of failure is minimized. However, the bandwidth limits the bit rate that can be carried, and optical systems, with their much higher carrier (optical) frequencies, have much greater bandwidth available. Further, because of the growing demand for secure, high-speed wireless communications, optical ISLs have a definite advantage over rf ISLs for data rates >1 Gbps. In addition, telescope apertures and optical equipment tends to be smaller and more compact than their rf counterparts [Wikipedia, 2023] (also, see Optical Communications and Intersatellite Links, undated, at www.wtec.org/loyola/satcom2/03_06.htm-22k-). For example, the optical beamwidth is typically 5 µrad (Maral et al., 2002), and Table 12.3 lists properties of some solid-state lasers.

Type	Wavelength, µm	Power	Beam Diameter, mm	Beam Divergence
GaAs/GaAlAs	0.78–0.905	1–40 mW Avg		10° × 35°
InGaAsP	1.1–1.6	1–10 mW		10° × 30° × 20° × 40°
Nd: YAG Pulsed	1.064	Up to 600 W Avg	1–10	0.3–20 mrad
Nd:YAG Diode pumped	1.064	0.5–10 mW	1–2	0.5–2.0 mrad
Nd:YAG (cw)	1.064	0.04–600 W	0.7–8	2–25 mrad

Notes: Al, aluminum; As, arsenide; cw, continuous wave; Ga, gallium; mm, millimeter; mrad, milliradian; µm, micron = 10^{-6} m; mW, milliwatt; Nd, neodymium; P, phosphorus; W, watt; YAG, yttrium-aluminum garnet.
Source: Extracted from Chen (1996).

TABLE 12.3 Solid State Lasers

Further, satellite laser communications is the fastest growing innovation in the space sector and earth's telecom industry, and NASA, in collaboration with private industry, is the sponsor for space-based laser technology development. The agency's Laser Communications Relay Demonstration, LCRD [NASA, 2023] will support digital communications over laser links with both LEO and Interplanetary spacecraft.

Especially, future manned space flight missions to the Moon and Mars will require laser communications. For example, near real-time face-to-face video isn't available now, but it's close to becoming a reality. On the Moon, it's possible to establish multiple, bi-directional HD video channels with only two-second latency. While it would be challenging to implement laser communications to watch a live Mars landing, nevertheless, implementing a Mars telecom orbiter is the ultimate goal to make relay services more reliable. Scaling laser communications for deep space destinations like Saturn or Jupiter will take more time to develop. Distant spacecraft will also have to battle the data latency issues that laser communications developers need to improve.

Concerning optical communications link analysis, the free-space loss given by Eq. (12.5a) is repeated here:

$$L_p = \left(\frac{1}{4\pi \left(\frac{r}{\lambda} \right)} \right)^2 , \text{ratio}$$

Here, wavelength is used in the equation as this is the quantity usually specified for a laser, and of course r and λ must be in the same units.

The intensity distribution of a laser beam generally follows what is termed a *Gaussian law*, for which the intensity falls off in an exponential manner in a direction transverse to the direction of propagation. The *beam radius* is where the transverse electric field component drops to $(1/e)$ of its maximum value, where $e \approx 2.718$. The diameter of the beam (twice the radius) gives the total beamwidth. The on-axis gain (like the antenna gain defined in Sec. 6.6 is given by (Maral et al., 2002)

$$G_T = \frac{32}{\theta_T^2} \tag{12.73}$$

where θ_T is the total beamwidth.

On the receive side, the telescope aperture gain is given by:

$$G_R = \left(\frac{\pi D}{\lambda} \right)^2 \tag{12.74}$$

where D is the effective diameter of the receiving aperture. The optical receiver receives some amount of optical power P_R. The energy in a photon is hc/λ where h is Plank's constant (6.6256 e-34 J-s) and c is the speed of light in vacuum (approximately 3.0e-08 m/s). For a received power P_R the number of photons received per second is therefore $P_R\lambda/hc$. The detection process consists of photons imparting sufficient energy to valence band electrons to raise these to the conduction band. The *quantum efficiency* of a photo-diode is the ratio (average number of conduction electrons generated)/(average number of photons received). Denoting the quantum efficiency by η, the average number of electrons released is $\eta P_R \lambda/hc$ and the photo current is:

$$I_{ph} = \frac{q\eta P_R \lambda}{hc} \tag{12.75}$$

where q is the electron charge. The *responsivity* of a photodiode is defined as the ratio of photo current to incident power. Denoting responsivity by R_0 and evaluating the constants in Eq. (12.73) gives

$$R_0 = \frac{\eta\lambda}{1.24} \text{ with } \lambda \text{ in } \mu m \tag{12.76}$$

The energy band gap of the semiconductor material used for the photodiode determines the wavelengths that it can respond to. The requirement in general is that the bandgap energy must be less than the photon energy, or $E_G < hc/\lambda$. On this basis it turns out that silicon is useful for wavelengths shorter than 1 μm. Germanium is useful at 1.3 μm and InSb and InAs at 1.55 μm.

The total current flowing in a photodiode consists of the actual current generated by the photons, plus what is termed the *dark current*. This is the current that flows even when no signal is present. Ideally it should be zero, but in practice it is of the order of a few nanoamperes, and it can contribute to the noise. Denoting the dark current by I_d the total current is

$$I = I_{ph} + I_d \tag{12.77}$$

This current is accompanied by *shot noise* (a name that is a hangover from vacuum tube days), the mean square spectral density of which is $(2qI)$ in A^2/Hz with I in amperes. This is for a diode without any internal amplification such as a PIN diode. An *avalanche photo diode* (APD) multiplies the signal current by a factor M and at the same time generates excess noise, represented by an *excess noise factor, F*, so that the mean square spectral density is $(2qI)MF$, where F increases with increases with M (see Jones, 1988, p. 240). Typical values are of the order $M = 100$, $F = 12$. For a PIN diode, $M = 1$ and $F = 1$.

The analysis to follow is based on that given in Jones (1988). The equivalent circuit for the input stage of an optical receiver is shown in Fig. 12.19, where R is the parallel combination of diode resistance, input load resistance and preamplifier input resistance, and C is the parallel combination of diode capacitance, preamplifier input capacitance and stray circuit capacitance. Resistance R generates thermal noise current, the spectral density of which is $4kT/R$ (k is Boltzmann's constant and T is the temperature, which may be taken as room temperature). The preamplifier transistor also generates shot noise, which when referred to the input has a spectral density $2q/I_{in}$ in A^2/Hz, where I_{in} is the transistor gate (JFET) or base (BJT) current in amperes. Using curly brackets {.} for

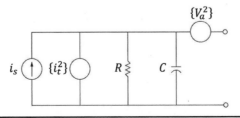

FIGURE 12.19   The equivalent input circuit for an optical receiver. $\{i_t^2\}$ is the mean square spectral density for the current noise source, and $\{v_a^2\}$ the mean square spectral density for the voltage noise source. i_s is the signal current.

the spectral density values (see Jones, 1988), the total noise current spectral density at the input is

$$\{i_t^2\} = 2qIM^2F + \frac{4kT}{R} + 2qI_{tr}A^2/\text{Hz} \tag{12.78}$$

The preamplifier also has a noise voltage component shown as v_a resulting from the shot noise in the drain or collector current. The mean square noise voltage spectral density is $2qI_{tr}/g_m^2\ V^2/\text{Hz}$ where I_{tr} is the transistor drain (JFET) or collector (BJT) current in amperes and g_m is the device transconductance in siemens. The total noise voltage spectral density at the input is therefore

$$\{v_n^2\} = \{i_t^2\}|Z_L|^2 + \frac{2qI_{tr}}{g_m^2}V^2/\text{Hz} \tag{12.79}$$

where Z_L is the impedance of R and C in parallel.

The average signal voltage is $MI_{ph}R$. Increasing R should result in an increase in signal to noise ratio since the signal voltage is proportional to R and the R component of noise current density is inversely proportional to R. However, the bandwidth of the RC input network is inversely proportional to R and therefore sets a limit to how large R can be. An equalizing filter, which has a transfer function, given by $H_{eq}(f) = 1 + j2\pi f RC$ can be included in the overall transfer function, which compensates for the input impedance frequency response over the signal bandwidth. The effective spectral density for the mean square noise voltage at the input is then $\{v_n^2\}|H(f)|^2$. The mean square noise voltage V_n^2 at the input is obtained by integrating this expression over the signal bandwidth. Only the result is given here:

$$V_n^2 = \left[\left(2qIM^2F + \frac{4kT}{R} + 2qI_{tr}\right)R^2 + \frac{2qI_{tr}}{g_m^2}\left(1 + \frac{(2\pi RCB)^2}{3}\right)\right]B \tag{12.80}$$

With the average signal voltage given by $V_s = MI_{ph}R$ the signal to noise ratio is

$$\frac{S}{N} = \frac{V_s^2}{V_n^2} = \frac{(MI_{ph}R)^2}{\left[\left(2qIM^2F + \frac{4kT}{R} + 2qI_{in}\right)R^2 + \frac{2qI_{tr}}{g_m^2}\left(1 + \frac{(2\pi RCB)^2}{3}\right)\right]B} \tag{12.81}$$

This can be simplified on dividing through by M^2R^2 to get:

$$\frac{S}{N} = \frac{I_{ph}^2}{\left[2qIF + \frac{4kT}{M^2R} + \frac{2qI_{in}}{M^2} + \frac{2qI_{tr}}{M^2g_m^2}\left(\frac{1}{R^2} + \frac{(2\pi CB)^2}{3}\right)\right]B} \tag{12.82}$$

Example 12.29 An optical receiver utilizes an APD for which $R_0 = 0.65$ A/W, $M = 100$, $F = 4$, $I_{in} = 0$, $g_m = 3000$ mS and $I_{tr} = 0.15$ mA. The dark current may be neglected. The input load consists of a 600 Ω resistor in parallel with a 10 pF capacitance. The signal bandwidth is 25 MHz, and equalization is employed. Calculate the resultant signal-to-noise ratio for an input signal power of 1 μW.

Solution   The photocurrent is $I_{ph} = 0.65\text{e-}06 = 0.65$ µA. The individual terms in the denominator are, with $I_d = 0$ and $I_{in} = 0$:

$$2qIF = 6.408e - 25 \text{ A}^2/\text{Hz}$$

$$\frac{4kT}{M^2R} = 0.027e - 25 \text{ A}^2/\text{Hz}$$

$$\frac{2qI_{tr}}{M^2 g_m^2}\left(\frac{1}{R^2} + \frac{(2\pi CB)^2}{3}\right) = 0.011e - 25 \text{ A}^2/\text{Hz}$$

$$\frac{S}{N} = \frac{(0.65e-06)^2 \times 1.0e+25}{(6.408+0.027+0.011)\times 25e+06} = 2.622e+04 \Rightarrow 44.2 \text{ dB}$$

12.14   Problems and Exercises

Note: In problems where temperature is not specified or is room temperature, assume a value of 290 K. In calculations involving antenna gain, assume an efficiency factor of 0.55.

12.1.   Give the decibel equivalents for the following quantities: (*a*) a power ratio of 30:1; (*b*) a power of 230 W in dBW; (*c*) a bandwidth of 36 MHz in dBHz; (*d*) a frequency ratio of 2 MHz/3 kHz in dB; (*e*) a temperature of 200 K in dBK; and (*f*) an antenna gain of 2.0 E + 04 in dBi.

12.2.   (*a*) Explain what is meant by EIRP. (*b*) A transmitter provides a power of 10 W into an antenna which has a gain of 46 dB. Calculate the EIRP in (*i*) watts; (*ii*) dBW.

12.3.   Calculate the boresight (peak) gain of a 3-m parabolic reflector antenna in power ratio and dBi at a frequency of (*a*) 6 GHz; (*b*) 14 GHz.

12.4.   Calculate the boresight (peak) gain in decibels and the effective area of a 30-m parabolic antenna at a frequency of 4 GHz.

12.5.   An antenna has a boresight (peak) gain of 46 dB at 12 GHz. Calculate its effective area.

12.6.   Calculate the effective area of a 10-ft parabolic reflector antenna at a frequency of (*a*) 4 GHz; (*b*) 12 GHz.

12.7.   The EIRP from a satellite is 49.4 dBW. Calculate (*a*) the power density at a ground station for which the range is 38,000 km and (*b*) the power delivered to a matched load at the input to the ground station receiver if the antenna gain is 50 dB for a downlink frequency of 4 GHz.

12.8.   Calculate the free-space loss as a power ratio and in decibels for transmission at frequencies of (*a*) 4 GHz, (*b*) 6 GHz, (*c*) 12 GHz, and (*d*) 14 GHz; for a range of 42,000 km.

12.9.   Repeat the calculation in Prob. 12.7*b* allowing for a fading margin of –1.0 dB and receiver feeder losses of –0.5 dB.

12.10.   Explain what is meant by (*a*) *antenna noise temperature*, (*b*) *amplifier noise temperature*, and (*c*) *system noise temperature* referred to input. A system operates with an antenna noise temperature of 40 K and an input amplifier noise temperature of 120 K. Calculate the available noise power density of the system referred to the amplifier input.

12.11.   Repeat Prob. 12.10; but now the apparent noise temperature at the antenna aperture is 40 K, but the antenna has an ohmic loss of –0.5 dB at a physical temperature = 290 K. Calculate the new noise power density of the system referred to the amplifier input.

12.12.   Two amplifiers are connected in cascade, each having a gain of 10 dB and a noise temperature of 200 K. Calculate (*a*) the overall gain and (*b*) the effective noise temperature referred to input.

12.13.   Explain what is meant by *noise factor*. What is the IEEE standard source temperature for which is noise factor defined?

12.14.   The noise factor of an amplifier is 7:1 power ratio. Calculate (*a*) the noise figure and (*b*) the equivalent noise temperature.

12.15.   An attenuator has an ohmic loss of –6 dB at a physical temperature = 290 K. Calculate (*a*) its noise figure and (*b*) its equivalent noise temperature referred to its input.

12.16.   An amplifier having a noise temperature of 200 K has a –4 dB attenuator connected at its input at a physical temperature = 290 K. Calculate the effective noise temperature referred to the attenuator input.

12.17.   A receiving system consists of an antenna having a noise temperature of 60 K, feeding directly into a LNA. The amplifier has a noise temperature of 120 K and a gain of 45 dB. The coaxial feeder between the LNA and the main receiver has an ohmic loss of –2 dB at a physical temperature = 290 K, and the main receiver has a noise figure of 9 dB. Calculate the system noise temperature referred to input.

12.18.   Explain why it is advantageous to place the LNA, of a receiving system, at the antenna end of the feeder cable. Give a numerical example for an antenna noise temperature of 20 K, a LNA with gain of 40 dB and noise figure = 1.5 dB, and a feeder cable with –2 dB ohmic loss at a physical temperature = 290 K.

12.19.   An antenna having a noise temperature of 35 K is connected through a feeder having –0.5 dB loss to an LNA. The LNA has a noise temperature of 90 K, and assume that the feeder has a physical temperature = 290 K. Calculate the system noise temperature referred to (*a*) the feeder input and (*b*) the LNA input. For an LNA gain = 30 dB, compare the output noise power density for the two cases, and explain the reason for the result.

12.20.   Explain what is meant by *carrier-to-noise ratio*. At the input to a receiver, the received carrier power is 400 pW, and the system noise temperature is 450 K. Calculate the carrier-to-noise density ratio in dBHz. Given that the bandwidth is 36 MHz, calculate the carrier-to-noise ratio in decibels.

12.21.   Explain what is meant by the G/T ratio of a satellite receiving system. A satellite receiving system employs a 5-m parabolic antenna operating at 12 GHz. The antenna noise temperature is 100 K, and the receiver front-end noise temperature is 120 K. Calculate $[G/T]$.

12.22.   In a satellite link the propagation loss is –200 dB. Margins and other losses account for another –3 dB. The receiver $[G/T]$ is 11 dB, and the [EIRP] is 45 dBW. Calculate the received $[C/N]$ for a system bandwidth of 36 MHz.

12.23.   A carrier-to-noise density ratio of 90 dBHz is required at a receiver having a $[G/T]$ ratio of 12 dB. Given that total losses in the link amount to –196 dB, calculate the [EIRP] required.

12.24.   Explain what is meant by *saturation flux density*. At 14 GHz, the power received by a 1.8-m parabolic antenna in the boresight direction is 250 pW. Calculate the power flux density (*a*) in W/m^2 and (*b*) in dBW/m^2 at the antenna.

12.25.   An earth station radiates an [EIRP] of 54 dBW at a frequency of 6 GHz. If the path loss = –196 dB, calculate the power flux density at the satellite receiving antenna.

12.26. A satellite transponder requires a saturation flux density of -110 dBW/m^2, operating at a frequency of 14 GHz. Calculate the earth station [EIRP] required if total losses (system and path) amount to -200 dB.

12.27. Explain what is meant by input BO. An earth station is required to operate at an [EIRP] of 44 dBW to produce saturation of the satellite transponder. If the transponder must be operated in a -10 dB input BO mode, calculate the new value of [EIRP] required.

12.28. Determine the carrier-to-noise density ratio at the satellite input for an uplink, which has the following parameters: operating frequency 6 GHz, saturation flux density $= -95$ dBW/m^2, input BO $= -11$ dB, satellite $[G/T] = -7$ dBK^{-1}, $[L_{ra}] = -0.5$ dB. (Tabulate the link budget values as shown in the text).

12.29. For an uplink the required $[C/N]$ ratio is 20 dB. The operating frequency is 30 GHz, and the bandwidth is 72 MHz. The satellite $[G/T]$ is 14.5 dBK^{-1}. Assuming operation with -11 dB input BO, calculate the saturation flux density. $[L_{ra}] = -1$ dB.

12.30. For the uplink in Prob. 12.29, the total losses amount to -218 dB. Calculate the earth station [EIRP] required.

12.31. An earth station radiates an [EIRP] of 54 dBW at 14 GHz from a 10-m parabolic antenna. The transmit feeder losses between the HPA and the antenna are -2.5 dB. Calculate the output of the HPA.

12.32. The following parameters apply to a satellite downlink: saturation [EIRP] 22.5 dBW, free-space loss -195 dB, other losses and margins -1.5 dB, earth station $[G/T]$ 37.5 dB/K. Calculate the $[C/N_o]$ at the earth station. Assuming an output BO of -6 dB is applied, what is the new value of $[C/N_o]$?

12.33. The output from a satellite TWTA is 10 W that is fed to a 1.2-m parabolic antenna operating at 12 GHz. Calculate the [EIRP].

12.34. The $[C/N]$ values for a satellite circuit are: uplink 25 dB, and downlink 15 dB. Calculate the overall $[C/N]$ value.

12.35. The required $[C/N]$ value at the ground station receiver is 22 dB, and the downlink $[C/N]$ is 24 dB. What is the minimum value of $[C/N]$ that the uplink can have in order that the overall value can be achieved?

12.36. A satellite circuit has the following parameters:

	Uplink, dB	Downlink, dB
[EIRP]	54	34
$[G/T]$	0	17
$[L_p]$	-200	-198
$[L_{ra}]$	-2	-2
$[L_{atm}]$	-0.5	-0.5
$[Lm]$	-0.5	-0.5

Calculate the overall $[C/N_o]$ value.

12.37. Explain how intermodulation noise originates in a satellite link and describe how it may be reduced. In a satellite circuit the carrier-to-noise ratios are uplink 25 dB; downlink 20 dB; intermodulation 13 dB. Calculate the overall carrier-to-noise ratio.

12.38. For the satellite circuit of Prob. 12.37, input BO is introduced that reduces the carrier-to-noise ratios by the following amounts: uplink −7 dB, downlink −2 dB, and improves the intermodulation by +11 dB. Calculate the new value of overall carrier-to-noise ratio.

12.39. As a follow-up to Example 12.28, calculate $[E_b/N_o]$, given that the coded bit rate is 25 Mbps.

12.40. Rework Example 12.29 to find the output [SNR] for a PIN diode, all other values remaining unchanged.

References

Andrew Antenna Co., Ltd. 1985. *1.8-Meter 12-GHz Receive-only Earth Station Antenna. Bulletin 1206A*. Whitby, Ontario, Canada.

CCIR Report 338-3. 1978. "Propagation Data Required for Line-of-Sight Radio Relay Systems." *14th Plenary Assembly*, vol. V, Kyoto.

Chen, C. L. 1996. *Elements of Optoelectronics and Fiber Optics*. Irwin, a Times Mirror Higher Education Group, Inc., Chicago.

Freeman, R. L. 1981. *Telecommunications Systems Engineering*. Wiley, New York.

Ha, T. T. 1990. *Digital Satellite Communications*. 2nd ed. McGraw-Hill, New York.

Hogg, D. C., and T. Chu. 1975. "The Role of Rain in Satellite Communications." *Proc. IEEE*, vol. 63, no. 9, pp. 1308–1331.

Jones, W. B. Jr. 1988. *Introduction to Optical Fiber Communication Systems*. Holt, Rinehart and Winston, Inc., Texas.

Lin, S. H., H. J. Bergmann, and M. V. Pursley. 1980. "Rain Attenuation on Earth-Satellite Paths: Summary of 10-Year Experiments and Studies." *Bell Syst. Tech. J.*, vol. 59, no. 2, February, pp. 183–228.

Maral, G., and M. Bousquet. 2002. *Satellite Communications Systems*. Wiley, New York.

Morgan, W. L. 1999. "Intersatellite Links." *Space Business International*, Quarter 1, pp. 9–12.

Motorola Satellite Communications Inc., 1992. Minor amendment to application before the FCC to construct and operate a low earth orbit satellite system in the RDSS uplink band File No.9-DSS-P91 (87) CSS-91-010.

NASA, 2023, https://www.nasa.gov/mission_pages/tdm/lcrd/index.html.

Pratt, Timothy, Charles Bostian, and Jeremy Allnutt. 2003. Satellite Communications. 2nd ed. Wiley, New York.

Ulaby, F., and D. G. Long. 2014. *Microwave Radiometric and Radar Remote Sensing*. University of Michigan Press, Ann Arbor, Michigan.

Webber, R. V., J. I. Strickland, and J. J. Schlesak. 1986. "Statistics of Attenuation by Rain of 13-GHz Signals on Earth-Space Paths in Canada." *CRC Report 1400*, Communications Research Centre, Ottawa, April.

Wikipedia, 2023, https://en.wikipedia.org/wiki/Laser_communication_in_space.

CHAPTER **13**

Interference

13.1 Introduction

With many telecommunications services using radio transmissions, interference between services can arise in several ways, and Fig. 13.1 shows in a general way the possible interference paths between services. In this figure, the *earth stations* are specifically associated with satellite circuits, and the *terrestrial stations* are specifically associated with ground-based microwave line-of-sight (LOS) circuits. The possible modes of interference shown are classified by the *International Telecommunications Union* (ITU, 1985) as follows:

A_1: terrestrial station transmissions, possibly causing interference to reception by an earth station

A_2: earth station transmissions, possibly causing interference to reception by a terrestrial station

B_1: space station transmission of one space system, possibly causing interference to reception by an earth station of another space system

B_2: earth station transmissions of one space system, possibly causing interference to reception by a space station of another space system

C_1: space station transmission, possibly causing interference to reception by a terrestrial station

C_2: terrestrial station transmission, possibly causing interference to reception by a space station

E: space station transmission of one space system, possibly causing interference to reception by a space station of another space system

F: earth station transmission of one space system, possibly causing interference to reception by an earth station of another space system

A_1, A_2, C_1, and C_2 are possible modes of interference between space and terrestrial services. B_1 and B_2 are possible modes of interference between stations of different space systems using separate uplink and downlink frequency bands, and E and F are extensions to B_1 and B_2 where bidirectional frequency bands are used.

The Radio Regulations (ITU, 1986) specify maximum limits on radiated powers (more strictly, on the distribution of energy spectral density) to reduce the potential interference to acceptable levels in most situations. However, interference may still occur in certain cases, and in this occurrence, *coordination*, between the affected telecommunications administrations, is required. Coordination may require both administrations to change or adjust some of the technical parameters of their systems.

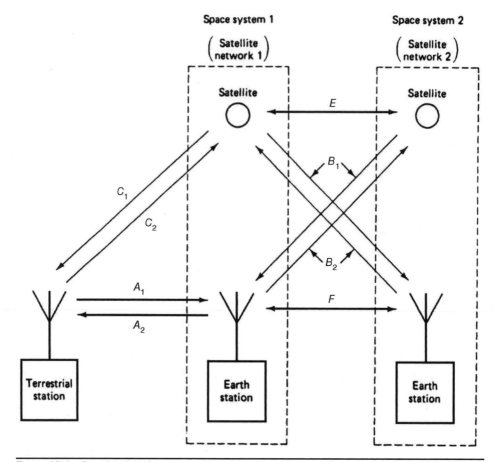

FIGURE 13.1 Possible interference modes between satellite circuits and a terrestrial station. (*Courtesy of CCIR Radio Regulations.*)

For geostationary satellites, interference modes B_1 and B_2 set a lower limit to the orbital spacing between satellites. To increase the capacity of the geostationary orbit, the *Federal Communications Commission* (FCC) in the United States has in recent years authorized a reduction in orbital spacing from 4° to 2° in the 6/4-GHz band. Some of the effects that this has on the B_1 and B_2 levels of interference are examined later in the chapter. It may be noted, however, that although the larger authorized operators are in general be able to meet the costs of technical improvements needed to offset the increased interference resulting from reduced orbital spacing, the same cannot be said for individually owned *television receive-only* (TVRO) installations (the "home satellite dish"), and these users have no recourse to regulatory control (Chouinard, 1984).

Interference with individually owned TVRO receivers also may occur from terrestrial station transmissions in the 6/4-GHz band. Although this may be thought of as an A_1 mode of interference, the fact that these home stations are considered by many broadcasting companies to be "pirates" means that regulatory controls to reduce interference are not applicable. Some steps that can be taken to reduce this form of interference are described in a publication by the Microwave Filter Company (1984).

It has been mentioned that the Radio Regulations place limits on the energy spectral density that may be emitted by an earth station. Energy dispersal is one technique employed to redistribute the transmitted energy more evenly over the transmitted bandwidth. This principle is described in more detail later in this chapter.

Intermodulation interference, briefly mentioned in Sec. 7.7.3, is a type of interference which can occur between two or more carriers using a common transponder in a satellite or a common high-power amplifier in an earth station. For all practical purposes, this type of interference can be treated as noise, as described in Sec. 12.11.

13.2   Interference between Satellite Circuits (B_1 and B_2 Modes)

A satellite circuit may suffer the B_1 and B_2 modes of interference shown in Fig. 13.1 from a few neighboring satellite circuits, the resultant effect being termed *aggregate interference*. Because of the difficulties of considering the range of variations expected in any practical aggregate, studies of aggregate interference have been quite limited, with most of the study effort going into what is termed *single-entry interference studies* (see Sharp, 1984a). As the name suggests, single-entry interference refers to the interference produced by a single interfering circuit on a neighboring circuit.

Interference may be considered as a form of noise, and as with noise, system performance is determined by the ratio of wanted to interfering powers, in this case the wanted carrier to the interfering carrier power or C/I ratio. The single most important factor controlling interference is the radiation pattern of the earth-station antenna. Comparatively large-diameter reflectors can be used with earth-station antennas, and hence narrow beamwidths can be achieved. For example, a 10-m antenna at 14 GHz has a −3-dB beamwidth of about 0.15°. This is very much narrower than the 2° to 4° orbital spacing allocated to satellites. To relate the C/I ratio to the antenna radiation pattern, it is necessary first to define the geometry involved.

Figure 13.2 shows the angles subtended by two satellites in geostationary orbit. The *orbital separation* is defined as the angle α subtended at the center of the earth, known as

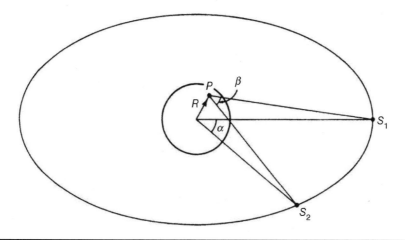

Figure 13.2   Geocentric angle α and the topocentric angle β.

the *geocentric angle*. However, from an earth station at point P the satellites appear to subtend an angle β. Angle β is referred to as the *topocentric angle*. In all practical situations relating to satellite interference, the topocentric and geocentric angles may be assumed equal, and in fact, making this assumption leads to an overestimate of the interference (Sharp, 1983).

However, this geometry is properly addressed in Sec. 3.2.1, Fig. 3.5, and Eq. (3.12) (repeated here). For example, as presented in Example 3.3, for an earth station located at 35°N × 100°W, the angle between two GEOSATS at 90°W and 92°W is $\theta = 2.27°$. Also a worst-case calculation, with an earth station on the equator, is presented in Example 13.1.

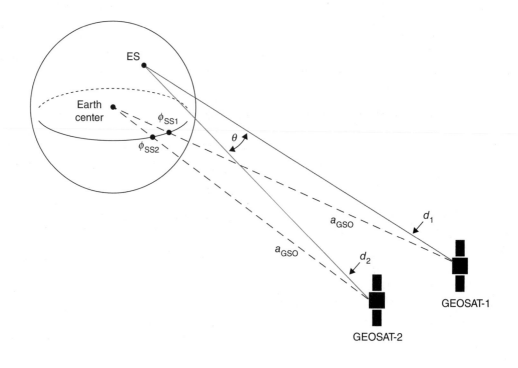

$$\theta = \arccos\left(\frac{\left(d_1^2 + d_2^2 - 2\alpha_{GSO}^2(1 - \cos(\phi_{ss1} - \phi_{ss2}))\right)}{2d_1 d_2}\right)$$

where d_1 and d_2 are the slant ranges to the two GEOSATs, a_{GSO} is the GEO radius, and ϕ_{ss1} and ϕ_{ss2} are the subsatellite point longitudes.

Example 13.1   For the earth station located at 0° × 91°W and its satellite GEOSAT-1 located at 90°W, find the off-boresight angle for an adjacent GEOSAT-2 located at an orbital location of 92°W.

Solution   First, using Eqs. (3.8) and (3.10), calculate the slant range to GEOSAT-1.

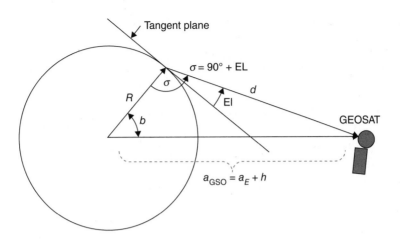

From Fig. 3.2b, the earth central angle "b" between the earth station and the subsatellite point is

$$b = \arccos(\cos B \cos \lambda_E)$$

where $B = \phi_E - \phi_{SS}$ (the delta-longitude between earth station and satellite), λ_E is the earth station latitude.

$$b = \arccos(\cos(1°)\cos(0°)) = 1°$$

Next using Eq. (3.10), calculate the slant range to GEOSAT-1 as

$$d_1 = \sqrt{6378.137^2 + 42164.2^2 - 2 \times 6378.137 \times 42164.2 \times \cos 1°} = 35{,}787.2 \text{ km}$$

By symmetry, for GEOSAT-2, the earth central angle $b_2 = -b_1$ and slant ranges are identical.
Next, using Eq. (3.12), the off-boresight angle of GEOSAT-2 is

$$\theta = \arccos\left(\frac{\left(d_1^2 + d_2^2 - 2\alpha_{GSO}^2(1 - \cos(\phi_{ss1} - \phi_{ss2}))\right)}{2d_1 d_2}\right)$$

Calculate the argument of the arccos

$$= \frac{(35{,}787.2^2 + 35{,}787.2^2 - 2 \times 42{,}164.2^2(1 - \cos 2°))}{(2 \times 35{,}787.2 \times 35{,}787.2)}$$

$$= 0.999154$$

and $\theta = \arccos(0.999154) = 0.04113 \text{ radians} = 2.36°$

So, using the orbital spacing (2°), for the off-boresight angle, is a conservative estimate of the interference.

Consider now S_1 as the wanted satellite and S_2 as the interfering satellite. An antenna has its main beam directed at S_1 and an off-axis component at angle θ directed at S_2. Angle θ is the same as the topocentric angle, which is assumed equal to the geocentric or orbital spacing angle. Therefore, when calculating the antenna sidelobe pattern, the orbital spacing angle may be used, as described in Sec. 13.2.4. Orbital spacing angles range from 2° to 4° in 0.5° intervals in the C-band.

In Fig. 13.3 the satellite circuit experiencing interference is that from earth station A via satellite S_1 to receiving station B. The B_1 mode of interference can occur from satellite S_2 into earth station B, and the B_2 mode of interference can occur from earth station C into

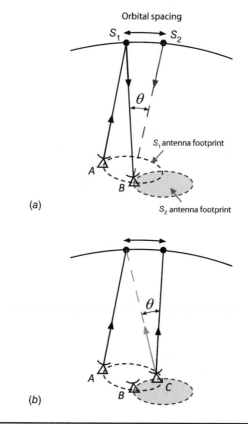

FIGURE 13.3 GEOSAT interference diagram, (a) B_1 mode and (b) B_2 mode.

satellite S_1. The total single-entry interference is the combined effect of these two modes. Because the satellites cannot carry very large antenna reflectors, the beamwidth is relatively wide, even for the so-called spot beams. For example, a 3.5-m antenna at 12 GHz has a beamwidth of about 0.5°, and the equatorial arc subtended by this angle is about 314 km. In interference calculations, therefore, the earth stations are assumed to be situated on the −3-dB contours of the satellite footprints, in which case the satellite antennas do not provide any gain discrimination between the wanted and the interfering carriers on either transmit or receive. However, when the geographic locations of the earth stations and the satellite antenna peak gains, boresight locations, and the effective reflector (lambda/diameter) are known, then the analysis presented in Sec. 6.13.1 may be used to perform an improved interference calculation.

13.2.1 Downlink (B_1 Mode)

Equation (12.6) may be used to calculate the wanted and interfering downlink carrier powers received by an earth station. Because engineers understand the concept of gain of a linear system, in this text edition, we choose to extend this concept to express losses as gains < 1.0 power ratio. In this way, the transfer function of cascaded blocks is simplified as the product of gains (some of which may be losses) for individual blocks. We feel that this is more intuitive than the alternative approach of treating losses as power ratios > 1.0 and then dividing by them in the linear transfer function. This topic is discussed in Chap. 12 and in App. G.

The desired carrier transmission from satellite-1 to earth station-B (shown as solid line in Fig. 13.3a) is

$$C_{1B} = P_{t,1} \, G_{t,1B} \, L_{p,1B} \, G_{r,1B} \, L_{\text{sys},1B} \qquad (13.1a)$$

and in dB form is

$$[C_{1B}] = [P_{t,1}] + [G_{t,1B}] + [L_{p,1B}] + [G_{r,1B}] + [L_{\text{sys},1B}] \qquad (13.1b)$$

When written in terms of the satellite EIRP, and by expanding $L_{\text{sys},1B}$, the received power is

$$C_{1B} = \text{EIRP}_1 \, G(\theta)_{1B} \, L_{\text{ta},1} \, L_{p,1B} \, L_{\text{atm},1} \, G_{r,1B} \, L_{\text{ra},B} \, L_{m,B} \qquad (13.2a)$$

and in dB form is

$$[C_{1B}] = [\text{EIRP}_1] + [G(\theta)_{1B}] + [L_{\text{ta},1}] + [L_{p,1B}] + [L_{\text{atm},1}] + [G_{r,1B}] + [L_{\text{ra},B}] + [L_{m,B}] \qquad (13.2b)$$

The first three terms are associated with GEOSAT-1, namely: $[\text{EIRP}]_1$, the equivalent isotropic radiated power; $[G(\theta)_{1B}]$, the LOS transmitting antenna gain loss; and $L_{\text{ta},1}$, feeder losses between the transmitter and antenna. The next two terms are associated with the free-space and atmosphere propagation, $L_{p,1B}$, is the free-space path loss and $L_{\text{atm},1}$ is the atmospheric attenuation. Finally, the last three terms are associated with the receiving earth station-B, namely: $G_{r,1B}$, the station-B peak receiving antenna gain; $L_{\text{ra},B}$ feeder losses between the receiver and antenna; and $L_{m,B}$ is the receiving antenna mispointing loss.

A similar equation, for the interfering carrier I, (shown as dashed line in Fig. 13.3a) is

$$I_{2B} = \text{EIRP}_2 \, G(\theta)_{2B} \, L_{\text{ta},2} \, L_{p,2B} \, L_{\text{atm},2} \, G(\theta)_{r,2B} \, L_{\text{ra},B} \qquad (13.3a)$$

and in dB form is

$$[I_{2B}] = [\text{EIRP}_2] + [G(\theta)_{2B}] + [L_{\text{ta},2}] + [L_{p,2B}] + [L_{\text{atm},2}] + [G_{r,2B}] + [L_{\text{ra},B}] \qquad (13.3b)$$

where the subscript "2" is for GEOSAT-2. Note that for the earth station-B, the antenna receiving gain ($G(\theta)_{r,2B}$) is the LOS gain in the direction of satellite-2, and there is no mispointing loss. Also, there may be an additional loss term $[Y]$, allowing for antenna polarization discrimination, when applicable.

Finally, taking the ratio C/I becomes

$$\frac{C}{I} = \frac{\text{EIRP}_1 \, G(\theta)_{1B} \, L_{\text{ta},1} \, L_{p,1B} \, L_{\text{atm},1} \, G_{r,1B} \, L_{\text{ra},B} \, L_{m,B}}{\text{EIRP}_2 \, G(\theta)_{2B} \, L_{\text{ta},2} \, L_{p,2B} \, L_{\text{atm},2} \, G(\theta)_{r,2B} \, L_{\text{ra},B} \, Y}$$

which reduces to

$$\frac{C}{I} = \frac{\text{EIRP}_1 \, G(\theta)_{1B} \, L_{\text{ta},1} \, G_{r,1B} \, L_{m,B}}{\text{EIRP}_2 \, G(\theta)_{2B} \, L_{\text{ta},2} \, G(\theta)_{r,2B} \, Y} \qquad (13.4a)$$

Converting to the dB form, the downlink becomes

$$\left[\frac{C}{I}\right]_D = [\Delta\text{EIRP}] + [\Delta G_t] + [\Delta L_{\text{ta}}] + [\Delta G_r] + [1/Y]_D \qquad (13.4b)$$

where $\Delta[\text{EIRP}] = [\text{EIRP}]_1 - [\text{EIRP}]_2$; $[\Delta G_t] = [G(\theta)_{1B}] - [G(\theta)_{2B}]$; $[\Delta L_{\text{ta}}] = [\Delta L_{\text{ta},1}] - [\Delta L_{\text{ta},2}]$; and $[\Delta G_r] = [G_{r,1B}] - [G(\theta)_{r,2B}]$

Example 13.2 The desired carrier [EIRP] from a satellite is 34 dBW, and the ground station receiving antenna gain is 44 dB in the desired direction and 24.47 dB toward the interfering satellite. The interfering satellite also radiates an [EIRP] of 34 dBW. Neglect the difference in $[\Delta L_{ta}]$, which is typically negligible and include a polarization discrimination is −4 dB. Determine the carrier-to-interference ratio at the ground receiving antenna.

Solution From Eq. (13.3)

$$\left[\frac{C}{I}\right]_D = (34 - 34) + (44 - 24.47) + (-4) = 23.3 \text{ dB}$$

13.2.2 Uplink (B_2 Mode)

In a similar manner, a result like Eq. (13.4) is derived for the uplink (see Fig. 13.3b). In this situation, however, it is desirable to work with the transmitter powers and the antenna transmit gains rather than the EIRPs of the two earth stations (A & C).

The desired carrier transmission is from earth station-A to GEOSAT-1, and this signal path is shown as a solid line in Fig. 13.3b. The received carrier power (linear units) is

$$C_{1A} = P_{t,A} G_{t,A} L_{m,A} L_{ta,A} L_{p,1A} L_{atm,1A} G(\theta)_{r,1A} L_{ra,1} \tag{13.5a}$$

and the resulting dB form is

$$[C_{1A}] = [P_{t,A}] + [G_{t,A}] + [L_{m,A}] + [L_{ta,A}] + [L_{p,1A}] + [L_{atm,1A}] + [G(\theta)_{r,1A}] + [L_{ra,1}] \tag{13.5b}$$

In a similar manner, the interference path is shown as a dashed line from earth station-C to GEOSAT-1 in Fig. 13.3b, and the linear form for the interference power is

$$I_{1C} = P_{t,C} G(\theta)_{t,1C} L_{ta,C} L_{p,1C} L_{atm,1C} G(\theta)_{r,1C} L_{ra,1} \tag{13.6a}$$

the antenna receiving gain $(G(\theta)_{r,2B})$ is the LOS gain in the direction of satellite-2, and there is no mispointing loss.

and in dB form

$$[I_{1A}] = [P_{t,C}] + [G(\theta)_{t,1C}] + [L_{ta,C}] + [L_{p,1C}] + [L_{atm,1C}] + [G(\theta)_{r,1C}] + [L_{ra,1}] \tag{13.6b}$$

So C/I becomes

$$\frac{C}{I} = \frac{P_{t,A} G_{t,A} L_{m,A} L_{ta,C} G(\theta)_{r,1A}}{P_{t,C} G(\theta)_{t,1C} L_{ta,C} G(\theta)_{r,1C} Y} \tag{13.7a}$$

and in the dB form

$$\left[\frac{C}{I}\right]_U \cong [\Delta P_t] + [\Delta G_t] + [\Delta G_r] + [1/Y]_U \tag{13.7b}$$

It is important to review the proper procedure to calculate the term $[1/Y]_U$. If Y is a power ratio < 1, then $[1/Y]$ expressed in dB = 10 log(1/Y) = positive number.

Example 13.3 Station A transmits at 24 dBW with an antenna gain of 54 dB, and station C transmits at 30 dBW. The off-axis gain in the S_1 direction is 24.47 dB, and the polarization discrimination is −4 dB. Calculate the [C/I] ratio on the uplink.

Solution First, it is important to review the proper procedure to calculate the term $[1/Y]_U$ dB. If Y is a power ratio < 1, then Y expressed in dB is

$$[Y] = 10 \log(Y) = \text{a negative number}$$

So, expressed as a power ratio, −4 dB = $10^{-4/10}$ = 0.398

Therefore, $\left[\dfrac{1}{Y}\right]_U = 10\log\left(\dfrac{1}{10^{-0.4}}\right) = 10\log(10^{+0.4}) = +4\text{ dB}$

So, $[1/-\text{dB}] = [+\text{dB}]$
Equation (13.6b) gives

$$\left[\dfrac{C}{I}\right]_D = (24-30)+(54-24.47)+4 = 27.53\text{ dB}$$

13.2.3   Combined [C/I] due to Interference on Both Uplink and Downlink

Interference may be considered as a form of noise, and assuming that the interference sources are statistically independent, the interference powers may be added directly to give the total interference at receiver B. The uplink and the downlink ratios are combined in the same manner described in Sec. 12.10 for noise, resulting in

$$\left(\dfrac{I}{C}\right)_{\text{tot}} = \left(\dfrac{I}{C}\right)_U + \left(\dfrac{I}{C}\right)_D \tag{13.8}$$

Here, power ratios must be used, not decibels, and the subscript "tot" denotes the total (combined ratio) at the input of station B receiver.

Example 13.4   Using the uplink and downlink values of $[C/I]$ determined in Examples 13.1 and 13.2, find the overall (total) ratio $[C/I]_{\text{tot}}$.

Solution   For the uplink, $[C/I] = 27.53$ dB gives $(I/C)_U = 0.001766$, and for the downlink, $[C/I] = 23.53$ dB gives $(I/C)_D = 0.004436$. Combining these gives

$$\left(\dfrac{I}{C}\right)_{\text{tot}} = 0.001766 + 0.004436 = 0.006202$$

Hence

$$\left[\dfrac{I}{C}\right]_{\text{tot}} = -10\log 0.006202 = 20.07\text{ dB}$$

13.2.4   Earth Station Antenna Gain Function

The earth station antenna radiation pattern (both transmit and receive) is divided into three regions: the main lobe region, the sidelobe region, and the transition region between the two. For interference calculations, the exact details of the antenna gain pattern are not required, and an envelope curve is used instead.

Figure 13.4 shows a sketch of the envelope pattern used by the FCC. The width of the main lobe and transition region depends on the antenna diameter in units of wavelength, and Fig. 13.4 is shown to provide only the general shape. The sidelobe gain function in units of dB is defined for different ranges of the off-boresight angle, θ given in units of degrees. The sidelobe gain function is

$$[G(\theta°)] = \begin{cases} 29 - 25\log\theta & 1° \le \theta \le 7° \\ +8 & 7° < \theta \le 9.2° \\ 32 - 25\log\theta & 9.2° < \theta \le 48° \\ -10 & 48° < \theta \le 180° \end{cases} \tag{13.9}$$

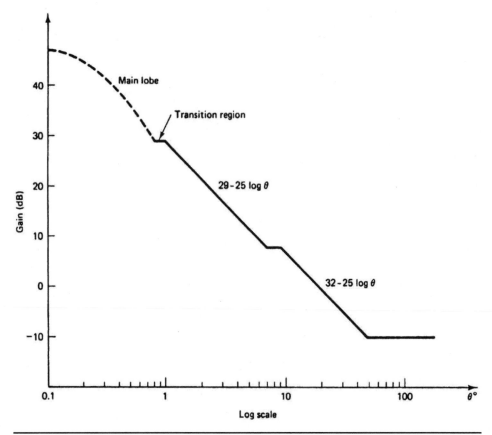

FIGURE 13.4 Earth-station antenna gain pattern used in FCC/OST R83-2, revised Nov. 30, 1984. (*Courtesy of Sharp, 1984b.*)

For the range of satellite orbital spacings presently in use, this sidelobe gain function is in the calculation of the interference levels.

Example 13.5 Determine the degradation in the downlink $[C/I]$ ratio, when satellite orbital spacing is reduced from 4° to 2°, and assuming that all other factors remain unchanged. FCC earth station antenna characteristics may be assumed.

Solution The decibel increase in interference is

$$(29 - 25 \log 2) - (29 - 25 \log 4) = 25 \log 2 = 7.5 \text{ dB}$$

The $[C/I]_D$ is degraded directly by this amount.

It should be noted that no simple relationship can be given for calculating the effect of reduced orbital spacing on the overall $[C/I]$. The separate uplink and downlink values must be calculated and combined as described in Sec. 13.2.3. Also, other telecommunications authorities specify antenna characteristics that differ from the FCC specifications (see CCIR Rep. 391-3, 1978).

13.2.5   Passband Interference

In the preceding section, the carrier-to-interference ratio at the receiver input is determined. However, the amount of interference reaching the detector depends on the bandwidth of frequency overlap between the interfering spectrum and the wanted channel passband.

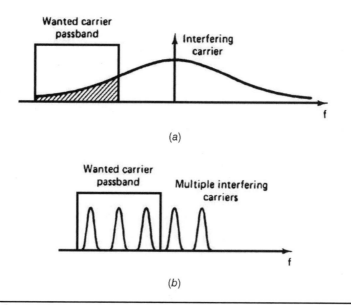

FIGURE 13.5 Power spectral density curves for (a) wideband interfering signal and (b) multiple interfering carriers.

Two situations can arise, as shown in Fig. 13.5. In Fig. 13.5a, partial overlap of the interfering signal spectra with the wanted passband is shown. The fractional interference is given as the ratio of the shaded area to the total area under the interference spectrum curve. This is denoted by the symbol Q (Sharp, 1983) or in decibels as $[Q]$. Where partial overlap occurs, Q is less than unity and $[Q]$ is a negative dB, and where the interfering spectrum coincides with the wanted passband, $[Q] = 0$ dB. So, for an accurate C/I estimation, the evaluation of Q usually must performed.

The second situation, illustrated in Fig. 13.5b, is where multiple interfering carriers are present within the wanted passband, such as with *single carrier per channel* (SCPC) operation discussed in Sec. 14.5. Here, Q represents the sum of the interfering carrier powers within the passband, and $[Q] > 0$ dB.

In the FCC report FCC/OST R83–2 (Sharp, 1983), Q values are computed for a wide range of interfering and wanted carrier combinations. Typical $[Q]$ values obtained from the FCC report are as follows: with the wanted carrier a TV/FM signal and the interfering carrier a similar TV/FM signal, $[Q] = 0$ dB; with SCPC/PSK interfering carriers, $[Q] = 27.92$ dB; and with the interfering carrier a wideband digital-type signal, $[Q] = -3.36$ dB.

The passband $[C/I]$ ratio is calculated using

$$\left(\frac{C}{I}\right)_{pb} = \left(\frac{C}{I}\right)_{tot}\left(\frac{1}{Q}\right), \text{ ratio} \tag{13.10a}$$

$$\left[\frac{C}{I}\right]_{pb} = \left[\frac{C}{I}\right]_{tot} + [1/Q], \text{ dB} \tag{13.10b}$$

The positions of these ratios in the receiver chain are illustrated in Fig. 13.6, where it can be seen that both $[C/I]_{ant}$ and $[C/I]_{pb}$ are predetection ratios, measured at RF or IF. Interference also can be measured in terms of the post-detector output, shown as $[S/I]$ in Fig. 13.6, and this is discussed in the following section.

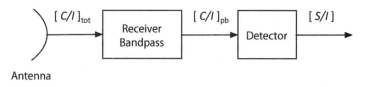

Figure 13.6 Carrier-to-interference ratios and signal-to-interference ratio.

13.2.6 Receiver Transfer Characteristic

In some situations a measure of the interference in the postdetection baseband, rather than in the IF or RF passband, is required. Baseband interference is measured in terms of baseband signal-to-interference ratio $[S/I]$. To relate $[S/I]$ to $[C/I]_{ant}$, a *receiver transfer characteristic* is introduced that considers the modulation characteristics of the wanted and interfering signals and the carrier frequency separation. Denoting the receiver transfer characteristic in decibels by [RTC], the relationship can be written as

$$\left(\frac{S}{I}\right) = \left(\frac{C}{I}\right)_{tot} (RTC) \tag{13.11a}$$

$$\left[\frac{S}{I}\right] = \left[\frac{C}{I}\right]_{tot} + [RTC], \text{ dB} \tag{13.11b}$$

The [RTC] is analogous to the receiver processing gain $[G_p]$ introduced in Sec. 9.6.3. Note that it is the $[C/I]$ at the receiver input, which is used, **not the passband value**, and the [RTC] considers any frequency offset. The [RTC] is always a positive number of decibels so that the baseband signal-to-interference ratio is greater than the carrier-to-interference ratio at the receiver input.

Calculation of [RTC] for various combinations of wanted and interfering carriers is very complicated and usually performed by computer simulation. As an example, taken from Sharp (1983), when the wanted carrier is TV/FM with a modulation index of 2.571 and the interfering carrier is TV/FM with a modulating index of 2.560, the carrier frequency separation being zero, the [RTC] is computed to be 31.70 dB. These computations are limited to low levels of interference (see Sec. 13.2.8).

13.2.7 Specified Interference Objectives

Although $[C/I]_{pb}$ or $[S/I]$ gives a measure of interference, ultimately, the effects of interference must be assessed in terms of what is tolerable to the end user. Such assessment usually relies on some form of subjective measurement. For TV, viewing tests are conducted, in which a mixed audience of experienced and inexperienced viewers (experienced from the point of view of assessing the effects of interference) assess the effects of interference on picture quality. By gradually increasing the interference level, a quality impairment factor can be established that ranges from 1 to 5. The five grades are defined as (Chouinard, 1984):

5. Imperceptible

4. Perceptible, but not annoying

3. Slightly annoying

2. Annoying

1. Very annoying

[S/I] objectives:

FDM/FM: 600 pW0p or [S/I] = 62.2 dB (62.2 dB). Reference: CCIR Rec. 466-3.

TV/FM (broadcast quality): [S/I] = 67 dB weighted (67 dB; 65.5 dB).

References: CCIR Rec. 483–1, 354–2, 567, 568.

CSSB/AM: (54.4 dB; 62.2 dB). No references quoted.

$[C/I]_{pb}$ objectives:

TV/FM (CATV): $[C/I]_{pb}$ = 20 dB (22 dB; 27 dB). Reference: ABC 62 FCC 2d 901 (1976).

Digital: $[C/I]_{pb}$ = [C/N] (at BER 10^{-6}) + 14 dB (20 to 32.2 dB). Reference: CCIR Rec. 523.

SCPC/PSK: (21.5 dB; 24 dB). No references quoted.

SCPC/FM: (21.2 dB; 23.2 dB). No references quoted.

SS/PSK: (11 dB; 0.6 dB). No references quoted.

TABLE 13.1 Summary of Single-Entry Interference Objectives Used in FCC/OST R83-2, May 1983

Acceptable picture quality requires a quality impairment factor of at least 4.2. Typical values of interference levels which result in acceptable picture quality are for broadcast TV, [S/I] = 67 dB; and for cable TV, $[C/I]_{pb}$ = 20 dB.

For digital circuits, the $[C/I]_{pb}$ is related to the *bit error rate* (BER) (see, e.g., CCIR Rec. 523, 1978). Values of the required $[C/I]_{pb}$ used by Sharp (1983) for different types of digital circuits range from 20 to 32.2 dB.

To give some idea of the numerical values involved, a summary of the objectives stated in the FCC single-entry interference program is presented in Table 13.1. In some cases the objective used differed from the reference objective, and the values used are shown in parentheses. In some entries in the table, noise is shown measured in units of pW0p. Here, the pW stands for picowatts. The 0 means that the noise is measured at a "zero-level test point," which is a point in the circuit where a test-tone signal produces a level of 0 dBm. The final p stands for psophometrically weighted noise, discussed in Sec. 9.6.6.

13.2.8 Protection Ratio

In CCIR Report 634–2 (1982), the *International Radio Consultative Committee* (CCIR) specifies the permissible interference level for TV carriers in terms of a parameter known as the *protection ratio*. The protection ratio is defined as the minimum carrier-to-interference ratio at the input to the receiver that results in "just perceptible" degradation of picture quality. The protection ratio applies only for wanted and interfering TV carriers at the same frequency, and it is equivalent to $[C/I]_{pb}$ evaluated for this situation. Denoting the quality impairment factor by Q_{IF} and the protection ratio in decibels by $[PR_0]$, the equation given in CCIR Report 634–2 (1982) is

$$[PR_0] = 12.5 - 20\log\left(\frac{Dv}{12}\right) - Q_{IF} + 1.1\, Q_{IF}^2 \tag{13.12}$$

Here Dv is the peak-to-peak deviation in megahertz.

Example 13.6 An FM/TV carrier is specified as having a modulation index of 2.571 and a top modulating frequency of 4.2 MHz. Calculate the protection ratio required to give a quality impairment factor of (*a*) 4.2 and (*b*) 4.5.

Solution   The peak-to-peak deviation is $2 \times 2.571 \times 4.2 = 21.6$ MHz. Applying Eq. (13.12) gives the following results:

(a) $[PR_0] = 12.5 - 20 \log \left(\dfrac{21.6}{12} \right) - 4.2 + 1.1 \times 4.2^2 = 22.6$ dB

(b) $[PR_0] = 12.5 - 20 \log \left(\dfrac{21.6}{12} \right) - 4.5 + 1.1 \times 4.5^2 = 25.2$ dB

It should be noted that the receiver transfer characteristic discussed in Sec. 13.2.6 was developed from the CCIR protection ratio concept (see Jeruchim and Kane, 1970).

13.3   Energy Dispersal

The power in a frequency-modulated signal remains constant, independent of the modulation index. When unmodulated, all the power is at the carrier frequency, and when modulated, the same total power is distributed among the carrier and the sidebands. At low modulation indices the sidebands are grouped close to the carrier, and the power spectral density, or wattage per unit bandwidth, is relatively high in that spectral region. At high modulation indices, the spectrum becomes widely spread, and the power spectral density relatively low.

Use is made of this property in certain situations to keep radiation within CCIR recommended limits. For example, to limit the A_2 mode of interference in the 1 to 15 GHz range for the fixed satellite service, CCIR Radio Regulations state, in part, that the earth station EIRP should not exceed 40 dBW in any 4-kHz band for $\Theta \leq 0°$ and should not exceed $40 + 3\Theta$ dBW in any 4-kHz band for $0° < \Theta \leq 5°$. The angle is the angle Θ of elevation of the horizon viewed from the center of radiation of the earth station antenna. It is positive for angles above the horizontal plane, as illustrated in Fig. 13.7a, and negative for angles below the horizontal plane, as illustrated in Fig. 13.7b.

For space stations transmitting in the frequency range 3400 to 7750 MHz, the limits are specified in terms of power flux density for any 4-kHz bandwidth. Denoting the angle of arrival as δ degrees, measured above the horizontal plane as shown in Fig. 13.7c, these limits are

- -152 dB (W/m^2) in any 4-kHz band for $0° \leq \delta \leq 5°$
- $-152 + 0.5\,(\delta - 5)$ dB (W/m^2) in any 4-kHz band for $5° < \delta \leq 25°$
- -142 dB (W/m^2) in any 4-kHz band for $25° < \delta \leq 90°$

Because the specification is in terms of power or flux density in any 4-kHz band, not the total power or the total flux density, a carrier may be within the limits when heavily frequency-modulated, but the same carrier with light frequency modulation may exceed the limits. An energy-dispersal waveform is a low-frequency modulating wave which is inserted below the lowest baseband frequency for the purpose of dispersing the spectral energy when the current value of the modulating index is low. In the INTELSAT system for FDM carriers, a symmetrical triangular wave is used, a different fundamental frequency for this triangular wave in the range 20 to 150 Hz being assigned to each FDM carrier. The rms level of the baseband is monitored, and the amplitude of the dispersal waveform is automatically adjusted to keep the overall frequency deviation within defined limits. At the receive end, the dispersal waveform is removed from the demodulated signal by lowpass filtering.

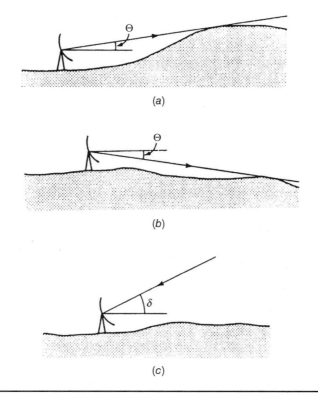

(a)

(b)

(c)

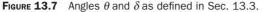

FIGURE 13.7   Angles θ and δ as defined in Sec. 13.3.

With television signals the situation is more complicated. The dispersal waveform, usually a sawtooth waveform, must be synchronized with the field frequency of the video signal to prevent video interference, so for the 525/60 standard, a 30-Hz wave is used, and for the 625/50 standard, a 25-Hz wave is used. If the TV signal occupies the full bandwidth of the transponder, known as *full-transponder television*, the dispersal level is kept constant at a peak-to-peak deviation of 1 MHz irrespective of the video level. In what is termed *half-transponder television*, where the TV carrier occupies only one-half of the available transponder bandwidth, the dispersal deviation is maintained at 1-MHz peak to peak when video modulation is present and is automatically increased to 2 MHz when video modulation is absent. At the receiver, video clamping is the most used method of removing the dispersal waveform.

Energy dispersal is effective in reducing all modes of interference but particularly that occurring between earth and terrestrial stations (A_2 mode) and between space and terrestrial stations (C_1). It is also effective in reducing intermodulation noise.

13.4   Coordination

When a new satellite network is in the planning stage, certain calculations must be made to ensure that the interference levels remain within acceptable limits. These calculations include determining the interference that is caused by the new system and interference it receives from other satellite networks.

In Sec. 13.2, procedures are outlined showing how interference may be calculated by considering modulation parameters and carrier frequencies of wanted and interfering systems. These calculations are very complex, and the CCIR uses a simplified method to determine whether *coordination* is necessary. As mentioned previously, where the potential for interference exists, the telecommunication administrations are required to coordinate the steps to be taken to reduce interference, a process referred to as *coordination*.

To determine whether coordination is necessary, the interference level is calculated assuming maximum spectrum density levels of the interfering signals and converted to an equivalent increase in noise temperature. The method is specified in detail in CCIR Report 454–3 (1982) for several possible situations. To illustrate the method, one specific situation where the existing and proposed systems operate on the same uplink and downlink frequencies, is explained here.

Figure 13.8a shows the two networks, R and R'. The method is described for network R' interfering with the operation of R. Satellite S' can interfere with the earth-station E, this being a B_1 mode of interference, and earth-station E' can interfere with the satellite S, this being a B_2 mode. Note that the networks need not be physically adjacent to one another.

13.4.1 Interference Levels

Consider first the interference B_1. This is illustrated in Fig. 13.8b. Let U_S represent the maximum power density transmitted from satellite S'. The units for U_S are W/Hz, or joules (J), and this quantity is explained in more detail shortly. Let the transmit gain of satellite S' in the direction of earth-station E be G'_S, and let G'_E be the receiving gain of earth-station E in the direction of satellite S'. The interfering spectral power density received by the earth station is therefore

$$[I_1] = [U_S] + [G'_S] + [G_E] + [L_D] \tag{13.13}$$

where L_D is the propagation loss for the downlink. The gain and loss factors are power ratios, and the brackets denote the corresponding decibel values as before. The increase in equivalent noise temperature at the earth-station receiver input can then be defined using Eq. (12.13) as

$$[\Delta T_E] = [I_1] + [1/k] = [I_1] + 228.6 \tag{13.14}$$

Here, k is Boltzmann's constant and $[k] = -228.6$ dBJ/K.

A similar argument can be applied to the uplink interference B_2, as illustrated in Fig. 13.8c, giving

$$[I_2] = [U_E] + [G'_E] + [G_S] + [L_U] \tag{13.15}$$

The corresponding increase in the equivalent noise temperature at the satellite receiver input is then

$$[\Delta T_S] = [I_2] + 228.6 \tag{13.16}$$

Here, U_E is the maximum power spectral density transmitted by earth station E', G'_E is the transmit gain of E' in the direction of S, G_S is the receive gain of S in the direction of E', and L_U is the uplink propagation loss.

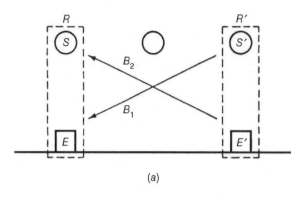

(a)

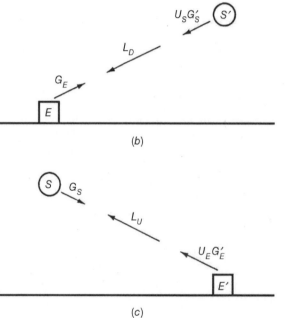

(b)

(c)

Figure 13.8 (a) Interference modes B_1 and B_2 from network R' into network R. (b) For the B_1 mode the interfering power in dBW/Hz is $[I_1] = [U_S] + [G_S'] + [G_E] + [L_U]$. (c) For the B_2 mode the interfering power in dBW/Hz is $[I_2] = [U_E] + [G_E'] + [G_S] + [L_U]$.

13.4.2 Transmission Gain

The effect of the equivalent temperature rise ΔT_S must be transferred to the earth-station E, and this is done using the transmission gain for system R, which is calculated for the situation shown in Fig. 13.9. Figure 13.9a shows the satellite circuit in block schematic form. U_E represents the maximum power spectral density transmitted by earth-station E, and G_{TE} represents the transmit gain of E in direction S. G_{RS} represents the receive gain of S in the direction E. The received power spectral density at satellite S is therefore

$$[U_{RS}] = [U_E] + [G_{TE}] + [G_{RS}] + [L_U] \tag{13.17}$$

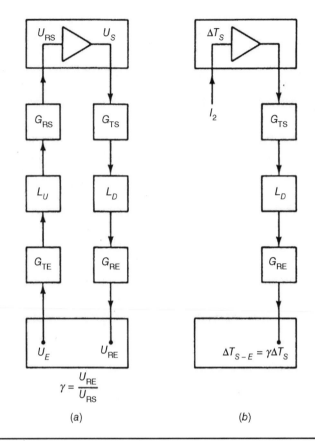

FIGURE 13.9 (a) Defining the transmission gain γ in Sec. 13.4.2. (b) Use of transmission gain to refer satellite noise temperature to an earth station.

In a similar way, with satellite S transmitting and earth-station E receiving, the received power spectral density at earth-station E is

$$[U_{RE}] = [U_S] + [G_{TS}] + [G_{RE}] + [L_D] \tag{13.18}$$

where U_S is the maximum power spectral density transmitted by S, G_{TS} is the transmit gain of S in the direction of E, and G_{RE} is the receive gain of E in the direction of S. It is assumed that the uplink and downlink propagation losses, L_U and L_D, are the same as those used for the interference signals.

The transmission gain for network R is then defined as

$$\gamma = \frac{U_{RE}}{U_{RS}} \tag{13.19a}$$

$$[\gamma] = [U_{RE}] + [1/U_{RS}] \tag{13.19b}$$

Note that this is the same transmission gain shown in Fig. 12.14.

Using the transmission gain, the interference I_2 at the satellite may be referred to the earth-station receiver as γI_2, and hence the noise-temperature rise at the satellite receiver input may be referred to the earth-station receiver input as $\gamma \Delta T_S$. This is illustrated in Fig. 13.9*b*. Expressed in decibel units, the relationship is

$$[\Delta T_{S-E}] = [\gamma] + [\Delta T_S] \tag{13.20}$$

13.4.3 Resulting Noise-Temperature Rise

The overall equivalent rise in noise temperature at earth-station E because of interference signals B_1 and B_2 is then

$$\Delta T = \Delta T_{S-E} + \Delta T_E \tag{13.21}$$

In this final calculation the dBK values must first be converted to degrees, which are then added to give the resulting equivalent noise-temperature rise at the earth-station E receive antenna output.

Example 13.7   Given that $L_U = -200$ dB, $L_D = -196$ dB, $G_E = G_E' = 25$ dB, $G_S = G_S' = 9$ dB, $G_{TE} = G_{RE} = 48$ dB, $G_{RS} = G_{TS} = 19$ dB, $U_S = U_S' = 1$ µJ, and $U_E' = 10$ µJ; calculate the transmission gain $[\gamma]$, the interference levels $[I_1]$ and $[I_2]$, and the equivalent temperature rise overall.

Solution   Using Eq. (13.14) gives

$$[U_{RS}] = [U_E] + [G_{TE}] + [G_{RS}] + [1/L_U]$$

$$[U_{RS}] = -50 + 48 + 19 + (+200) = -50 + 48 + 19 + 200 = -183 \text{ dBJ}$$

Using Eq. (13.15) gives

$$[U_{RE}] = -60 + 19 + 48 + (-196) = -189 \text{ dBJ}$$

Therefore,

$$[\gamma] = -189 + (+183) = -6 \text{ dB}$$

From Eq. (13.10),

$$[I_1] = -60 + 9 + 25 + (-196) = -222 \text{ dBJ}$$

From Eq. (13.12),

$$[I_2] = -50 + 25 + 9 - 200 = -216 \text{ dBJ}$$

From Eq. (13.11),

$$[\Delta T_E] = -222 + 228.6 = 6.6 \text{ dBK or } \Delta T_E = 4.57 \text{ K}$$

From Eq. (13.13),

$$[\Delta T_S] = -216 + 228.6 = 12.6 \text{ dBK}$$

From Eq. (13.17),

$$[\Delta T_{S-E}] = -6 + 12.6 = 6.6 \text{ dBK or } \Delta T_{S-E} = 4.57 \text{ K}$$

The resulting equivalent noise-temperature rise at the earth-station E receive antenna output is 4.57 + 4.57 = 9.14 K

13.4.4 Coordination Criterion

CCIR Report 454–3 (1982) specifies that the equivalent noise-temperature rise should be no more than 4 percent of the equivalent thermal noise temperature of the satellite link. The equivalent thermal noise temperature is defined in the CCIR Radio Regulations, App. 29.

As an example, the CCIR Recommendations for FM Telephony allows up to 10,000 pW0p total noise in a telephone channel. The abbreviation pW0p stands for picowatts at a zero-level test point, psophometrically weighted, as already defined in connection with Table 13.1. The 10,000-pW0p total includes a 1000-pW0p allowance for terrestrial station interference and 1000 pW0p for interference from other satellite links. Thus the thermal noise allowance is 10,000 − 2000 = 8000 pW0p. Four percent of this is 320 pW0p. Assuming that this is over a 3.1-kHz bandwidth, the spectrum density is 320/3100 or approximately 0.1 pJ0p (pW0p/Hz). In decibels, this is −130 dBJ. This is output noise, and to relate it back to the noise temperature at the antenna, the overall gain of the receiver from antenna to output, including the processing gain, discussed in Sec. 9.6.3, must be known. For illustration purposes, assume that the gain is 90 dB, so the antenna noise is −130 − 90 = −220 dBJ. The noise-temperature rise corresponding to this is −220 + 228.6 = 8.6 dBK. Converting this to kelvins gives 7.25 K.

13.4.5   Noise Power Spectral Density

The concept of noise power spectral density was introduced in Sec. 12.5 for a flat frequency spectrum. Where the spectrum is not flat, an average value for the spectral density can be calculated. To illustrate this, consider a simplified spectrum of input noise (Fig. 13.10) for an analog signal at the input to a typical multiple SCPC satellite transponder. The maximum spectrum density is flat at typical value of 3×10^{-20} W/Hz from 0 to 2 MHz and then slopes linearly down to zero over the range from 2 to 8 kHz.

The noise power in any given bandwidth is calculated as the area under the curve, the width of which is the value of the bandwidth. Thus, for the first 2 kHz, the noise power is 3×10^{-20} W/Hz; so the total noise over this bandwidth = 6×10^{-17} W. Next, from 2 to 8 kHz, the noise power is $3 \times 10^{-20} \times (8 - 2) \times 1000/2 = 9 \times 10^{-17}$ W. Thus, the total power is therefore 15×10^{-17} W, and the average spectral density is 15,000/8000 = 1.88×10^{-20} W/Hz.

The noise power spectral density over the worst 4-kHz bandwidth must include the highest part of the curve and is therefore calculated for the 0- to 4-kHz band. The power over this band is seen to be the area of the rectangle 3 W/Hz × 4 kHz minus the area of the triangle shown dashed in Fig. 13.10, which equals 11×10^{-17} W, and the spectral density is 2.75×10^{-20} W/Hz.

The units for spectral power density are often stated as watts per hertz (W/Hz), which is descriptive of how the power spectral density is calculated. In terms of

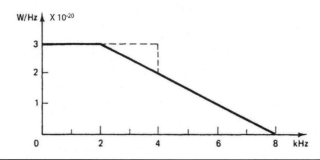

Figure 13.10   Typical transponder input noise power spectral density for a narrow band SCPC signal.

fundamental units, watts are equivalent to joules per second and hertz to cycles per second or simply seconds^{-1}. Since cycles are a dimensionless quantity, then 1 W/Hz is equivalent to 1 J. This is verified by Eq. (12.13) for noise power spectral density.

13.5 Problems and Exercises

13.1. Briefly describe the modes of interference that can occur in a satellite communications system. Distinguish carefully between satellite and terrestrial modes of interference.

13.2. Define and explain the difference between topocentric angles and geocentric angles as applied to satellite communications. Two geostationary satellites have an orbital spacing of 4°. Calculate the topocentric angle subtended by the satellites, measured (*a*) from the midpoint between the subsatellite points and (*b*) from either of the subsatellite points.

13.3. Westar IV is located at 98.5°W and Telstar at 96°W. The coordinates for two earth stations are 104°W, 36°N and 90°W, 32°N. By using the look angle and range formulas given in Sec. 3.2, calculate the topocentric angle subtended at each earth station by these two satellites.

13.4. Explain what is meant by *single-entry interference*. Explain why it is the radiation pattern of the earth station antennas, not the satellite antennas, which governs the level of interference.

13.5. A geostationary satellite employs a 3.5-m parabolic antenna at a frequency of 12 GHz. Calculate the half-power beamwidth and the spot diameter on the equator.

13.6. Calculate the half-power beamwidth for an earth station antenna operating at 14 GHz. The antenna utilizes a parabolic reflector of 3.5-m diameter. Compare the distance separation of satellites at 2° spacing with the diameter of the beam at the −3-dB points on the geostationary orbit.

13.7. Compare the increase in interference levels expected when satellite orbital spacing is reduced from 4° to 2° for earth station antenna sidelobe patterns of (*a*) 32 − 25 log θ dB and (*b*) 29 − 25 log θ dB.

13.8. A satellite circuit operates with an uplink transmit power of 28.3 dBW and an antenna gain of 62.5 dB. A potential interfering circuit operates with an uplink power of 26.3 dBW. Assuming a −4 dB polarization discrimination figure and earth station sidelobe gain function of 32 − 25 log θ dB, calculate the [*C*/*I*] ratio at the satellite for 2° satellite spacing.

13.9. The downlink of a satellite circuit operates at a satellite [EIRP] of 35 dBW and a receiving earth station antenna gain of 59.5 dB. Interference is produced by a satellite spaced 3°, its [EIRP] also being 35 dBW. Calculate the [*C*/*I*] ratio at the receiving antenna, assuming −6 dB polarization discrimination. The sidelobe gain function for the earth station antenna is 32 − 25 log θ.

13.10. A satellite circuit operates with an earth station transmit power of 30 dBW and a satellite [EIRP] of 34 dBW. This circuit causes interference with a neighboring circuit for which the earth-station transmit power is 24 dBW, the transmit antenna gain is 54 dB, the satellite [EIRP] is 34 dBW, and the receive earth station antenna gain is 44 dB. Calculate the carrier-to-interference ratio at the receive station antenna. Antenna sidelobe characteristics of 32 − 25 log θ dB may be assumed, with the satellites spaced at 2° and polarization isolation of −4 dB on uplink and downlink.

13.11. Repeat Prob. 13.10 for a sidelobe pattern of 29 − 25 log θ.

13.12. A satellite TV/FM circuit operates with an uplink power of 30 dBW, an antenna gain of 53.5 dB, and a satellite [EIRP] of 34 dBW. The destination earth station has a receiver antenna gain of 44 dB. An interfering circuit has an uplink power of 11.5 dBW and a satellite [EIRP] of 15.7 dBW.

Given that the spectral overlap is $[Q] = 3.8$ dB, calculate the passband $[C/I]$ ratio. Assume polarization discrimination figures of -4 dB on the uplink and 0 dB on the downlink and an antenna sidelobe pattern of $32 - 25 \log \theta$. The satellite spacing is $2°$.

13.13.   Repeat Prob. 13.12 for an interfering carrier for which the uplink transmit power is 26 dBW, the satellite [EIRP] is 35.7 dBW, and the spectral overlap figure is -3.36 dB.

13.14.   An FDM/FM satellite circuit operates with an uplink transmit power of 11.9 dBW, an antenna gain of 53.5 dB, and a satellite [EIRP] of 19.1 dBW. The destination earth station has an antenna gain of 50.5 dB. A TV/FM interfering circuit operates with an uplink transmit power of 28.3 dBW and a satellite [EIRP] of 35 dBW. Polarization discrimination figures are -6 dB on the uplink and 0 dB on the downlink. Given that the receiver transfer characteristic for wanted and interfering signals is [RTC] = 60.83 dB, calculate the baseband $[S/I]$ ratio for an antenna sidelobe pattern of $29 - 25 \log \theta$. The satellite spacing is $2°$.

13.15.   Repeat Prob. 13.14 for an interfering circuit operating with an uplink transmit power of 27 dBW, a satellite [EIRP] of 34.2 dBW, and [RTC] = 37.94 dB.

13.16.   Explain what is meant by *single-entry interference objectives*. Show that an interference level of 600 pW0p is equivalent to an $[S/I]$ ratio of 62.2 dB.

13.17.   For the wanted circuit in Probs. 13.12 and 13.13, the specified interference objective is $[C/I]_{pb} = 22$ dB. Is this objective met?

13.18.   For broadcast TV/FM, the permissible video-to-noise objective is specified as $[S/N] = 53$ dB. The CCIR recommendation for interference is that the total interference from all other satellite networks should not exceed 10 percent of the video noise and that the single-entry interference should not exceed 40 percent of this total. Show that this results in a single-entry objective of $[S/I] = 67$ dB.

13.19.   Explain what is meant by *protection ratio*. A TV/FM carrier operates at a modulation index of 2.619, the top modulating frequency being 4.2 MHz. Calculate the protection ratio required for quality factors of (*a*) 4.2 and (*b*) 4.5. How do these values compare with the specified interference objective of $[C/I]_{pb} = 22$ dB?

13.20.   Explain what is meant by *energy dispersal* and how this may be achieved.

13.21.   Explain what is meant by *coordination* in connection with interference assessment in satellite circuits.

References

CCIR Recommendation 523. 1978. "Maximum Permissible Levels of Interference in a Geostationary Satellite Network in the Fixed Satellite Service Using 8-bit PCM Encoded Telephony Caused by Other Networks of This Service." *14th Plenary Assembly*, vol. IV, Kyoto.

CCIR Report 391-3. 1978. "Radiation Diagrams of Antennae for Earth Stations in the Fixed Satellite Service for Use in Interference Studies and for the Determination of a Design Objective." *14th Plenary Assembly*, vol. IV, Kyoto.

CCIR Report 454-3. 1982. "Method of Calculation to Determine Whether Two Geostationary-Satellite Systems Require Coordination." *15th Plenary Assembly*, vol. IX, part 1, Geneva.

CCIR Report 634-2. 1982. "Maximum Interference Protection Ratio for Planning Television Broadcast Systems." *Broadcast Satellite Service (Sound and Television)*, vols. X and IX, part 2, Geneva.

Chouinard, G. 1984. "The Implications of Satellite Spacing on TVRO Antennas and DBS Systems." *Canadian Satellite User Conference*, Ottawa.

ITU. 1985. *Handbook on Satellite Communications* (FSS). Geneva.

ITU. 1986. *Radio Regulations*. Geneva.

Jeruchim, M. C., and D. A. Kane. 1970. *Orbit/Spectrum Utilization Study*, vol. IV. General Electric Doc. No. 70SD4293, December 31.

Microwave Filter Company, Inc. 1984. "TI and TVROs: A Brief Troubleshooting Guide to Suppressing Terrestrial Interference at 3.7–4.2 GHz TVRO Earth Stations."

Sharp, G. L. 1983. Reduced Domestic Satellite Orbital Spacings at 4/6 GHz. FCC/OST R83-2, May.

Sharp, G. L. 1984a. "Reduced Domestic Satellite Orbit Spacing." *AIAA Communications Satellite Systems Conference*, Orlando, FL, March 18–22.

CHAPTER 14

Satellite Access

14.1 Introduction

A transponder channel aboard a satellite may be fully loaded by a single transmission from an earth station. This is referred to as a *single access* mode of operation. It is also possible, and more common, for a transponder to be loaded by several carriers. These may originate from several earth stations geographically separate, and each earth station may transmit one or more of the carriers. This mode of operation is termed *multiple access*. The need for multiple access arises because more than two earth stations may be within the service area of a satellite. Even so-called spot beams from satellite antennas cover areas several hundred miles across.

The two most used methods of multiple access are *frequency-division multiple access* (FDMA) and *time-division multiple access* (TDMA). These are analogous to frequency-division multiplexing (FDM) and time-division multiplexing (TDM) described in Chaps. 9 and 10. However, multiple access and multiplexing are different concepts, where modulation (and hence multiplexing) is essentially a transmission feature, and multiple access is essentially a traffic feature.

A third category of multiple access is *code-division multiple access* (CDMA). In this method, each signal is associated with a particular code that is used to spread the signal in frequency and/or time. Such signals are received simultaneously at an earth station, but by using the key to the code, the station can recover the desired signal by means of correlation. The other signals occupying the transponder channel appear very much like random noise to the correlation decoder.

Multiple access also may be classified by the way in which circuits are assigned to users (*circuits* in this context implies one communication channel through the multiple-access transponder). Circuits may be *preassigned*, which means they are allocated on a fixed or partially fixed basis to certain users. These circuits are therefore not available for general use. Preassignment is simple to implement but is efficient only for circuits with *continuous heavy* traffic.

An alternative to preassignment is *demand-assigned multiple access* (DAMA). In this method, all circuits are available to all users and are assigned according to the demand. DAMA results in more efficient overall use of the circuits but is more costly and complicated to implement.

Both FDMA and TDMA can be operated as preassigned or demand assigned systems. CDMA is a random-access system, there being no control over the timing of the access or of the frequency slots accessed.

These multiple-access methods refer to the way in which a single *transponder* channel is utilized. A satellite carries many transponders, and normally each covers a different frequency channel, as shown in Figs. 7.13 and 7.16. This provides a form of FDMA to the whole satellite. It is also possible for transponders to operate at the same frequency but to be connected to different spot-beam antennas. These allow the satellite to be accessed by earth stations widely separated geographically but transmitting on the same frequency. This is termed *frequency reuse*. This method of access is referred to as *space-division multiple access* (SDMA). It should be kept in mind that each spot beam may itself be carrying signals in one of the other multiple-access formats.

14.2   Single Access

With single access, a single modulated carrier occupies the whole of the available bandwidth of a transponder. Single-access operation is used on heavy-traffic routes and requires large earth station antennas such as the class A antenna shown in Fig. 8.7. As an example, Telesat Canada provides heavy route message facilities, with each transponder channel being capable of carrying 960 one-way voice circuits on an FDM/FM carrier, as illustrated in Fig. 14.1. The earth station employs a 30-m-diameter antenna and a parametric amplifier, which together provide a minimum $[G/T]$ of 37.5 dB/K.

14.3   Preassigned FDMA

Frequency slots may be preassigned to analog and digital signals, and to illustrate the method, analog signals in the FDM/FM/FDMA format are considered first. As the acronyms indicate, the signals are frequency-division multiplexed, frequency modulated (FM), with FDMA to the satellite. In Chap. 9, FDM/FM signals are discussed. Recall that the voice-frequency (telephone) signals are first SSBSC amplitude modulated onto voice carriers to generate the single sidebands needed for the FDM. For illustration, each earth station is assumed to transmit a 60-channel supergroup. Each 60-channel supergroup is then frequency modulated onto a carrier which is then upconverted to a frequency in the satellite uplink band.

Figure 14.2 shows the situation for three earth stations: one in Ottawa, one in New York, and one in London. All three earth stations access a single satellite transponder channel simultaneously, and each communicates with both others. Thus, it is assumed that the satellite receive and transmit antenna beams are *global*, encompassing all three earth

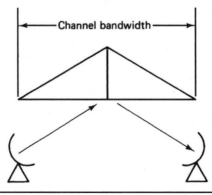

FIGURE 14.1   Heavy route message (frequency modulation—single access). (*Courtesy of Telesat Canada, 1983.*)

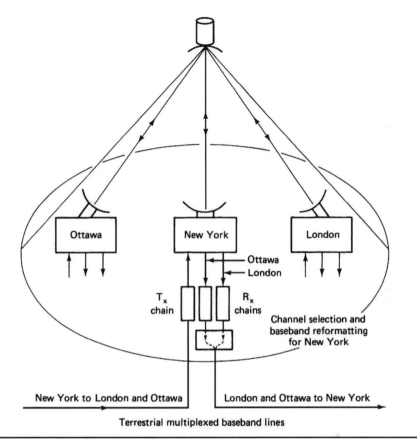

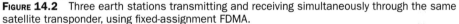

FIGURE 14.2 Three earth stations transmitting and receiving simultaneously through the same satellite transponder, using fixed-assignment FDMA.

stations. Each earth station transmits one uplink carrier modulated with a 60-channel supergroup and receives two similar downlink carriers.

The earth station at New York is shown in more detail. One transmit chain is used, and this carries telephone traffic for both Ottawa and London. On the receive side, two receive chains must be provided, one for the Ottawa-originated carrier and one for the London-originated carrier. Each of these carriers has a mixture of traffic, and in the demultiplexing unit, only those telephone channels intended for New York are passed through. These are remultiplexed into an FDM/FM format which is transmitted out along the terrestrial line to the New York switching office. This earth-station arrangement should be compared with that shown in Fig. 8.6.

Figure 14.3 shows a hypothetical frequency assignment scheme for the hypothetical network of Fig. 14.2. Uplink carrier frequencies of 6253, 6273, and 6278 MHz are shown for illustration purposes. For the satellite transponder arrangement of Figs. 7.15 and 7.16, these carriers are translated down to frequencies of 4028, 4048, and 4053 MHz (i.e., the corresponding 4-GHz-band downlink frequencies) and sent to a transponder of the satellite. Typically, a 60-channel FDM/FM carrier occupies 5 MHz of transponder bandwidth, including guard bands. A total frequency allowance of 15 MHz is therefore required for the

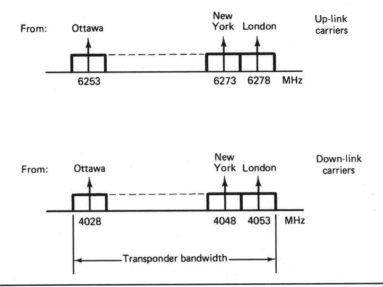

FIGURE 14.3   Transponder channel assignments for the earth stations shown in Fig. 14.2.

three stations, and each station receives all the traffic. The remainder of the transponder bandwidth may be unused, or it may be occupied by other carriers, which are not shown.

As an example of preassignment, suppose that each station can transmit up to 60 voice circuits and that 40 of these are preassigned to the New York–London route. If these 40 circuits are fully loaded, additional calls on the New York–London route are blocked even though there may be idle circuits on the other preassigned routes.

Telesat Canada operates medium-route message facilities utilizing FDM/FM/FDMA. Figure 14.4 shows how five carriers may be used to support 168 voice channels.

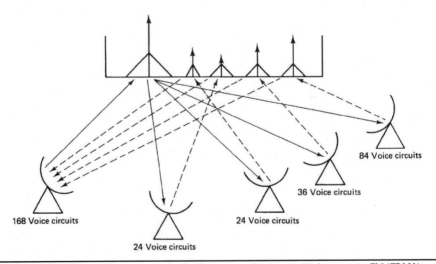

FIGURE 14.4   Medium route message traffic (frequency-division multiple access, FM/FDMA). (*Courtesy of Telesat Canada, 1983.*)

The earth station that carries the full load has a $[G/T]$ of 37.5 dB/K, and the other four have $[G/T]$ of 28 dB/K.

Preassignment also may be made based on a *single channel per carrier* (SCPC). This refers to a single voice (or data) channel per carrier, not a transponder channel, which may in fact carry some hundreds of voice channels by this method. The carriers may be frequency modulated or phase-shift modulated, and an earth station may be capable of transmitting one or more SCPC signals simultaneously.

Figure 14.5 shows the INTELSAT SCPC channeling scheme for a 36-MHz transponder. The transponder bandwidth is subdivided into 800 channels each 45-kHz wide. The 45 kHz, which includes a guard band, is required for each digitized voice channel, which utilizes *quadrature phase-shift keying* (QPSK) modulation. The channel information signal may be digital data or *pulse-code modulation* (PCM) voice signals (see Chap. 10). A pilot frequency is transmitted for the purpose of frequency control, and the adjacent channel slots on either side of the pilot are left vacant to avoid interference. The scheme therefore provides a total of 798 one-way channels or up to 399 full-duplex voice circuits. In duplex operation, the frequency pairs are separated by 18.045 MHz, as shown in Fig. 14.5.

The frequency tolerance relative to the assigned values is within ±1 kHz for the received SCPC carrier and must be within ±250 Hz for the transmitted SCPC carrier (Miya, 1981). The pilot frequency is transmitted by one of the earth stations designated as a primary station. This provides a reference for *automatic frequency control* (AFC) (usually using phase-locked-loops) of the transmitter frequency synthesizers and receiver local oscillators. In the event of failure of the primary station, the pilot frequency is transmitted from a designated backup station.

An important feature of the INTELSAT SCPC system is that each channel is voice-activated. This means that on a two-way telephone conversation, only one carrier is operative at any one time. Also, in long pauses between speech, the carriers are switched off. It has been estimated that for telephone calls, the one-way utilization time is 40 percent of the call duration. Using voice activation, the average number of carriers being amplified at any one time by the transponder traveling-wave tube (TWT) is reduced. For a given level of intermodulation distortion (see Secs. 7.7.3 and 12.11), the TWT power output per FDMA carrier therefore can be increased.

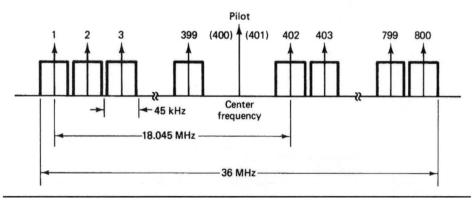

FIGURE 14.5 Channeling arrangement for INTELSAT SCPC system.

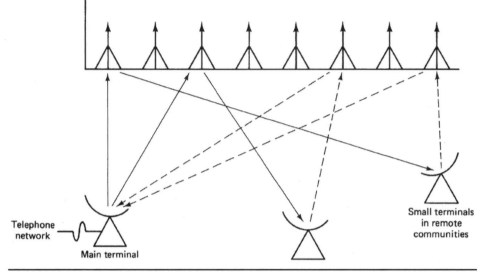

Figure 14.6 Thin route message traffic (single channel per carrier, SCPC/FDMA). (*Courtesy of Telesat Canada, 1983.*)

SCPC systems are widely used on lightly loaded routes, this type of service being referred to as a *thin route service*. It enables remote earth stations in sparsely populated areas to connect into the national telephone network in a reasonably economical way. A main earth station is used to make the connection to the telephone network, as illustrated in Fig. 14.6. The Telesat Canada Thin Route Message Facilities provide up to 360 two-way circuits using PSK/SCPC (PSK = *phase-shift keying*). The remote terminals operate with 4.6-m-diameter antennas with $[G/T]$ values of 19.5 or 21 dB/K. Transportable terminals are also available, one of these being shown in Fig. 14.7. This is a single-channel station that uses a 3.6-m antenna and comes complete with a desktop electronics package that can be installed on the customers' premises.

Figure 14.7 Transportable message station. (*Courtesy of Telesat Canada, 1983.*)

14.4   Demand-Assigned FDMA

In the demand-assigned mode of operation, the transponder frequency bandwidth is subdivided into several channels. A channel is assigned to each carrier in use, giving rise to the single-channel-per-carrier mode of operation discussed in the preceding section. As in the preassigned access mode, carriers may be frequency modulated with analog information signals, these being designated FM/SCPC, or they may be phase modulated with digital information signals, these being designated as PSK/SCPC.

Demand assignment may be carried out in several ways. In the polling method, a master earth station continuously polls all the earth stations in sequence, and if a *call request* is encountered, frequency slots are assigned from the pool of available frequencies. The polling delay with such a system tends to become excessive as the number of participating earth stations increases.

Instead of using a polling sequence, earth stations may request calls through the master earth station as the need arises. This is referred to as *centrally controlled random access*. The requests go over a digital order wire, which is a narrowband digital radio link or a circuit through a satellite transponder reserved for this purpose. Frequencies are assigned, if available, by the master station, and when the call is completed, the frequencies are returned to the pool. If no frequencies are available, the blocked call requests may be placed in a queue, or a second call attempt may be initiated by the requesting station.

As an alternative to centrally controlled random access, control may be exercised at each earth station, this being known as *distributed control random access*. A good illustration of such a system is provided by the Spade system operated by INTELSAT on some of its satellites.

This is described in the following section.

14.5   Spade System

The word *Spade* is a loose acronym for SCPC pulse-code-modulated multiple-access demand-assignment equipment. Spade was developed by Comsat for use on the INTELSAT satellites (see, e.g., Martin, 1978) and is compatible with the INTELSAT SCPC preassigned system described in Sec. 14.3. However, the distributed-demand assignment facility requires a *common signaling channel* (CSC). This is shown in Fig. 14.8.

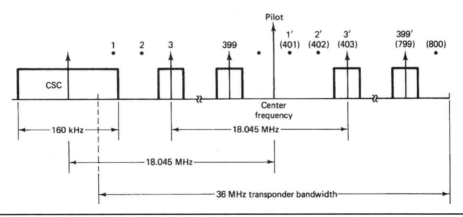

Figure 14.8 Channeling scheme for the Spade system.

The CSC bandwidth is 160 kHz, and its center frequency is 18.045 MHz below the pilot frequency, as shown in Fig. 14.8. To avoid interference with the CSC, voice channels 1 and 2 are left vacant, and to maintain duplex matching, the corresponding channels 1' and 2' are also left vacant. Recalling from Fig. 14.5 that channel 400 also must be left vacant, this requires that channel 800 be left vacant for duplex matching. Thus, six channels are removed from the total of 800, leaving a total of 794 one-way or 397 full-duplex voice circuits, the frequencies in any pair being separated by 18.045 MHz, as shown in Fig. 14.8. (An alternative arrangement is shown in Freeman, 1981.)

All the earth stations are permanently connected through the CSC. This is shown diagrammatically in Fig. 14.9 for six earth stations A, B, C, D, E, and F. Each earth station has the facility for generating any one of the 794 carrier frequencies using frequency synthesizers. Furthermore, each earth station has a memory containing a list of the frequencies currently available, and this list is continuously updated through the CSC. To illustrate the procedure, suppose that a call to station F is initiated from station C in Fig. 14.9. Station C first selects a frequency pair at random from those currently available on the list and signals this information to station F through the CSC. Station F must acknowledge, through the CSC, that it can complete the circuit. Once the circuit is established, the other earth stations are instructed, through the CSC, to remove this frequency pair from the list.

The round-trip time between station C initiating the call and station F acknowledging it is about 600 ms. During this time, the two frequencies chosen at station C may be assigned to another circuit. In this event, station C receives the information on the CSC update and immediately chooses another pair at random, even before hearing from station F.

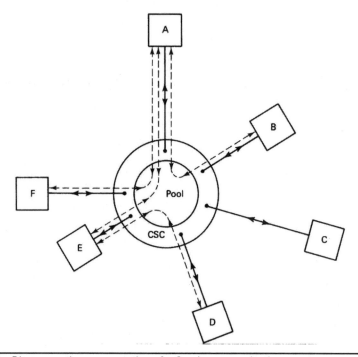

FIGURE 14.9   Diagrammatic representation of a Spade communications system.

Once a call has been completed and the circuit disconnected, the two frequencies are returned to the pool, the information again being transmitted through the CSC to all the earth stations.

As well as establishing the connection through the satellite, the CSC passes signaling information from the calling station to the destination station, in the example above from station C to station F. Signaling information in the Spade system is routed through the CSC rather than being sent over a voice channel. Each earth station has an equipment called the *demand assignment signaling and switching* (DASS) unit that performs the functions required by the CSC.

Some type of multiple access to the CSC must be provided for all the earth stations using the Spade system. This is quite separate from the SCPC multiple access of the network's voice circuits. TDMA, described in Sec. 14.7.8, is used for this purpose, allowing up to 49 earth stations to access the common signaling channel.

14.6   Bandwidth-Limited and Power-Limited TWT Amplifier Operation

Since a transponder has a total bandwidth B_{TR}, this can impose a limitation on the number of carriers that can access the transponder in an FDMA mode. For example, if there are K carriers each of bandwidth B, then the best that can be achieved is $K = B_{TR}/B$. Any increase in the transponder EIRP does not improve on this, and so the system is said to be *bandwidth-limited*. Likewise, for digital systems, the bit rate is determined by the bandwidth, which is limited to some maximum value by B_{TR}.

Power limitation occurs where the EIRP is insufficient to meet the $[C/N]$ requirements, as shown by Eq. (12.38). The signal bandwidth is approximately equal to the noise bandwidth, and if the EIRP is below a certain level, the bandwidth must be correspondingly reduced to maintain the $[C/N]$ at the required value. These limitations are discussed in more detail in the following two sections.

14.6.1   FDMA Downlink Analysis

To see the effects of output backoff, which results with FDMA operation, consider the overall carrier-to-noise ratio given as

$$\left(\frac{N}{C}\right)_{tot} = \left(\frac{N}{C}\right)_{U} + \left(\frac{N}{C}\right)_{D} + \left(\frac{N}{C}\right)_{IM} \tag{14.1}$$

A minimum value of carrier-to-noise ratio is required, as specified in the system design. This is denoted by the subscript REQ. The overall C/N must be at least as great as the required value, a condition which can therefore be stated as

$$\left(\frac{N}{C}\right)_{REQ} \geq \left(\frac{N}{C}\right)_{tot} \tag{14.2}$$

Note that because the noise-to-carrier ratio (rather than the carrier-to-noise ratio) is involved, the total value must be less than or equal to the required value. Using Eq. (14.1), the condition can be rewritten as

$$\left(\frac{N}{C}\right)_{REQ} \geq \left(\frac{N}{C}\right)_{U} + \left(\frac{N}{C}\right)_{D} + \left(\frac{N}{C}\right)_{IM} \tag{14.3}$$

With FDMA, backoff is usually utilized to reduce the intermodulation noise to an acceptable level, and as shown in Sec. 12.10, the uplink noise contribution is usually negligible. Thus, the expression can be approximated by

$$\left(\frac{N}{C}\right)_{REQ} \geq \left(\frac{N}{C}\right)_{D}$$

or

$$\left(\frac{C}{N}\right)_{REQ} \leq \left(\frac{C}{N}\right)_{D} \tag{14.4}$$

Consider the situation where each carrier of the FDMA system occupies a bandwidth B and has a downlink power denoted by $[EIRP]_D$. Equation (12.53) gives

$$\left(\frac{C}{N}\right)_{D} = EIRP_D \left(\frac{G}{T}\right)_D (L_p)_D (L_{sys})_D \left(\frac{1}{k}\right)\left(\frac{1}{B_N}\right) \tag{14.5a}$$

and when converted to dB becomes

$$\left[\frac{C}{N}\right]_{D} = [EIRP]_D + \left[\frac{G}{T}\right]_D + [LOSSES]_D + [1/k] + [1/B] \tag{14.5b}$$

where it is assumed that $B_N \approx B$ *(however for digital modulation $B_N = R_{sym}$)*, and

$$[LOSSES]_D = [L_{sys}] + [L_p] .$$

As discussed in Sec. 12.4, we structure our link equations as a product of terms in units of power ratio. In Eq. (14.5a), the product "kB" appears in the denominator, but we restructure the equation to multiply by $(1/k)*(1/B)$. Further, we treat losses as gains <1 that are multiplicative in the link equation. These losses are associated with multiple sources such as: free-space spreading loss (L_p), and other losses combined under the system losses (L_{sys}), which includes; atmospheric absorption loss, antenna mispointing losses, feeder losses for both the transmitter and receiver, etc. In Eq. (14.5b), after conversion to dB, these losses are a summation of negative dB's in the term "$[LOSSES]_D$."

Equation (14.5b) can be written in terms of the required carrier-to-noise ratio as

$$\left[\frac{C}{N}\right]_{REQ} \leq [EIRP]_D + \left[\frac{G}{T}\right]_D + [LOSSES]_D + [1/k] + [1/B] \tag{14.6}$$

To set up a reference level, consider first single-carrier operation. The satellite has a saturation value of EIRP and a transponder bandwidth B_{TR}, both of which are assumed fixed. With single-carrier access, no backoff is needed, and Eq. (14.6) becomes

$$\left[\frac{C}{N}\right]_{REQ} \leq [EIRP]_S + \left[\frac{G}{T}\right]_D + [LOSSES]_D + [1/k] + [1/B_{TR}] \tag{14.7}$$

or

$$\left[\frac{C}{N}\right]_{REQ} - \left[EIRP_S\right] - \left[\frac{G}{T}\right]_D - [LOSSES]_D - [1/k] - \left[1/B_{TR}\right] \leq 0 \qquad (14.8)$$

If the system is designed for single-carrier operation, then the equality sign applies, and the reference condition is

$$\left[\frac{C}{N}\right]_{REQ} - \left[EIRP_S\right] - \left[\frac{G}{T}\right]_D - [LOSSES]_D - [1/k] - \left[1/B_{TR}\right] = 0 \qquad (14.9)$$

Consider now the effect of power limitation imposed by the need for backoff. Suppose the FDMA access provides for M-carriers that share the output power equally, and each requires a bandwidth B. The output power for each of the FDMA carriers is

$$[EIRP]_D = \left[EIRP_S\right] + [BO]_0 + [1/M] \qquad (14.10)$$

where $[BO]_0$ is a *negative* dB.

The transponder bandwidth B_{TR} is shared between the carriers, but not all B_{TR} can be utilized because of the power limitation. Let α represent the fraction of the total bandwidth occupied, such that

$$B = \frac{\alpha B_{TR}}{M}$$

or in terms of dB

$$[B] = [\alpha] + \left[B_{TR}\right] + [1/M] \qquad (14.11)$$

Substituting these relationships in Eq. (14.6) gives

$$\left[\frac{C}{N}\right]_{REQ} \leq \left[EIRP_S\right] + [BO]_0 + \left[\frac{G}{T}\right]_D + [LOSSES]_D + [1/k] + \left[1/B_{TR}\right] + [1/\alpha] \quad (14.12)$$

It is noted that the [M] term cancels out, and the expression can be rearranged as

$$\left[\frac{C}{N}\right]_{REQ} - \left[EIRP_S\right] - \left[\frac{G}{T}\right]_D - [LOSSES]_D - [1/k] - \left[1/B_{TR}\right] \leq [BO]_0 + [1/\alpha] \quad (14.13)$$

But as shown by Eq. (14.9), the left-hand side is equal to zero if the single carrier access is used as reference, and hence

$$0 \leq [BO]_0 + [1/\alpha] \text{ or } -[1/\alpha] \leq [BO]_0 \qquad (14.14)$$

The best that can be achieved is to make $[1/\alpha] = [BO]_0$, and since the backoff is a negative number of decibels, $[\alpha]$ must be positive, or equivalently, α is fractional. The following example illustrates the limitation imposed by backoff.

Example 14.1   A satellite transponder has a bandwidth of 36 MHz and a saturation EIRP of 27 dBW. The earth-station receiver has a [G/T] of 30 dB/K, and the total link losses are −196 dB. The transponder is accessed by FDMA carriers each of 3-MHz bandwidth, and an output backoff of −6 dB is employed.

(a) Calculate the downlink carrier-to-noise ratio for single-carrier operation,
(b) Determine the number of carriers that can be accommodated in the FDMA system, and
(c) Compare this with the number that can be accommodated if no backoff is needed.

The carrier-to-noise ratio determined for single-carrier operation may be taken as the reference value, and it may be assumed that the uplink noise and intermodulation noise are negligible.

Solution   Transponder bandwidth: B_{TR} = 36 MHz, therefore [B_{TR}] = 75.56 dBHz
Carrier bandwidth: B = 3 MHz, therefore [B] = 64.77 dBHz
  For single carrier operation B_{TR} can replace B in Eq. (14.5) and [$EIRP_S$] can replace [$EIRP$]$_D$ to give

$$\left[\frac{C}{N}\right]_D = \left[EIRP_S\right] + \left[\frac{G}{T}\right]_D + \left[LOSSES\right]_D + \left[1/k\right] + \left[1/B_{TR}\right]$$

$$= 27 + 30 - 196 + 228.6 - 75.56 = 14.0 \text{ dB}$$

Setting [α] = [BO]$_0$ gives [α] = −6 dB and hence from Eq. (14.11)

$$[M] = [\alpha] + [B_{TR}] + [1/B]$$

$$= -6 + 75.56 - 64.77 = 4.79 \text{ dB}$$

Hence

$$M = 10^{4.79/10} = \text{Integer} (3.013) = 3 \text{ carriers}$$

If backoff is not required, the number of carriers which can be accommodated is given by

$$B_{TR}/B = 36/3 = 12$$

Example 14.2   Very Small Aperture Earth Terminals (VSATs), operating at Ku-band, in a Mesh network access a single GEOSAT transponder using SCPC FDMA in a bandwidth limited mode, with a guard band of 30 kHz between channels. Each terminal transmits digital data at a bit rate of 128 kbps using BPSK with RRC filters with a roll-off factor ρ = 0.40. The required end-to-end link BER $\leq 10^{-6}$ and the BPSK demodulation implementation loss = −1.5 dB.

(a) How many VSATs can be supported if the transponder bandwidth B_{TR} = 36 MHz?

Solution   First we calculate the RF bandwidth required to transmit the bit rate. For BPSK, the bit rate (R_b) equals the digital symbol rate (R_{sym}), and the using Eq. (10.22), the RF bandwidth

$$B_{RF} = (1 + \rho)R_{sym} = 1.4 * 128e+03 = 179.20 \text{ kHz,}$$

and including the guard band, the total SCPC bandwidth

$$B = 179.20 + 30 = 209.20 \text{ kHz}$$

the number of VSATs supported is

$$\#VSAT = INT\left(\frac{B_{TR}}{B}\right) = INT\left(\frac{36.0e+06}{209.2e+03}\right) = 172$$

Next, the system requires that all VSATs maintain identical $(C/N_0)_{up}$ in clear-sky operation, but the TWTA must be operated in a backoff mode to minimize the generation of 3rd order intermodulation noise (3IM). Using a typical TWTA power transfer characteristic given in the following table, find a suitable clear-sky operating point.

TWTA Power Transfer Characteristic for Multi-Carrier Inputs (from Ha, Table 5.3)

BO$_i$, dB	BO$_o$, dB	C/3IM, dB	BO$_i$, dB	BO$_o$, dB	C/3IM, dB
0	−2.0	8.51	−10	−5.21	21.96
−1	−1.84	9.44	−11	−5.97	23.81
−2	−1.8	10.47	−12	−6.79	25.74
−3	−1.89	11.6	−13	−7.64	27.77
−4	−2.09	12.8	−14	−8.57	29.89
−5	−2.39	14.1	−15	−9.44	32.1
−6	−2.8	15.49	−16	−10.37	34.4
−7	−3.29	16.97	−17	−11.31	36.79
−8	−3.86	18.54	−18	−12.24	39.27
−9	−4.5	20.21	−19	−13.17	41.84

For the uplink @ 14 GHz, selecting a BO$_i$ = −9.0 dB yields an acceptable $[(C/N)_{IM}]$ = 20.21 dB and converted to $[(C/N_0)_{IM}]$ = 20.21 dB + $[R_b]$ = 20.21 + 51.07 = 71.28 dBHz.

Using a VSAT transmit terminal with: EIRP = 37.18 dBW, a GEOSAT transponder with: G/T = 16.34 dB/K, a path loss = −206.76 dB, system losses = −2.77 dB; and $[1/k]$ = 228.6 dB, results in a $(C/N_0)_{up}$ = 72.59 dBHz.

Next, calculate the $(C/N_0)_{dn}$.

For the GEOSAT transponder with a downlink @ 12 GHz, the TWTA has a P_{sat} = 80 W $\Rightarrow$ 19.0 dBW, and for the selected BO$_i$ = −9.0 dB, the corresponding BO$_o$ = −4.50 dB from the above TWTA table. Thus, the total multicarrier transmitter output power is

$$[P_{tot}] = [P_{sat}] + [BO_o] = 19.0 + (−4.5) = 14.5 \text{ dBW} \Rightarrow 28.39 \text{ W,}$$

which is equally shared by 172 VSATs, and the single carrier transmitted power is

$$P_t = 28.39/172 = 0.17 \text{ W} \Rightarrow −7.8 \text{ dBW.}$$

Using a transmitting antenna gain of 43.0 dB, results in a GEOSAT [EIRP/carrier] = 36.70 dBW. For the receiving VSAT terminal the (G/T) = 20.85 dB, and the path loss = −205.29 dB and the system losses = −1.22 dB. Using Eq. (12.42) produces a

$$[(C/N_0)_{dn}] = 73.62 \text{ dBHz,}$$

(b) Calculate the $(C/N_0)_{tot}$ for clear-sky, the P_e, and the link rain margin.
Using Eq. (14.1) the $(N_0/C)_{tot}$ is calculated below

Calc $(C/N_0)_{tot}$	Ratio	dBHz
$(C/N_0)_{up}$	1.81E+07	72.59
$(N_0/C)_{up}$	5.52E-08	
$(C/N_0)_{dn}$	2.30+07	73.62
$(N_0/C)_{dn}$	4.35E-08	
$(C/N_0)_{3IM}$	1.34E+07	71.28
$((N_0)/C)_{3IM}$	7.45E-08	
$(N_0/C)_{tot}$	1.73E-07	
$(C/N_0)_{tot}$	5.78E+06	67.62

Using Eq. (10.24) the probability of bit error is

$$P_e = \frac{1}{2} \operatorname{erfc}\left(\sqrt{\left(\frac{E_b}{N_0}\right)_{\text{eff}}}\right), \text{ errors/bit}$$

where $\left[\left(\frac{E_b}{N_0}\right)_{\text{eff}}\right] = \left[\left(\frac{E_b}{N_0}\right)_{\text{tot}}\right] + [\text{Imp Loss}]$

Using an [implementation Loss] = −1.5 dB, then

$$\left[\left(\frac{C}{N_0}\right)_{\text{eff}}\right] = 67.62 + (-1.5) = 66.12 \text{ dB} \Rightarrow 4.097\text{e}+06$$

Recognizing that $\dfrac{E_b}{N_0} = \dfrac{C/N_0}{R_b} = \dfrac{4.097e+06}{128e+03} = 31.96$ and the P_e calculation is given below

Calc the P_e (BPSK)	Ratio	dB
$[(C/N_0)_{\text{eff}}] = [(C/N_0)_{\text{tot}}] + [\text{Imp-Loss}]$	4.09E+06	66.12
$(E_b/N_0)_{\text{eff}} = (C/N_0)_{\text{eff}}/R_b$	31.96	
$P_e = 0.5*\text{ERFC}(\text{sqrt}(E_b/N_0))$	6.39E-16	

Next using an iterative solution, $[(C/N_0)_{\text{eff}}] = 61.60$ dB is the required threshold value that yields the specified BER $\sim 10^{-6}$ for BPSK modulation. So the clear sky $(C/N_0)_{\text{tot}}$ results in a +4.52 dB rain margin above.

Effect of rain attenuation on $(C/N_0)_{\text{tot}}$
Referring to Sec. 12.9, the Rain Attenuation Exceedance for 0.010% of 1 year from Table 4.6 is −6.22 dB for 12 GHz (downlink) and −8.58 for 14 GHz (uplink). Since the probability of rain occurring simultaneously on both the uplink and downlink is negligible, we perform the uplink and downlink analysis separately.

Effect of rain attenuation on $(C/N_0)_{\text{dn}}$
As discussed in Sec. 12.9.1, rain degrades the $(C/N_0)_{\text{dn}}$ in two ways, namely; by carrier attenuation and by increased rain emission noise. For the downlink and clear-sky, assume that the system noise temperature $T_{\text{sys}} = 150$ K.

(b) What is the P_e during rain on the downlink?

Solution   Calculations are given in the tables below.

Rain on Downlink	Ratio	dB
Rain attenuation @ 12 GHz	0.2388	−6.22
$T_{\text{sys clear sky}}$	150	
Tap_rain = (274*(1 − Rain_atten))	208.6	
$T_{\text{sys\_rain}}$ = Tap_rain + $T_{\text{sys\_clear}}$	358.6	
Delta-N = $T_{\text{sys\_rain}}/T_{\text{sys\_clear-sky}}$	2.390	3.78
$(C/N_0)_{\text{dn}} = (C/N_0)_{\text{clear}} + [\text{atten}] - [\text{delta-N}]$		63.61

Calc Rain: $(C/N_0)_{tot}$ at VSAT	Ratio	dB
$(C/N_0)_{up}$	1.81E+07	72.59
$(N_0/C)_{up}$	5.51E-08	
$(C/N_0)_{dn}$	2.30E+06	63.61
$(N_0/C)_{dn}$	4.35E-07	
$(C/N_0)_{3IM}$	1.34E+07	71.28
$(N_0/C)_{3IM}$	7.44E-08	
$(N_0/C)_{tot}$	5.65E-07	
$(C/N_0)_{tot}$	1.77E+06	62.48

Calc the P_e (BPSK)	Ratio	dB
$[(C/N)_{eff}] = [(C/N)_{tot}] + [Imp\text{-}Loss]$	1.253E+06	60.98
$(E_b/N_0)_{eff}$	9.79	
$P_e = 0.5*ERFC(sqrt(E_b/N_0))$	4.82E-06	

Note that the link fail to meet the BER spec.

(c) What is the P_e during rain on the uplink?

Solution For the uplink, the system noise does not change during rain (see Example 12.21); so only the carrier attenuation reduces the $(C/N_0)_{up}$. From Table 4.6, the Rain Attenuation Exceedance for 0.010% of one-year is −8.58 for 14 GHz. So,

$$\left[\left(\frac{C}{N_0}\right)_{up\text{-}rain}\right] = \left[\left(\frac{C}{N_0}\right)_{up\text{-}clear\text{-}sky}\right] + [Rain\ Atten]$$

$$= 72.59 + (-8.58) = 64.01\ dB$$

Furthermore, this attenuation increases the BO_i (larger negative dB) to the transponder TWTA; which affects both the $(C/N_0)_{dn}$ and the $(C/3IM_o)$. To perform this calculation, it is necessary to use the TWTA power transfer characteristic table given above.

The new $[BO_i] = [(BO_i)_{clear\text{-}sky}] + [Rain\ Atten] = -9.0 + (-8.58) = -17.58\ dB$

Interpolating the TWTA Table, the new $BO_o = -11.90\ dB$, which is a change in BO_o

$$delta\text{-}BO_o = -11.90 - (-4.50) = -7.40\ dB,$$

so the downlink carrier to noise density becomes

$$\left[\left(\frac{C}{N_0}\right)_{dn\text{-}rain}\right] = \left[\left(\frac{C}{N_0}\right)_{dn\text{-}clear\text{-}sky}\right] + [delta - BO_o]$$

$$= 73.62 + (-7.40) = 66.22\ dB$$

and from the table, the new $(C/3IM) = 89.34\ dB$

The calculation of the $[(C/N)_{tot}]$ and the P_e are tabulated below.

Calc $(C/N_0)_{tot}$ for Rain on Uplink	Ratio	dB
$(C/N_0)_{up}$	2.515E+06	64.01
$(N_0/C)_{up}$	3.98E-07	
$(C/N_0)_{dn}$	4.18E+06	66.22
$(N_0/C)_{dn}$	2.39E-07	
$(C/N_0)_{3IM}$	8.59E+08	89.34
$(N_0/C)_{3IM}$	1.16E-09	
$(N_0/C)_{tot}$	6.38E-07	
$(C/N_0)_{tot}$	1.57E+06	61.95

Calc the P_e (BPSK)	Ratio	dB
$[(C/N_0)_{eff}] = [(C/N_0)_{tot}] + [Imp\text{-}Loss]$	1.11E+06	60.45
$(E_b/N_0)_{eff}$	8.6723	
$P_e = 0.5*ERFC(sqrt(E_b/N_0))$	1.56E-05	

Note that the BER also fails to meet the spec ($P_e \leq 10^{-6}$ errors/bit), when he rain occurs on the uplink. To mitigate this problem, it is possible to switch to QPSK modulation and add half-rate forward error correction (FEC) to improve the BER. This example follows.

Example 14.3   For this case, the same 172 VSATs using SCPC/FDMA can access the same satellite transponder. The only changes made are in the VSAT terminal, which transmits digital data at a bit rate of 256 kbps (128 kbps digital data + 128 kbps FEC) using QPSK modulation. Further, the FEC corrects some of the bit errors; so it is equivalent to having a higher E_b/N_0 value, which is implemented as a processing gain (power ratio > 1). So, the effective carrier to noise density ratio can be expressed as

$$[(C/N_0)_{eff}] = [(C/N_0)_{tot}] + [\text{implementation loss}] + [\text{FEC processing gain}], dB$$

(a) Calculate the $(C/N_0)_{tot}$ for clear-sky, the P_e, and the link rain margin?

Solution   Using QPSK the symbol rate $R_{sym} = R_b/2$, and for BPSK the symbol rate $R_{sym} = R_b$. For this problem the R_b has doubled compared to the previous example, but the symbol rate has not changed. Therefore, the $B_{RF} = (1+\rho)R_{sym}$, and the number of VSAT's is the same. As a result, the clear-sky $(C/N_0)_{tot}$ is the same, but the P_e calculation is different because, for the same $[(C/N_0)_{eff}]$, the E_b/N_0 is one-half for QPSK (compared with BPSK), and because we are using FEC, the processing gain improves the value of E_b/N_0.

From Ex. 14.2 (a); the $[(C/N_0)_{tot}] = 67.62$ dBHz, the [Imp Loss] $= -1.5$ dB, and the [FEC gain] $= +6$ dB.

$$[(C/N_0)_{eff}] = [(C/N_0)_{tot}] + [\text{implementation loss}] + [\text{FEC processing gain}] \, dB,$$

$$[(C/N_0)_{eff}] = 67.62 + (-1.5) + 6 = 72.12 \, dB \Rightarrow 1.629E+07, \text{ and}$$

$$\frac{E_b}{N_0} = \frac{1.629E+07}{(2*128E+03)} = 63.61, \text{ and}$$

$$P_e = 0.5*ERFC\left(\sqrt{\frac{E_b}{N_0}}\right) = 8.28e\text{-}30, \text{ errors/bit}$$

The threshold value for BER $\leq 10^{-6}$ is $[(C/N_0)_{eff}] = 64.61$ dB (+3 dB higher than BPSK). So the rain margin $= 72.12 - 64.61 = +7.51$ dB

Next, consider the Effect of rain attenuation on $(C/N_0)_{tot}$

(b) What is the P_e during rain on the downlink?

Solution   For this case, the calculation of $[(C/N_0)_{tot-rain}] = 63.61$ dB is the same as Ex. 14.2, and the $[(C/N_0)_{eff}] = [(C/N_0)_{tot-rain}] + $ [implementation loss] $+$ [FEC processing gain] dB, where $[(C/N)_{tot}] = 62.48$ dB, [Imp Loss] $= -1.5$ dB and [FEC gain] $= +6$ dB, and the $[(C/N)_{eff}] = 62.48$ dB $+ (-1.5) + 6 = 66.98$ dB $\Rightarrow 4.989e+06$, and

$$\frac{E_b}{N_0} = \frac{4.989e+06}{(2*128e+03)} = 19.49, \text{ and}$$

$$P_e = 0.5 * \text{ERFC}\left(\sqrt{\frac{E_b}{N_0}}\right) = 2.15e\text{-}10 \text{ errors/bit}$$

which now meets the BER spec!

(c) What is the P_e during rain on the uplink?

Solution   For this case, the calculation of $[(C/N_0)_{tot-rain}] = 61.95$ dB is the same as Ex. 14.2, and the $[(C/N_0)_{eff}] = [(C/N_0)_{tot-rain}] + $ [implementation loss] $+$ [FEC processing gain] dB, where $[(C/N_0)_{tot}] = 61.95$ dB, [Imp Loss] $= -1.5$ dB and [FEC gain] $= +6$ dB, and the $[(C/N)_{eff}] = 61.95$ dB $+ (-1.5) + 6 = 66.45$ dB $\Rightarrow 4.419e+06$, and

$$\frac{E_b}{N_0} = \frac{4.429e+06}{(2*125e+03)} = 17.26, \text{ and}$$

$$P_e = 0.5 * \text{ERFC}\left(\sqrt{\frac{E_b}{N_0}}\right) = 2.10e\text{-}9 \text{ errors/bit}$$

14.7   TDMA

With TDMA, only one carrier uses the transponder at any one time, and therefore, inter-modulation products, which result from the nonlinear amplification of multiple carriers, are absent. This leads to one of the most significant advantages of TDMA, which is that the TWT can be operated at maximum power output or saturation level.

Because the signal information is transmitted in bursts, TDMA is only suited to digital signals. Digital data can be assembled into burst format for transmission and reassembled from the received bursts using digital buffer memories.

Figure 14.10 illustrates the basic TDMA concept, in which the stations transmit bursts in sequence. Burst synchronization is required, and in the system illustrated in Fig. 14.10, one station is assigned solely for the purpose of transmitting *reference bursts* to which the others can be synchronized. The time interval from the start of one reference burst to the next is termed a *frame*. A frame contains the reference burst R and the bursts from the other earth stations, these being shown as A, B, and C in Fig. 14.10.

Figure 14.11 illustrates the basic principles of burst transmission for a single channel. Overall, the transmission appears continuous because the input and output bit rates are continuous and equal. However, within the transmission channel, input bits are temporarily stored and transmitted in bursts. Since the time interval between bursts is the frame time T_F, the required buffer capacity is

$$M = R_b T_F \tag{14.15}$$

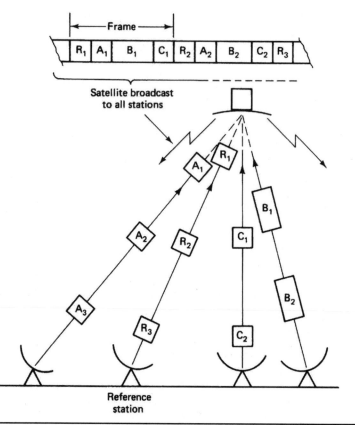

FIGURE 14.10   Time-division multiple access (TDMA) using a reference station for burst synchronization.

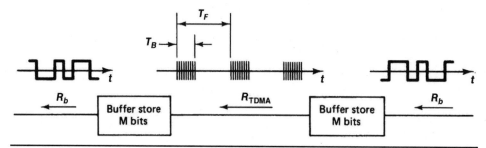

FIGURE 14.11   Burst-mode transmission linking two continuous-mode streams.

The buffer memory fills up at the input bit rate R_b during the frame time interval. Then, these M bits are transmitted as a burst in the next frame without any break in continuity of the input. The M bits are transmitted in the burst time T_B, and the *transmission rate*, which is equal to the burst bit rate, is

$$R_{\mathrm{TDMA}} = \frac{M}{T_B} = R_b \frac{T_F}{T_B} \qquad (14.16)$$

This is also referred to as the *burst rate*, but note that this refers to the instantaneous bit rate *within* a burst (not the number of bursts per second, which is simply equal to $1/T_B$). The *average* bit rate for the burst mode is simply M/T_F, which is equal to the input and output rates.

The frame time T_F increases the overall propagation delay. For example, in the simple system illustrated in Fig. 14.11, even if the actual propagation delay between transmit and receive buffers is assumed to be zero, the receiving side still has to wait a time T_F before receiving the first transmitted burst. In a geostationary satellite system, the actual propagation delay is a significant fraction of a second, and excessive delays from other causes must be avoided. This sets an upper limit to the frame time, although with current technology other factors restrict the frame time to well below this limit. The frame period is usually chosen to be a multiple of 125 µs, which is the standard sampling period used in PCM telephony systems, since this ensures that the PCM samples can be distributed across successive frames at the PCM sampling rate.

Figure 14.12 shows some of the basic units in a TDMA ground station, which for discussion purposes is labeled earth station A. Terrestrial links coming into earth station A carry digital traffic addressed to destination stations, labeled B, C, X. It is assumed that the bit rate is the same for the digital traffic on each terrestrial link. In the units labeled *terrestrial interface modules* (TIMs), the incoming continuous-bit-rate signals are converted into the intermittent-burst-rate mode. These individual burst-mode signals are *time-division multiplexed* in the time division multiplexer (MUX) so that the traffic for each destination station appears in its assigned time slot within a burst.

Certain time slots at the beginning of each burst are used to carry timing and synchronizing information. These time slots collectively are referred to as the *preamble*. The complete burst containing the preamble and the traffic data is used to phase modulate the *radio frequency* (RF) carrier. Thus, the composite burst which is transmitted at RF consists of several time slots, as shown in Fig. 14.13. These are described in more detail shortly.

The received signal at an earth station consists of bursts from all transmitting stations arranged in the frame format shown in Fig. 14.13. The RF carrier is converted to *intermediate frequency* (IF), which is then demodulated. A separate preamble detector provides timing information for transmitter and receiver along with a carrier synchronizing signal for the phase demodulator, as described in the following section. In many systems, a station receives its own transmission along with the others in the frame, which can then be used for burst-timing purposes.

A reference burst is required at the beginning of each frame to provide timing information for the *acquisition* and *synchronization* of bursts (these functions are described further in Sec. 14.7.4). In the INTELSAT international network, at least two reference stations are used, one in the East and one in the West. These are designated *primary* reference stations, one of which is further selected as the *master primary*. Each primary station is duplicated by a *secondary* reference station, making four reference stations in all. The fact that all the reference stations are identical means that anyone can become the master primary. All the system timing is derived from the high-stability clock in the master primary, which is accurate to 1 part in 10^{11} (Lewis, 1982). A clock on the satellite is locked to the master primary, and this acts as the clock for the other participating earth stations. The satellite clock provides a constant frame time, but the participating earth stations must make corrections for variations in the satellite range, since the transmitted bursts from all the participating earth stations must reach the satellite in synchronism. Details of the timing requirements are found in Spilker (1977).

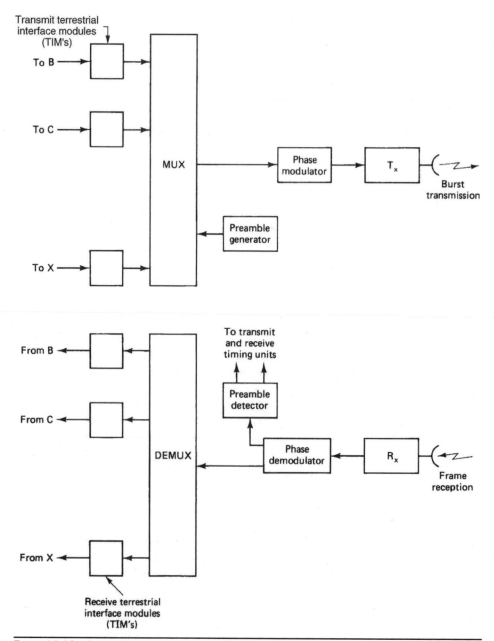

Figure 14.12 Some of the basic equipment blocks in a TDMA system.

In the INTELSAT system, two reference bursts are transmitted in each frame. The first reference burst, which marks the beginning of a frame, is transmitted by a master primary (or a primary) reference station, and it contains the timing information needed for the acquisition and synchronization of bursts. The second reference burst, which is transmitted by a secondary reference station, provides synchronization but not acquisition

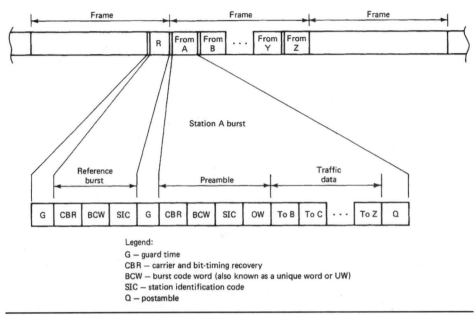

Figure 14.13 Frame and burst formats for a TDMA system.

information. The secondary reference burst is ignored by the receiving earth stations unless the primary or master primary station fails.

14.7.1 Reference Burst

The reference burst that marks the beginning of a frame is subdivided into time slots or channels used for various functions. These differ in detail for different networks, but Fig. 14.13 shows some of the basic channels that are usually provided. These can be summarized as follows:

Guard time (G). A guard time is necessary between bursts to prevent the bursts from overlapping. The guard time varies from burst to burst depending on the accuracy with which the various bursts can be positioned within each frame.

Carrier and bit-timing recovery (CBR also known as: CBTR or CCR). To perform coherent demodulation of the phase-modulated carrier, as described in Secs. 10.7 and 10.8, a coherent carrier signal must first be recovered from the burst. An unmodulated carrier wave is provided during the first part of the CBR time slot. This is used as a synchronizing signal for a local oscillator at the receiver detector, which then produces an output coherent with the carrier wave. The carrier in the subsequent part of the CBR time slot is modulated by a known phase-change sequence which enables the bit timing to be recovered. Accurate bit timing is needed for the operation of the sample-and-hold function in the detector circuit (see Figs. 10.18 and 10.26). Carrier recovery is described in more detail in Sec. 14.7.3.

Burst code word (BCW). (Also known as a *unique word.*) This is a binary word, a copy of which is stored at each earth station. By comparing the incoming bits in a burst with the stored version of the BCW, the receiver can detect when a group of

received bits matches the BCW, and this in turn provides an accurate time reference for the burst position in the frame. A known bit sequence is also carried in the BCW, which enables the phase ambiguity associated with coherent detection to be resolved.
Station identification code (SIC). This identifies the transmitting station.

Figure 14.14 shows the makeup of the reference bursts used in some of the INTELSAT networks. The numbers of symbols and the corresponding time intervals allocated to the various functions are shown. In addition, to the channels already described, a

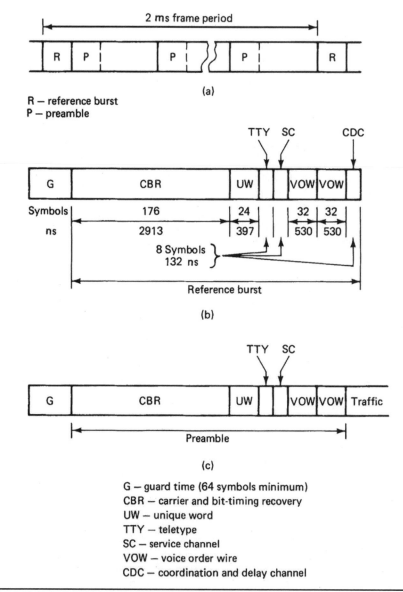

G — guard time (64 symbols minimum)
CBR — carrier and bit-timing recovery
UW — unique word
TTY — teletype
SC — service channel
VOW — voice order wire
CDC — coordination and delay channel

FIGURE 14.14 (a) INTELSAT 2-m427s frame; (b) composition of the reference burst *R*; (c) composition of the preamble *P*. (QPSK modulation is used, giving 2 bits per symbol. Approximate time intervals are shown.)

coordination and delay channel or CDC (sometimes referred to as the *control and delay channel*) is provided. This channel carries the identification number, of the earth station being addressed, and various codes used in connection with the acquisition and synchronization of bursts, at the addressed earth station. It is also necessary for an earth station to know the propagation time delay to the satellite to implement burst acquisition and synchronization. In the INTELSAT system, the propagation delay is computed from measurements made at the reference station and transmitted to the earth station in question through the coordination and delay channel.

The other channels in the INTELSAT reference burst are the following:

TTY: telegraph order-wire channel, used to provide text communications between earth stations.

SC: service channel which carries various network protocol and alarm messages.

VOW: voice-order-wire channel used to provide voice communications between earth stations. Two VOW channels are provided.

14.7.2 Preamble and Postamble

The *preamble* is the initial portion of a traffic burst that carries information like that carried in the reference burst. In some systems the channel allocations in the reference bursts and the preambles are identical. No traffic is carried in the preamble. In Fig. 14.13, the only difference between the preamble and the reference burst is that the preamble provides an *orderwire* (OW) channel.

For the INTELSAT format shown in Fig. 14.14, the preamble differs from the reference burst in that it does not provide a CDC. Otherwise, the two are identical.

As with the reference bursts, the preamble provides a carrier and bit-timing recovery channel and a burst-code-word channel for burst-timing purposes. The burst code word in the preamble of a traffic burst is different from the burst code word in the reference bursts, which enables the two types of bursts to be identified.

In certain phase detection systems, the phase detector must be allowed time to recover from one burst before the next burst is received by it. This is termed *decoder quenching*, and a time slot, referred to as a *postamble*, is allowed for this function. The postamble is shown as Q in Fig. 14.13. Many systems are designed to operate without a postamble.

14.7.3 Carrier Recovery

A factor, which must be considered with TDMA is that the various bursts in a frame lack coherence so that carrier recovery must be repeated for each burst. This applies to the traffic as well as the reference bursts. Where the carrier recovery circuit employs a phase-locked loop such as shown in Fig. 10.25, a problem known as *hangup* can occur. This arises when the loop moves to an unstable region of its operating characteristic. The loop operation is such that it eventually returns to a stable operating point, but the time required to do this may be unacceptably long for burst-type signals.

One alternative method utilizes a narrowband tuned circuit filter to recover the carrier. An example of such a circuit for *quadrature phaseshift keying* (QPSK), taken from Miya (1981), is shown in Fig. 14.15. The QPSK signal, which has been downconverted to a standard IF of 140 MHz, is quadrupled in frequency to remove the modulation, as described in Sec. 10.7. The input frequency must be maintained at the resonant frequency of the tuned circuit, which requires some form of automatic frequency control.

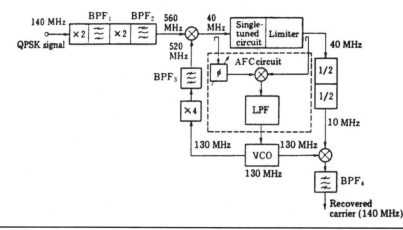

Figure 14.15 An example of carrier recovery circuit with a single-tuned circuit and AFC. (*Courtesy of Miya, 1981.*)

Because of the difficulties inherent in working with high frequencies, the output frequency of the quadrupler is downconverted from 560 to 40 MHz, and the AFC is applied to the *voltage controlled oscillator* (VCO) used to make the frequency conversion. The AFC circuit is a form of *phase-locked loop* (PLL), in which the phase difference between input and output of the single-tuned circuit is held at zero, which ensures that the 40-MHz input remains at the center of the tuned circuit response curve. Any deviation of the phase difference from zero generates a control voltage, which is applied to the VCO in such a way as to bring the frequency back to the required value.

Interburst interference may be a problem with the tuned-circuit method because of the energy stored in the tuned circuit for any given burst. Avoidance of interburst interference requires careful design of the tuned circuit (Miya, 1981) and possibly the use of a postamble, as mentioned in the Sec. 14.7.2.

Other methods of carrier recovery are discussed in Gagliardi (1991).

14.7.4 Network Synchronization

Network synchronization is required to ensure that all bursts arrive at the satellite in their correct time slots. As mentioned previously, timing markers are provided by the reference bursts, which are tied to a highly stable clock at the reference station and transmitted through the satellite link to the traffic stations. At any given traffic station, detection of the unique word (or burst code word) in the reference burst signals the *start of receiving frame* (SORF), the marker coinciding with the last bit in the unique word.

It is desirable to have the highly stable clock located aboard the satellite because this eliminates the variations in propagation delay arising from the uplink for the reference station, but this is impractical because of weight and space limitations. However, the reference bursts retransmitted from the satellite can be treated, for timing purposes, as if they originated from the satellite (Spilker, 1977).

The network operates what is termed a *burst time plan*, a copy of which is stored at each earth station. The burst time plan shows each earth station where the receive bursts intended for it are relative to the SORF marker. This is illustrated in Fig. 14.16. At earth station A the SORF marker is received after some propagation delay t_A, and the burst time plan tells station A that a burst intended for it follows at time T_A after the

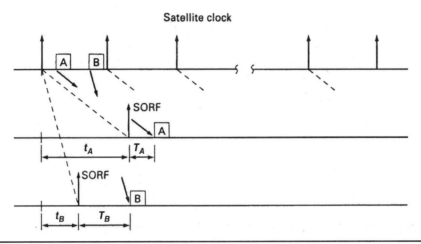

Figure 14.16   Start of receive frame (SORF) marker in a time burst plan.

SORF marker received by it. Likewise, for station B, the propagation delay is t_B, and the received bursts start at T_B after the SORF markers received at station B. The propagation delays for each station differ, but typically they are approximately 120 ms each.

The burst time plan also shows a station when it must transmit its bursts to reach the satellite in the correct time slots. A major advantage of the TDMA mode of operation is that the burst time plan is essentially under software control so that changes in traffic patterns can be accommodated much more readily than is the case with FDMA, where modifications to hardware are required. Against this, implementation of the synchronization is a complicated process.

Corrections must be included for changes in propagation delay, which result from the slowly varying position of the satellite (see Sec. 7.4). In general, the procedure for transmit timing control has two stages. First, there is the need for a station just entering, or reentering after a long delay, to acquire its correct slot position, this being referred to as *burst position acquisition*. Once the time slot has been acquired, the traffic station must maintain the correct position, this being known as *burst position synchronization*.

Open-loop timing control.   This is the simplest method of transmit timing. A station transmits at a fixed interval following reception of the timing markers, according to the burst time plan, and sufficient guard time is allowed to absorb the variations in propagation delay. The burst position error can be large with this method, and longer guard times are necessary, which reduces frame efficiency (see Sec. 14.7.7). However, for frame times longer than about 45 ms, the loss of efficiency is less than 10 percent. In a modified version of the open-loop method known as *adaptive open-loop timing*, the range is computed at the traffic station from orbital data or from measurements, and the traffic earth station makes its own corrections in timing to allow for the variations in the range. It should be noted that with open-loop timing, no special acquisition procedure is required.

Loopback timing control.   *Loopback* refers to the fact that an earth station receives its own transmission, from which it can determine range. It follows that the loopback method can only be used where the satellite transmits a global or regional beam encompassing all the earth stations in the network. Several methods are available for the acquisition process (see, e.g., Gagliardi, 1991), but basically, these all require some form of ranging to be carried out so that a close estimate of the slot position can be acquired. In one method, the traffic station transmits a low-level burst consisting of the preamble only. The power level is 20 to 25 dB below the normal operating level (Ha, 1990) to prevent interference with other bursts, and the short burst is swept through the frame until it is observed to fall within the assigned time slot for the station. The short burst is then increased to full power, and fine adjustments in timing are made to bring it to the beginning of the time slot. Acquisition can take up to about 3 s in some cases. Following acquisition, the traffic data can be added, and synchronization can be maintained by continuously monitoring the position of the loopback transmission with reference to the SORF marker. The timing positions are reckoned from the last bit of the unique word in the preamble (as is also the case for the reference burst). The loopback method is also known as *direct closed-loop feedback*.

Feedback timing control.   Where a traffic station lies outside the satellite beam containing its own transmission, loopback of the transmission does not of course occur, and some other method must be used for the station to receive ranging information. Where the synchronization information is transmitted back to an earth station from a distant station, this is termed *feedback closed-loop control*. The distant station may be a reference station, as in the INTELSAT network, or it may be another traffic station which is a designated *partner*. During the acquisition stage, the distant station can feed back information to guide the positioning of the short burst, and once the correct time slot is acquired, the necessary synchronizing information can be fed back on a continuous basis.

Figure 14.17 illustrates the feedback closed-loop control method for two earth stations *A* and *B*. The SORF marker is used as a reference point for the burst transmissions. However, the reference point which denotes the *start of transmit frame* (SOTF) must be

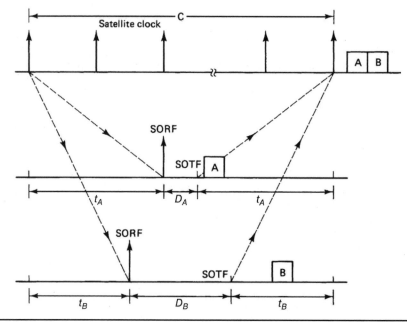

FIGURE 14.17   Timing relationships in a TDMA system. SORF, start of receive frame; SOTF, start of transmit frame.

delayed by a certain amount, shown as D_A for earth station A and D_B for earth-station B. This is necessary so that the SOTF reference points for each earth station coincide at the satellite transponder, and the traffic bursts, which are transmitted at their designated times after the SOTF, arrive in their correct relative positions at the transponder, as shown in Fig. 14.17. The total time delay between any given satellite clock pulse and the corresponding SOTF is a constant, shown as C in Fig. 14.17. C is equal to $2t_A + D_A$ for station A and $2t_B + D_B$ for station B. In general, for earth station i, the delay D_i is determined by

$$2t_i + D_i = C \qquad (14.17)$$

In the INTELSAT network, $C = 288$ ms.

For a truly geostationary satellite, the propagation delay t_i is constant. However, as shown in Sec. 7.4, station-keeping maneuvers are required to keep a geostationary satellite at its assigned orbital position, and hence this position can be held only within certain tolerances. For example, in the INTELSAT network, the variation in satellite position can lead to a variation of up to ± 0.55 ms in the propagation delay (INTELSAT, 1980). To minimize the guard time needed between bursts, this variation in propagation delay must be considered in determining the delay, D_i, required at each traffic station. In the INTELSAT network, the D_i numbers are updated every 512 frames, which is a period of 1.024 s, based on measurements and calculations of the propagation delay times made at the reference station. The D_i numbers are transmitted to the earth stations through the CDC channel in the reference bursts. (It should be noted that the open-loop synchronization described previously amounts to using a constant D_i value.)

The use of traffic burst preambles along with reference bursts to achieve synchronization is the most common method, but at least one other method, not requiring preambles, has been proposed by Nuspl and de Buda (1974). It also should be noted that there are certain types of "packet satellite networks," for example, the basic Aloha system (Rosner, 1982), which are closely related to TDMA, in which synchronization is not used.

14.7.5 Unique Word Detection

The *unique word* (UW) or *burst code word* (BCW) is used to establish burst timing in TDMA. Figure 14.18 shows the basic arrangement for detecting the UW. The received bit stream is passed through a shift register which forms part of a correlator. As the bit stream moves through the register, the sequence is continuously compared with a stored version of the UW. When correlation is achieved, indicated by a high output from the threshold detector, the last bit of the UW provides the reference point for timing purposes. It is important therefore to know the probability of error in detecting the UW. Two possibilities must be considered. One, termed the *miss probability*, is the probability of the correlation detector failing to detect the UW even though it is present in the bit stream. The other, termed the *probability of false alarm*, is the probability that the correlation detector misreads a sequence as the UW. Both are examined in turn.

Miss probability. Let E represent the maximum number of errors allowed in the UW of length N bits, and let I represent the actual number of errors in the UW as received. The following conditions apply:

When $I \le E$, the detected sequence is declared to be the UW.

When $I > E$, the detected sequence N is declared not to be the UW; that is, the unique word is missed.

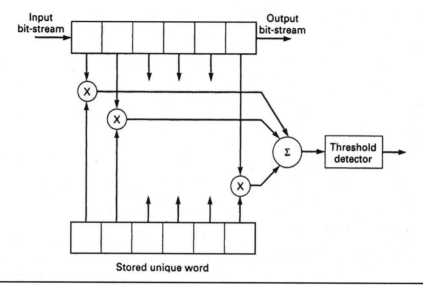

FIGURE 14.18   Basic arrangement for detection of the unique word (UW).

Let p represent the average probability of error in transmission [the *bit error rate* (BER)]. The probability of receiving a sequence N containing I errors in any one arrangement is

$$p_I = p^I(1-p)^{N-1} \qquad (14.18)$$

The number of combinations of N things taken I at a time, usually written as $_NC_I$, is given by

$$_NC_I = \frac{N!}{I!(N-I)!} \qquad (14.19)$$

The probability of receiving a sequence of N bits containing I errors is therefore

$$P_I = {_NC_I}p_I \qquad (14.20)$$

Now since the UW is just such a sequence, Eq. (14.20) gives the probability of a UW containing I errors. The condition for a miss occurring is that $I > E$, and therefore, the miss probability is

$$P_{\text{miss}} = \sum_{I=E+1}^{N} P_I \qquad (14.21)$$

Written out in full, this is

$$P_{\text{miss}} = \sum_{I=E+1}^{N} \frac{N!}{I!(N-I)!} p^I(1-p)^{N-I} \qquad (14.22)$$

Equation (14.22) gives the *average* probability of missing the UW even though it is present in the shift register of the correlator. Note that because this is an average probability, it is not necessary to know any specific value of I.

Example 14.4 Determine the miss probability for these following values: $N = 40$, $E = 5$, $p = 10^{-3}$

Solution

$$P_{miss} = \sum_{I=6}^{40} \frac{40!}{I!(40-I)!} \, 10^{-3I}(1-10^{-3})^{40-I}$$

$$= 3.7 \times 10^{-12}$$

False detection probability. Consider now a sequence of N bits which is not the UW. However, this sequency can be interpreted as the UW even if it differs from it by E bit positions. Further, let I represent the actual number of bit positions by which the random sequence does differ from the UW. Thus, E represents the number of acceptable "bit errors" considered from the point of view of the UW, although they may not be errors in the message they represent. Likewise, I represents the actual number of "bit errors" considered from the point of view of the UW, although they may not be errors in the message they represent. As before, the number of combinations of N things taken I at a time is given by Eq. (14.19), and hence the number of words acceptable as the UW is

$$W = \sum_{I=0}^{E} {}_N C_I \tag{14.23}$$

The number of words that can be formed from a random sequence of N bits is 2^N, and on the assumption that all such words are equiprobable, the probability of receiving any one word is 2^{-N}. Hence the probability of a false detection is

$$P_F = 2^{-N} W \tag{14.24}$$

Written out in full, this is

$$P_F = 2^{-N} \sum_{I=0}^{E} \frac{N!}{I!(N-I)!} \tag{14.25}$$

Notice that because this is an average probability, it is not necessary to know a specific value of I. Also, in this case, the BER does not enter the calculation.

Example 14.5 Determine the probability of false detection for the following values: $N = 40$, $E = 5$

Solution

$$P_F = 2^{-40} \sum_{I=0}^{5} \frac{40!}{I!(40-I)!} = 6.9e\text{-}07$$

From Examples 14.4 and 14.5, it is seen that the probability of a false detection is much higher than the probability of a miss, and this is true in general. In practice, once frame synchronization has been established, a time window can be formed around the expected time of arrival for the UW such that the correlation detector is only in operation for the window period. This greatly reduces the probability of false detection.

14.7.6 Traffic Data

The traffic data immediately follow the preamble in a burst. As shown in Fig. 14.13, the traffic data subburst is further subdivided into time slots addressed to the individual destination stations. Any given destination station selects only the data in the time slots intended for that station. As with FDMA networks, TDMA networks can be operated with both preassigned and demand assigned channels, and examples of both types are given shortly. The greater the fraction of frame time that is given over to traffic, the higher is the efficiency. The concept of *frame efficiency* is discussed in the following section.

14.7.7 Frame Efficiency and Channel Capacity

The frame efficiency is a measure of the fraction of frame time used for the transmission of traffic. *Frame efficiency* may be defined as

$$\text{Frame efficiency} = \eta_F = \frac{\text{traffic bits}}{\text{total bits}} \tag{14.26}$$

Alternatively, this can be written as

$$\eta_F = 1 - \frac{\text{overhead bits}}{\text{total bits}} \tag{14.27}$$

In these equations, bits per frame are implied. The overhead bits consist of the sum of the preamble, the postamble, the guard intervals, and the reference-burst bits per frame. The equations may be stated in terms of symbols rather than bits, or the actual times may be used.

For a fixed overhead, Eq. (14.27) shows that a longer frame, or greater number of total bits, results in higher efficiency. However, longer frames require larger buffer memories and add to the propagation delay. Synchronization also may be made more difficult, keeping in mind that the satellite position is varying with time. A lower overhead also leads to higher efficiency, but again, reducing synchronizing and guard times may result in more complex equipment being required.

Example 14.6 Calculate the frame efficiency for an INTELSAT frame given the following information:

Total frame length = 120,832 symbols

Traffic bursts per frame = 14

Reference bursts per frame = 2

Guard interval = 103 symbols

Solution From Fig. 14.14, the preamble symbols add up to

$$P = 176 + 24 + 8 + 8 + 32 + 32 = 280$$

With addition of the CDC channel, the reference channel symbols add up to

$$R = 280 + 8 = 288$$

Therefore, the overhead symbols are

$$OH = 2 \times (103 + 288) + 14 \times (103 + 280) = 6144 \text{ symbols}$$

Therefore, from Eq. (14.27),

$$\eta_F = 1 - \frac{6144}{120832} = 0.949$$

The voice-channel capacity of a frame, which is also the voice-channel capacity of the transponder being accessed by the frame, can be found from a knowledge of the frame efficiency and the bit rates. Let R_b be the bit rate of a voice channel, and let there be a total of n voice channels shared between all the earth stations accessing the transponder. The total incoming *traffic* bit rate to a frame is nR_b. The traffic bit rate of the frame is $\eta_F R_{TDMA}$, and therefore

$$nR_b = \eta_F R_{TDMA}$$

or

$$n = \frac{\eta_F R_{TDMA}}{R_b} \tag{14.28}$$

Example 14.7 Calculate the voice-channel capacity for the INTELSAT frame in Example 14.2, given that the voice-channel bit rate is 64 kb/s and that QPSK modulation is used. The frame period is 2 millisec.

Solution The number of symbols per frame is 120,832, and the frame period is 2 millisec. Therefore, the symbol rate is 120,832/2e-03 = 60.416 Msps. QPSK modulation utilizes 2 bits per symbol, and therefore, the transmission bit rate is $R_{TDMA} = 60.416 \times 2 = 120.832$ Mbps.

Using Eq. (14.28) and the number of voice channels is

$$n = \text{INT}\left(\frac{0.949 \times 120.832e+06}{64e+03}\right) = 1792$$

14.7.8 Preassigned TDMA

An example of a preassigned TDMA network is the CSC for the Spade network described in Sec. 14.5. The frame and burst formats are shown in Fig. 14.19. The CSC can accommodate up to 49 earth stations in the network plus one reference station, making a maximum of 50 bursts in a frame.

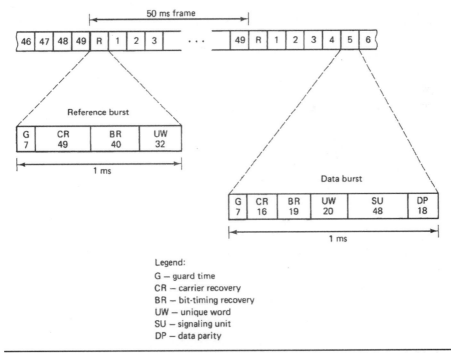

Figure 14.19 Frame and bit formats for the common signaling channel (CSC) used with the Spade system. (*Data from Miya, 1981.*)

All the bursts are of equal length. Each burst contains 128 bits and occupies a 1-millisec time slot. Thus, the bit rate is 128 kb/s. As discussed in Sec. 14.5, the frequency bandwidth required for the CSC is 160 kHz.

The *signaling unit* (SU) shown in Fig. 14.19 is that section of the data burst which is used to update the other stations on the status of the frequencies available for the SCPC calls. It also carries the signaling information, as described in Sec. 14.5.

Another example of a preassigned TDMA frame format is the INTELSAT frame shown in simplified form in Fig. 14.20. In the INTELSAT system, preassigned and demand-assigned voice channels are carried together, but for clarity, only a preassigned traffic burst is shown. The traffic burst is subdivided into time slots, termed *satellite channels* in the INTELSAT terminology, and there can be up to 128 of these in a

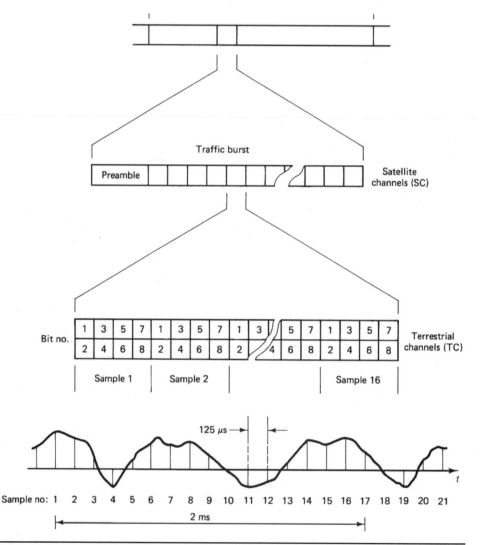

FIGURE 14.20 Preassigned TDMA frame in the INTELSAT system.

traffic burst. Each satellite channel is further subdivided into 16 time slots termed *terrestrial channels*, each terrestrial channel carrying one PCM sample of an analog telephone signal. QPSK modulation is used, and therefore, there are 2 bits per symbol as shown. Thus, each terrestrial channel carries 4 symbols (or 8 bits). Each satellite channel carries $4 \times 16 = 64$ symbols, and at its maximum of 128 satellite channels, the traffic burst carries 8192 symbols.

As discussed in Sec. 10.3, the PCM sampling rate is 8 kHz, and with 8 bits per sample, the PCM bit rate is 64 kb/s. Each satellite channel can accommodate this bit rate. Where input data at a higher rate must be transmitted, multiple satellite channels are used. The maximum input data rate which can be handled is $128(SC) \times 64$ kb/s = 8.192 Mbps.

The INTELSAT frame is 120,832 symbols or 241,664 bits long. The frame period is 2 ms, and therefore, the burst bit rate is 120.832 Mbps.

As mentioned previously, preassigned and demand-assigned voice channels can be accommodated together in the INTELSAT frame format. The demand-assigned channels utilize a technique known as *digital speech interpolation* (DSI), which is described in the following section.

The preassigned channels are referred to as *digital noninterpolated* (DNI) channels.

14.7.9 Demand-Assigned TDMA

With TDMA, the burst and subburst assignments are under software control, compared with hardware control of the carrier frequency assignments in FDMA. Consequently, compared with FDMA networks, TDMA networks have more flexibility in reassigning channels, and the changes can be made more quickly and easily.

Several methods are available for providing traffic flexibility with TDMA. The burst length assigned to a station may be varied as the traffic demand varies. A central control station may be employed by the network to control the assignment of burst lengths to each participating station. Alternatively, each station may determine its own burst-length requirements and assign these in accordance with a prearranged network discipline.

As an alternative to burst-length variation, the burst length may be kept constant and the number of bursts per frame used by a given station varied as demand requires. In one proposed system (CCIR Report 708, 1982), the frame length is fixed at 13.5 ms. The basic burst time slot is 62.5 µs, and stations in the network transmit information bursts varying in discrete steps over the range 0.5 ms (8 basic bursts) to 4.5 ms (72 basic bursts) per frame. Demand assignment for speech channels takes advantage of the intermittent nature of speech, as described in the following section.

14.7.10 Speech Interpolation and Prediction

Because of the intermittent nature of speech, a speech transmission channel lies inactive for a considerable fraction of the time it is in use. Several factors contribute to this. The talk-listen nature of a normal two-way telephone conversation means that transmission in any one direction occurs only about 50 percent of the time. In addition, the pauses between words and phrases may further decrease this to about 33 percent. If further allowance is made for "end party" delays such as the time required for a party to answer a call, the average fraction of the total connect time may drop to as low as 25 percent. The fraction of time a transmission channel is active is known as the *telephone load activity factor*, and for system design studies, the value of 0.25 is recommended by *Comité Consutatif Internationale Télégraphique et Téléphonique* (CCITT), although higher values are also used (Pratt and Bostian, 1986). The point is that for a significant fraction of the time the channel is available for other transmissions, and advantage is taken of this in the form of demand assignment known as *digital speech interpolation*.

Digital speech interpolation may be implemented in one of two ways, these being digital *time assignment speech interpolation* (digital TASI) and *speech predictive encoded communications* (SPEC).

Digital TASI. The traffic-burst format for an INTELSAT burst carrying demand-assigned channels and preassigned channels is shown in Fig. 14.21. As mentioned previously, the demand-assigned channels utilize digital TASI, or what is referred to in the INTELSAT nomenclature as DSI, for *digital speech interpolation*. These are shown by the block labeled "interpolated" in Fig. 14.21. The first satellite channel (channel 0) in this block is an assignment channel, labeled *DSI-AC*. No traffic is carried in the assignment channel; it is used to transmit channel assignment information as described shortly.

Figure 14.22 shows in outline the DSI system. Basically, the system allows N terrestrial channels to be carried by M satellite channels, where $N > M$. For example, in the INTELSAT arrangement, $N = 240$ and $M = 127$.

On each incoming terrestrial channel, a speech detector senses when speech is present, the intermittent speech signals being referred to as *speech spurts*. A speech spurt lasts on average about 1.5 s (Miya, 1981). A control signal from the speech detector is sent to the channel assignment unit, which searches for an empty TDMA buffer. Assuming that one is found, the terrestrial channel is assigned to this satellite channel, and the speech spurt is stored in the buffer, ready for transmission in the DSI subburst. A delay is inserted in the speech circuit, as shown in Fig. 14.22, to allow some time for the assignment process to be completed. However, this delay cannot exactly compensate for the assignment delay, and the initial part of the speech spurt may be lost. This is termed a *connect clip*.

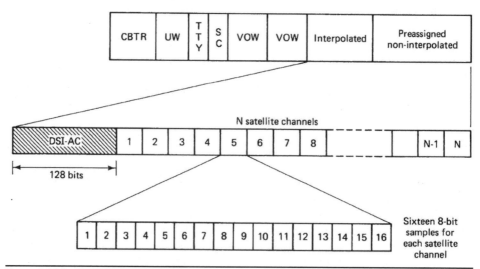

FIGURE 14.21 INTELSAT traffic burst structure. (*Courtesy of INTELSAT, 1983. With permission.*)

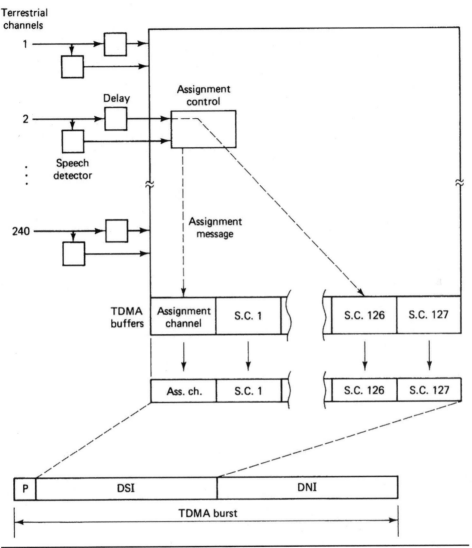

Figure 14.22   Digital speech interpolation. DNI, digital noninterpolation; DSI, digital speech interpolation.

In the INTELSAT system an intermediate step occurs where the terrestrial channels are renamed *international channels* before being assigned to a satellite channel (Pratt and Bostian, 1986). For clarity, this step is not shown in Fig. 14.22.

At the same time as an assignment is made, an assignment message is stored in the assignment channel buffer, which informs the receive stations which terrestrial channel is assigned to which satellite channel. Once an assignment is made, it is not interrupted, even during pauses between spurts, unless the pause times are required for another DSI channel. This reduces the amount of information needed to be transmitted over the assignment channel.

At the receive side, the traffic messages are stored in their respective satellite-channel buffers. The assignment information ensures that the correct buffer is read out to the corresponding terrestrial channel during its sampling time slot. During speech pauses when the channel has been reassigned, a low-level noise signal is introduced at the receiver to simulate a continuous connection.

It has been assumed that a free satellite channel can be found for any incoming speech spurt, but of course there is a finite probability that all channels are occupied, and the speech spurt is lost. Losing a speech spurt in this manner is referred to as *freeze-out*, and the freeze-out fraction is the ratio of the time the speech is lost to the average spurt duration. It is found that a design objective of 0.5 percent for a freeze-out fraction is satisfactory in practice. This means that the probability of a freeze-out occurring is 0.005.

Another source of signal mutilation is the *connect clip* mentioned earlier. Again, it is found in practice that clips longer than about 50 ms are very annoying to the listener. An acceptable design objective is to limit the fraction of clips that are equal to or greater than 50 ms to a maximum of 2 percent of the total clips. In other words, the probability of encountering a clip that exceeds 50 ms is 0.02.

The *DSI gain* is the ratio of the number of terrestrial channels to number of satellite channels, or N/M. The DSI gain depends on the number of satellite channels provided as well as the design objectives stated earlier. Typically, DSI gains somewhat greater than 2 can be achieved in practice.

Speech Predictive Encoded Communications (SPEC).   The block diagram for the SPEC system is shown in Fig. 14.23 (Sciulli and Campanella, 1973). In this method, the

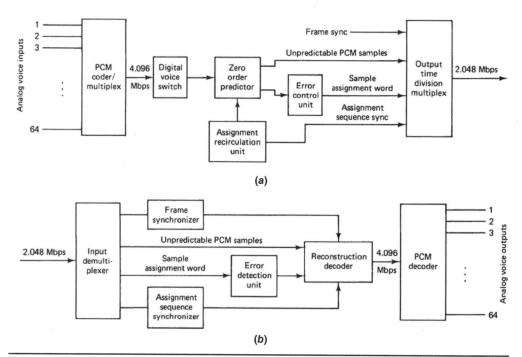

FIGURE 14.23   (a) SPEC transmitter; (b) SPEC receiver. (*Courtesy of Sciulli and Campanella, 1973. Copyright 1973, IEEE.*)

incoming speech signals are converted to a PCM multiplexed signal using 8 bits per sample quantization. With 64 input lines and sampling at 125 μs, the output bit rate from the multiplexer is $8 \times 64/125 = 4.096$ Mbps.

The digital voice switch following the PCM multiplexer is timeshared between the input signals. It is voice-activated to prevent transmission of noise during silent intervals. When the zero-order predictor receives a new sample, it compares it with the previous sample for that voice channel, which it has stored, and transmits the new sample only if it differs from the preceding one by a predetermined amount. These new samples are labeled *unpredictable PCM samples* in Fig. 14.23a.

For the 64 channels a 64-bit assignment word is also sent. A logic 1 in the channel for the assignment word means that a new sample was sent for that channel, and a logic 0 means that the sample was unchanged. At the receiver, the sample assignment word either directs the new (unpredictable) sample into the correct channel slot, or it results in the previous sample being regenerated in the reconstruction decoder. The output from this is a 4.096-Mbps PCM multiplexed signal that is demultiplexed in the PCM decoder.

By removing the redundant speech samples and silent periods from the transmission link, a doubling in channel capacity is achieved. As shown in Fig. 14.23, the transmission is at 2.048 Mbps for an input-output rate of 4.096 Mbps.

An advantage of the SPEC method over the DSI method is that freeze-out does not occur during overload conditions. During overload, sample values which should change may not. This effectively leads to a coarser quantization and therefore an increase in quantization noise. This is subjectively more tolerable than freeze-out.

14.7.11 Downlink Analysis for Digital Transmission

As mentioned in Sec. 14.6, the transponder power output and bandwidth both impose limits on the system transmission capacity. With TDMA, TWT backoff is not generally required, which allows the transponder to operate at saturation. One drawback arising from this is that the uplink station must be capable of saturating the transponder, which means that even a low-traffic-capacity station requires comparatively large power output compared with what is required for FDMA. This point is considered further in Sec. 14.7.12.

As with the FDM/FDMA system analysis, it is assumed that the overall carrier-to-noise ratio is essentially equal to the downlink carrier-to-noise ratio. With a power-limited system this C/N ratio is one of the factors that determines the maximum digital rate, as shown by Eq. (10.24), which is rewritten as

$$[R_b] = \left[\frac{C}{N_0}\right] - \left[\frac{E_b}{N_0}\right] \tag{14.29}$$

The $[E_b/N_0]$ ratio is determined by the required BER, as shown in Fig. 10.22, and described in Sec. 10.6.4. For example, for a BER of 10^{-5} an $[E_b/N_0]$ of 9.6 dB is required. If the rate R_b is specified, then the $[C/N_0]$ ratio is determined, as shown by Eq. (14.29), and this value is used in the link budget calculations as required by Eq. (12.57). Alternatively, if the $[C/N_0]$ ratio is fixed by the link budget parameters as given by Eq. (12.57), the bit rate is then determined by Eq. (14.29).

The bit rate is also constrained by the IF bandwidth. As shown in Sec. 10.6.3, the ratio of bit rate to IF bandwidth is given by

$$\frac{R_b}{B_{IF}} = \frac{m}{1+\rho}$$

where $m = 1$ for *binary phase-shift keying* (BPSK) and $m = 2$ for QPSK and ρ is the rolloff factor. The value of 0.2 is commonly used for the rolloff factor, and therefore, the bit rate for a given bandwidth becomes

$$R_b = \frac{mB_{IF}}{1.2} \tag{14.30}$$

Example 14.8   Using Eq. (12.57), a downlink $[C/N_0]$ of 87.3 dBHz is calculated for a TDMA circuit that uses QPSK modulation. A BER of 10^{-5} is required. Calculate the maximum transmission rate. Also calculate the IF bandwidth required assuming a rolloff factor of 0.2.

Solution   From Fig. 10.23 which is applicable for QPSK and BPSK. $[E_b/N_0] = 9.65$ dB for a BER of 10^{-5}. Hence

$$[R_b] = 87.3 - 9.65 = 77.65 \text{ dB}$$

This is equal to 58.21 Mbps.
    For QPSK $m = 2$ and using Eq. (14.30), we have

$$B_{IF} = 58.21 \times 1.2/2 = 34.9 \text{ MHz}$$

From Example 14.8, it is seen that an EIRP, that results in a $[C/N_0]$ of 87.3 dBHz at the receiving ground station, is near optimum in that the transponder bandwidth (36 MHz) is almost fully occupied.

14.7.12   Comparison of Uplink Power Requirements for FDMA and TDMA

With FDMA, the modulated carriers at the input to the satellite are retransmitted from the satellite as a combined frequency-division-multiplexed signal. Each carrier retains its modulation, which may be analog or digital. For this comparison, digital modulation is assumed. The modulation bit rate for each carrier is equal to the input bit rate (adjusted as necessary for *forward error correction* [FEC]). The situation is illustrated in Fig. 14.24a, where for simplicity, the input bit rate R_b is assumed to be the same for each earth station. The [EIRP] is also assumed to be the same for each earth station.

    With TDMA, the uplink bursts that are displaced in time from one another are retransmitted from the satellite as a combined time-division-multiplexed signal. The uplink bit rate is equal to the downlink bit rate in this case, as illustrated in Fig. 14.24b. As described in Sec. 14.7, compression buffers are needed to convert the input bit rate R_b to the transmitted bit rate R_{TDMA}.

    Because the TDMA earth stations must transmit at a higher bit rate compared with FDMA, a higher [EIRP] is required, as can be deduced from Eq. (10.32), which states that

$$\left[\frac{C}{N_0}\right] = \left[\frac{E_b}{N_0}\right] + [R]$$

where $[R]$ is equal to $[R_b]$ for an FDMA uplink and $[R_{TDMA}]$ for a TDMA uplink.

    For a given BER the $[E_b/N_0]$ ratio is fixed as shown by Fig. 10.22. Hence, assuming that $[E_b/N_0]$ is the same for the TDMA and the FDMA uplinks, an increase in $[R]$ requires a corresponding increase in $[C/N_0]$. If the TDMA and FDMA uplinks operate with the

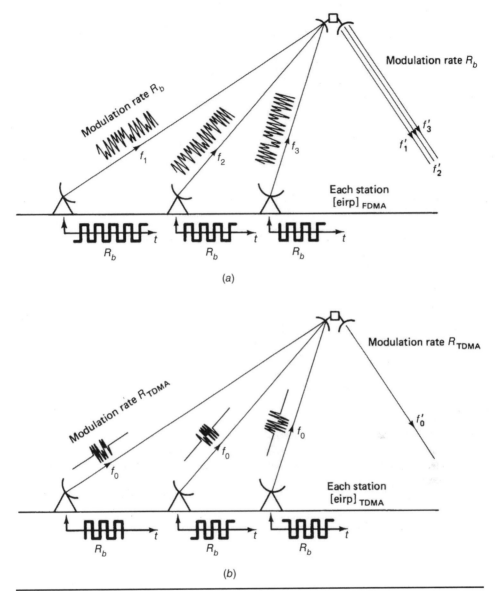

FIGURE 14.24 (a) FDMA network; (b) TDMA network.

same [LOSSES] and satellite $[G/T]$, Eq. (12.42) shows that the increase in $[C/N_0]$ can be achieved only through an increase in the earth station [EIRP], and therefore

$$[\text{EIRP}]_{\text{TDMA}} - [\text{EIRP}]_{\text{FDMA}} = [R_{\text{TDMA}}] - [R_b] \tag{14.31}$$

For the same earth-station antenna gain in each case, the decibel increase in earth station transmit power for TDMA compared with FDMA is

$$[P]_{\text{TDMA}} - [P]_{\text{FDMA}} = [R_{\text{TDMA}}] - [R_b] \tag{14.32}$$

Example 14.9   A 14-GHz uplink operates with transmission losses and margins totaling −212 dB and a satellite $[G/T] = 10$ dB/K. The required uplink $[E_b/N_0]$ is 12 dB. (*a*) Assuming FDMA operation and an earth-station uplink antenna gain of 46 dB, calculate the earth-station transmitter power needed for transmission of a T1 baseband signal. (*b*) If the downlink transmission rate is $[R_{TDMA}] = 74$ dB, calculate the uplink power increase required for TDMA operation.

Solution   (*a*) From Sec. 10.4 the T1 bit rate is 1.544 Mbps or $[R] = 62$ dB.

Using the $[E_b/N_0] = 12$ dB value specified, Eq. (10.32) gives

$$\left[\frac{C}{N_0}\right] = 12 + 62 = 74 \text{ dBHz}$$

From Eq. (12.42),

$$[\text{EIRP}] = \left[\frac{C}{N_0}\right] - \left[\frac{G}{T}\right] - [\text{LOSSES}] - 228.6$$

$$= 74 - 10 + 212 - 228.6$$

$$= 47.4 \text{ dBW}$$

Hence the transmitter power required is

$$[P] = 47.4 - 46 = 1.4 \text{ dBW} \Rightarrow 1.38 \text{ W}$$

(*b*) With TDMA operation the rate increase is

$$[R_{TDMA}] - [R_b] = 74 - 62 = 12 \text{ dB}.$$

All other factors being equal, the earth station [EIRP] must be increased by this amount, and hence

$$[\text{delta-}P] = 1.4 + 12 = 13.4 \text{ dBW} \Rightarrow 21.88 \text{ W}$$

Example 14.10   Consider the Ku-band VSAT mesh network discussed in Example 14.2, but now operating using TDMA. For comparison purposes, the goal is to convert the network without changes to the GEOSAT transponder and only necessary hardware changes to the VSAT earth stations. After performing this analysis we compare the relative performance of these two networks.

As discussed above, the most significant impact of changing from FDMA to TDMA is the increase in the VSAT EIRP, which we choose to implement by using the existing antennas and to increase the transmitter to higher power pulsed transmitter. Even though the peak transmitter power increases by two orders of magnitude, the duty cycle is less than 1 percent, so that the average power is basically unchanged. Of course, the filters need to be wide-band RRC filters with a roll-off factor $\rho = 0.40$, but this is not a major cost driver for the conversion. Each terminal transmits the same digital data at a bit rate of 128 kbps with the BPSK demodulation, and the link BER $\leq 10^{-6}$ with a BPSK demodulation implementation loss = −1.5 dB.

For SCPC-FDMA the VSAT uplink EIRP = 37.18 dBW, which comprise a peak antenna gain = 44.18 dB and the transmit power = −7.0 dBW $\Rightarrow 0.2$ W. For TDMA, to drive the transponder to saturation, requires a $P_t = 16.0$ dBW $\Rightarrow 40$ W, so the new VSAT EIRP = 16.0 + 44.18 = 60.18 dBW. With this change, the TDMA uplink $(C/N_0)_{up} = 72.59$ dBHz is the same as FDMA.

For SCPC FDMA, the transponder 80 W TWTA operates at a $BO_o = -4.5$ dB that provides a total multi-carrier output power = 28.4 W with a $[(C/N_0)_{3IM}] = 71.28$ dBHz. For TDMA, the TWTA operates with a single carrier input and an output power = 80 W, with no 3IM interference generated. The resulting GEOSAT EIRP = 62.02 dBW, and a $[(C/N_0)_{dn}] = 73.64$ dBHz. All the other link parameters are identical, so the overall $[(C/N_0)_{tot}] = 73.64$ dBHz, and this is compared to the SCPC FDMA $[(C/N_0)_{tot}] = 67.62$ dBHz (+6 dB improvement for TDMA). This is the result of TDMA operating at a $BO_o = 0$ dB and the single carrier operation eliminating the 3IM interference. Of course the GEOSAT downlink is much greater than required to meet the BER specification, but this is the result of not changing the transponder design.

TDMA Frame Design (BPSK: symbol = bit)

The TDMA frame consists of one traffic burst for 171 VSATs and a single reference earth station burst (172 total as before). Assume that the reference burst = 221 bits and the traffic preambles = 195 bits, and all bursts have ≥ 1 microsec guard band.

Each burst fills the transponder B_{TR} = 36 MHz transponder with a burst rate

$$R_{TDMA} = \frac{B_{TR}}{(1+\rho)} = \frac{36.0e+06}{1.4} = 25.714 \text{ Mbps}$$

and the corresponding time to transmit one bit = $1/R_{TDMA}$ = 3.8888e-08 sec/bit

The next decision is the selection of frame time T_F. Because of the large number of VSATs, it is desirable to increase T_F to improve the TDMA efficiency, and increasing the number of PCM samples/ burst also improves the efficiency.

So, we select an arbitrary integer (80) PCM samples/frame @ 125 microsec/sample.

$$T_F = 80 * 125 \text{ microsec} = 10.0 \text{ millisec,}$$

Thus, the traffic data buffer size $M = T_F R_b = 10.0e-03 * 128e+03 = 1280$ bits, which must be transmitted each frame (with no carry-over). In addition, we must account for the overhead (OH) of 195 symbols, which makes the total bits/burst = 1475 bits/burst. The time to transmit a single traffic burst

$$T_B = 1475 * 3.8889e-08 = 57.361 \text{ microsec}$$

One reference burse is T_{Ref} = 221 * 3.8889e-08 = 8.59 microsec

The total time required for traffic and reference burst is = 171 * 57.36e-06 + 8.59e-06 = 9.817 millisec

Time available for guard time = 10.0 − 9.817 = 0.183 millisec = 183 microsec divided by (171 + 1) = 183/171 = 1.06 microsec, which satisfies the specification.

(*a*) What is the clear sky $[C/N_0]$, the corresponding BER, and the rain margin?

Solution The (C/N_0) link analysis for clear sky is presented below.

Calc Clear-air: $(C/N_0)_{tot}$ at VSAT	Ratio	dB
$(C/N_0)_{up}$	1.81E+07	72.59
$(N_0/C)_{up}$	5.53E-08	
$(C/N_0)_{dn}$	1.59E+08	81.94
$(N_0/C)_{dn}$	6.39E-09	
$(N_0/C)_{tot}$	6.17E-08	
$(C/N_0)_{tot}$	1.62E+07	72.10

Calc the P_e	Ratio	dB
$[(C/N_0)_{eff}] = [(C/N_0)_{tot}] + [\text{Imp-Loss}]$	1.147E+07	70.60
$(E_b/N_0)_{eff}$	89.70	
$P_e = 0.5*ERFC(sqrt(E_b/N_0))$	3.63E-41	

Using an iterative solution the threshold for BER ~ 10^{-6} is $[C/N_0]_{eff}$ = 61.60 dBHz
Therefore, the rain margin = $[C/N_0]_{clear-sky}$ − [threshold] = 70.60 − 61.60 = +9.0 dB

(b) What is the BER ≥ 99.99% of 1 year?

Solution Referring to Sec. 12.9, the Rain Attenuation Exceedance for 0.010% of 1 year from Table 4.6 is −6.22 dB for 12 GHz (downlink) and −8.58 for 14 GHz (uplink). Since the probability of rain occurring simultaneously on both the uplink and downlink is negligible, we treat the uplink and downlink separately.

Effect of rain attenuation on $(C/N_0)_{dn}$
As discussed in Sec. 12.9.1, rain degrades the $(C/N_0)_{dn}$ in two ways, namely; by carrier attenuation and by increased rain emission noise. For the downlink and clear-sky, assume that the system noise temperature $T_{sys} = 150$ K.

Rain Atten on the Downlink @ 12 GHz	0.23878113	−6.22
$T_{sys\_clear-sky}$, K	150	
Tap_rain = (274*(1 − Rain_atten)), K	208.6	
T_{sys_rain} = Tap_rain + $T_{sys\_clear-sky}$	358.6	
Delta-N = $T_{sys\_rain}/T_{sys\_clear-sky}$	2.39	3.78
$(C/N_0)_{dn}$ = $(C/N_0)_{clear}$ + [atten] − [delt-N]		71.94

Calc Rain: $(C/N_0)_{tot}$ at VSAT	Ratio	dB
$(C/N_0)_{up}$	1.81E+07	72.57
$(N_0/C)_{up}$	5.53E-08	
$(C/N_0)_{dn}$	1.56E+07	71.94
$(N_0/C)_{dn}$	6.40E-08	
$(N_0/C)_{tot}$	1.19E-07	
$(C/N_0)_{tot}$	8.38E+06	69.23

Calc the P_e (BPSK)	Ratio	dB
$[(C/N_0)_{eff}]$ = $[(C/N_0)_{tot}]$ + [Imp-Loss]	5.932E+06	67.74
$(E_b/N_0)_{eff}$	46.34	
P_e = 0.5*ERFC(sqrt(E_b/N_0))	3.07E-22	

(c) What is the P_e during rain on the uplink?

Solution For the uplink, the system noise does not change during rain (see Example 12.21); so only the carrier attenuation reduces the $(C/N_0)_{up}$. From Table 4.6, the Rain Attenuation Exceedance for 0.010% of 1 year is −8.58 for 14 GHz. So,

$$\left[\left(\frac{C}{N_0}\right)_{up-rain}\right] = \left[\left(\frac{C}{N_0}\right)_{up-clear-sky}\right] + [\text{Rain Atten}]$$

$$= 72.57 + (-8.58) = 63.99 \text{ dB}$$

Furthermore, this attenuation increases the BO_i to the transponder TWTA, which affects both the $(C/N_0)_{dn}$. To perform this calculation, it is necessary to use the single carrier TWTA power transfer characteristic table given below.

TWTA Transfer Function for Single Carrier (from Ha, Table 5.3)

BO$_i$, dB	BO$_o$, dB	BO$_i$, dB	BO$_o$, dB
0	0	−10	−6.6
−1	−0.15	−11	−7.56
−2	−0.4	−12	−8.53
−3	−0.8	−13	−9.51
−4	−1.45	−14	−10.5
−5	−2.2	−15	−11.5
−6	−3	−16	−12.5
−7	−3.85	−17	−13.5
−8	−4.73	−18	−14.5
−9	−5.65	−19	−15.5

The new $[\text{BO}_i]$ = [Rain Atten] = −8.58 dB, and interpolating the above TWTA table, the new BO_o = −5.29 dB, so the downlink carrier to noise density becomes

$$\left[\left(\frac{C}{N_0}\right)_{\text{dn-rain}}\right] = \left[\left(\frac{C}{N_0}\right)_{\text{dn-clear-sky}}\right] + \left[\text{BO}_o\right]$$

$$= 81.94 + (-5.29) = 76.65 \text{ dB}$$

The calculation of the $[(C/N)_{\text{tot}}]$ and the P_e are tabulated below.

Calc (C/N$_0$)$_{\text{tot}}$ for Rain on Uplink	Ratio	dB
$(C/N_0)_{\text{up}}$	2.506E+06	63.99
$(N_0/C)_{\text{up}}$	3.99E-07	
$(C/N_0)_{\text{dn}}$	4.63E+07	76.65
$(N_0/C)_{\text{dn}}$	2.16E-08	
$(N_0/C)_{\text{tot}}$	4.21E-07	
$(C/N_0)_{\text{to}}$	2.38E+06	63.76

Calc the P_e (BPSK)	Ratio	dB
$[(C/N)_{\text{eff}}] = [(C/N)_{\text{tot}}] + [\text{Imp-Loss}]$	1.68E+06	62.26
$(E_b/N_0)_{\text{eff}}$	13.1486	
$P_e = 0.5*\text{ERFC}(\text{sqrt}(E_b/N_0))$	1.46E-07	

Performance comparison for VSAT mesh network

Both FDMA and TDMA have achievable solutions to meeting the VSAT network requirements; however for FDMA the VSATs had to switch to QPSK with FEC to meet the 99.99 percent link availability specification during rainfall (for separate scenarios: rain on uplink and rain on downlink). On the other hand, the TDMA design is "overkill" because the assumptions prevented changing the GEOSAT transponder. If this constraint is relaxed, a lower EIRP system can be designed that significantly reduces the complexity and cost for the TDMA system in space and on the ground.

Moreover, for small satellite business systems, it is desirable to be able to operate with relatively small earth stations, which suggests that FDMA should be the mode of operation. But TDMA permits more efficient use of the satellite transponder by eliminating the need for backoff. This suggests that it might be worthwhile to operate a hybrid system in which FDMA is the uplink mode of operation, with the individual signals converted to a time-division-multiplexed format in the transponder before being amplified by the TWTA. This allows the transponder to be operated at saturation as in TDMA. Such a hybrid mode of operation requires the use of a signal-processing transponder as discussed in the next section.

14.8   On-Board Signal Processing for FDMA/TDM Operation

As seen in the preceding section, for small earth stations carrying digital signals at relatively low data rates, there is an advantage to be gained in terms of earth station power requirements by using FDMA. On the other hand, TDMA signals make more efficient use of the transponder because backoff is not required.

Market studies in the late 1980s (e.g., Stevenson et al., 1984) predicted that *customer premises services* (CPS) would make up a significant portion of the satellite demand in the 2000s. For CPS, multiplexed digital transmission are used at the T1 rate to provides for most of the popular services, such as voice, data, and videoconferencing, but specifically excludes standard television signals. Modern customer premises services have operated in the FDMA/TDM mode of operation mentioned in the preceding section, and this mode requires the use of *signal-processing transponders or regenerative digital repeaters* (see Sec. 12.12) in which the FDMA uplink signals are converted to the TDM format for retransmission on the downlink. It also should be noted that the use of signal processing transponders "decouples" the uplink from the downlink. This is important because it allows the performance of each link to be optimized independently of the other, and two signal-processing methods are presented next.

The first approach is illustrated in the simplified block schematic of Fig. 14.25*a*, where the individual uplink carriers (at the satellite) are selected by frequency filters and detected in the normal manner, i.e., PSK demodulation. The baseband signals are then combined in the baseband processor, where they are converted to a time-division-multiplexed format for remodulation onto a downlink carrier. More than one downlink carrier may be provided, but only one is shown for simplicity. The disadvantages of this approach are those of excessive size, weight, and power consumption, because the circuitry must be duplicated for each input carrier.

However, the disadvantages associated with processing each carrier separately can be avoided by means of *group processing*, in which the input FDMA signals are demultiplexed as a group in a single processing circuit, as illustrated in Fig. 14.25*b*. Initial feasibility studies, conducted in the early 1990s, found that technology was not available to build spaceborne digital-type group processors using *very high speed integrated circuits* (VHSICs). However, the use of an analog *surface acoustic wave* (SAW) Fourier transformer were feasible. (Atzeni et al., 1975; Hays et al., 1975; Hays and Hartmann, 1976; Maines and Paige, 1976; Nud and Otto, 1975). In today's technology both approaches are quite feasible.

In its basic form, the SAW device is a transformer of electromagnetic energy into a surface acoustic wave of a piezoelectric dielectric substrate. Because the propagation velocity of the acoustic wave is much lower than that of an electromagnetic wave, the SAW device exhibits useful delay characteristics. In addition, the electrodes are readily shaped to provide a wide range of transfer characteristics, which makes it a powerful signal-processing component. SAW devices are used extensively in today's cell phones and other passive RF electronics components such as delay lines, bandpass filters, and

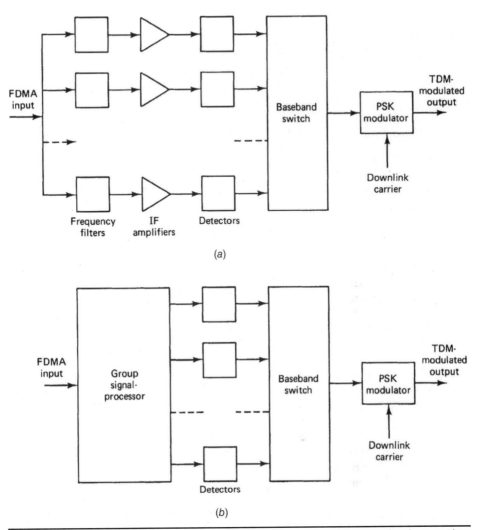

(a)

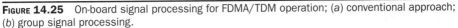

(b)

Figure 14.25 On-board signal processing for FDMA/TDM operation; (a) conventional approach; (b) group signal processing.

the key component in a unit known as a Fourier transformer. The unique property of the Fourier transformer is that the output signal is a time analog of the frequency spectrum of the input signal. When the input is a group of FDMA carriers, the output is an analog of the FDMA frequency spectrum. This allows the FDMA signals to be demultiplexed in real time by means of a commutator switch, which eliminates the need for the separate frequency filters required in the conventional analog approach. Once the signals have been separated in this way, the original modulated-carrier waveforms are recovered using SAW inverse Fourier transformers.

In a practical transformer, continuous operation can only be achieved by repetitive cycling of the transformation process. As a result, the output is periodic, and the observation interval is chosen to correspond to the desired spectral interval. Also, the periodic

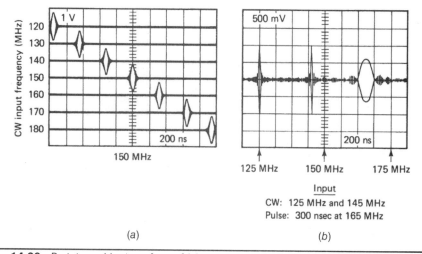

FIGURE 14.26   Prototype chirp transform of (a) seven successive CW input signals and (b) three simultaneous RF input signals, namely: two CW and a chirp pulse. (*Courtesy of Hays and Hartmann, 1976. Copyright 1976, IEEE.*)

interval over which transformation takes place results in a broadening of the output "pulses," which represent the FDMA spectra.

So, repetitive operation of the Fourier transformer at a rate equal to the data bit rate produces a suitable repetitive output. The relative positions of the output pulses remain unchanged, as they are fixed by the frequencies of the FDMA carriers, and the PSK modulation on the individual FDMA carriers appears in the phase of the carriers within each output pulse. Thus, the FDMA carriers are converted to a pulsed TDM signal, but further signal processing is required before this can be retransmitted as a TDM signal.

Figure 14.26 shows the output obtained from a practical Fourier transformer for various RF input signals. For Fig. 14.26a the input is seven *continuous-wave* (CW) signals applied in succession that produces an output of pulses corresponding to the line spectra for these waves.

The broadening of the lines is a result of the finite time gate over which the Fourier transformer operates. It is important to note that the horizontal axis in Fig. 14.26 is a time axis on which the equivalent frequency points are indicated. Figure 14.26b shows the output obtained with three simultaneous inputs, two CW waves and one chirped-pulsed carrier wave. Again, the output contains two pulses corresponding to the CW signals and a time function which has the shape of the spectrum for the pulsed wave (Hays and Hartmann, 1976). A detailed account of SAW devices is found in Morgan (1985) and in the IEEE Proceedings (1976).

14.9   Satellite-Switched TDMA

More efficient utilization of satellites in the geostationary orbit is achieved using antenna spot beams. The use of spot beams is also referred to as *space-division multiplexing*. Further improvements can be realized by switching the antenna interconnections in synchronism with the TDMA frame rate, this being known as *satellite-switched* TDMA (SS/TDMA).

Figure 14.27a shows in simplified form the SS/TDMA concept (Scarcella and Abbott, 1983). Three antenna beams are used, with each beam serving two earth stations, and a

3×3 satellite switch matrix permits the antenna interconnections to be made on a switched basis. A *switch mode* connectivity arrangement is shown in Fig. 14.27b. For three antenna beams, six modes are required, but for N beams, $N!$ modes are required for full interconnectivity. Full interconnectivity means that the signals carried in each beam are transferred to each of the other beams at some time in the switching sequence. This includes the loopback connection, where signals are returned along the same beam, enabling intercommunications between stations within a beam. Of course, the uplink and downlink microwave frequencies are different.

Because of high-isolation between beams, the same frequency plan for uplinks and downlinks can be used (e.g., 14 and 12 GHz in the Ku band). To simplify the satellite switch

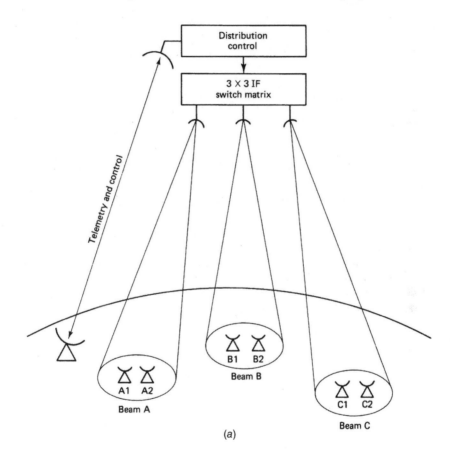

	Output					
Input	Mode 1	Mode 2	Mode 3	Mode 4	Mode 5	Mode 6
A	A	A	B	C	B	C
B	B	C	A	A	C	B
C	C	B	C	B	A	A

(b)

Figure 14.27   (a) Satellite switching of three spot beams; (b) connectivities or modes.

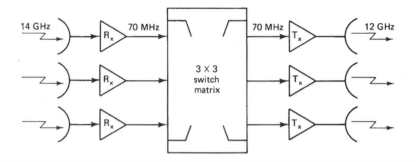

Figure 14.28   Switch matrix in the R.F. link.

design, the switching is carried out at the intermediate frequency that is common to uplinks and downlinks. The simplified block schematic for the 3×3 system is shown in Fig. 14.28.

A mode *pattern* is a repetitive sequence of satellite switch modes, also referred to as SS/TDMA frames. Successive SS/TDMA frames need not be identical, because there is some redundancy between modes. For example, in Fig. 14.27b, beam *A* interconnects with beam *B* in modes 3 and 5, and thus not all modes need be transmitted during each SS/TDMA frame. However, for full interconnectivity, the *mode pattern* must contain all modes.

All stations within a beam receive all the TDM frames transmitted in the downlink beam. Each frame is a normal TDMA frame consisting of bursts, addressed to different stations in general. As mentioned, successive frames may originate from different transmitting stations and therefore have different burst formats. The receiving station in a beam recovers the bursts addressed to it in each frame.

The two basic types of switch matrix are the *crossbar matrix* and the *rearrangeable network*. The crossbar matrix is easily configured for the *broadcast mode*, in which one station transmits to all stations. The broadcast mode with the rearrangeable network-type switch is more complex, and this can be a deciding factor in favor of the crossbar matrix (Watt, 1986). The schematic for a 3×3 crossbar matrix is shown in Fig. 14.29, which also shows input beam *B* connected in the broadcast mode.

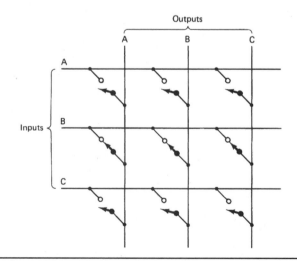

Figure 14.29 3×3 crossbar matrix switch, showing input *B* connected in the broadcast mode.

The switching elements may be ferrites, pin diodes, or transistors. The dual-gate FET appears to offer significant advantages over the other types and is considered by some to be the most promising technology (Scarcella and Abbott, 1983).

Figure 14.30 shows how a 3×3 matrix switch may be used to reroute traffic. Each of the ground stations U, V, and W accesses a separate antenna on the satellite and carries traffic destined for the downlink beams X, Y, and Z. The switch is controlled

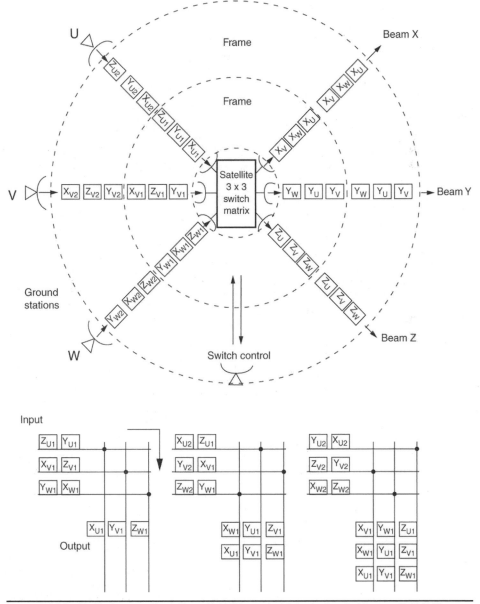

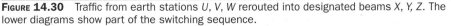

FIGURE 14.30 Traffic from earth stations U, V, W rerouted into designated beams X, Y, Z. The lower diagrams show part of the switching sequence.

from a ground control station, and the switching sequence for the frame labeled with subscript 1 is shown in the lower part of the figure.

The schematic for a 4×4 matrix switch as used on the European Olympus satellite is shown in Fig. 14.31 (Watt, 1986). This arrangement is derived from the crossbar matrix.

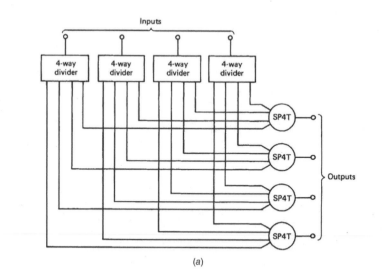

(a)

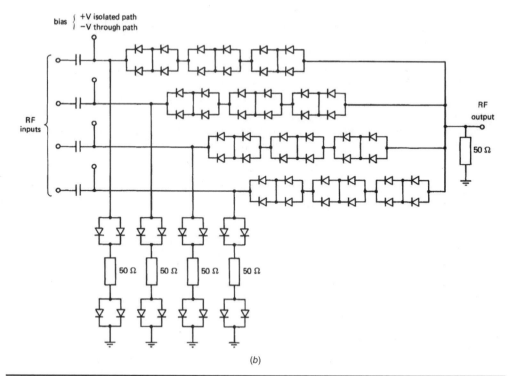

(b)

Figure 14.31 (a) 4×4 switch matrix; (b) circuit diagram of redundant SP4T switch element. (*Courtesy of Watt, 1986; reprinted with permission of IEE, London.*)

It permits broadcast mode operation, but does not allow more than one input to be connected to one output. Diodes are used as switching elements, and as shown, diode quads are used which provide redundancy against diode failure. It is recognized that satellite-switched TDMA adds to the complexity of the on-board equipment and to the synchronization requirements.

Use of multiple antenna beams can also be used for *space-division multiple access* (SDMA). Both beam switching, and the use of phased adaptive arrays (see Sec. 6.19) have been studied for mobile applications. An analysis and comparison of antenna beam switching and phased adaptive arrays is found in Zaharov (2001).

14.10 Code-Division Multiple Access

With CDMA the individual carriers may be present simultaneously within the same RF bandwidth, but each carrier carries a unique code waveform (in addition to the information signal) that allows it to be separated from all the others at the receiver. The carrier is modulated in the normal way by the information waveform and then is further modulated by the code waveform to spread the spectrum over the available RF bandwidth. Many of the key properties of CDMA rely on this spectrum spreading, and the systems employing CDMA are also known as *spread-spectrum multiple access* (SSMA). Care must be taken not to confuse the SS here with that for satellite switched (SS/TDMA) used in the Sec. 14.9.

While CDMA can be used with analog and digital signals (see Dixon, 1984), current satellite usage is predominantly digital. For illustration purposes, a polar *non-return-to-zero* (NRZ) waveform denoted by $p(t)$ (see Fig. 10.2) is used for the information signal, and BPSK modulation (see Sec. 10.6.1) is assumed. The code waveform $c(t)$ is also a polar NRZ signal, as sketched in Fig. 14.32. What is called *bits* in an information waveform are called *chips* for the code waveform, and the chip rate is usually much greater than the information bit rate.

The pulses (chips) in the code waveform vary randomly between *logic-1* and *logic-0*, and the randomness is an essential feature of spread-spectrum systems, and more is said about this shortly. The code signal may be applied as modulation in the same way as the information signal so that the BPSK signal carries both the information signal $p(t)$ and the code signal $c(t)$. This method is referred to as *direct-sequence spread spectrum* (DS/SS).

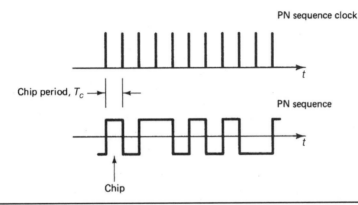

FIGURE 14.32   PN binary sequence. One element is known as a chip.

Other techniques are also used to spread the spectrum, such as frequency hopping, but for satellite communications, the approach is the DS/SS method.

14.10.1 Direct-Sequence Spread Spectrum

In Fig. 14.33, $p(t)$ is an NRZ binary information signal, and $c(t)$ is a NRZ binary code signal. These two signals form the inputs to a multiplier (balanced modulator), the output of which is proportional to the product $p(t)c(t)$. This product signal is applied to a second balanced modulator, the output of which is a BPSK signal at the carrier frequency. For clarity, it is assumed that the carrier is the uplink frequency, and hence the uplink carrier is described by

$$e_U(t) = c(t)p(t)\cos\omega_U t \tag{14.33}$$

The corresponding downlink carrier is

$$e_D(t) = c(t)p(t)\cos\omega_D t \tag{14.34}$$

At the receiver, an identical $c(t)$ generator is synchronized to the $c(t)$ of the downlink carrier. This synchronization is carried out in the *acquisition and tracking block*. With $c(t)$ a polar NRZ type waveform, and with the locally generated $c(t)$ exactly in synchronism with the transmitted $c(t)$, the product $c^2(t) = 1$. Thus the output from the multiplier is

$$c(t)e_D(t) = c^2(t)p(t)\cos\omega_D t = p(t)\cos\omega_D t \tag{14.35}$$

This is identical to the conventional BPSK signal given by Eq. (10.20), and hence detection proceeds in the normal manner.

14.10.2 The Code Signal c(t)

The code signal $c(t)$ carries a binary code that has special properties needed for successful implementation of CDMA. The binary symbols used in the codes are referred to as *chips* rather than *bits* to avoid confusion with the information bits also present. Chip generation is controlled by a clock, and the chip rate, in chips per second, is given by the clock speed. Denoting the clock speed by R_{ch}, the chip period is the reciprocal of the clock speed:

$$T_{ch} = \frac{1}{R_{ch}} \tag{14.36}$$

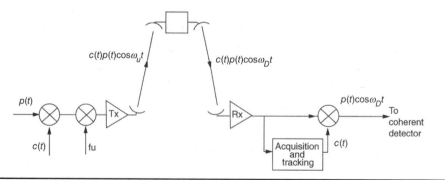

FIGURE 14.33 A basic CDMA system.

The waveform $c(t)$ is periodic, in that each period is a repetition of a given sequence of N chips. The sequence itself exhibits random properties, which are described shortly. The periodic time for the waveform is

$$T_N = NT_{ch} \tag{14.37}$$

The codes are generated using binary shift registers and associated linear logic circuits. The circuit for a three-stage shift register that generates a sequence of $N = 7$ chips is shown in Fig. 14.34a. Feedback occurs from stages 1 and 3 as inputs to the exclusive OR gate. This provides the input to the shift register, and the chips are clocked through at the clock rate R_{ch}. The generator starts with all stages holding binary 1s, and the following states are as shown in the table in Fig. 14.34. Stage 3 also provides the binary output sequence. The code waveform generated from this code is shown in Fig. 14.34b.

Such codes are known as *maximal sequence* or *m-sequence* codes because they utilize the maximum length sequence that can be generated. For Fig. 14.34a the maximum length sequence is 7 chips as shown. In general, the shift register passes through all states (all combinations of 1s and 0s in the register) except the all-zero state when generating a maximal sequence code. Therefore, a code generator employing an *n*-stage shift register can generate a maximum sequence of N chips, where

$$N = 2^n - 1 \tag{14.38}$$

The binary 1s and 0s are randomly distributed such that the code exhibits noise-like properties. However, there are certain deterministic features described below, and the codes are more generally known as *PN codes*, which stands for *pseudo-noise codes* (or *PRN pseudo-random noise codes*).

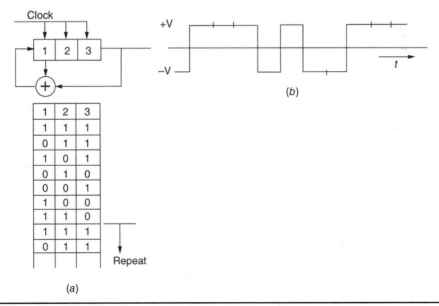

(b)

1	2	3
1	1	1
0	1	1
1	0	1
0	1	0
0	0	1
1	0	0
1	1	0
1	1	1
0	1	1

Repeat

(a)

FIGURE 14.34 Generation of a 7-chip maximal sequence code.

1. The number of binary 1s is given by

$$\text{No. of 1s} = \frac{2^n}{2} \tag{14.39}$$

and the number of binary 0s is given by

$$\text{No. of 0s} = \frac{2^n}{2} - 1 \tag{14.40}$$

The importance of this relationship is that when the code uses $+V$ volts for a binary 1 and $-V$ volts for a binary 0, the dc offset is close to zero. Since there is always one more positive chip than negative, the dc offset is given by

$$\text{dc offset} = \frac{V}{N} \tag{14.41}$$

The dc offset determines the carrier level relative to the peak value; that is, the carrier is suppressed by amount $1/N$ for BPSK. For example, using a code with $n = 8$ with BPSK modulation, the carrier is suppressed by $1/255$ or -48 dB.

2. The total number of maximal sequences that can be generated by an n-stage shift register (and its associated logic circuits) is given by

$$S_{\text{max}} = \frac{\phi(N)}{n} \tag{14.42}$$

Here, $\phi(N)$ is known as *Euler's* ϕ-*function*, which gives the number of integers in the range 1, 2, 3 . . ., $N - 1$, that are relatively prime to N [N is given by Eq. (14.38)]. Two numbers are relatively prime when their greatest common divisor is 1. A general formula for finding $\phi(N)$ is (see Ore, 1988)

$$\phi(N) = N\left(\frac{p_1 - 1}{p_1}\right)\cdots\left(\frac{p_r - 1}{p_r}\right) \tag{14.43}$$

where $p_1 \ldots p_r$ are the prime factors of N. For example, for $n = 8$, $N = 2^8 - 1 = 255$. The prime factors of 255 are 3, 5, and 17, and hence

$$\phi(255) = 255\left(\frac{2}{3}\right)\left(\frac{4}{5}\right)\left(\frac{16}{17}\right) = 128$$

The total number of maximal sequences that can be generated by an eight-stage code generator is therefore

$$S_{\text{max}} = \frac{128}{8} = 16$$

As a somewhat simpler example, consider the case when $n = 3$. In this instance, $N = 7$. There is only one prime factor, 7 itself, and therefore

$$\varphi(7) = 7 \cdot \frac{6}{7} = 6$$

and

$$S_{\text{max}} = \frac{\varphi(7)}{3} = 2$$

In this case there are only two distinct maximal sequences.

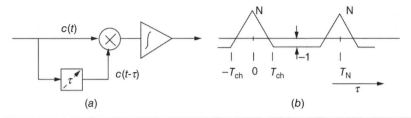

FIGURE 14.35 (a) Generating the autocorrelation function; (b) the autocorrelation waveform.

One of the most important properties of $c(t)$ is its *autocorrelation function*. The autocorrelation function is a measure of how well a time-shifted version of the waveform compares with the unshifted version. Figure 14.35a shows how the comparison may be made. The $c(t)$ waveform is multiplied with a shifted version of itself, $c(t - \tau)$, and the output is averaged (shown by the integrator). The average, of course, is independent of time t (the integrator integrates out the time-t dependence), but it depends on the time lead or lag introduced by τ. When the waveforms are coincident, $\tau = 0$, and the average output is a maximum, which for convenience are normalized to 1. Any shift in time, advance or delay, away from the $\tau = 0$ position results in a decrease in output voltage. A property of m-sequence code waveforms is that the autocorrelation function decreases linearly from the maximum value (unity in this case) to a negative level $1/N$, as shown in Fig. 14.35b. The very pronounced peak in the autocorrelation function provides the chief means for acquiring acquisition and tracking so that the locally generated m-sequence code can be synchronized with the transmitted version.

14.10.3   Acquisition and Tracking

One form of acquisition circuit that makes use of the autocorrelation function is shown in Fig. 14.36. The output from the first multiplier is

$$e(t) = c(t - \tau)c(t)p(t) \cos \omega_D t$$
$$= c(t - \tau)c(t) \cos [\omega_D t + \varphi(t)] \tag{14.44}$$

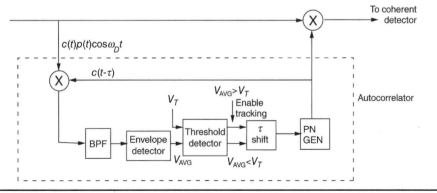

FIGURE 14.36 Acquisition of a carrier in a CDMA system.

Here, the information modulation, which is BPSK, is shown as $\varphi(t)$ so that the effect of the following *bandpass filter* (BPF) on the amplitude can be more clearly seen. The BPF has a passband centered on ω_D, wide with respect to the information modulation but narrow with respect to the code signal. It performs the amplitude-averaging function on the code signal product (see Maral and Bousquet, 1998). The averaging process can be illustrated as follows. Consider the product of two cosine terms and its expansion:

$$\cos\omega t\cos(\omega t-\delta)=\frac{1}{2}\{\cos[\omega t+(\omega t-\delta)]+\cos[\omega t-(\omega t-\delta)]\}$$

$$=\frac{1}{2}[\cos(2\omega t-\delta)+\cos\delta] \tag{14.45}$$

The BPF rejects the high-frequency component, leaving only the average component $0.5\cos(\delta)$. This signal is analogous to the $c(t)c(t-\tau)$ term in Eq. (14.44). The envelope detector following the BPF produces an output proportional to the envelope of the signal, that is, to the average value of $c(t)c(t-\tau)$. This is a direct measure of the autocorrelation function. When it is less than the predetermined threshold V_T required for synchronism, the time shift τ is incremented. Once the threshold has been reached or exceeded, the system switches from acquisition mode to tracking mode.

One form of tracking circuit, the *delay lock loop*, is shown in Fig. 14.37. Here, two correlators are used, but the local signal to one is advanced by half a chip period relative

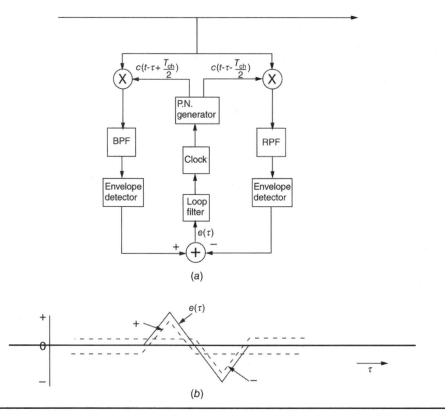

(a)

(b)

Figure 14.37   (a) The delay lock loop; (b) the waveform at the adder.

to the desired code waveform, and the other is delayed by the same amount. The outputs from the correlators are subtracted, and this difference signal provides the control voltage for the VCO that drives the shift register clock. With the control voltage at the zero crossover point, the locally generated code signal is in phase with the received code signal. Any tendency to drift out of phase changes the VCO in such a way as to bring the control voltage back to the zero crossover point, thus maintaining synchronism.

The acquisition and tracking circuits also attempt to correlate the stored version of $c(t)$ at the receiver with all the other waveforms being received. Such correlations are termed *cross-correlations*. It is essential that the cross-correlation function does not show a similar peak as the autocorrelation, and this requires careful selection of the spreading functions used in the overall system (see, for example, Dixon, 1984).

14.10.4 Spectrum Spreading and Despreading

In Sec. 10.6.3 the idea of bandwidth for PSK modulation is introduced. In general, for a BPSK signal at a bit rate R_b, the main lobe of the power-density spectrum occupies a bandwidth extending from $f_c - R_b$ to $f_c + R_b$. This is sketched in Fig. 14.38a. A similar result applies when the modulation signal is $c(t)$, the power-density spectrum being as sketched in Fig. 14.38b. Because $c(t)$ exhibits periodicity, the spectrum density is ideally a line function, and Fig. 14.38b shows the envelope of the spectrum. The spectrum shows the *power density* (watts per Hertz) in the signal. For constant carrier power, it follows that if a signal is forced to occupy a wider bandwidth, its spectrum density is reduced. This is a key result in CDMA systems. In all direct-sequence spread-spectrum systems, the chip rate is very much greater than the information bit rate, or $R_{ch} \gg R_b$. The bandwidth is determined mainly by R_{ch} so that the power density of the signal described by Eq. (14.34) is spread over the bandwidth determined by R_{ch}. The power density is reduced by approximately the ratio of R_{ch} to R_b.

Assuming that acquisition and tracking have been accomplished, $c(t)$ in the receiver (Fig. 14.33) performs in effect a *despreading function* that it restores the spectrum of the wanted signal to what it was before the spreading operation in the transmitter. This is

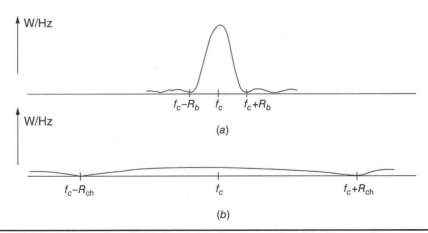

FIGURE 14.38 Spectrum for a BPSK signal: (*a*) without spreading, (*b*) with spreading.

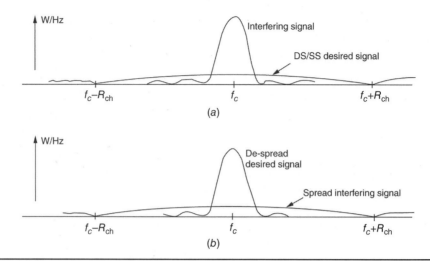

FIGURE 14.39   (a) Spectrum of an interfering, nonspread signal along with the spread desired signal; (b) the effect of the despreading operation on the desired signal resulting in spread of the interferer.

also how the spread-spectrum technique can reduce interference. Figure 14.39a shows the spectra of two signals, an interfering signal that is not part of the CDMA system and that has not been spread, and the desired DS/SS received signal. Following the despreading operation for the desired signal, its spectrum is restored as described previously. The interfering signal, however, is simply multiplied by the $c(t)$ signal, which results in it being spread.

14.10.5   CDMA Throughput

The maximum number of channels in a CDMA system can be estimated as follows: It is assumed that the thermal noise is negligible compared with the noise resulting from the overlapping channels, and for comparison purposes, it is assumed that each channel introduces equal power P_R into the receiver. For a total of K channels, $K - 1$ of these produce noise, and assuming that this is evenly spread over the noise bandwidth B_N of the receiver, the noise density, in W/Hz, is

$$N_0 = \frac{(K-1)P_R}{B_N}$$

(14.46)

Let the information rate of the wanted channel to be R_b; then, from Eq. (10.30),

$$E_b = \frac{P_R}{R_b}$$

(14.47)

Hence the bit energy to noise density ratio is

$$\frac{E_b}{N_0} = \frac{B_N}{(K-1)R_b}$$

(14.48)

The noise bandwidth at the BPSK detector is approximately equal to the IF bandwidth as given by Eq. (10.21), but using the chip rate

$$B_N \cong B_{IF} = (1 + \rho)R_{ch} \qquad (14.49)$$

where ρ is the rolloff factor of the filter. Hence the bit energy to noise density ratio becomes

$$\frac{E_b}{N_0} = \frac{(1+\rho)R_{ch}}{(K-1)R_b} \qquad (14.50)$$

As pointed out in Chap. 10, the probability of bit error is usually a specified objective, and this determines the E_b/N_0 ratio, for example, through Fig. 10.23. The number of channels is therefore

$$K = 1 + (1 + \rho)\frac{R_{ch}N_0}{R_b E_b} \qquad (14.51)$$

The processing *gain* G_p is basically the ratio of power density in the unspread signal to that in the spread signal. Since the power density is inversely proportional to bandwidth, an approximate expression for the processing gain is

$$G_p = \frac{R_{ch}}{R_b} \qquad (14.52)$$

Hence

$$K = 1 + (1 + \rho)G_p\frac{N_0}{E_b} \qquad (14.53)$$

Example 14.11   The code waveform in a CDMA system spreads the carriers over the full 36 MHz bandwidth of the channel, and the rolloff factor for the filtering is 0.4. The information bit rate is 64 kb/s, and the system uses BPSK. Calculate the processing gain in decibels. Given that the BER must not exceed 10^{-5}, give an estimate of the maximum number of channels that can access the system.

Solution

$$R_{ch} = \frac{B_{IF}}{1+\rho} = \frac{36e+06}{1.4} = 25.7e+06 \text{ chips/s}$$

Hence the processing gain is

$$G_p = \frac{25.7e+06}{64e+03} = 401.56$$

From Fig. 10.24 for $P_e = 10^{-5}$, $[E_b/N_0] = 9.6$ dB approximately. This is a power ratio of 9.12, and from Eq. (14.53),

$$K = 1 + \frac{1.4 \times 401.56}{9.12} = \text{Integer}\,(61.6) = 61$$

The *throughput efficiency* is defined as the ratio of the total number of bits per unit time that can be transmitted with CDMA to the total number of bits per unit time that can be transmitted with single access and no spreading. For K accesses as determined earlier, each at bit rate R_b, the total bits per unit time is KR_b. A single access can utilize the full bandwidth, and hence its transmission rate as determined by Eq. (10.21) is

$$R_T = \frac{B_{IF}}{1+\rho} \qquad (14.54)$$

This is the same as the chip rate, and hence the throughput is

$$\eta = \frac{KR_b}{R_T}$$

$$= \frac{KR_b}{R_{ch}} \qquad (14.55)$$

$$= \frac{K}{G_p}$$

Using the values obtained in Example 14.11 gives a throughput of 0.15, or 15 percent. This should be compared with the frame efficiency for TDMA (see Example 14.6), where it is seen that the throughput efficiency can exceed 90 percent.

CDMA offers several advantages for satellite networking, especially where VSAT-type terminals are involved. These are:

1. The beamwidth for VSAT antennas is comparatively broad and therefore can be subject to interference from adjacent satellites. The interference rejection properties of CDMA through spreading are of considerable help here.
2. Multipath interference, for example, that resulting from reflections, can be avoided provided the time delay of the reflected signal is greater than a chip period and the receiver locks onto the direct wave.
3. Synchronization between stations in the system is not required (unlike TDMA, where synchronization is a critical feature of the system). This means that a station can access the system at any time.
4. Degradation of the system (reduction in E_b/N_0) is gradual with an increase in number of users. Thus additional traffic can be accommodated if some reduction in performance is acceptable. The main disadvantage is the low throughput efficiency.

14.11   Problems and Exercises

14.1. Explain what is meant by a *single access* in relation to a satellite communications network. Give an example of the type of traffic route where single access is used.

14.2. Distinguish between *preassigned* and *demand-assigned traffic* in relation to a satellite communications network.

14.3. Explain what is meant by FDMA, and show how this differs from FDM.

14.4. Explain what the abbreviation SCPC stands for. Explain in detail the operation of a preassigned SCPC network.

14.5. Explain what is meant by *thin route service*. What type of satellite access is most suited for this type of service?

14.6. Briefly describe the ways in which demand assignment may be carried out in an FDMA network.

14.7. Explain in detail the operation of the Spade system of demand assignment. What is the function of the common signaling channel?

14.8. Explain what is meant by *power-limited* and *bandwidth-limited operation* as applied to an FDMA network. In an FDMA scheme the carriers utilize equal powers and equal bandwidths, the bandwidth in each case being 5 MHz. The transponder bandwidth is 36 MHz. The saturation EIRP for the downlink is 34 dBW, and an output backoff of −6 dB is employed. The downlink losses are −201 dB, and the destination earth station has a G/T ratio of 35 dBK^{-1}. Determine the

[C/N] value assuming this is set by single carrier operation. Also determine the number of carriers which can access the system, and state, with reasons, whether the system is power limited or bandwidth limited.

14.9. A satellite transponder has a saturation EIRP of 25 dBW and a bandwidth of 27 MHz. The transponder resources are shared equally by several FDMA carriers, each of bandwidth 3 MHz, and each requiring a minimum EIRP of 12 dBW. If –7 dB output backoff is required, determine the number of carriers that can be accommodated.

14.10. In some situations it is convenient to work in terms of the carrier-to-noise temperature. Show that $[C/T] = [C/N_0] + [k]$. The downlink losses for a satellite circuit are –196 dB. The earth station $[G/T]$ ratio is 35 dB/K, and the received $[C/T]$ ratio is –138 dBW/K. Calculate the satellite [EIRP].

14.11. The earth-station receiver in a satellite downlink has an FM detector threshold level of 10 dB and operates with a 3 dB threshold margin. The emphasis improvement figure is 4 dB, and the noise-weighting improvement figure is 2.5 dB. The required $[S/N]$ at the receiver output is 46 dB. Calculate the receiver processing gain. Explain how the processing gain determines the IF bandwidth.

14.12. A 252-channel FM/FDM telephony carrier is transmitted on the downlink specified in Prob. 14.10. The peak/rms ratio factor is 10 dB, and the baseband bandwidth extends from 12 to 1052 kHz. The voice-channel bandwidth is 3.1 kHz. Calculate the peak deviation, and hence, using Carson's rule, calculate the IF bandwidth.

14.13. Given that the IF bandwidth for a 252-channel FM/FDM telephony carrier is 7.52 MHz and that the required $[C/N]$ ratio at the earth-station receiver is 13 dB, calculate (a) the $[C/T]$ ratio and (b) the satellite [EIRP] required if the total losses amount to –200 dB and the earth-station $[G/T]$ ratio is 37.5 dB/K.

14.14. Determine how many carriers can access an 80-MHz transponder in the FDMA mode, given that each carrier requires a bandwidth of 6 MHz, allowing for –6.5 dB output backoff. Compare this number with the number of carriers possible without backoff.

14.15. (a) Analog television transmissions may be classified as *full-transponder* or *half-transponder transmissions*. State what this means in terms of transponder access. (b) A composite TV signal (video plus audio) has a top baseband frequency of 6.8 MHz. Determine for a 36-MHz transponder the peak frequency deviation limit set by (1) half-transponder and (2) full transponder transmission.

14.16. Describe the general operating principles of a TDMA *network*. Show how the transmission bit rate is related to the input bit rate.

14.17. Explain the need for a reference burst in a TDMA system.

14.18. Explain the function of the preamble in a TDMA traffic burst. Describe and compare the channels carried in a preamble with those carried in a reference burst.

14.19. What is the function of (a) the burst-code word and (b) the carrier and bit-timing recovery channel in a TDMA burst?

14.20. Explain what is meant by (a) *initial acquisition* and (b) *burst synchronization* in a TDMA network. (c) The nominal range to a geostationary satellite is 42,000 km. Using the station-keeping tolerances stated in Sec. 7.4 in connection with Fig. 7.10, determine the variation expected in the propagation delay.

14.21. (*a*) Define and explain what is meant by *frame efficiency* in relation to TDMA operation. (*b*) In a TDMA network the reference burst and the preamble each requires 560 bits, and the nominal guard interval between bursts is equivalent to 120 bits. Given that there are eight traffic bursts and one reference burst per frame and the total frame length is equivalent to 40,800 bits, calculate the frame efficiency.

14.22. Given that the frame period is 2 ms and the voice-channel bit rate is 64 kb/s, calculate the equivalent number of voice channels that can be carried by the TDMA network specified in Prob. 14.21.

14.23. Calculate the frame efficiency for the CSC shown in Fig. 14.19.

14.24. (*a*) Explain why the frame period in a TDMA system is normally chosen to be an integer multiple of 125 μs. (*b*) Referring to Fig. 14.20 for the INTELSAT preassigned frame format, show that there is no break in the timing interval for sample 18 when this is transferred to a burst.

14.25. Show that, all other factors being equal, the ratio of uplink power to bit rate is the same for FDMA and TDMA. In a TDMA system the preamble consists of the following slots, assigned in terms of number of bits: bit timing recovery 304; unique word 48; station identification channel 8; order wire 64. The guard slot is 120 bits, the frame reference burst is identical to the preamble, and the burst traffic is 8192 bits. Given that the frame accommodates 8 traffic bursts, calculate the frame efficiency. The traffic is preassigned PCM voice channels for which the bit rate is 64 kb/s, and the satellite transmission rate is nominally 60 Mbps. Calculate the number of voice channels which can be carried.

14.26. In comparing design proposals for multiple access, the two following possibilities were considered: (1) uplink FDMA with downlink TDM and (2) uplink TDMA with downlink TDM. The incoming baseband signal is at 1.544 Mbps in each case and the following table shows values in dB:

	Uplink	Downlink
$[E_b/N_0]$	12	12
$[G/T]$	10	19.5
[LOSSES]	−212	−210
[EIRP]	—	48
Transmit Antenna Gain $[G_T]$	45.8	—

Determine (*a*) the downlink TDM bit rate and (*b*) the transmit power required at the uplink earth station for each proposal.

14.27. A TDMA network utilizes QPSK modulation and has the following symbol allocations: guard slot 32; carrier and bit timing recovery 180; burst code word (unique word) 24; station identification channel 8; order wire 32; management channel (reference bursts only) 12; service channel (traffic bursts only) 8. The total number of traffic symbols per frame is 115,010, and a frame consists of two reference bursts and 14 traffic bursts. The frame period is 2 ms. The input consists of PCM channels each with a bit rate of 64 kb/s. Calculate the frame efficiency and the number of voice channels that can be accommodated.

14.28. For the network specified in Prob. 14.27 the BER must be at most 10^{-5}. Given that the receiving earth-station $[G/T]$ value is 30 dB and total losses are −200 dB, calculate the satellite [EIRP] required.

14.29. Discuss briefly how demand assignment may be implemented in a TDMA network. What is the advantage of TDMA over FDMA in this respect?

14.30. Define and explain what is meant by the terms *telephone load activity factor* and *digital speech interpolation*. How is advantage taken of the load activity factor in implementing digital speech interpolation?

14.31. Define and explain the terms *connect clip* and *freeze-out* used in connection with digital speech interpolation.

14.32. Describe the principles of operation of a SPEC system, and state how this compares with digital speech interpolation.

14.33. Determine the bit rate that can be transmitted through a 36-MHz transponder, assuming a rolloff factor of 0.2 and QPSK modulation.

14.34. On a satellite downlink, the $[C/N_0]$ ratio is 86 dBHz and an $[E_b/N_0]$ of 12 dB is required at the earth station. Calculate the maximum bit rate that can be transmitted.

14.35. FDMA is used for uplink access in a satellite digital network, with each earth station transmitting at the T1 bit rate of 1.544 Mbps. Calculate (*a*) the uplink $[C/N_0]$ ratio required to provide a $[E_b/N_0] = 14$ dB ratio at the satellite and (*b*) the earth station [EIRP] needed to realize the $[C/N_0]$ value. The satellite $[G/T]$ value is 8 dB/K, and total uplink losses amount to –210 dB.

14.36. In the satellite network of Prob. 14.35, the downlink bit rate is limited to a maximum of 74.1 dB, with the satellite TWT operating at saturation. A –5 dB output backoff is required to reduce intermodulation products to an acceptable level. Calculate the number of earth stations that can access the satellite on the uplink.

14.37. The [EIRP] of each earth station in an FDMA network is 47 dBW, and the input data are at the T1 bit rate with 7/8 FEC added. The downlink bit rate is limited to a maximum of 60 Mbps with –6 dB output backoff applied. Compare the [EIRP] needed for the earth stations in a TDMA network utilizing the same transponder.

14.38. (*a*) Describe the general features of an on-board signal processing transponder that allows a network to operate with FDMA uplinks and a TDMA downlink. (*b*) In such a network, the overall BER must not exceed 10^{-5}. Calculate the maximum permissible BER of each link, if each link contributes equally to the overall value.

14.39. Explain what is meant by *full interconnectivity* in connection with satellite switched TDMA. With four beams, how many switch modes are required for full interconnectivity?

14.40. Identify all the redundant modes in Fig. 14.27.

14.41. The shift register in an *m*-sequence generator has seven stages. Calculate the number of binary 1s and 0s. The code is used to generate a NRZ polar waveform at levels +1 V and –1 V. Calculate the dc offset and the carrier suppression in decibels that can be achieved when BPSK is used.

14.42. The shift register in an *m*-sequence generator has 10 stages. Calculate the length of the *m*-sequences. Determine the prime factors for N and, hence, the total number of maximal length sequences that can be produced.

14.43. As shown in Sec. 14.10.2, an *m*-sequence generator having a three-stage shift register can generate a total of two maximal sequences, and Fig. 14.34 shows one of these. Draw the corresponding circuit for the other sequence, and the waveform.

14.44. Draw accurately to scale the autocorrelation function over one complete cycle for the waveform shown in Fig. 14.34. Assume $V = 1$ V and $T_{ch} = 1$ ms.

14.45. Draw accurately to scale the autocorrelation function over one complete cycle for the waveform determined in Prob. 14.43. Assume $V = 1$ V, and $T_{ch} = 1$ ms.

14.46. Describe in your own words how signal acquisition and tracking are achieved in a DS/SS system.

14.47. An m-sequence generator having a three-stage shift register can generate a total of two maximal sequences. Neatly sketch the cross-correlation function for the two m-sequences.

14.48. Explain the principle behind spectrum spreading and despreading and how this is used to minimize interference in a CDMA system.

14.49. The IF bandwidth for a CDMA system is 3 MHz, the rolloff factor for the filter being 1. The information bit rate is 2.4 kb/s, and an $[E_b/N_0]$ of 11 dB is required for each channel accessing the CDMA system. Calculate the maximum number of accesses permitted.

14.50. Determine the throughput efficiency for the system in Prob. 14.49.

14.51. Show that when K is large such that the first term, unity, on the right hand side of Eq. (14.53) can be neglected, the throughput efficiency is independent of the processing gain. Hence, plot the throughput efficiency as a function of $[E_b/N_0]$ for the range 7 to 11 dB.

14.52. A military secure telephone system uses a Ku-band switched-beam satcom system from a LEOSAT with independent spot beams (see Fig. 14.27). CDMA is the multiple access method for groups of terminals within each of its multiple antenna beams. The terminals generate and receive compressed digital voice signals with a bit rate of 9.6 kbps. The signals are transmitted and received at a chip rate of 2.0 Mbps as BPSK modulated DS-CDMA. The ComSat transmits 10 simultaneous CDMA signals within each antenna beam. Assume the following specs: $(C/N)_{up} = 10.0$ dB; SATCOM transponder: EIRP $= 18.0$ dBW, path loss $= -186.8$ dB, (C/3IM) $= 6.0$ dB, $L_{sys} = -3$ dB; and Earth terminal: $G/T = 7.0$ dB. (*a*) What is the symbol rate for a digital voice signal? (*b*) Assuming RRC filters with a roll-off $= 0.4$, what is the downlink thermal noise power in dBW? (*c*) What is the downlink CDMA interference noise power in dBW? (*d*) What is the downlink CDMA interference noise power in dBW? (*e*) What is the downlink 3IM noise power in dBW? (*f*) What is the $(C/(N+I))_{tot}$ in dB at the input to the correlation receiver before despreading? (*g*) Assuming theoretical processing gain, what is the S/N for a digital voice signal after despreading? (*h*) Assuming an implementation margin of -0.8 dB in the receiver, what is the P_e in errors/bit? and (*i*) What is the BET in errors/sec for a digital voice signal?

14.53. Consider the military secure telephone system (given in Prob. 14.52). If a user is simultaneously illuminated by two spot beams (20 CDMA signals), (*a*) what is new S/N for a digital voice signal after despreading? (*b*) Assuming an implementation margin of -0.8 dB in the receiver, what is the P_e in errors/bit? and (*c*) What is the BET in errors/sec for a digital voice signal?

References

Atzeni, C., G. Manes, and L. Masotti. 1975. "Programmable Signal Processing by Analog Chirp-Transformation using SAW Devices." *Ultrasonics Symposium Proceedings*, IEEE, New York.

CCIR Report 708 (mod I). 1982. "Multiple Access and Modulation Techniques in the Fixed Satellite Service." *14th Plenary Assembly*, Geneva.

Dixon, R. C. 1984. *Spread Spectrum Systems.* Wiley, New York.

Freeman, R. L. 1981. *Telecommunications Systems Engineering.* Wiley, New York.

Gagliardi, R. M. 1991. *Satellite Communications,* 2d ed. Van Nostrand Reinhold, New York.

Ha, T. T. 1990. *Digital Satellite Communications,* 2d ed. McGraw-Hill, New York.

Hays, R. M., W. R. Shreve, D. T. Bell, Jr., L. T. Claiborne, and C. S. Hartmann. 1975. "Surface Wave Transform Adaptable Processor System." *Ultrasonics Symposium Proceedings,* IEEE, New York.

Hays, Ronald M., and C. S. Hartmann. 1976. "Surface Acoustic Wave Devices for Communications." *Proc. IEEE,* vol. 64, no. 5, May, pp. 652–671.

IEEE Proceedings. 1976. Special Issue on Surface Acoustic Waves. May.

INTELSAT. 1980. "Interfacing with Digital Terrestrial Facilities." ESS-TDMA-1-8, p. 11.

Lewis, J. R. 1982. "Satellite Switched TDMA. Colloquium on the Global INTELSAT VI Satellite System." *Digest No. 1982/76,* IEEE, London.

Maines, J. D., and E. G. S. Paige. 1976. "Surface Acoustic Wave Devices for Signal Processing Applications." *Proc. IEEE,* vol. 64, no. 5, May.

Maral, G., and M. Bousquet. 1998. *Satellite Communications Systems.* Wiley, New York.

Martin, J. 1978. *Communications Satellite Systems.* Prentice-Hall, Englewood Cliffs, NJ.

Miya, K. (ed.). 1981. *Satellite Communications Technology.* KDD Engineering and Consulting, Inc., Japan.

Morgan, D. P. 1985. *Surface-Wave Devices for Signal Processing.* Elsevier, New York.

Nudd, G. R., and O. W. Otto. 1975. "Chirp Signal Processing Using Acoustic Surface Wave Filters." *Ultrasonics Symposium Proceedings,* IEEE, New York.

Nuspl, P. P., and R. de Buda. 1974. "TDMA Synchronization Algorithms." *IEEE EASCON Conference Record,* Washington, DC.

Ore, O. 1988. *Number Theory and Its History.* Dover Publications, New York.

Pratt, T., C. W. Bostian, and J. E. Allnutt. 2003. *Satellite Communications.* Wiley, New York.

Rosner, R. D. 1982. *Packet Switching.* Lifetime Learning Publications, New York.

Scarcella, T., and R. V. Abbott. 1983. "Orbital Efficiency Through Satellite Digital Switching." *IEEE Communications Magazine,* vol. 21, no. 3, May.

Sciulli, J. A., and S. J. Campanella. 1973. "A Speech Predictive Encoding Communication System for Multichannel Telephony." *IEEE Transactions on Communications,* vol. Com-21, no. 7, July.

Spilker, J. J. 1977. *Digital Communications by Satellite.* Prentice-Hall, Englewood Cliffs, NJ.

Stevenson, S., W. Poley, L. Lekan, and J. Salzman. 1984. "Demand for Satellite-Provided Domestic Communications Services to the Year 2000." NASA Tech. Memo. 86894, November.

Watt, N. 1986. "Multibeam SS-TDMA Design Considerations Related to the Olympus Specialised Services Payload." *IEE Proc.,* vol. 133, part F, no. 4, July.

Zaharov, V. F. C., M. Gutierrez, 2001. "Smart Antenna Application for Satellite Communications Systems with Space Division Multiple Access." *Journal of Radio Electronics* N2 2001, at http://jre.cplire.ru/jre/feb01/1/text.html#fig1.

CHAPTER **15**

Satellites in Networks

15.1   Introduction

Wireless communication networks are going through a revolution to meet the requirements of the *"Internet of Everything."* Future networks will have to carry the communication and computation services and provide security for a broad variety and demanding deluge of data to an ever-increasing number of devices driving the need for broadband Internet connectivity everywhere. Driven by the growing demand for broadband Internet and communication, satellite networks have developed rapidly in the last decade and are on the cusp of exponential growth. Groups of LEO satellites, with on-board processor (OBP) devices, are being launched weekly, adding to the mega-constellations covering the earth, and are extending the terrestrial IP network to more and more regions involving satellites in modern networks.

Modern networks are converging on packet-switching networks. Satellite networks can connect user terminals to each other, they can connect user terminals to the Internet, or be a backbone connecting networks together. Thus, these networks provide broadband access to services and applications like television, telephony, and Internet connection between terrestrial networks, where physical cables and radio cannot economically reach. In addition, satellites networks can extend these services to vehicles, aircraft, ships, and space.

A typical satellite network architecture, Fig. 15.1, consists of a space segment, ground segment, control and management segment, and user segment. The space segment provides user access and data transmission with different satellites organized in LEO, MEO, GEO or layered combinations. Routing, adaptive access control, and spot-beam management functions occur in this segment. The ground segment manages, via satellite gateways (SG) and satellite terminals (ST) enabled through points of presence (PoP) in the backbone network, the interconnection of the space segment with other public or private networks. The control and management segment composed of the network control centers (NCCs) and network management centers (NMCs) manages the system operation by establishing, monitoring and releasing of connections, admission control, and resource allocation functions. The user segment through direct access to satellite networks or indirect access through terrestrial access points is responsible for providing satellite-based network services to end-users fixed or mobile devices.

In telecommunications work, a network is a connection of two or more devices such as telephones, computers, switches, and printers, and of particular interest to this chapter are *broadband networks*. The word *broadband* in the context of telecommunication networks means that the network is designed to carry voice, data, video, and image type signals. The *Internet* is a broadband network of interconnected networks. This mixture

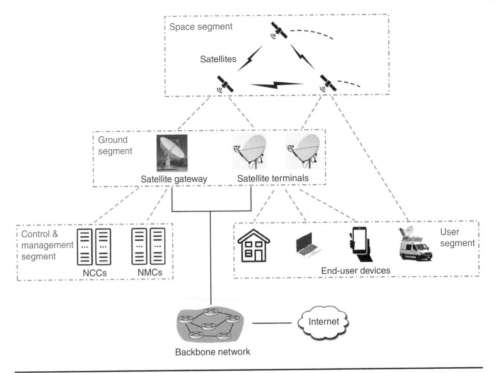

FIGURE 15.1   Typical Network for Satellite Internet.

of signals can also be carried by a method known as *asynchronous transfer mode* (ATM). In this chapter the word *network* refers to a broadband network.

A key feature of broadband networks is the way in which information is assembled into *packets,* which can be transmitted on an "as needed" basis. This packet-switching data network (structured data) differs from previous methods where information was (and still is in many cases) transmitted as a continuous signal (unstructured data).

A packet may be transferred through a network in one of two ways, *connection oriented* and *connectionless oriented.* In connection-oriented transfer, the path between sender and receiver is established before the information is transferred, and the information packets are transmitted in sequence. In connectionless oriented transfer, a path may not be established before sending the packets, and these may not be sent in sequence. The independent packets have the flexibility to reroute, if there is failure or congestion, and may follow different paths through the network. As an analogy, the connection-oriented transfer can be compared to the exchange of information that takes place during a normal telephone conversation (i.e., you initiate the connection by dialing and then wait for the connection to go through before you start communicating). On the other hand, the connectionless-oriented transfer is like sending information through the postal mail services (i.e., you put your written communication in an envelope along with a to address and return address before posting it in a mailbox).

As originally defined, connectionless packet (IP *datagrams*) transmission designed for data (such as computer output and email) is unsuited for voice transmission since the variable sized packets are too large and unpredictable. With mixed packets of voice and data, the large data packets can introduce unacceptable delays in the reassembly of the time-dependent voice packets, which affects the quality of service (QoS). The ATM, which uses relatively small fixed sized packets, known as *cells*, is designed specifically for the integration of telecommunications and data networks using a fast packet switching connection-oriented approach to manage QoS and bandwidth resources to carry broadband traffic (voice, data, video, images). The ATM is described in Sec. 15.4 and provides a legacy telecommunication solution to QoS.

15.2   Bandwidth

Bandwidth is a key concept in any telecommunications network. Originally, the word bandwidth was used to define a range of frequencies, and it applied mainly to analog circuits and signals; see, for example, Sec. 9.6.2. As shown in Sec. 10.5, digital signals also require bandwidth, and for this case, bandwidth is directly proportional to the bit rate. For example, as shown by Eq. (10.19), under certain conditions the bandwidth of a T1 signal is equal numerically to the bit rate, and with broadband signals, it is common practice to state the bandwidth as a bit rate.

The term *wideband* is also encountered in telecommunications networks; for example, the bandwidth at the input to a satellite receiver is wideband, as described in Sec. 7.7.1. Although "wideband" and "broadband" appear to be similar in meaning, each is used in quite specific contexts, wideband referring to the frequency range of signals and systems, and broadband as a bit rate as encountered in networks.

15.3   Network Basics

A data network is composed of hardware, software, and protocols, and the *topology* of a network refers to the way in which these network devices are connected. Common layouts are bus, star, ring, mesh, fully connect and tree. Satellite networking is usually configured in star or mesh topology allowing transmission of point-to-point, point-to-multipoint, and multipoint-multipoint connections. A *star* is where several connections radiate out from a central hub or switch and is used in traditional broadcast services or internet connections via satellite although it is not optimal for real-time telephony or video-conferencing because of the two-hop transmissions. A *mesh* is where each node is connected to an arbitrary number of neighboring nodes, such that there is at least one connected path between any node. A *node* is where several connections meet at a common point within the network. Two types of nodes are encountered, *terminal nodes* and *communicating nodes* (see, e.g., Walrand, 1991). A terminal node, as the name suggests is where terminal equipment such as computers and telephones, connect into the network. Communicating nodes are interconnected nodes within the network where various switching and routing functions are carried out. Nodes in general are complex structures, and to ensure the smooth flow of information through the node, the equipment at the node must operate according to well defined rules and conventions that define how data should be formatted, transmitted, and received, or what is referred to as a *protocol*. To reduce complexity, the overall protocol for a node is structured in distinct *layers*, where each encapsulated layer has its own protocol (i.e., a set of rules) that is independent of the protocol in other layers.

Figure 15.2 shows how users A and B might communicate through a three-layer protocol. Each layer interfaces with the layers immediately adjacent, above, and below.

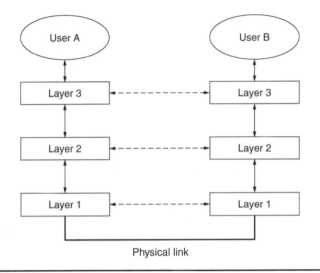

Figure 15.2   Network layers.

It is assumed that the information to be sent from A to B is in binary form. This may be a bit stream as generated, for example, by pulse code modulation for voice (see Sec. 10.3), or it could be bits stored in a digital file such as an email message. Originally, "digital networks" in the context used here were separate from the "telephone network" and the information to be transferred was called *data*, since it was mostly computer-generated. However, as mentioned in the introduction, present-day networks carry voice, data, video, and image type signals, all in digital form. Layer 3 interfaces with user A and may, for example, rearrange the bits into *packets*. There must be *interface control information* exchanged between user A and layer 3, but once the message information has been transferred to layer 3, the packet arrangement in layer 3 does not depend on the user A. Layer 3 also interfaces with layer 2. Again, some interface control information is required to ensure a smooth transfer of the message information, now in packets in this example. Layer 2 might, for example, add more bits to the message to provide an address, or perhaps some form of error control, but this can be done without reference to layer 3. Layer 2 interfaces with layer 1, which is always known as the *physical layer* in network literature. The protocol in the physical layer takes care of such matters as specifying the type of connectors to be used, the type of modulation to be employed, and the nature of the physical link. In the case of satellites, the physical link is the equipment and properties of the uplink and downlink.

In Fig. 15.2, the dotted connections show a *virtual connection* between peer layers, each layer on user A's side is virtually connected to the corresponding layer on user B's side. However, the only physical connection is that through layer 1. Peer layers must exchange information to ensure that the layer protocol is being followed, but such information is only of use to the two peer layers.

Some of the networks in common use are:

Personal area network (PAN).   Connects devices within a person's space, for example bluetooth connection between smartwatch and smartphone.

Local area network (LAN).   Connects devices that are close geographically, for example in the same building.

Campus area network (CAN).   Designed for operation in a campus such as a military campus or the campus of an educational establishment.

Metropolitan area network (MAN).   Connects LANs of a geographic region (i.e., city).

Radio access network (RAN).   Connects devices to the cloud using radio waves, for example a smartphone to cell tower.

Wide Area Network (WAN).   Connects devices that are well-separated geographically, for example long-distance optical or radio links may be used for the connections.

In this chapter, the use of satellite communications using the Internet Protocol Suite (connectionless-oriented), and using the ATM (connection oriented) are examined.

15.4   Asynchronous Transfer Mode

When information is transferred through a network it passes through several stages. The information is assembled into *packets*, and in broadband networks the packets are multiplexed to form a single bit stream. Then there is the actual transmission of signals from node to node, and at the nodes the signals undergo some form of switching. The complete process of passing information from source to destination is referred to as a *transfer mode*. With the ATM, the packets originating from an individual user do not have to be transmitted at periodic intervals, which is called *asynchronous*. For comparison, the time division multiplexing, described in Sec. 10.4, is a form of *synchronous* transmission, whereby the ATM "packets" are known as *cells*; and they can be given time slots on demand as and when required.

15.4.1   ATM Layers

The protocol for ATM was created by several standards organizations:

The International Telecommunication Union—Telecommunications Standardization Sector (ITU-T)

The American National Standards Institute (ANSI)

European Telecommunications Standards Institute (ETSI)

The ATM Forum

The layers for ATM are shown in Fig. 15.3.

- *Physical layer.*   The lowest layer is the physical layer (corresponding to layer 1 in Fig. 15.2). The physical layer has two sublayers, the *physical medium sublayer*, which deals with such matters as bit alignment, line codes to be used and bit timing plus conversions of electrical, optical and radio signals if necessary. Although the cells are transmitted asynchronously, bit timing is required for correct reception of the bit stream. This is provided at the physical medium sublayer. The other, (upper) sublayer is the *transmission convergence sublayer*. Certain transmission systems require the bit stream to be framed (an example of framing is the T1 system shown in Fig. 10.12). Packing cells into a frame, and unpacking them on receive, is carried out in the transmission convergence sublayer. The transmission conversion sublayer also provides the mechanism

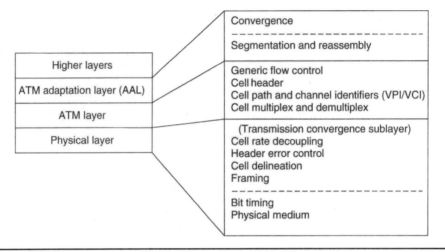

Figure 15.3   ATM layers.

for delineating cells, that is identifying the boundaries of the cell. Also, for certain types of bursty data such as computer output where there may be relatively long idle periods. "Idle" cells are inserted to maintain timing on transmit, and removed on reception at the transmission convergence sublayer (*cell decoupling*). Cell header error control is implemented at this transmission sublayer. The cell header is described in detail shortly.

- *ATM layer.* Cells from various sources (voice, data, video, images) are multiplexed, that is, arranged in sequence for transmit, and are separated (demultiplexed) into their respective channels on receive. Also, at this layer a cell header is added, which provides path and channel identification. The function of the header is key to ATM operation, and this is discussed in detail shortly.

- *ATM adaptation layer (AAL).*   The adaptation layer is divided into two sublayers. The *convergence sublayer* (CS) where the service requirements for the multimedia (voice, data, video, image) are established, and the *segmentation and reassembly sublayer* where the incoming information is segmented into cells, and the outgoing information is reassembled into its original format. The incoming bit streams for the various media are quite different in bit rate and burstiness. The ATM adaptation layers are:

  - *AAL-1* is used for *constant bit rate* (CBR) applications, and is designed for voice and data is sent over circuit facilities such as T1 (described in Sec. 10.4). As shown below, 48 octets in a cell are reserved for the payload, but AAL-1 uses one of these octets for overhead, leaving 47 octets for the information payload.

  - *AAL-2* is used for *variable bit rate* (VBR) such as video and voice that use compression techniques. Again, part of the payload, from 1 to 3 octets is used for certain overhead functions, leaving 45 to 47 octets for the information payload.

- *AAL-3/4*—Originally there was a separate AAL-3 but it was found that this could be combined with AAL-4. In this way AAL-3/4 was established. It is connectionless oriented (see Sec. 15.3) and is used for VBR data. An overhead of four octets is included in the payload field leaving 44 octets for the information payload. It should be noted that all the other AALs are connection-oriented (see Sec. 15.3).

- *AAL-5* is used for VBR and is connection-oriented. The payload does not contain any overhead, so the information payload makes full use of the available 48 octets.

There is also an ATM adaptation layer known as the null layer, and designated *AAL-0*. This is used where the information is already assembled in cell format.

15.4.2   ATM Networks and Interfaces

Two basic types of ATM networks are in general use—*public* and *private*. The public ATM network is designed to provide connections between any two subscribers, rather in the way that the public telephone network does. A private ATM network is one set up by an organization to provide ATM communications between the various parts of the organization. A private ATM network can exist independently of the public network, and the methods of interfacing to these also differ in general.

A few different interfaces are encountered in an ATM network, and these are denoted as follows:

- *User network interface* (UNI).   This is the interface between an end user and an ATM network. It occurs at the entry point for a user. A UNI may be public or a private.

- *Network node interface* (NNI).   This is the interface between two nodes in a network, which may be public or private. A private NNI is usually indicated by PNNI in the literature.

- *Network network interface*.   This is also abbreviated NNI and is an interface between two ATM networks. The interface between two private networks is indicated by PNNI. The context should make clear in both instances whether the NNI refers to network-node or network-network interface. A private network must interface with the public network through a public UNI.

15.4.3   The ATM Cell and Header

The standard ATM cell is made up of 53 octets, where an octet is a sequence of 8 bits. The 8-bit sequence is also known as a *byte*, but it should be noted that in computer terminology a byte can consist of other than 8 bits (see also Sec. 15.7). The cell structure shown in Fig. 15.4*a* is seen to consist of a 5 octet header, and a 48 octet payload. The header contains several *fields*, which provide the information necessary to guide the cell through the network. These fields are described later, but as seen in Fig. 15.4, the header for cells at the point of entry to a network (at a UNI) differs slightly from the header for cells traversing an NNI. The header fields summarized in Fig. 15.4 are:

GFC   This is the *generic flow control* field. Its function is to provide control and metering of the data flow before it enters the network. No flow control is exercised once the cell has entered the network. The 4 bits in the header are reallocated, only in the UNI, as described under VPI.

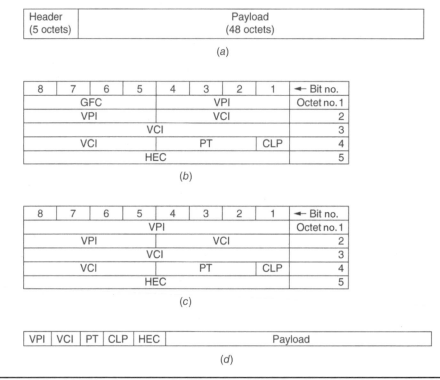

Figure 15.4 (a) ATM cell structure; (b) UNI header; (c) NNI header; (d) header and payload bit stream.

VCI This is the *virtual channel identifier* field. A virtual channel carries a single stream of cells, which may be a mixture of voice, video, image, and data. From the point of view of the users, the different signals appear to have their own channels, and hence these are called *virtual channels*. Each channel is identified by its VCI.

VPI This is the *virtual path identifier* field. Virtual channels can be "bundled" together to form what is termed a *virtual path*. With binary coding it should be kept in mind that all the information forms a single serial bit stream, which is transmitted over a physical link (e.g., optical fiber and satellite). The VPI enables the cells to be grouped into separate (but virtual) paths over the physical link. Switching within the network can be carried out by switching the virtual paths, without the need to switch the channels separately. As shown in Fig. 15.4, the GFC field is not required for cells once they are in the network, and the 4 bits for this field are reassigned to the VPI field, which enables the number of virtual paths within the network to be increased.

CLP This is the *cell loss priority* field. It consists of a single bit, a 1 for a low priority cell, meaning that the cell can be discarded in the event of congestion. A 0 indicates high priority, meaning that the cell should only be discarded if it cannot be delivered.

PT This is the *payload type* field. It consists of three bits. Bit 1 ATM user-to-user (AAU) bit indicating packet boundaries AAL5. Bit 2 explicit forward congestion indication (EFCI), if 1 then network congestion experienced. Bit 3 is used to indicate data cells from cells of operation, administration, and maintenance (OMA).

HEC   This is the *header error control* field. The HEC field enables single bit errors in the header (including the HEC field) to be corrected, and double bit errors to be detected (but not corrected). The receiver is normally in the error correction mode. If a single error is detected, it is corrected, and the cell transmitted onward. However, after detection and correction of a single error, the receiver automatically switches to the error detection mode. Of course, if more than one error is detected, it automatically switches to error detection mode, and the cell is discarded. Thus, in error detection mode, *no errors* are corrected, and cells with errors are discarded. Once a cell with an error-free header is detected, the receiver automatically switches back to error-correction mode. The rationale behind this approach is discussed at length in Goralski (1995). Error control is applied only to the header (the data in the payload may have its own error-control coding). ATM was originally used for transmission over low *bit error rate* (BER) links such as optical fiber, where the bit error probability can be as low as 10^{-11}. Unfortunately, over satellite links the BER can be as high as 10^{-2} that often occur in bursts, which require special error-control coding (see Secs. 11.3.3 and 11.6). The HEC field is placed at the end of the header, as shown in Fig. 15.4*d,* where it functions also as a marker for cell position.

15.4.4   ATM Switching

Keep in mind that the octets in the header form a serial bit stream when being transmitted, as shown in Fig. 15.4*d.* A cell is switched through a network on a path determined by the VPI and VCI fields. At each node in the network, a cell is switched from one physical link to another, and two basic types of ATM switches are encountered. In one type, known as an ATM *digital cross connect switch* (DCS), the VPI field in a cell is overwritten by a new number on switching, while the VCI is left unchanged. This enables several virtual channels to be switched together, thereby, speeding-up the switching process. The functioning of the DCS is illustrated in Fig. 15.5*a,* where for simplicity only the VPIs and VCIs are shown. These have been given hypothetical numbers for illustration purposes, for example 200/12 indicates a VCI of 200 and a VPI of 12. The look-up table in the DCS determines the new VPI for the virtual paths, and the physical output link to which the paths must be switched. Different suppliers offer different types of switching stages (the switch fabric). For ease of identification the cells have been labeled P, Q, R, S, T, U in Fig. 15.5.

The second basic type of switch is referred to as an ATM switch. This differs from the DCS in that both the VPI and VCI fields are overwritten with new numbers on switching. Figure 15.5*b* illustrates the functioning of an ATM switch, where the VPI and the VCI of cells are both overwritten by new numbers based on the look-up table. Again, for ease of identification the cells have been labeled V, W, X, Y, Z. Although the virtual channels and virtual paths are formed over a single physical link, it is common practice to show the virtual channels bundled into a virtual path as in Fig. 15.5*c.*

From Fig. 15.4, it is seen that the VPI field at the UNI has 8 bits, and therefore 2^8 or 256 virtual paths can be supported. At the NNI the VPI field is increased to 12 bits, allowing 2^{12} or 4096 virtual paths to be supported. The VCI has 16 bits and therefore in theory the number of virtual channels is 2^{16} or 65536. This in practice is limited to 64000 (see Russell, 2000).

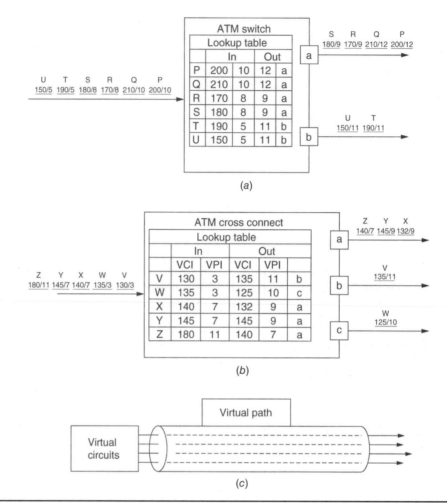

FIGURE 15.5 (a) ATM switch; (b) ATM cross connect; (c) virtual circuits and path.

15.4.5 Permanent and Switched Virtual Circuits

A virtual path may be set up on a permanent basis. The switching identifiers are preset so that the path does not change and is available for use, as required. This is referred to as a *permanent virtual circuit*, abbreviated PVC (note the use of the word "circuit" to describe this). A path may also be established anew each time a connection is required. Thus, users are disconnected when the transfer is finished, and a new connection is required if another session is requested. This is described as a *switched virtual circuit* (SVC)—the term *switch* being used in analogy with telephone usage. The terms PVC and SVC can also be applied to virtual channels.

15.4.6 ATM Bandwidth

A major advantage claimed for ATM is that it can provide bandwidth on demand, also known as *flexible bandwidth allocation*. The concept of bandwidth and its relationship to bit rate is described in Sec. 15.2; and with the transmission of digital signals over networks, the significant parameter is the bit rate. It is common practice, therefore, to state bandwidth as a bit rate; and the maximum bit rate that a physical link can support is referred to as the *speed* of the link. Denoting the speed by S, the *maximum* bandwidth available for the payload can be calculated simply as S multiplied by the ratio of payload bits per cell to total bits per cell. The payload has 48 octets, and hence the payload bits per cell is $8 \times 48 = 384$. An ATM cell has a total of 53 octets; and therefore, there are a total of $8 \times 53 = 424$ bits per cell. Unfortunately, the situation is somewhat more complicated as, in certain cases, the 48 octets contain some overhead bits; but this is ignored for the present calculation. If a single user makes use of all the cells and these are transmitted without a break at speed S, then the payload bandwidth would be:

$$BW = S \times \frac{384}{424} \tag{15.1}$$

For example, at a speed of 30 Mbps the payload bandwidth is approximately 27 Mbps.

Although it is possible for all the cells to be assigned to a single channel, it is much more likely that the cells are distributed over many channels, that is, several signals are *multiplexed* into the ATM transmission. This is one of the strengths of ATM, namely; the ability to accommodate multiple signals, in general each requiring a different bandwidth. The bandwidth used by a signal can be calculated from knowledge of the cell distribution. In practice, cells may be transmitted in *frames*. Although the cells are asynchronous (i.e., not necessarily distributed periodically over the frames), for ATM, the frames themselves are transmitted synchronously at 8000 frames per second (see Mackenzie 1998), which is termed the *synchronous transmission module-1* (STM-1). If one cell per frame is assigned to one user, and since each cell carries 384 payload bits, the corresponding bandwidth is:

$$B_{\text{payload}} = 384 \times F \tag{15.2}$$

where F is the number of frames per second. For example, using the value $F = 8000$ frames/s, the payload bandwidth for one cell per frame is:

$$B_{\text{payload}} = 384 \times 8000 = 3.072 \text{ Mbps}$$

The flexibility of ATM arises because a cell may be sent only once every nth frame, or more than one cell per frame may be sent. For example, suppose that a user cell is sent once every 48th frame and that the frame rate is 8000 frames/s, then the corresponding payload bandwidth is:

$$\frac{3.072}{48} = 0.064 \text{ Mbps} = 64 \text{ kbps}$$

As shown by Eq. (10.10), this is the bit rate for a single PCM signal.

If more than one cell per frame is allocated to a user, say C cells per frame then the corresponding payload bandwidth is

$$B_C = B_{\text{payload}} \times C = 384 \times F \times C$$

There is of course a limit to the number of cells that can fit in a frame. If S is the speed of the link in bits/s and F the number of frames/s, then the bits per frame is S/F. Each cell has 424 bits (the header bits must be included here), and hence maximum number of cells per frame is

$$C_{\text{max}} = \frac{S}{424 \times F} \tag{15.3}$$

Substituting this for C in the B_C equation and simplifying gives:

$$BW_{C_{max}} = \frac{384}{424}S \qquad (15.4)$$

This, of course, is just the Eq. (15.1) arrived at earlier for the maximum payload bandwidth for a single user making use of all the cells.

Voice, television, and videoconferencing all require a CBR service (essentially a fixed bandwidth). These signals are sensitive to variations in cell delay, so the cells for these services must be spaced at regular (periodic) intervals. Other services such as email and file transfer are not sensitive to variations in cell delay, and cells can be interspersed as space permits. This is called VBR service, and the use of ATM for CBR and VBR is illustrated in Fig. 15.6.

15.4.7 Quality of Service

In addition to flexible bandwidth allocation, ATM offers a guaranteed QoS, which differs for various applications. For example, voice and video are sensitive to cell delay and variations in cell delay, but may be able to tolerate some loss of cells. On the other hand, digital data is not affected by cell delay (within limits), but it cannot tolerate cell loss. When a connection is requested by a user, the network is first tested to ensure that the guaranteed QoS can be provided, otherwise the user request is not accepted. In fact, there are five service classes based on bit rate, and these are summarized in Table 15.1.

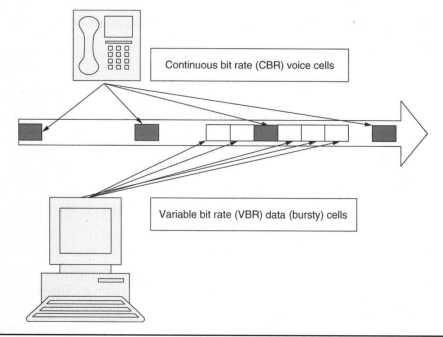

FIGURE 15.6   The use of ATM for CBR and VBR.

Service Class	Quality of Service Parameter
Constant bit rate (CBR)	This class is used for emulating circuit switching. The cell rate is constant with time, and applications are sensitive to cell-delay variation. Telephone traffic, videoconferencing, television are all examples of CBR transmissions.
Variable bit rate–non real (VBR–NRT)	Traffic can be sent at a variable rate, which depends time on the availability of user information. Statistical multiplexing* is used to optimize network resources. Multimedia is an example of VBR-NRT transmission.
Variable bit rate–real time (VBR–RT)	Similar to VBR–NRT, but designed for applications that are sensitive to cell delay. Application examples are voice with speech activity detection and interactive compressed video.
Available bit rate (ABR)	Provides rate-based flow control, the rate depending on the cell congestion present in the network. Examples of use are file transfer and email.
Unspecified bit rate (UBR)	A "catch all" class, widely used for TCP/IP (see Sec. 15.6)

Notes: With statistical multiplexing the sum of the peak input rates exceeds the capacity of the link, but the assumption is made that sources are statistically independent and are unlikely to send peak rates at the same time.

Source: With minor modifications from Web ProForum Tutorials, ©The International Engineering Consortium, http://www.iec.org.

TABLE 15.1 ATM Service Classes

A few parameters are used in the measurement of ATM performance, and these are shown in Table 15.2.

15.5 ATM over Satellite

The ATM, described in the previous sections is used in terrestrial networks, to carry a mixture of signals, for example voice, data, video, and images. A natural development is to extend the ATM networks to include satellite links, which bring the services of ATM to remote or isolated users and which provide broadcast facilities. Satellite links have the additional advantage of enabling LANs that are widely separated geographically to be linked together, thus forming a WAN. ATM over satellite is usually abbreviated as SATM in the literature (for satellite ATM). Satellites may be incorporated into ATM networks in several ways, some of which are described here, but there are problems unique to satellite links that must be addressed.

The first of these is the BER. ATM was originally designed to operate over optical fiber links where bit errors are randomly distributed and the probability of bit error is

Technical Parameter	Definition
Cell loss ratio (CLR)	Percentage of cells not delivered to their destination because of congestion or buffer overflow.
Cell transfer delay (CTD)	The delay experienced by a cell between entry and exit points of a network
Cell delay variation (CDV)	A measure of the variance of the cell transfer delay.
Peak cell rate (PCR)	The maximum cell rate at which a user transmit.
Sustained cell rate (SCR)	The average transmission rate measured over a long period, of the order of the connection lifetime.
Burst tolerance (BT)	Determines the maximum burst that can be sent at the peak rate.

Source: With minor modifications from Web ProForum Tutorials, ©The International Engineering Consortium, http://www.iec.org.

TABLE 15.2 ATM Technical Parameters

comparatively low, on the order of 10^{-10}. In satellite links, the BER is generally much higher than this (see Figs. 10.22 and 11.8), and more importantly, the bit errors can occur in bursts. In laboratory tests involving MPEG-2 signals it has been shown (Ivancic et al., 1997; 1998) that a BER of better than 10^{-8} is required for the most stringent ATM applications. Such values may be achieved using concatenated convolution and *Reed-Solomon* (R-S) *forward error correction* (FEC) (see Sec. 11.6).

The problem, with applying FEC to ATM signals overall, is that error correction is being applied to channels that might be tolerant of a higher BER, such as voice, and thus is a waste of resources. A number of commercial error-control schemes for ATM over satellite are available. For example, the Ericsson CLA-2000 ATM Link Accelerator (Ericsson, 2003) uses adaptive R-S and interleaving mechanisms, resulting in a *cell loss ratio* (CLR) and a *cell error ratio* (CER) of 10^{-10} or better (see Table 15.2). Adaptive means that the coding scheme is adjusted to suit the application (e.g., voice or data) and that it dynamically adapts to the link conditions. Thereby, on clear days it uses less FEC overhead, which results in a 7 percent increase in bandwidth.

Another approach is to use the LANET protocol, where LANET stands for *Limitless ATM Network*, a product of Yurie-Lucent. Adaptive R-S coding is applied to both the payload and the PTI and CLP fields in the header (Akyildiz et al., 1998). Also, Yurie has implemented a redundancy coding scheme for the circuit (channel) addresses in the header. Multiple addresses are provided and are chosen such that the most probable error occurrences change a given address to another permissible address within the group for the channel.

Delay and variance of delay (termed *delay jitter*) also present problems not normally found in terrestrial networks. For geostationary satellites the one-way propagation delay is about 250 milliseconds. Variations in delay, termed *delay jitter*, can be more of a problem for delay-sensitive channels such as voice and video, and some form of buffering is needed to minimize the delay jitter. The use of *low earth orbit* (LEO) and *medium earth orbit* (MEO) satellites reduces the propagation delay, although other problems are then introduced. LEO and MEO networks are discussed shortly.

ATM satellite networks can be broadly classified as *bent pipe architecture* and *on-board processing architecture*. With the "bent pipe" architecture, the satellite acts as a conduit between two earth stations, which may be fixed or mobile. With the *on-board processing* (usually abbreviated as OBP) architecture, the satellite can be considered a network node in the sky where ATM switches form part of the transponder in the satellite, which allows for on-board switching based on the VPI and VCI fields. This may be incorporated with the OBP and satellite beam switching described in Secs. 14.8 and 14.9. The use of OBP provides greater connectivity, reduces transmission delay, and permits the use of smaller and cheaper user terminals; but of course, all this is at the cost of a more complex satellite structure.

The simplest situation is the bent pipe *relay architecture* illustrated in Fig. 15.7. This makes use of geostationary satellites, where the satellite link can be thought of as replacing a terrestrial link between two fixed points. The satellite link is likely to operate at a lower bit rate than the terrestrial ATM services connecting through it, and some rate adaptation is necessary. This is provided by the modems shown in Fig. 15.7. The *ATM link accelerator* (ALA), shown in Fig. 15.7, provides the adaptive error-control coding mentioned earlier. Cell switching (using the VPI and VCI fields) is carried out at the ATM end stations, not in the satellite. Also, signaling (e.g., call set up and teardown) takes place between the ground ATM switches. The relay designation applies only to "bent pipe" satellites, and is not applicable to OBP satellites.

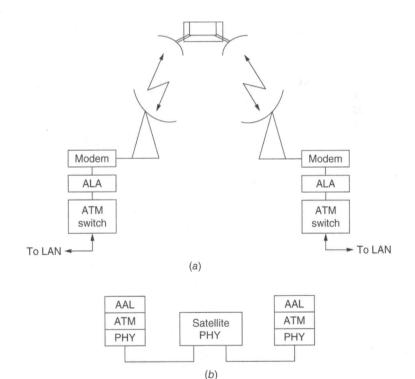

Figure 15.7   (a) "Bent pipe" satellite relay; (b) layer architecture.

The broad classifications, "bent pipe" and OBP defined earlier can be further classi-fied as *network access*, and *network interconnect*. The distinction here is that network access provides satellite access for ATM end-users, who may connect to each other and to ATM networks; network interconnect provides for satellite interconnection of ATM networks. Access and interconnect can be used with fixed and mobile users and networks (a mobile network for example could be a group of users on a ship or plane), where mobile operations must be supported. Figure 15.8 shows the bent pipe access arrange-ment for fixed end-users and an ATM network in a star network topology. Satellites can be interconnected using *Inter-satellite links* (ISL). By connecting multiple satellites, we can optimize earth station hops, which reduces earth to space spectrum traffic and extends service to regions that do not have gateway earth stations enroute. In general, the access and interconnect modes of operation may utilize a combination of *geostation-ary earth orbiting satellites* (GEOs), LEO, and MEO satellites.

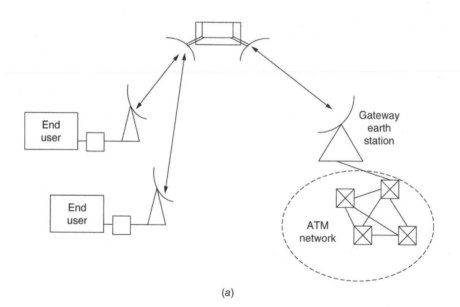

(a)

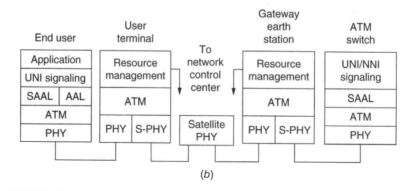

(b)

FIGURE 15.8   (a) Fixed network access; (b) layer architecture.

The most complex arrangement, which applies only to OBP satellites, is called *full mesh*. Here the satellites themselves form an ATM network in space, in which the full complement of ATM functions operate. This includes traffic switching, flow and congestion control, connection setup and teardown, and quality of service requirements.

Where mobile operations must be supported, mobility management functions are provided by a *mobility enhanced UNI layer* at the end-user (M+UNI) and by a mobility enhanced UNI/NNI layer (M+UNI/NNI) at the ATM switch. Mobility management includes *location management*, (authentication, registration, paging, roaming, and routing) and *handover*, required when the mobile units move from one location to another.

An extensive overview of satellite architectures is found in Toh and Li (1998), a summary of which is given in Table 15.3.

Network	Bent Pipe	On Board Processing (OBP)
Relay		
	Point to point linkage between two fixed ATM users.	Not applicable.
Access		
Fixed ATM	UNI at user terminal; UNI/NNI at GES. No mobility support.	Media access required. UNI between user and satellite. No mobility support.
Mobile ATM	Mobility enhanced UNI at user terminal and NNI between GES and ATM network.	Mobility support provided by the mobility enhanced switching at terrestrial ATM and at satellite.
Interconnect		
Fixed ATM	High speed interconnection between fixed ATM networks PNNI, B-ICI, or public UNI between GESs and ATM networks. No mobility support.	ATM switch on board satellite acts as intermediate node. Supports NNI signaling, cell switching, and multiplexing. No mobility support if GEOs used.
Mobile ATM	High speed interconnections between mobile and fixed ATM networks and between two mobile ATM networks. Mobility enhanced NNI between GESs and networks.	High speed interconnections between mobile and fixed ATM networks, and between two mobile ATM networks. Mobility enhanced NNI between GESs and networks.
Full mesh		
SATM	Not applicable.	Satellites form an ATM network in space, which supports all of the above scenarios.

Notes: ATM, asynchronous transfer mode; B-ICI, broadband inter-carrier interface; GES, gateway earth station; NNI, network node interface; PNNI, private network node interface; SATM, satellite ATM (network); UNI, user network interface.

TABLE 15.3 Satellite ATM Architectures

A LEO satellite constellation has many advantages over GEO satellites for networking. Because of the shorter ranges involved, the propagation delay is much smaller, there are lower power requirements both on the satellite and user end. Also, there is more spectrum bandwidth available because of available channel assignments for millimeter waves or optical wavelengths, as well as frequency reuse among spotbeams. These advantages are offset by frequent *handover* in the link-layer and network-layer due to high satellite speed and different orbital planes. A LEO satellite at 500 km altitude travels at 7.6 km/s and takes around 95 minutes to orbit the earth resulting in a handover approximately every 5 minutes. If the handover fails, then the links are lost, which can result in the termination of the connection-oriented services. The timing of the handover is crucial, if it is too early then unnecessary handoffs can occur, and if it is too late, the chance of forced termination increases. The handover can be triggered by multiple factors including signal level strength, distance measurement, traffic congestion, political geographic boundaries, and others.

Figure 15.9 shows *spotbeam, satellite,* and *ISL handover* are three link-layer handover categories which occur, when one or more links between communication endpoints change, because of the dynamic nature of LEO satellites. Consider Fig. 15.9*a* when the users endpoint cross the boundary between neighboring spotbeams of a satellite, an *intra-satellite* or spotbeam handover occurs. Otherwise Fig. 15.9*b* or *c, shows satellite* or *inter-satellite handover* occurring, when a satellite's existing connection of a user's attachment point is transferred to a different satellite. For the latter case, an *ISL handover* occurs when two interconnected neighboring satellites interplane ISL's approach disconnection, and then the ongoing connections must be rerouted that occurs because of changing viewing angles and/or distance between satellites.

A network-layer handover is when one of the communications satellite or user endpoint changes its routable address due to change of coverage area at the satellite or user terminal. A network or *higher-layer* handover is needed to migrate the existing connections. This all adds to the complexity of the on-board requirements. Figure 15.10*a* is a block schematic of a typical ATM LEO satellite network (Todorova, 2002). User terminals may access the LEO satellite system directly, or they may require an *adaptation unit* (like the modem shown in Fig. 15.7). The public ATM network accesses the LEO satellite system through *gateways* (as in Fig. 15.8). Decentralized network management is proposed in the system described by Todorova (2002), where overall management is provided by the *network control center* (NCC), but certain functions are carried out by the satellite ATM switches (S-ATM). The management information is carried in signaling channels termed *management virtual channels* (mVCs). For added security a stand-by

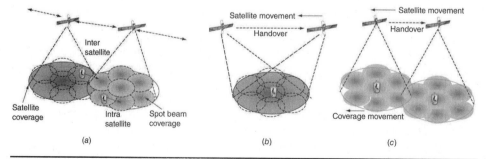

FIGURE 15.9 Example of (*a*) spotbeam, (*b*) satellite and (*c*) ISL handover scenarios for LEO/MEO satellites.

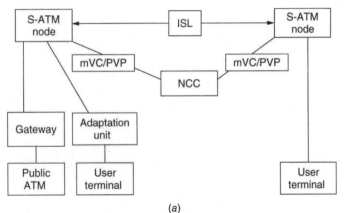

(a)

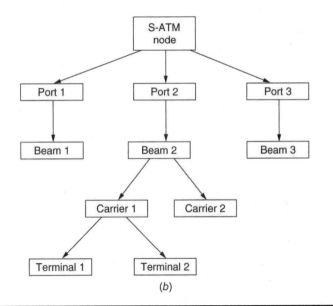

(b)

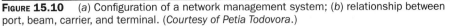

FIGURE 15.10 (a) Configuration of a network management system; (b) relationship between port, beam, carrier, and terminal. (*Courtesy of Petia Todovora.*)

path, in the form of a *permanent virtual path* (PVP), is provided in the event that the signaling channel should fail.

Figure 15.10b is a block schematic showing the relationship between port, beam, carrier, and terminal (Todorova and Nguyen, 2001). As shown, the ports from the ATM switch connect to individual beams. The beams may be multicarrier, two being shown in Fig. 15.9, and each carrier can be received by more than one earth terminal.

15.6 The Internet

The Internet is a global system of interconnected networks providing various information and communication facilities that use the Internet protocol suite (TCP/IP) to communicate between networks and devices (see Fig. 15.11). It does not have its own

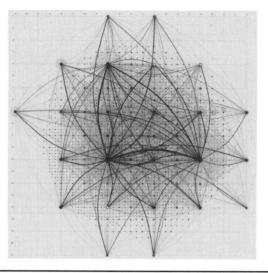

Figure 15.11   The Internet: a global system of interconnected networks. (*Courtesy of Wikepedia.org*)

physical structure; rather, it makes use of existing physical networks, the copper wires, optical fibers, wireless, and satellite links, owned by private, public, academic, business, governments, of local to global scale. Although there is no identifiable structure, access to the Internet follows well-defined rules. Users connect to *Internet service providers* (ISPs), who in turn connect through *routers* to *network service providers* (NSPs), who complete the connections to other users and to *servers*. Servers are computers dedicated to the purpose of providing information to the Internet. They run specialized software for each type of Internet application. These include the world wide web (WWW), electronic mail (email), voice over internet protocol (VoIP), multicast and content distribution (RTP), long-distance computing (Telnet), domain name system (DNS) and file transfers (FTP). *Routers* are computers that form part of the communications net and that route or direct data packets along the best available path between networks.

Although there is no central management or authority for the Internet, its extraordinarily rapid growth has meant that some control must be exercised over what is permitted. A summary of the controlling groups is shown in Fig. 15.12*a*. A description of the groups is found in Leiner et al. (2000) and Mackenzie (1998).

The WWW is probably the most widely used application on the Internet. The evolution and growth of the WWW has been rather like that of the Internet itself, with no central authority but still with a general structure that attempts to regulate what happens. The WWW Consortium, referred to as W3C, was founded in October 1994 (Jacobs, 2000). W3C oversees several special interest groups, as shown in Fig. 15.12*b*, and coordinates its efforts with the Internet Engineering Task Force (IETF) and with other standards bodies. Details of the W3C are found in Jacobs (2000).

15.7   Internet Layers

In a satellite network the uplink and downlink between satellite and earth stations and the ISL links form the *physical layer* in a data communication system. By *data communications* is meant communications between computers and peripheral equipment. The signals are

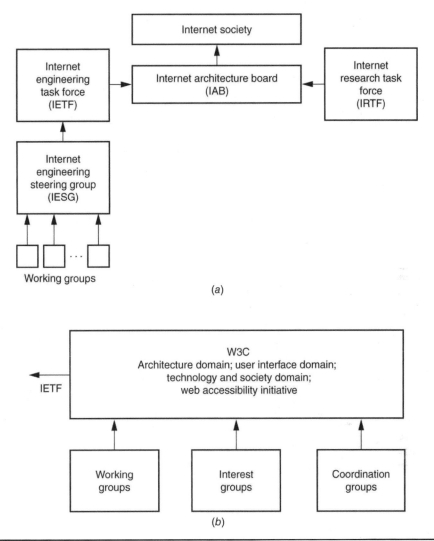

FIGURE 15.12   (a) Internet groups; (b) World Wide Web groups.

digital, and although digital signals are covered in Chap. 10, the satellite links must be able to accommodate the special requirements imposed by networks. The terminology used in networks is highly specialized, and some of these terms are explained here to provide the background needed to understand the satellite aspects (see also Sec. 15.3). The Internet, of course, is a data communication system using TCP/IP protocol suite.

The data is transmitted in *packets across different network technologies*. Many separate functions must be performed in packet transmission, such as packet addressing, routing, and coping with packet congestion. The modern approach is to assign each function to a layer in what is termed the *network architecture*. This has already been discussed in connection with ATM, as shown in Fig. 15.2. The layers are conceptual in the sense that they may consist of software or some combination of software and hardware. In the case of

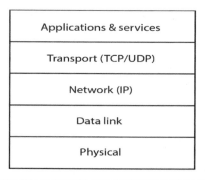

FIGURE 15.13   Layered structure for TCP/IP. (*Courtesy of Feit, 1997.*)

the Internet, the network architecture is referred to as the TCP/IP model, although there are protocols other than TCP/IP contained in the model. The layered structure is shown in Fig. 15.13. A brief description of these layers is included to familiarize the reader with some of the terms used in network communications, although the TCP layer is of most interest in this chapter.

- *Physical layer*   This covers such items as the physical connectors, signal format, modulation, and the uplink and downlink in a satellite communications system.

- *Data-link layer*   The function of this layer is to organize the digital data into blocks as required by the physical layer and seemingly free of undetected transmission errors to the network layer. For example, if the physical layer uses ATM technology, as described in Sec. 15.3, the data is organized into cells. Digital transmission by satellite frequently uses TDMA, as described in Chap. 14, and satellite systems may transmit Internet data over ATM. Thus the data-link layer must organize the data into a suitable format to suit the physical-layer technology. In the terrestrial Internet, the data link converts the data into frames. The network interface card (NIC) receives and transmits the signal usually on Ethernet wire. The MAC address uniquely identifies each NIC and is used to deliver packets to other nodes on the local network. A switch's primary responsibility is to facilitate communication within networks. The data-link layer and the physical layer are closely interrelated, and it can be difficult sometimes to identify the interface between these two layers (Mackenzie, 1998).

- *Network layer*   This is strictly an IP layer best effort service based on the datagram. The Internet packets are passed along the Internet from the source host to the next router or gateway and finally to the destination host with no guarantee of QoS. No exact path is laid out beforehand, and the IP layers in the routers must provide the destination address from a routing table or software process for the next leg of the journey so to speak. This destination address is part of the IP header attached to the packet along with the source address and data from the transport layer which in totality is called the *datagram*. The problems of lost packets or packets arriving out of sequence are not a concern of the IP layer, and for this reason, the IP layer is called *connectionless* (i.e., it does not require a connection to be established before sending a packet on). These problems are taken care of by the transport layer.

- *Transport layer* Two sets of protocol are provided in this layer. Transmission control protocol (TCP) and user datagram protocol (UDP) provide ports or sockets for applications to send and receive data from applications on other hosts across the Internet. For TCP, information is passed back and forth between transport layers, which controls the information flow. TCP provides reliable message delivery and ensures that data is not damaged, lost, duplicated, or delivered out of order to the receiving process. In the early days of the Internet, when traffic was comparatively light, these problems could be handled even where satellite transmissions were involved. With the enormous increase in traffic on the present-day Internet, these problems require special solutions where satellite systems are used, which are discussed in later sections. The TCP layer is termed *connection-oriented* (compared with the connectionless service mentioned above) because the sender and receiver must be in communication with each other to implement the protocol, which allows TCP to provide a reliable service. For example, there are situations where a simple standalone message may need to be sent, which does not require the more complex TCP. For these types of messages, the UDP is used. The UDP provides a connectionless service, similar to IP. This is a best-effort service which is useful for real-time applications because retransmission of any packet may cause more problems than lost packets. The UDP header adds the port numbers for the source and destination applications.

- *Application layer* Classical Internet application protocols include HTTP for WWW, FTP for file transfer, SMTP for email, telnet for remote login, DNS for domain name service, real-time services RTP, RTCP, RTSP, and active web services.

The Internet is in the process of moving from Internet protocol version 4 (IPv4) to IPv6 to provide extensions and solutions to address QoS, security, mobility, multicast, and more address space.

The term *packet* has been used somewhat loosely up to this point. A more precise terminology for the Protocol Data Unit (PDU) used for packets at the various layers are shown in Fig. 15.14 along with a sample ethernet frame. At the application level the packet is simply referred to as *data*. The packet comprising the TCP header and the application data is a *TCP segment*. The packet comprising the UDP header, and the data is a *UDP message.* The packet comprising the IP header, the TCP or UDP header, and the data is an *IP datagram*. Finally, the packet comprising the data-link frame header, the frame trailer (used for error control), and the IP datagram is a *frame*. The wrapping of each higher layer PDU into the next layer down as the data moves from the Application layer down the layer stack until reaching the data link layer is called *encapsulation*. *Decapsulation* reverses this process at the receiving host by removing each layers PDU and passing the result to the next layer starting at the data link layer until the application layer receives the data sent.

The beauty of the IP datagram is that it provides an end-to-end network layer service to the layers above that can then be carried across different types of networks, while the IP datagram stays the same. The data link and physical layer all manage the different types of network technologies such as ATM, WiFi, ethernet, 5G mobile, satellite, and future space networks.

It should be noted here that the preceding definitions are those used in the version IPv4. IPv6 is a more recent version being brought on-stream in which the IP datagram is in fact called an *IP packet*.

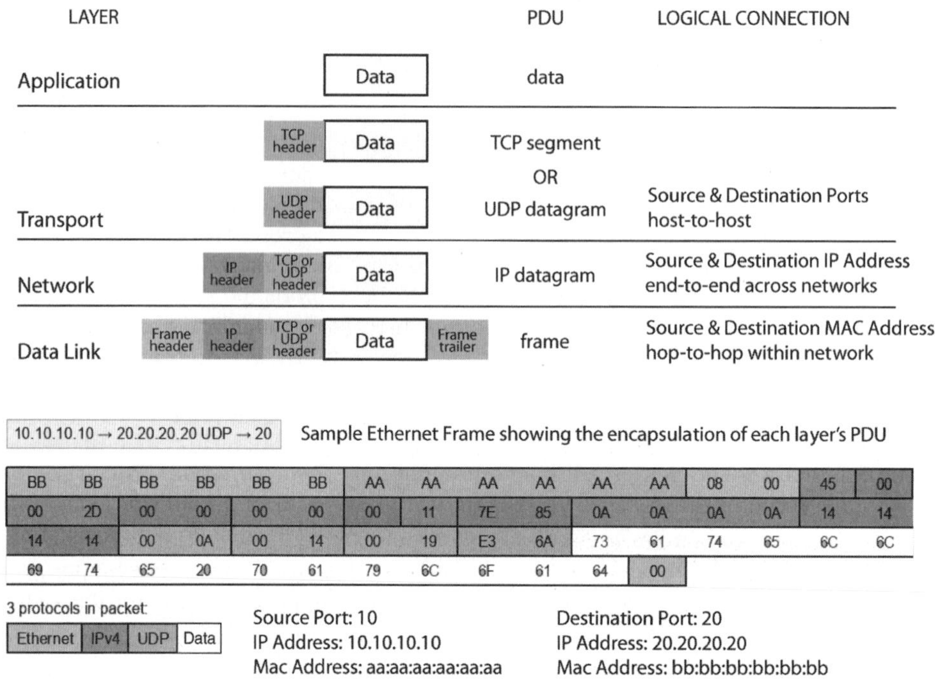

LAYER		PDU	LOGICAL CONNECTION

Application — Data — data

Transport — TCP header / Data — TCP segment
OR
UDP header / Data — UDP datagram — Source & Destination Ports host-to-host

Network — IP header / TCP or UDP header / Data — IP datagram — Source & Destination IP Address end-to-end across networks

Data Link — Frame header / IP header / TCP or UDP header / Data / Frame trailer — frame — Source & Destination MAC Address hop-to-hop within network

10.10.10.10 → 20.20.20.20 UDP → 20   Sample Ethernet Frame showing the encapsulation of each layer's PDU

BB	BB	BB	BB	BB	BB	AA	AA	AA	AA	AA	AA	08	00	45	00
00	2D	00	00	00	00	00	11	7E	85	0A	0A	0A	0A	14	14
14	14	00	0A	00	14	00	19	E3	6A	73	61	74	65	6C	6C
69	74	65	20	70	61	79	6C	6F	61	64	00				

3 protocols in packet:
Ethernet IPv4 UDP Data

Source Port: 10
IP Address: 10.10.10.10
Mac Address: aa:aa:aa:aa:aa:aa

Destination Port: 20
IP Address: 20.20.20.20
Mac Address: bb:bb:bb:bb:bb:bb

FIGURE 15.14 Protocol Data Unit (PDU) terminology.

Some of the units used in data transmission are:

- *Byte* Common usage has established the byte (symbol B) as a unit of 8 bits, and this practice is followed here. It should be noted, however, that in computer terminology, a byte can sometimes refer to a unit other than 8 bits. Further, an 8-bit unit may be called an *octet*.

- *Kilobyte* A kilobyte (symbol kB) is 1024 bytes. Transmission rates may be stated in kilobytes per second or kbps.

- *Megabyte* The megabyte (symbol MB) is 1024 kilobytes. Transmission rates may be stated in megabytes per second or Mbps.

The TCP/IP suite is shown in Fig. 15.15, and an excellent detailed description of these protocols are found in Feit (1997). The present text is concerned more with the special enhancements needed on TCP/IP for successful satellite transmission.

15.8 IP Addressing and Routing

TCP/IP has an Internet addressing scheme that allows users and applications to identify specific networks and hosts with that to communicate. This IP address is used to route data through the network. IP addresses are assigned to networks, subnetworks, and hosts, and special addresses can be used for broadcasting and for local loopback. Each network attached to the public Internet has a unique subnetwork address and

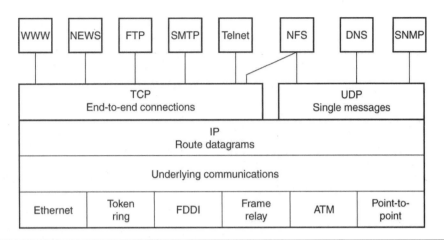

FIGURE 15.15 The TCP/IP suite. (*Courtesy of Feit, 1997.*)

each host has a unique address in its network. Hosts in two different networks communicate through bridges, routers, or gateways using the IP addresses. Within a network, nodes communicate using low-level hardware addresses. Routers exchange information using routing protocols to discover the topology of the network and calculate the best route to be used for forwarding packets to their destinations. Address resolution protocol (ARP) is used to find the mapping between the IP address and hardware address of hosts in the local network-. Reverse address resolution protocol (RARP) is the protocol used to solve the IP address given the hardware address. Each router in the network has a routing table showing the next router or default router to forward packets to for all known destinations. If there is not a route the packet is dropped. A domain or *autonomous system* (AS) is a part of the Internet owned or managed by a common policy or organization (for example a company's LAN). Interior Border Gateway Protocol (IBGP) is a routing protocol to exchange routing information and reachability updates among the routers within an AS. IBGP makes routing decisions by considering a combination of route attributes, including AS path length, next-hop reachability, local preference, and optionally weight and local administrative policies. These decisions are made to optimize routing within the autonomous system while avoiding loops and ensuring efficient traffic flow. Border gateway protocol (BGP) is a standardized IP routing protocol designed to exchange routing information between autonomous systems on the global Internet. It makes routing decisions based on paths, network policies, or rule-sets configured by a network administrator.

15.9   IP Routing in Satellite Networks

A Router is a network device whose primary purpose is to facilitate communication between networks. Each interface on a router creates a network boundary. Routers operate at Layer 3 enabling end-to-end delivery of a packet across networks. A router maintains a Routing Table that contains paths to all the networks a Router knows how to reach. There are many ways a Router can learn of a network and populate its Routing Table. All packets received for networks not in the Routing Table are silently dropped. Routing is a well known and researched problem on terrestrial networks with decades of optimization and innovation designed to maintain synchronous end-to-end connections across networks. When network topology, connection links or link costs change, routing protocols must recalculate routes and update the routing tables of all routers in the domain until the routers all agree they have the same topological information (*convergence*). Taking a long time to reach convergence (slow convergence problem) can cause packets to be forwarded in a loop between routers (routing loops) or packets to be forwarded to a router that is unable to forward to the destination (black holes). Factors that can contribute to slow convergence include: large networks can take more time to propagate routing updates throughout the network; complexity of the routing protocol; frequency of network changes; high level of link failures or congestion. LEO and multitier satellite networks by their nature have all these factors that can contribute to slow convergence. They have groups of thousands of nodes travelling fast in different planes in opposite directions with frequent link-layer and network-layer handovers causing massive system wide changing of topology and network metrics. The use of existing terrestrial routing protocols as they are, would never reach convergence from all the routing shocks. Satellite networks require a different approach to routing between the satellite networks and the Internet and is an active area of research.

Solutions using dynamic satellite routing are being researched using the fact that a satellite's orbital motion is periodic and its real-time positions, satellite link connectivity, ground links, and real-time metrics are predictable. Using the predictions, the management plane, the control plane and the forwarding plane of the network can be modeled and adaptively improved to ensure a stable and optimal end-to-end path between satellite pairs. Single-layer and multiple-layer satellite networks incorporating LEOs, MEOs, and GEOs can be modeled as a dynamic network model and analyzed. These new satellite mesh networks composed of thousands of OBP satellites that are not static but are growing and shrinking as more satellites are added or are expiring require a dynamic and flexible management scheme. One such solution can be implemented using software-defined network (SDN) technologies and multiprotocol label switching (MPLS).

SDN is a hierarchical network architecture that decouples and separates the tightly coupled TCP/IP data plane and control plane thus providing a scheme to dynamically and flexibly manage the network, see Figure 15.16. Open southbound APIs are designed for centralized control of switches and routers in the network by the SDN controller. Using network virtualization technology to abstract the physical network and its components, the binding relationship between the infrastructure layer and application layer are severed. The network is no longer bound by the hardware architecture thereby allowing flexible networking for on-demand applications with customization and efficient orchestration of network resources including continuous update.

The SDN control layer interacts with the infrastructure layer through the Southbound API and with the application layer through the Northbound API. The control layer is the core of SDN and includes the link discovery module, the topology manger module, the decision making module, and the flow manager module. Link Layer Discovery Protocol (LLDP) and Broadcast Domain Discovery Protocol (BDDP) are used by the link discovery module to build the network topology. In real-time, the topology manager collects and manages the working status and connection of the SDN switches and updates the network topology once the connection status changes. The decision making module, using the real-time view of the global network, can make optimal decisions for data forwarding behaviors. The flow manager module distributes the flow table actively

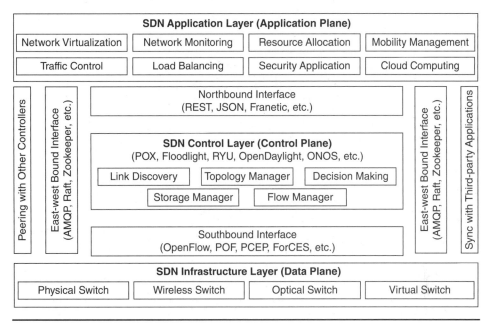

FIGURE 15.16   Software defined network structure.

or passively. An active approach distributes pre-configured flow tables to the SDN switches. In a passive approach the SDN switches request a new flow table entry, for any unmatched data packets in their flow tables, as they occur.

More than one SDN controller, managing different sub-networks, can be used to manage large complex networks, and they are connected with the east-west bound interfaces along with third-party applications. The controllers can be in a peer-to-peer or leader-follower configuration.

SDN infrastructure layer focuses on receiving and forwarding packets through different ports with physical switches that work across multiple layers. This process is based on the data forwarding decisions from the SDN controller that are stored in the SDN switch flow table. The SDN controller maintains the connection between the switch and controller. Hierarchical flow tables are supported and outdated flow table entries are discarded automatically. A single controller has centralized management of many different switches with global network status. Because of network virtualization techniques, the switches can be a distributed virtual multilayer switch, which lowers the cost and moves diagnosis, measurement and operation into intelligent automated remote maintenance and which is especially helpful when the switch is flying in space.

SDN controller placement is crucial and must maximize reliability between the control and data plane and minimize the communication delay between the controller and switches while economizing deployment costs. Because of volatility of the space segment, special attention is needed in link and node failures along with out-of-band control communication reliability. One solution is to have the GEO network as the control plane and LEO as the data plane.

MPLS is a packet-based label switching (forwarding) technology that directs data from one node to the next using labels. It separates packet forwarding (data forwarding plane)

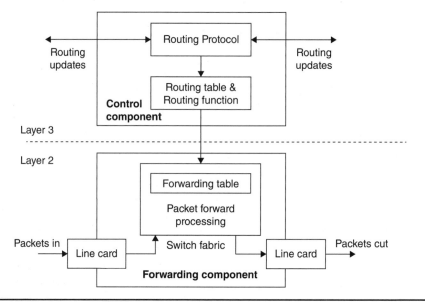

FIGURE 15.17 Functional view of MPLS components.

from routing (control plane). It is different from IP forwarding because the forwarding is done based on labels, which identify established paths between endpoints (similar to ATM virtual circuits VPI/VCI) and not based on IP addresses. Labels are normally local to a single data link, between adjacent routers, and have no global significance (as opposed to an IP address). It is multiprotocol as it supports multiple protocols such as Internet protocol (IP), ATM, T1, frame relay and digital subscriber line (DSL) network protocols. MPLS layers lie between the data link (layer 2) and the network layer (layer 3) or often referred to as layer 2.5 or shim protocol as shown in Fig. 15.17.

An MPLS header containing one or more labels (a label stack) are prefixed onto packets. Each entry in the label stack contains a 32-bit MPLS label with four fields: *label value (20 bits)* contains the label or router alert label; *traffic class (3 bits)* used for QoS priority and explicit congestion notification; *bottom of stack (1 bit)* indicate current label is the last in the stack; *TTL (8 bits)* time to live field as shown in Fig. 15.18.

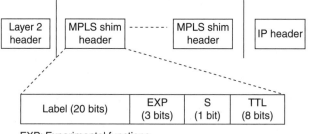

EXP: Experimental functions
S: Level of stack indicator, 1 indicates the bottom of the stack
TTL: Time to live

FIGURE 15.18 MPLS header structure.

MPLS networks contain label switch routers (LSR) and label edge routers (LER). The LSR operates in the middle of the MPLS network and receives a labeled packet, indexes into the label forwarding information base (LFIB) using the label and replaces it with the next hop (label) on the label-switched path (LSP) and then forwards the updated packet. The LER operates on the edge of the MPLS network acting as an entry point (pushing label onto packet) to the MPLS network and an exit point (popping label from packet) from the MPLS network. When forwarding an IP datagram, on entry the LER determines the appropriate label to affix and then forwards it into the MPLS domain. On exiting the MPLS domain, the LER strips the label and forwards the IP packet using IP rules. A LSP can be defined as the sequence of LSRs a packet must traverse on an end-to-end connection. Labels may be distributed between LERs and LSRs using the label distribution protocol (LDP) or resource reservation protocol (RSVP).

At any particular router a PUSH, POP, or SWAP of the label takes place. By viewing the label as an encoded instruction, MPLS can be viewed as an architectural framework that decouples transport from service, since there is no packet inspection (IP addresses are not exposed).

Figure 15.19 depicts a diagram depicting two unidirectional MPLS Label-Switched Paths (LSPs) named PE1→PE4 and PE4→PE2. When an IPv4 H1→H3 (10.1.12.10→ 10.2.34.30) packet arrives at PE1 it is determined that H3 is reachable through PE4, so PE1 places the packet in the PE1→PE4 LSP. It does so by inserting a new MPLS header between the IPv4 and the Ethernet headers of the H1→H3 packet. This header contains MPLS label 1000001, which is locally significant to P1. In MPLS terminology, this operation is a label push. Finally, PE1 sends the packet to P1 that receives the packet and inspects and removes the original MPLS header. Then, P1 adds a new MPLS header with label 1000002, which is locally significant to P2, and sends the packet to P2. This MPLS operation is called a label swap. P2 receives the packet, inspects and removes the MPLS header, and then sends the plain IPv4 packet to PE4. This MPLS operation is

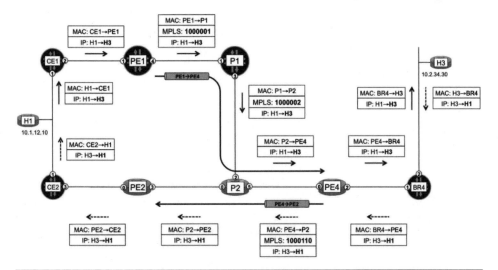

FIGURE 15.19   MPLS Label-Switched Paths PE1→PE4 and PE4→PE2 example diagram.

called a label pop. PE4 receives the IPv4 packet without any MPLS headers. This is fine because PE4 speaks BGP and is aware of all the IPv4 routes, so it knows how to forward the packet toward its destination.

The return packet of H3→H1 travels from PE4 to PE2 in a shorter LSP where only two MPLS operations take place: label push at PE4 and label pop at P2. There is no label swap.

Label stacking can also be viewed as a sequence of instructions. The label stack can contain the predetermined path (set of instructions) a packet would traverse for an end-to-end connection at a particular point in time, and we can determine these connections in advance using an algorithm that takes advantage of the fact that satellites follow predetermined paths. If we discretize the future network in smaller moments of time, we can view the chaotic dynamic topology of LEOs as a set of static networks and route packets using a shortest path/cost algorithm accordingly as shown in Fig. 15.20.

This described process can be accomplished using a centralized control scheme based on SDN technologies to design the routing mechanism, network management and label flow control. The SDN controller completes the routing calculation. By combining inter-satellite and terrestrial user links, a private MPLS protocol can carry and exchange multilayer network protocols. The SDN controller handles routing protocols such as open shortest path first (OSPF). These routing protocols can also be adjusted in real time, based on the time-varying topology, link status, and current traffic. They configure network address mapping tables for tags, satellites, and terminals; and they delete MPLS and LDP switching paths.

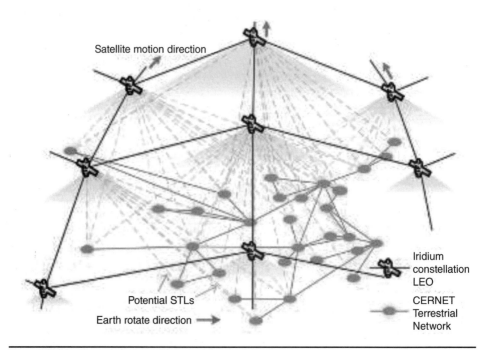

FIGURE 15.20 Snapshot view of future network state.

15.10   The TCP Link

A *virtual communications link* exists between corresponding layers in a network. The header in the TCP segment (see Fig. 15.21) carries instructions that enable communication between the send and receive TCP layers. Of course, the communication must pass through the other layers and along the physical link, but only the TCP layers act on the TCPs contained in the segment header. There is no direct physical link between the TCP layers, and for this reason, it is called a *virtual link*.

The send and receive TCP layers have buffer memories (usually just called *buffers*). The receive buffer holds incoming data, while they are being processed. The send buffer holds data until they are ready for transmission, and it also holds copies of data already sent, until it receives an acknowledgment that the original has been received correctly. The *receive window* is the amount of receive buffer space available at any given time. This changes as the received data is processed and removed from the buffer. The receive TCP layer sends an *acknowledgment* (ACK) signal to the send TCP layer, when it has cleared data from its buffer, and the ACK signal also provides an update on the current size of the receive window.

The send TCP layer keeps track of the amount of data in transit and, therefore, unacknowledged. It can calculate the amount of receive buffer space remaining, allowing for the data in transit. This remaining buffer space represents the amount of data that can still be sent and is termed the *send window*. The send TCP layer also sets a *time-out period*, and failure to receive an ACK signal within this period results in a duplicate packet being sent. On terrestrial networks, the probability of bit error (see Chap. 10) is extremely low, and *congestion* is the most likely reason for loss of ACK signals. Because a network carries traffic from many sources, traffic congestion can occur. The IP layer of the TCP/IP discards packets when congestion occurs, and hence, the corresponding ACK signals from the TCP layer do not get sent. Rather than continually resending packets, the send station reduces its rate of transmission, this being known as *congestion control*. A *congestion window* is applied, which starts at a size of one segment for a new connection. The window is doubled in size for each ACK received until it reaches a maximum value determined by the number of failed ACKs experienced. For normal operation, the congestion window grows equal to the receive window. The congestion window increases slowly at first, but as each doubling takes effect, the size increases exponentially. This controlling mechanism is known as *slow start*. If congestion sets in, it is evidenced by an increase in the failure to receive ACKs, and the send TCP reverts to the slow start.

15.11   Satellite Links and TCP

Although satellite links have formed part of the Internet from its beginning, the rapid expansion of the Internet and the need to introduce congestion control have highlighted certain performance limitations imposed by the satellite links. Before discussing these, it should be pointed out that the increasing demand for Internet services may well be met best with satellite direct-to-home links, and many companies are actively engaged in setting up just such systems. In the ideal case, the virtual link between TCP layers should not be affected by the physical link, and certainly the TCP is so well established that it would be undesirable (some would say unacceptable) to modify it to accommodate peculiarities of the physical link. The factors that can adversely affect TCP performance over satellite links are as follows:

Bit error rate   Satellite links have a higher BER (see Chaps. 10 and 12) than the terrestrial links forming the Internet. Typically, the satellite link BER without error-control coding is around 10^{-6}, whereas a level of 10^{-8} or lower is needed for successful TCP transfer (Chotikapong and Sun, 2000). The comparatively low BER on terrestrial links means that most packet losses are the result of congestion, and the TCP send layer is programmed to act on this assumption. When packets are lost because of high BER, therefore, as they might on satellite links, the TCP layer assumes that congestion is at fault and automatically invokes the congestion control measures. As a result, this slows the throughput.

Round-trip time (RTT)   The round-trip time (RTT) of interest here is the time interval that elapses between sending a TCP segment and receiving its ACK. With geostationary (GEO) satellites, the round-trip propagation path is ground station-to-satellite-to-ground station and back again. The range from ground station to the satellite (see Chap. 3) is on the order of 40,000 km, and therefore, the propagation path for the round trip is $4 \times 40{,}000 = 160{,}000$ km. The propagation delay is therefore $160{,}000 \times 10^3/(3 \times 10^8) = 0.532$ s. This is just the space propagation delay. The total round-trip time must consider the propagation delays on the terrestrial circuits and the delays resulting from signal processing. For order of magnitude calculations, an RTT value of 0.55 s would be appropriate. The send TCP layer must wait this length of time to receive the ACKs, and of course, it cannot send new segments until the ACKs are received, which is going to slow the throughput. The send TCP timeout period is also based on the RTT, and may be unduly lengthened. Also, with interactive applications, such as Telnet, this delay is highly undesirable.

Bandwidth-delay product (BDP)   The RTT is also used in determining an important factor known as the *bandwidth delay product* (BDP). The delay part of this refers to the RTT, since a sender must wait this amount of time for the ACK before sending more data. The bandwidth refers to the channel bandwidth. As shown in Chaps. 10 and 12, bandwidth and bit rate are directly related. In network terminology, the bandwidth is usually specified in bytes per second (or multiples of this), where it is understood that 1 byte is equal to 8 bits. For example, a satellite bandwidth of 36 MHz carrying a Binary Phase-shift keying BPSK signal could handle a bit rate given by Eq. (14.30) as 30 Mbps. This is equivalent to 3.75×10^6 Bps or about 3662 kbps. If the sender transmits at this rate, the largest packet it can send within the RTT of 0.55 s is $3662 \times 0.55 = 2014$ kB approximately. This is the BDP for the two-way satellite channel. The channel is sometimes referred to as a pipeline, and one that has a high BDP, as a long fat pipe. Now the receive TCP layer uses a 16-bit word to notify the send TCP layer of the size of the receive window it is going to use. Allowing 1 byte for certain overheads, the biggest segment size that can be declared for the receive window is $2^{16} - 1 = 65{,}535$ bytes, or approximately 64 kB. (Recall that 1 kB is equal to 1024 bytes.) This falls well short of the 2014 kilobytes set by the BDP for the channel, and thus the channel is very underutilized.

Variable round-trip time   Where lower earth orbiting satellites are used such as those in LEOs and MEOs, the propagation delays are much less than that for the GEO. The slant range to LEOs is typically on the order of a few thousand kilometers at most, and for MEOs, a few tens of thousands kilometers. The problem with these orbits is not so much the absolute value of delay as the variability. Because these satellites are not geostationary, the slant range varies, and for continuous communications there is the need for intersatellite links, which also adds to the delay and the variability. For example, for LEOs, the delay can vary from a few to about 80 millis. Whether or not this has an impact on TCP performance is currently an open question (RFC-2488).

Asymmetric use For economic reasons equipment used for sending to satellites may be asymmetrical which may have an impact on TCP performance.

Intermittent connectivity In non-GSO orbit, TCP connections may be handover to another satellite or routed to a different ground station which may cause packet or connection loss or delay.

15.12 Enhancing TCP over Satellite Channels Using Standard Mechanisms (RFC-2488)

In keeping with the objective that, where possible, the TCP itself should not be modified to accommodate satellite links, the *Request for Comments 2488* (RFC-2488) describes in detail several ways in which the performance over satellite links can be improved. These are summarized in Table 15.4. The first two mechanisms listed do not require any changes to the TCP. The others do require extensions to the TCP. As always, any extensions to the TCP must maintain compatibility with networks that do not employ the extensions. Brief descriptions of the mechanisms are included, but the reader is referred to the *Requests for Comments* (RFCs) for full details (see Sec. 15.11).

MTU stands for *maximum transmission unit*, and *Path MTU-Discovery* is a method that allows the sender to find the largest packet and, hence, largest TCP segment size that can be sent without fragmentation. The congestion window is incremented in

Mechanism	Use	RFC-2488 Section	Where Applied
Path MTU-Discovery	Recommended	3.1	Sender
FEC	Recommended	3.2	Link
TCP congestion control			
Slow start	Required	4.1.1	Sender
Congestion avoidance	Required	4.1.1	Sender
Fast retransmit	Recommended	4.1.2	Sender
Fast recovery	Recommended	4.1.2	Sender
TCP large windows			
Window scaling	Recommended	4.2	Sender and receiver
PAWS	Recommended	4.2	Sender and receiver
RTTM	Recommended	4.2	Sender and receiver
TCP SACKS	Recommended	4.4	Sender and receiver

TABLE 15.4 Summary of Objectives in RFC-2488

segments; hence, larger segments allow the congestion window to increment faster in terms of number of bytes carried. There is a delay involved in implementing Path MTU-Discovery, and of course, there is the added complexity. Overall, however, it improves the performance of TCP over satellite links.

Forward error correction (FEC)   *Lost packets, whether from transmission errors or* congestion, are assumed by the TCP to happen because of congestion, which means that congestion control is implemented, with its resulting reduction in throughput. Although there is ongoing research into ways of identifying the mechanisms for packet loss, the problem remains. Application of FEC (as described in Chap. 11) therefore should be used where possible.

Slow start and congestion avoidance   These strategies have already been described in Sec. 15.8, along with the problems introduced by long RTTs. Slow start and congestion avoidance control the number of segments transmitted, but not the size of the segments. Using Path MTU-Discovery as described earlier can increase the size, and hence the data throughput is improved.

Fast retransmit and fast recovery   From the nature of the ACKs received, the fast retransmit algorithm enables the sender to identify and resend a lost segment before its timeout expires. Since the data flow is not interrupted by timeouts, the sender can infer that congestion is not a problem, and the fast recovery algorithm prevents the congestion window from reverting to slow start. The fast retransmit algorithm can only respond to one lost segment per send window. If there is more than one, the others trigger the slow start mechanism.

TCP large windows   As shown in Sec. 15.9 in connection with the BDP, the receive window size is limited by the address field to 64 kilobytes maximum. By introducing a window scale extension into the TCP header, the address field can be effectively increased to 32 bits. Allowing for certain overheads, the maximum window size that can be declared is $2^{30} = 1$ gigabyte (again keeping in mind that 1 gigabyte = 1024^3 bytes). The window size and hence the scale factor can be set locally by the receive TCP layer. Note, however, that the TCP extension must be implemented at the sender and the receiver.

The two mechanisms *protection against wrapped sequence* (PAWS) and *round-trip time measurement* (RTTM) are extensions that should be used with large windows. Maintaining steady traffic flow and avoiding congestion require a current knowledge of the RTT, which can be difficult to obtain with large windows. By including a *time stamp* in the TCP header, the RTT can be measured. Another problem that arises with large windows is that the numbering of old sequences can overlap with new, a condition known as *wrap-around*. The protection against wrapped sequences is an algorithm that also makes use of the time stamp. These algorithms are described fully in RFC-1323.

SACK stands for *selective acknowledgment* and is a strategy that enables the receiver to inform the sender of all segments received successfully. The sender then needs to resend only the missing segments. The strategy should be used where multiple segments may be lost during transmission, such as, for example, in a satellite link, since clearly, retransmission of duplicate segments over long delay paths would seriously reduce the throughput. Full details of SACK are found in RFC-2018.

15.13   Requests for Comments

The rapid growth of the Internet resulted, in large part, from the free and open access to documentation provided by network researchers. The ideas and proposals of researchers are circulated in memos called *requests for comments* (RFCs). They can be accessed on the world wide web at a few sites, for example, http://www.rfceditor.org/. Following is a summary of some of the RFCs that relate specifically to satellite links and have been referred to in Sec. 15.5:

- *RFC-2760, Ongoing TCP Research Related to Satellites, February 2000. Abstract*: This document outlines possible TCP enhancements that may allow TCP to better utilize the available bandwidth provided by networks containing satellite links. The algorithms and mechanisms outlined have not been judged to be mature enough to be recommended by the IETF. The goal of this document is to educate researchers as to the current work and progress being done in TCP research related to satellite networks.

- *RFC-2488, Enhancing TCP Over Satellite Channels Using Standard Mechanisms, January 1999. Abstract*: The TCP provides reliable delivery of data across any network path, including network paths containing satellite channels. While TCP works over satellite channels, there are several IETF standardized mechanisms that enable TCP to more effectively utilize the available capacity of the network path. This document outlines some of these TCP mitigations. At this time, all mitigations discussed in this document are IETF standards track mechanisms (or are compliant with IETF standards).

- *RFC-2018, TCP Selective Acknowledgment Options, October 1996. Abstract*: TCP may experience poor performance when multiple packets are lost from one window of data. With the limited information available from cumulative acknowledgments, a TCP sender can only learn about a single lost packet per round-trip time. An aggressive sender could choose to retransmit packets early, but such retransmitted segments may have already been received successfully. A SACK mechanism, combined with a selective repeat retransmission policy, can help to overcome these limitations. The receiving TCP sends back SACK packets to the sender informing the sender of data that have been received. The sender can then retransmit only the missing data segments. This memo proposes an implementation of SACK and discusses its performance and related issues.

- *RFC-1323, TCP Extensions for High Performance, May 1992. Abstract*: This memo presents a set of TCP extensions to improve performance over large BDP paths and to provide reliable operation over very high-speed paths. It defines new TCP options for scaled windows and timestamps, which are designed to provide compatible interworking with TCPs that do not implement the extensions. The timestamps are used for two distinct mechanisms: RTTM and PAWS. Selective acknowledgments are not included in this memo. This memo combines and supersedes RFC-1072 and RFC-1185, adding additional clarification and more detailed specification. App. C of RFC-1323 summarizes the changes from the earlier RFCs.

- *RFC-1072, TCP Extensions for Long Delay Paths, October 1988. Status of this memo*: This memo proposes a set of extensions to the TCP to provide efficient operation over a path with a high BDP. These extensions are not proposed as an Internet standard currently. Instead, they are intended as a basis for further experimentation and research on TCP performance. Distribution of this memo is unlimited.

15.14   Split TCP Connections

The TCP provides end-to-end connection. This means that the TCP layers at the sender and receiver are connected through a virtual link (see Sec. 15.3) so that such matters as congestion control and regulation of data flow can be carried out without intervention of intermediate stages. It is to preserve this end-to-end connection that many of the extensions to TCP described in the preceding section have been introduced.

If, however, it is assumed that the end-to-end connectivity can be split, new possibilities are opened for the introduction of satellite links as part of the overall Internet. Figure 15.14 shows one possible arrangement (Ghani and Dixit, 1999). Breaking the network in this way is termed *spoofing*. This refers to the fact that the TCP source thinks it is connected to the TCP destination, whereas the *interworking unit* (IWU) performs a protocol conversion. In Fig. 15.14, TCP Reno refers to the TCP with extensions: slow start, congestion avoidance, fast retransmit, fast recovery, support for large windows, and delayed ACKs. At the IWU the data are transferred from the TCP Reno to the data-link protocol. As shown in Fig. 15.21, any one of several link layer protocols may be used. At the destination end, the IWU performs the conversion back to TCP Reno.

One approach developed at Roke Manor Research, Ltd. (West and McCann, 2000) illustrates some of the possibilities and problems associated with splitting. The two key issues are setup and teardown (West and McCann, 2000). To illustrate the process, consider a connection being set up between host A and host B. In setting up the connection, host A sends a synchronizing segment, labeled SYN in the TCP/IP scheme, which

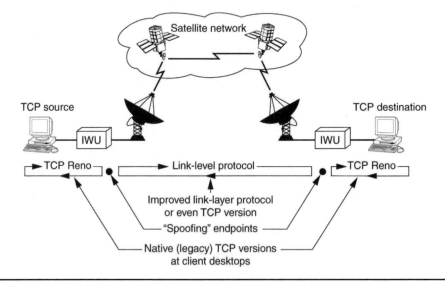

FIGURE 15.21   TCP/IP satellite link spoofing configuration. (*Courtesy of Ghani and Dixit, 1999. Copyright, 1999 IEEE.*)

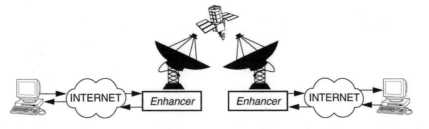

FIGURE 15.22 Physical architecture incorporating enhancer technology. (*Courtesy of West and McCann, 2000.*)

specifies certain protocols to be followed. Host B responds with its own SYN, which contains its protocol requirements, also an ACK signal that carries the number to be used by host A for the first data byte it sends. Host A then responds with an ACK signal that carries the number to be used by host B for the first data byte it sends. The three signals, SYN, SYN/ACK, and ACK, constitute, what is known as a *three-way handshake*.

A three-way handshake is also used to close (teardown). Suppose host B wishes to close. It sends a *final segment* (FIN). Host A acknowledges the FIN with an ACK and follows this with its own FIN. Host B acknowledges the FIN with an ACK. On connections with long RTTs, these three-way handshakes are very time-consuming.

Figure 15.22 shows the system developed at Roke Manor. The *enhancers* perform the same function as the IWUs in Fig. 15.6 in that they terminate the Internet connections and do not require any modifications to the TCP/IP. A propriety protocol is used over the satellite link. Figure 15.23 illustrates a situation where both setup and teardown are spoofed. In this illustration, host B refuses the connection, but host A receives the RESET signal too late. The spoofed FIN ACK tells host A that the data transfer was successful.

Figure 15.24 shows a more appropriate strategy. In this case, the FIN sent by host A is not spoofed. Since host B has refused the connection, host A receives no FIN, ACK

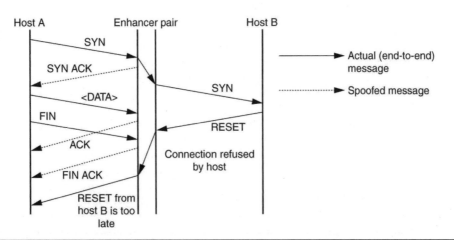

FIGURE 15.23 Connection establishment and closing. (*Courtesy of West and McCann, 2000.*)

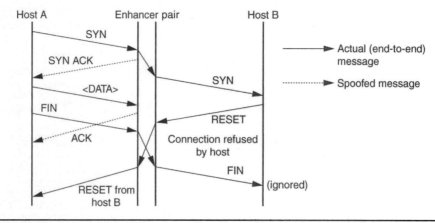

FIGURE 15.24   Improved connection close. (*Courtesy of West and McCann, 2000.*)

back from host B. Therefore, host A can infer from this that there was an error. While this is not as good as a regular TCP/IP connection, which would have reported a failure to connect on the first SYN, the system does adjust to the error and removes the spoofing on the setup on a second try.

15.15   Asymmetric Channels

The term *asymmetry* applies in two senses to an Internet connection. It can refer to the data flow, which is often asymmetric in nature. A short request being sent for a Web page and the returned Web page may be a much larger document. Also, the acknowledgment packets sent on the return or reverse link are generally shorter than the TCP segments sent on the forward link. Values of 1500 bytes for data segments on the forward link and 40 bytes for ACKs on the reverse link are given in RFC-2760.

Asymmetry is also used to describe the physical capacities of the links. For small earth stations (e.g., VSATs), transmit power and antenna size (in effect, the EIRP) limit the uplink data rate, which therefore may be much less than the downlink data rate. Such asymmetry can result in ACK congestion. Again, using some values given in RFC-2760, for a 1.5 Mbps data link a reverse link of less than 20 kbps can result in ACK congestion. The levels of asymmetry that lead to ACK congestion are readily encountered in VSAT networks that share the uplink through multiple access.

In some situations, the reverse link may be completed through a terrestrial circuit, as shown in Fig. 15.25 (Ghani and Dixit, 1999). Here, the TCP source is connected to the satellite uplink through an IWU as before. The downlink signal feeds the small residential receiver, which is a receive-only earth station. An IWU on the receive side converts the data to the TCP format and sends them on to the destination. The ACK packets from the TCP destination are returned to the TCP source through a terrestrial network. As pointed out in RFC-2760, the reverse link capacity is limited not only by its bandwidth but also by queue lengths at routers, which again can result in ACK congestion. Some of the proposed methods of handling asymmetry problems and ongoing research are described in Ghani and Dixit (1999).

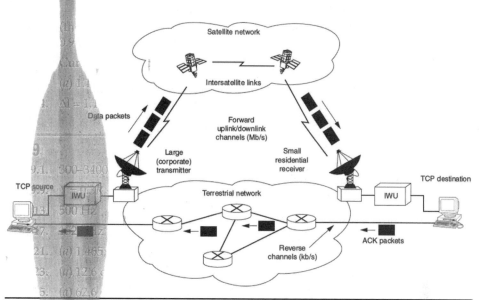

FIGURE 15.25 Asymmetric reverse ACK channel configuration. (*Courtesy of Ghani and Dixit, 1999. Copyright, 1999 IEEE.*)

15.16   Mobility Management

One approach to maintaining connections in satellite networks is to view it as a similar problem as a mobile network. The IETF introduced the mobile Internet protocol (MIP) (RFC5944) and proxy mobile Internet protocol version 6 (PMIPv6) (RFC6475) to provide mobility management for terrestrial IP networks. These IP-based mobility management protocols strive to maintain the TCP connection between a mobile node (MN) which is the device being moved and a static access point (AP) while minimizing the effects of location changes. By using two IP addresses, fixed home address (HA) and a care-of address (CoA), the transport layer connections (i.e., TCP) to the fixed HA are maintained as a home agent on the home network tunnels (encapsulation) the IP packets to a foreign agent on a foreign network that forwards the IP packet to the attached MN on that foreign network. This is similar to how postal mail forwarding works using temporary change of address. Mail addressed to the original home address is received at the home address local post office that uses the change of address information (foreign address) to encapsulate and reroute the mail to the foreign address serviced by the foreign addresses local post office that then delivers the mail to the physical foreign address.

15.17   Proposed Systems

Most of the currently employed satellites operate in what is called a "bent pipe" mode, that is, they relay the data from one host to another without any onboard processing. Also, many of the problems with using geostationary satellites for Internet traffic arise because of the long propagation delay and the resulting high delay-bandwidth product. In some of the newer satellite systems, use is made of LEO and MEO satellites to cut down on the propagation delay time. Also, onboard signal processing is used in many instances that may result in the TCP/IP protocol being exchanged for propriety protocols over the satellite links. The satellite part of the network may carry the IP over ATM. The Ka-band, which covers from 27 to 40 GHz, is used in many (but not all) of these newer systems. Wider bandwidths are available for carriers in the Ka-band compared with those in the Ku-band. A survey of some of these broadband systems are found in Farserotu and Prasad (2000), but company changes in the form of mergers and takeovers occur that can drastically alter the services a company may offer.

15.18   Problems and Exercises

15.1. Write brief notes on the defining features of a broadband network.

15.2. Explain the difference between a wideband network and a broadband network. In network terminology, bandwidth is usually given in terms of bit rate, while in radio systems bandwidth is given in terms of frequency. Show how these concepts are connected.

15.3. Explain briefly what is meant by a *network protocol*. By referring to any of the standard publications showing the *open systems interconnection* (OSI) model, show how the ATM layer model, of Fig. 15.2, relates to the OSI model. (A comprehensive description of the OSI model can be found on the Web site of Wikipedia, the free encyclopedia).

15.4. Briefly describe the difference between synchronous and asynchronous transfer modes. One author (Ramteke, 1994) points out that *synchronous transmission mode* (STM) uses asynchronous multiplexing, and ATM is transmitted using synchronous multiplexing (SONET). Explain how this is.

15.5. Describe the different types of interfaces and their functions in an ATM network.

15.6. With a BER of 10^{-8} the probability of a cell being discarded is about 10^{-13} and the probability of cells with undetected errors getting through is about 10^{-20} (Goralski, 1995, p. 140). On average, how many cells would be transmitted before a discard occurs? Is the probability of a cell with undetected errors getting through of any concern?

15.7. Briefly describe the difference between an ATM digital cross connect switch, and an ATM switch.

15.8. Briefly describe the difference between permanent virtual circuits and switched virtual circuits.

15.9. Explain what is meant by *bandwidth on demand* and how this is achieved in ATM.

15.10. A single user makes use of all the cells in an ATM transmission at a speed of 50 Mbps. What is the payload bandwidth?

15.11. An ATM system transmits 8000 frames per second with a maximum of 10 cells per frame. What is the speed of the link?

15.12. What type of signals are sensitive to variations in cell delay, and how does ATM accommodate these signals?

15.13. State the essential differences between the five classes of service offered with ATM, and the applications for which each class would be used. What is meant by quality of service?

15.14. What are the main technical parameters used in measuring ATM performance?

15.15. Discuss briefly the main problems encountered in incorporating satellite links in ATM networks, and the steps taken to overcome these.

15.16. Explain what is meant by a "bent pipe" system in connection with satellite communications.

15.17. Describe the main distinguishing features between satellite relay, satellite access, and satellite interconnect, in connection with ATM over satellite.

15.18. What are the main advantages and disadvantages in using LEO satellites compared with GEO satellites in ATM networks?

15.19. Write brief notes on the Internet and the WWW showing the relationship between them.

15.20. Describe the layered architecture that applies to networks, and indicate which of these layers contain (*a*) the IP and (*b*) the TCP. Explain what is meant by protocol.

15.21. Describe how *connectionless* and *connection-oriented* layers differ.

15.22. Describe what is meant by a *packet* in the different layers of a network.

15.23. Binary transmission occurs at a rate of 5000 bytes per second. What is this in bits per second?

15.24. Binary transmission occurs at a rate of 25 kilobytes per second. What is this in kilobits per second?

15.25. Binary transmission occurs at a rate of 30 megabytes per second. What is this in megabits per second?

15.26. What is meant by a *virtual link* in a network?

15.27. Define and explain what is meant by (*a*) *receive window*, (*b*) *send window*, and (*c*) *timeout* with reference to the Internet.

15.28. Explain what is meant by *congestion* and *slow start* in relation to Internet traffic.

15.29. In the early days of the Internet, Internet traffic could be sent over satellite links without any difficulty. State briefly the changes that have taken place that introduce difficulties and that require special attention.

15.30. Explain what is meant by RTT. Given that the uplink range to a satellite is 38,000 km and the downlink range is 40,000 km, calculate the RTT.

15.31. The overall bandwidth for a satellite link is 36 MHz, and the filtering is raised-cosine with a rolloff factor of 0.2. BPSK modulation is used. The range to the satellite is 39,000 km in both directions. Calculate the size of the largest packet that could be in transit if the BDP was the limiting factor. Assume that the RTT is equal to the space propagation delay.

15.32. A LEO satellite is used for Internet transmissions. The orbit can be assumed circular at an altitude of 800 km. Assuming that the minimum usable angle of elevation of the satellite is 10° (the angle above the horizon), calculate the approximate values of maximum and minimum RTTs. A spherical earth of uniform mass and radius of 6371 km may be assumed.

15.33. Repeat Prob. 15.32 for a circular MEO at altitude 20,000 km.

15.34. Briefly discuss the enhancements covered in RFC-2488 and how these might improve the performance of Internet traffic over satellite links.

15.35. By accessing the RFC site on the WWW, find the latest version of RFC-2760. Comment on this.

15.36. Explain what is meant by *split TCP connections* and why these might be considered undesirable for Internet use.

15.37. Explain what is meant by *asymmetric channels*. Describe how asymmetric channels may be incorporated in Internet connections via satellites.

References

Akyildiz, I. F., I. Joe, H. Driver, and Y. L. Ho. 1998. "A New Adaptive FEC Scheme for Wireless Networks." *Proc. IEEE Milcom'98*, Boston, MA, October.

Chotikapong, Y., and Z. Sun. 2000. "Evaluation of Application Performance for TCP/IP via Satellite Links." *IEE (London) Aerospace Group Seminar: Satellite Services and the Internet*, London. 17 February.

Ericsson. 2003. CLA-2000™/ATM Link Accelerator™, at http://www.componedex. com/ products/cla_atm.htm

Farserotu, J., and R. Prasad. 2000. "A Survey of Future Broadband Multimedia Satellite Systems, Issues and Trends." *IEEE Commun. Mag.*, vol. 38, no. 6, pp. 128–133, June.

Feit, S. 1997. *TCP/IP*, 2d ed. McGraw-Hill, New York.

Ghani, N., and S. Dixit. 1999. "TCP/IP Enhancements for Satellite Networks." *IEEE Commun. Mag.*, vol. 37, no. 7, pp. 64–72, July. Goralski, W. J. 1995. *Introduction to ATM Networking*. McGraw-Hill, New York.

Hayt, W. H., and J. E. Kemmerly 1978. *Engineering Circuit Analysis*. McGraw-Hill, New York.

Ivancic, W. D., B. D. Frantz, and M. J. Spells. 1998. "MPEG-2 Over Asynchronous Transfer Mode (ATM) Over Satellite Quality of Service (QoS) Experiments: Laboratory Tests." *NASA/TM/1998 206535*, September.

Ivancic, W. D., D. E. Brooks and B. D. Frantz. 1997. "ATM Quality of Service Tests for Digitized Video Using ATM Over Satellite: Laboratory Tests." *NASA/TM/107421*, July.

Jacobs, I. 2000. W3C. At http://www.w3.org/Consortium/, March.

Leiner, B. M., V. G. Cerf, D. D. Clark, R. E. Khan, L. Kleinrock, D. C. Lynch, J. Postel, L. G. Roberts, and S. Wolff. 2000. A Brief History of the Internet, at http://www.isoc.org/ internet-history/brief.html, revised 14 April.

Mackenzie, L. 1998. *Communications and Networks*. McGraw-Hill, New York.

Ramteke, T. 1994. *Networks*. Prentice Hall, New Jersey.

Russell, T. 2000. *Telecommunications Protocols*. 2d ed. McGraw-Hill, New York.

Todorova, P. 2002. Network Management in ATM LEO Satellite Networks. Proceedings of the 35th Annual Hawaii International Conference on System Sciences. (Google search: Todorova network management ATM)

Todorova, P., and H. N. Nguyen. 2001. "On-board Buffer Architectures for Low Earth (LEO) Satellite ATM Systems." *IEEE GLOBECOM 2001*, November 25–29, 2001, San Antonio, Texas.

Toh, C., and V. O. K. Li, 1998. "Satellite ATM Network Architectures: An Overview." *IEEE Network*, vol. 12, no. 5, pp. 61–71, September/October. Walrand, J. 1991. *Communication Networks*. Aksen Associates, Homewood, IL.

Web ProForum Tutorials of the International Engineering Consortium, at http://www.lec.org.

West, M., and S. McCann. 2000. "Improved TCP Performance over Long-Delay and Error-Prone Links." *IEE (London) Aerospace Group Seminar, Satellite Services and the Internet*, London. 17 February.

CHAPTER 16

Direct Broadcast Satellite (DBS) Television

16.1 Introduction

Satellites provide *broadcast* transmissions in the fullest sense of the word, because antenna footprints can be made to cover large areas of the earth. The idea of using satellites to provide direct transmissions into the home has been around for many years, and the services provided are known generally as *direct broadcast satellite* (DBS) services. Broadcast services include audio, television, and Internet services. Direct broadcast television, which is digital TV, is the subject of this chapter.

A comprehensive overview covering the early years of DBS in Europe, the United States, and other countries is given in Prichard and Ogata (1990). Some of the regulatory and commercial aspects of European DBS are found in Chaplin (1992), and the U.S. market is discussed in Reinhart (1990). Reinhart defines three categories of U.S. DBS systems, shown in Table 1.4. Of interest to the topic of this chapter is the high power category, the primary intended use of which is for DBS.

16.2 Orbital Spacing

From Table 1.4 it is seen that the orbital spacing is 9° for the high-power satellites, so adjacent satellite interference is considered nonexistent. The DBS orbital positions along with the transponder allocations for the United States are shown in Fig. 16.1. It should be noted that although the DBS services are spaced by 9°, *clusters of satellites* occupy the nominal orbital positions. For example, the following satellites are located at 119°W longitude: EchoStar VI, launched on July 14, 2000; EchoStar IV, launched on May 8, 1998; EchoStar II, launched on September 10, 1996; and EchoStar I, launched on December 28, 1995 (source: http://www.dishnetwork.com/).

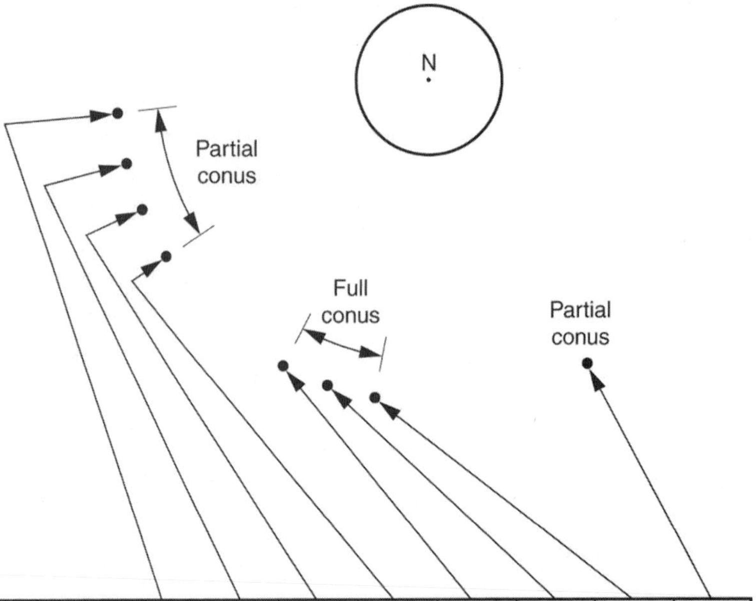

Satellite location	175°	166°	157°	148°	119°	110°	101°	61.5°
Company	Transponder totals							
Continental		11						11
DBSC	11							11
DirecTV			27				27	
Dominion		8						8
Echostar				24				
Echostar/Directsat	11	11			21	1		
MCI						28		
TCI/Tempo					11			
USSB				8		3	5	
Unassigned	10	2	5					2

FIGURE 16.1   DBS orbital positions for the United States.

16.3   Power Rating and Number of Transponders

From Table 1.4 it can be seen that satellites primarily intended for DBS have a higher [EIRP] than for the other categories, being in the range 51 to 60 dBW. At a *Regional Administrative Radio Council* (RARC) meeting in 1983, the value established for DBS was 57 dBW (Mead, 2000). Transponders are rated by the power output of their high-power amplifiers. Typically, a satellite may carry 32 transponders. If all 32 are in use, each operates at the lower power rating of 120 W. By doubling up the high-power amplifiers, the number of transponders is reduced by half to 16, but each transponder operates at the higher power rating of 240 W. The power rating has a direct bearing on the bit rate that can be handled, as described in Sec. 16.8.

16.4   Frequencies and Polarization

The frequencies for DBSs vary from region to region throughout the world, although these are generally in the Ku-band. For region 2 (see Sec. 1.2), Table 1.4 shows that for high-power satellites, the primary use of which is for DBS, the uplink frequency range is 17.3 to 17.8 GHz, and the downlink range is 12.2 to 12.7 GHz. The medium-power satellites listed in Table 1.4 also operate in the Ku-band at 14 to 14.5 GHz uplink and 11.7 to 12.2 GHz downlink. The primary use of these satellites, however, is for point-to-point applications, with additional use in the DBS service. In this chapter only the high-power satellites intended primarily for DBS are discussed.

The available bandwidth (uplink and downlink) is 500 MHz, and 32 transponder channels, each of bandwidth 24 MHz, can be accommodated. The bandwidth is sometimes specified as 27 MHz, but this includes a 3-MHz guard band allowance. Therefore, when calculating bit-rate capacity, the 24 MHz value is used, as shown in Sec. 16.5. The total of 32 transponders requires the use of both *right-hand circular polarization* (RHCP) and *left-hand circular polarization* (LHCP) to permit frequency reuse, and guard bands are inserted between channels of a given polarization. The DBS frequency plan for region 2 is shown in Fig. 16.2.

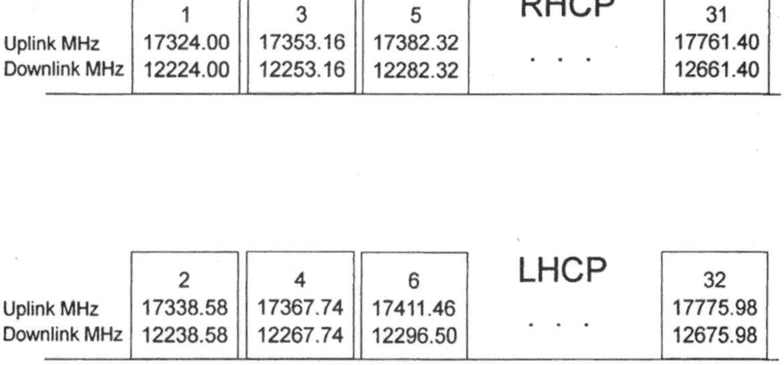

	1	3	5	RHCP		31
Uplink MHz	17324.00	17353.16	17382.32	· · ·		17761.40
Downlink MHz	12224.00	12253.16	12282.32			12661.40

	2	4	6	LHCP		32
Uplink MHz	17338.58	17367.74	17411.46	· · ·		17775.98
Downlink MHz	12238.58	12267.74	12296.50			12675.98

FIGURE 16.2   The DBS frequency plan for Region 2.

16.5 Transponder Capacity

The 24-MHz bandwidth of a transponder can carry only one analog television channel. However, to be commercially viable, DBS television [also known as *direct-to-home* (DTH) television] requires many more channels, and this requires a move from analog to digital television. Digitizing the audio and video components of a television program allows *signal compression* to be applied, which greatly reduces the bandwidth required. The signal compression used in DBS is a highly complex process, and only a brief overview is given here of the process. Before doing this, an estimate of the bit rate that can be carried in a 24-MHz transponder is made.

From Eq. (10.22), the symbol rate that can be transmitted in each bandwidth is

$$R_{sym} = \frac{B_{IF}}{1+\rho} \tag{16.1}$$

Thus, with a bandwidth of 24 MHz and allowing for a rolloff factor of 0.2, the symbol rate is

$$R_{sym} = \frac{24 \times 10^6}{1+0.2} = 20 \times 10^6 \text{ symbols/s}$$

Satellite digital television uses QPSK. Thus, using $M = 4$ in Eq. (10.3) gives $m = 2$, and the bit rate from Eq. (10.5) is

$$R_b = 2 \times R_{sym} = 40 \text{ Mbps}$$

This is the bit rate that can be carried in the 24-MHz channel using QPSK.

16.6 Bit Rates for Digital Television

The bit rate for digital television depends very much on the picture format. One way of estimating the uncompressed bit rate is to multiply the number of pixels in a frame by the number of frames per second, and multiply this by the number of bits used to encode each pixel. The number of bits per pixel depends on the color depth per pixel, for example 16 bits per pixel gives a color depth of $2^{16} = 65,536$ colors. Using the HDTV format having a pixel count per frame of 1920×1080 and a refresh rate of 30 frames per second as shown in Table 16.1, the estimated bit rate is 995 Mbps. (A somewhat different estimate is sometimes used, which allows for 8 bits for each of the three primary colors. This results in a bit rate of approximately 1.49 Gbps for this version of HDTV).

From Table 16.1 it is seen that the uncompressed bit rate ranges from 118 Mbps for standard definition television at the lowest pixel resolution to 995 Mbps for high definition TV at the highest resolution. As a note of interest, the broadcast raster for studio-quality television, when digitized according to the international CCIR-601 television standard, requires a bit rate of 216 Mbps (Netravali and Lippman, 1995).

A single DBS transponder must carry somewhere between four and eight TV programs to be commercially viable (Mead, 2000). The programs may originate from a variety of sources, for example film, analog TV, and videocassette. Before transmission, these must all be converted to digital, compressed, and then *time-division multiplexed* (TDM). This TDM baseband signal is applied as QPSK modulation to the uplink carrier reaching a given transponder.

No.	Format Type	Name	Aspect Ratio	Resolution Pixels	Frames per Second	Uncompressed Bit Rate, mbps
1	SDTV	480i	4:3	640 × 480	30	148
2	EDTV	480p	4:3	640 × 480	24	118
3	EDTV	480p	4:3	640 × 480	30	148
4	EDTV	480p	4:3	640 × 480	60	295
5	EDTV	480i	4:3	704 × 480	30	162
6	EDTV	480p	4:3	704 × 480	24	130
7	EDTV	480p	4:3	704 × 480	30	162
8	EDTV	480p	4:3	704 × 480	60	324
9	EDTV	480i	16:9	704 × 480	30	162
10	EDTV	480p	16:9	704 × 480	24	130
11	EDTV	480p	16:9	704 × 480	30	162
12	EDTV	480p	16:9	704 × 480	60	324
13	HDTV	720p	16:9	1280 × 720	24	334
14	HDTV	720p	16:9	1280 × 720	30	442
15	HDTV	720p	16:9	1280 × 720	60	885
16	HDTV	1080i	16:9	1920 × 1080	30	995
17	HDTV	1080p	16:9	1920 × 1080	24	796
18	HDTV	1080p	16:9	1920 × 1080	30	995

Notes: ATSC, Advanced Television Systems Committee; HDTV, high-definition television; EDTV, enhanced definition television; I, interlaced scanning; p, progressive scanning; SDTV, standard definition television.
Sources: Booth (1999); www.timefordvd.com, 2004.

TABLE 16.1 ATSC Television Formats

The compressed bit rate, and hence the number of channels that are carried, depends on the type of program material. Talk shows where there is little movement require the lowest bit rate, while sports channels with lots of movement require comparatively large bit rates. Typical values for SDTV are in the range of 4 Mbps for a movie channel, 5 Mbps for a variety channel, and 6 Mbps for a sports channel (from MPEG and DSS Technical Notes (v0.3) by C. Fogg, 1995). Compression is carried out to *Moving Pictures Expert Group* (MPEG) standards.

16.7 MPEG Compression Standards

MPEG is a group within the *International Standards Organization and the International Electrochemical Commission* (ISO/IEC) that undertook the job of defining standards for the transmission and storage of moving pictures and sound. The standards are concerned only with the bit stream syntax and the decoding process, not with how encoding and decoding might be implemented. Syntax covers matters such as bit rate, picture resolution, time frames for audio, and the packet details for transmission. The design of hardware for the encoding and decoding processes is left to the equipment manufacturer. Comprehensive descriptions of the MPEG can be found at http://www.mpeg.org and in Sweet (1997) and Bhatt et al. (1997). The MPEG standards currently available are MPEG-1, MPEG-2, MPEG-4, and MPEG-7. For a brief explanation of the "missing numbers," see Koenen (1999).

In DBS systems, MPEG-2 is used for video compression. As a first or preprocessing step, the analog outputs from the red (R), green (G), and blue (B) color cameras are converted to a luminance component (Y) and two chrominance components (Cr) and (Cb). This is like the analog NTSC arrangement shown in Fig. 9.7, except that the coefficients of the matrix **M** (the 3×3 matrix) are different. In matrix notation, the equation relating the three primary colors to the Y, Cr, and Cb components is

$$\begin{bmatrix} Y \\ Cr \\ Cb \end{bmatrix} = \begin{bmatrix} 0.299 & 0.587 & 0.114 \\ -0.168736 & -0.331264 & 0.5 \\ 0.5 & -0.418688 & -0.081312 \end{bmatrix} \begin{bmatrix} R \\ G \\ B \end{bmatrix} \tag{16.2}$$

The Y, Cr, and Cb analog signals are sampled in the digitizer shown in Fig. 16.3. It is an observed fact that the human eye is less sensitive to resolution in the color components (Cr and Cb) than the luminance (Y) component. This allows a lower sampling rate to be used for the color components. This is referred to as *chroma subsampling*, and it represents one step in the compression process. MPEG-2 uses 4:2:0 sampling, which is described next.

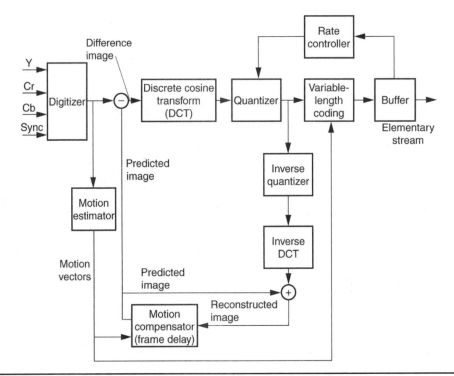

FIGURE 16.3 MPEG-2 encoder paths. (*Courtesy of Bhatt et al., 1997. IEEE.*)

Sampling is usually indicated by the ratios Y:U:V where Y represents the luminance (or luma) sampling rate, U the Cb sampling rate, and V the Cr sampling rate. The values for YUV are normalized to a value of 4 for Y, and ratios commonly encountered with digital TV are 4:4:4, 4:2:2 and 4:2:0. These are explained next.

4:4:4 means that the sampling rates of Y, Cb, and Cr are equal. Each pixel has three digital words, one for each of the component signals. If the words are 8 bits (commonly called a byte, but see Sec. 15.2) then each pixel is encoded in 3 bytes.

4:2:2 means that the Cb and Cr signals are sampled at half the rate of the Y signal component. Every two pixels have two bytes for the Y signal, one byte for the Cb signal and one byte for the Cr signal, resulting in 4 bytes for the 2-pixel block.

4:2:0 means that Cb and Cr are sampled at half the Y sampling rate, but they are sampled only on alternate scan lines. Thus vertical as well as horizontal resolution is reduced by half. A 2×2 pixel block has 6 bytes: 4 bytes for Y, 1 byte for Cb, and 1 byte for Cr.

Following the digitizer, difference signals are formed, and the *discrete cosine transform* (DCT) block converts these to a "spatial frequency" domain. The familiar Fourier transform transforms a time signal $g(t)$ to a frequency domain representation $G(f)$, allowing the signal to be filtered in the frequency domain. Here, the variables are time t and frequency f. In the DCT situation, the input signals are functions of the x (horizontal) and y (vertical) space coordinates, $g(x, y)$. The DCT transforms these into a domain of new variables u and v $G(u, v)$. The variables are called *spatial frequencies* in analogy with the time-frequency transform. It should be noted that $g(x, y)$ and $G(u, v)$ are *discrete functions*. In the quantizer following the DCT transform block, the discrete values of $G(u, v)$ are quantized to predetermined levels. This reduces the number of levels to be transmitted and therefore provides compression. The components of $G(u, v)$ at the higher spatial frequencies represent finer spatial resolution. The human eye is less sensitive to resolution at these high spatial frequencies; therefore, they can be quantized in much coarser steps. This results in further compression. (This step is analogous to the nonlinear quantization discussed in Sec. 10.3.)

Compression is also achieved through *motion estimation*. Frames in MPEG-2 are designated I, P, and B frames, and motion prediction is achieved by comparing certain frames with other frames. The I frame is an independent frame, meaning that it can be reconstructed without reference to any other frames. A P (for previous) frame is compared with the previous I frame, and only those parts that differ because of movement need to be encoded. The comparison is carried out in sections called *macroblocks* for the frames. A macroblock consists of 16×16 pixels. A B (for bidirectional) frame is compared with the previous I or P frame and with the next P frame. This obviously means that frames must be stored for the forward comparison to take place. Only the changes resulting from motion are encoded, which provides further compression. An estimate of the compression required can be made by assuming a value of 200 Mbps for the uncompressed bit rate for SDTV (see Table 16.1), and taking 5 Mbps as typical of that for a TV channel, the compression needed is on the order $200/5 = 40:1$. The 5 Mbps includes audio and data, but these should not take more than about 200 kbps. Audio compression is discussed later in this section.

The whole encoding process relies on digital decision-making circuitry and is computationally intensive and expensive. The decoding process is much simpler because the rules for decoding are part of the syntax of the bit stream. Decoding is carried out in the *integrated receiver decoder* (IRD) unit. This is described in Sec. 16.8.

In DBS systems, MPEG-1 is used for audio compression, and as discussed earlier, MPEG-2 is used for video compression. Both MPEG standards cover audio and video, but MPEG-1 video is not designed for DBS transmissions. MPEG-1 audio supports mono and two-channel stereo only, which is considered adequate for DBS systems currently in use. MPEG-2 audio supports multichannel audio in addition to mono and stereo. It is fully compatible with MPEG-1 audio, so the IRDs, which carry MPEG-2 decoders, have little trouble being upgraded to work with DBS systems transmitting multichannel audio.

The need for audio compression can be seen by considering the bit rate required for high-quality audio. The bit rate is equal to the number of samples per second (the sampling frequency f_s) multiplied by the number of bits per sample n:

$$R_b = f_s \times n \tag{16.3}$$

For a stereo CD recording, the sampling frequency is 44.1 kHz, and the number of bits per sample is 16:

$$R_b = 44.1 \times 10^3 \times 16 \times 2 = 1411.2 \text{ kbps}$$

The factor 2 appears on the right-hand side because of the two channels in stereo. This bit rate, approximately 1.4 Mbps, represents too high a fraction of the total bit rate allowance per channel, and hence the need for audio compression. Audio compression in MPEG exploits certain *perceptual phenomena* in the human auditory system. It is known that a loud sound at one frequency can mask a less intense sound at a nearby frequency. For example, consider a test conducted using two tones, one at 1000 Hz, which is called the *masking tone*, and the other at 1100 Hz, the *test tone*. Starting with both tones at the same level, say, 60 dB above the threshold of hearing, if now the level of the 1000-Hz tone is held constant while reducing the level of the 1100-Hz tone, a point is reached where the 1100-Hz tone becomes inaudible. The 1100-Hz tone is said to be *masked* by the 1000-Hz tone. Assume for purposes of illustration that the test tone becomes inaudible when it is 18 dB below the level of the masking tone. This 18 dB is the *masking threshold*. It follows that any noise below the masking threshold also is masked.

For the moment, assuming that only these two tones are present, then it can be said that the *noise floor* is 18 dB below the masking tone. If the test-tone level is set at, say, 6 dB below the masking tone, then of course it is 12 dB above the noise floor. This means that the signal-to-noise ratio for the test tone need to be no better than 12 dB. Now in a *pulse-code modulated* (PCM) system the main source of noise is that arising from the quantization process (see Sec. 10.3). It can be shown (see Roddy and Coolen, 1995) that the signal-to-quantization noise ratio is given by

$$\left(\frac{S}{N}\right)_q = 2^{2n} \tag{16.4}$$

where n is the number of bits per sample. In decibels this is

$$\left[\frac{S}{N}\right]_q = 10 \log 2^{2n}$$

$$\cong 6n \text{ dB} \tag{16.5}$$

This shows that increasing n by 1 bit increases the signal-to-quantization noise ratio by 6 dB. Another way of looking at this is to say that a 1-bit decrease in n increases the quantization noise by 6 dB. In the example above where 12 dB is an adequate signal-to-noise ratio, Eq. (16.5) shows that only 2 bits are needed to encode the 1100-Hz tone (i.e., the levels are quantized in steps represented by 00, 01, 10, 11). By way of contrast, the CD samples taken at a sampling frequency of 44.1 kHz are quantized using 16 bits to give a signal-to-quantization noise ratio of 96 dB.

Returning to the example of two tones the audio signal is not consist of two single tones but is a complex signal covering a wide spectrum of frequencies. In MPEG-1, two processes take place in parallel, as illustrated in Fig. 16.4. The filter bank divides the spectrum of the incoming signal into subbands. In parallel with this the spectrum is analyzed to permit identification of the masking levels. The masking information is passed to the quantizer, which then quantizes the subbands according to the noise floor.

The masking discussed so far is referred to as *frequency masking* for the reasons given earlier. It is also an observed phenomenon that the masking effect lasts for a short period after the masking signal is removed. This is termed *temporal masking,* and it allows further compression in that it extends the time for which the reduction in quantization applies. There are many more technical details to MPEG-1 than can be covered here, and the reader is referred to Mead (2000), which contains a detailed analysis of MPEG-1. The MPEG Web page—at http://www.mpeg.org/MPEG/audio.html—also provides several articles on the subject. The compressed bit rate for MPEG-1 audio used in DBS systems is 192 kbps.

MPEG-4 (part 10) was developed jointly by the *Video Coding Experts Group* (VCEG) of the *International Telecommunication Union* (ITU), Telecommunication Standardization Sector (ITU-T) which uses the designation H.264, and the MPEG of the ISO/IEC. As noted in Sullivan et al. (2004), this version of MPEG is known by at least six different names (H.264, H.26L, ISO/IEC 14496-10, JVT, MPEG-4 AVC, and MPEG-4 Part 10) and the abbreviation AVC is commonly used to denote *advanced video coding.* Following the usage in Sullivan et al., it is denoted here by H.264/AVC.

Areas of application include video telephony, video storage and retrieval (DVD and hard disk), digital video broadcast, and others. In general terms, MPEG-4 provides many features not present with other compression schemes, such as interactivity for viewers, where objects within a scene can be manipulated, but from the point

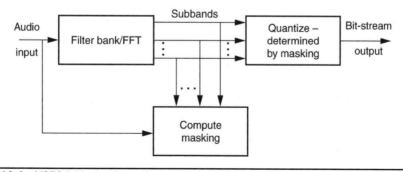

FIGURE 16.4   MPEG-1 block schematic.

of view of satellite television, the major advantage is the reduction in bit rate offered. About a 2:1 reduction in bit rate, on average is achievable with H.264/AVC compared with MPEG-2, and in July 2004, an amendment known as *fidelity range extensions* (FRExt, amendment 1) was added to H.264/AVC that can provide a reduction of as much as 3:1 in certain situations (Sullivan et al., 2004). FRExt supports 4:2:2 and 4:4:4 sampling.

As with MPEG-2 the analog outputs from the red (R), green (G), and blue (B) color cameras are converted to a luminance component (Y) and two chrominance components (Cr) and (Cb) but with a different **M** matrix, this being:

$$\begin{bmatrix} Y \\ Cr \\ Cb \end{bmatrix} = \begin{bmatrix} 0.2126 & 0.587 & 0.0722 \\ -0.119977 & -0.331264 & 0.523589 \\ 0.561626 & -0.418688 & -0.051498 \end{bmatrix} \begin{bmatrix} R \\ G \\ B \end{bmatrix} \tag{16.6}$$

It follows that any format conversion requires a matrix recalculation.

H.264/AVC takes advantage of the increases in processing power available from computer chips, but at the cost of more expensive equipment, both for the TV broadcaster and the consumer. As with MPEG-2, frames are compared for changes through comparing macroblocks of 16×16 pixels, but H.264/AVC also allows for comparisons of submacroblocks of pixel groups 16×8, 8×16, 8×8, 8×4, 4×8, and 4×4. At present it is not backward compatible with MPEG-2, which may present a problem with some high definition TV (see Sec. 16.13).

16.8 Forward Error Correction (FEC)

Because of the highly compressed nature of the DBS signal, there is little redundancy in the information being transmitted, and bit errors affect the signal much more severely than in a noncompressed bit stream. As a result, FEC is a must. Concatenated coding is used (see Sec. 11.6). The outer code is a Reed-Solomon code that corrects for block errors, and the inner code is a convolution code that corrects for random errors. The inner decoder utilizes the Viterbi decoding algorithm. These FEC bits, of course, add overhead to the bit stream. Typically, for a 240-W transponder (see Sec. 16.3), the bit capacity of 40 Mbps (see Sec. 16.5) may have a payload of 30 Mbps and coding overheads of 10 Mbps. The lower-power 120-W transponders require higher coding overheads to make up for the reduction in power and have a payload of 23 Mbps and coding overheads of 17 Mbps. More exact figures are given in Mead (2000) for DirecTV, where the overall code rates are given as 0.5896 for the 120-W transponder and 0.758 for the 240-W transponder.

Mead (2000) has shown that with FEC there is a very rapid transition in BER for values of $[E_b/N_0]$ between 5.5 and 6 dB. For $[E_b/N_0]$ values greater than 6 dB, the BER is negligible, and for values less than 5.5 dB, the BER is high enough to render the system useless.

The use of turbo codes, and LDPC codes (Sec. 11.11) currently being introduced for high definition TV provide a much greater increase in transponder capacity. As mentioned in Sec. 11.11.1, the Digital Video Broadcast S2 standard (DVB-S2) employs LDPC as the inner code in its FEC arrangement (Breynaert, 2005; Yoshida, 2003), and the DVB-RCS plans to use turbo codes (talk Satellite, 2004).

16.9   The Home Receiver Outdoor Unit (ODU)

The home receiver consists of two units—an outdoor unit and an indoor unit. Commercial brochures refer to the complete receiver as an IRD. Figure 16.5 is a block schematic for the ODU. This is like the outdoor unit shown in Fig. 8.1. The downlink signal, covering the frequency range 12.2 to 12.7 GHz, is focused by the antenna into the receive horn. The horn feeds into a polarizer that can be switched to pass either left-hand circular or right-hand circular polarized signals. The low-noise block that follows the polarizer contains a *low-noise amplifier* (LNA) and a downconverter. The function of the LNA is described in Sec. 12.5. The downconverter converts the 12.2- to 12.7-GHz band to 950 to 1450 MHz, a frequency range better suited to transmission through the connecting cable to the indoor unit.

The antenna usually works with an offset feed (see Sec. 6.14), and a typical antenna structure is shown in Fig. 16.6. It is important that the antenna have an unobstructed view of the satellite cluster to that it is aligned. Alignment procedures are described in Sec. 3.2.

The size of the antenna is a compromise among many factors but typically is around 18 in. (46 cm) in diameter. A small antenna is desirable for several reasons. Small antennas are less intrusive visually and are less subject to wind loading. In manufacture, it is easier to control surface irregularities, which can cause a reduction in gain by scattering the signal energy. The reduction can be expressed as a function

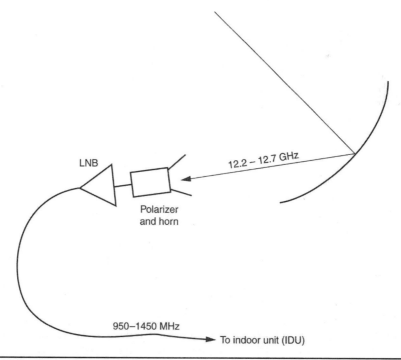

FIGURE 16.5   Block schematic for the outdoor unit (ODU).

Figure 16.6   A typical DBS antenna installation.

of the *root-mean-square* (rms) deviation of the surface, referred to an *ideal parabolic surface*. The reduction in gain is given by (see Baylin and Gale, 1991)

$$\eta_{\text{rms}} = e^{8.8\,\sigma/\varphi\lambda} \tag{16.7}$$

where σ is the rms tolerance in the same units as λ, the wavelength. For example, at 12 GHz (wavelength 2.5 cm) and for an rms tolerance of 1 mm, the gain is reduced by a factor

$$\eta_{\text{rms}} = e^{8.8 \times 0.1/2.5} = 0.7$$

This is a reduction of about 1.5 dB.

The isotropic power gain of the antenna is proportional to $(D/\lambda)^2$, as shown by Eq. (6.28), where D is the diameter of the antenna. Hence, increasing the diameter increases the gain (less any reduction resulting from rms tolerance), and in fact, many equipment manufacturers provide signal-strength contours showing the size of antenna that is best for a given region. Apart from the limitations to size stated earlier, it should be noted that at any given DBS location there are *clusters* of satellites, as described in Sec. 16.2. The beamwidth of the antenna must be wide enough to receive from all satellites in the cluster. For example, the Hughes DBS-1 satellite, launched on December 18, 1993, is located at 101.2°W longitude; DBS-2, launched on August 3, 1994, is at 100.8°W longitude; and DBS-3, launched on June 3, 1994, is at 100.8°W longitude. There is a spread of plus or minus 0.2° about the nominal 101°W position. The –3-dB beamwidth is given as approximately 70 λ/D, as shown by Eq. (6.29). A 60-cm dish at 12 GHz has a –3-dB beamwidth of approximately $70 \times 2.5/60 = 2.9°$, which is adequate for the cluster.

16.10   The Home Receiver Indoor Unit (IDU)

The block schematic for the IDU is shown in Fig. 16.7. The transponder frequency bands shown in Fig. 16.2 are downconverted to be in the range 950 to 1450 MHz, but of course, each transponder retains its 24-MHz bandwidth. The IDU must be able to receive any of the 32 transponders, although only 16 of these are available for a single polarization. The tuner selects the desired transponder. It should be recalled that the carrier at the center frequency of the transponder is QPSK modulated by the bit stream, which itself may consist of four to eight TV programs TDM. Following the tuner, the carrier is demodulated, the QPSK modulation being converted to a bit stream. Error correction is carried out in the decoder block labeled FEC^{-1}. The demultiplexer following the FEC^{-1} block separates the individual programs, which are then stored in buffer memories for further processing (not shown in the diagram). This further processing includes such things as conditional access, viewing history of *pay-per-view* (PPV) usage, and connection through a modem to the service provider (for PPV billing purposes). A detailed description of the IRD is found in Mead (2000).

16.11   Downlink Analysis

The main factor governing performance of a DBS system is the $[E_b/N_0]$ of the downlink. The downlink performance for clear-sky conditions can be determined using the method described in Sec. 12.8 and illustrated in Example 16.1 that follows. The

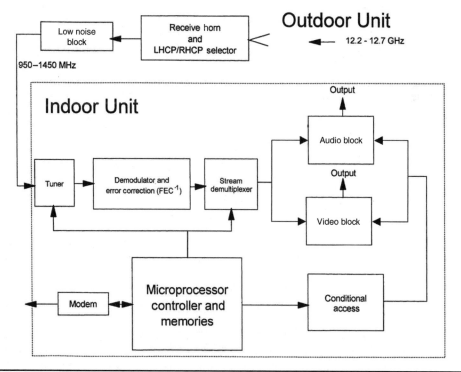

Figure 16.7   Block schematic for the indoor unit (IDU).

effects of rain can be calculated using the procedure given in Sec. 12.9.2 and illustrated in Example 16.2 that follows. In calculating the effects of rain, use is made of Fig. 16.8, which shows the regions (indicated by letters) tabulated along with rainfall in Table 16.2.

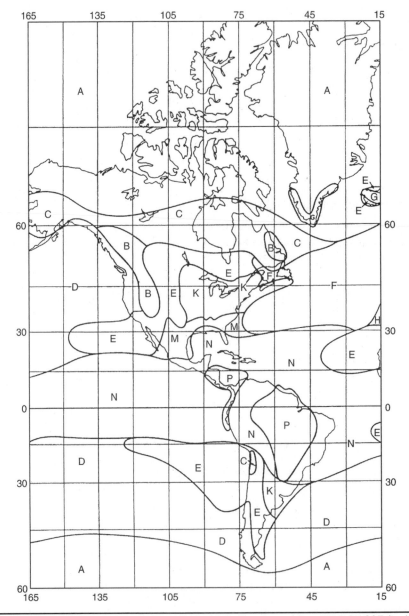

FIGURE 16.8 Rain climatic zones. Refer to Table 16.2. (*Courtesy of Rec. ITU-R PN.837-1, with permission from the copyright holder ITU. Sole responsibility for the reproduction rests with the author. The complete volume of the ITU material from which the material is extracted can be obtained from the International Telecommunication Union, Sales and Marketing Service, Place des Nations–CH-1211, Geneva 20, Switzerland.*)

Percentage of Time	A	B	C	D	E	F	G	H	J	K	L	M	N	P	Q
1.0	<0.1	0.5	0.7	2.1	0.6	1.7	3	2	8	1.5	2	4	5	12	24
0.3	0.8	2	2.8	4.5	2.4	4.5	7	4	13	4.2	7	11	15	34	49
0.1	2	3	5	8	6	8	12	10	20	12	15	22	35	65	72
0.03	5	6	9	13	12	15	20	18	28	23	33	40	65	105	96
0.01	8	12	15	19	22	28	30	32	35	42	60	63	95	145	115
0.003	14	21	26	29	41	54	45	55	45	70	105	95	140	200	142
0.001	22	32	42	42	70	78	65	83	55	100	150	120	180	250	170

Source: Table 16.2 and Fig. 16.8 reproduced from ITU Recommendation ITU-R PN.837-1 (1994), with permission.

TABLE 16.2 Rainfall Intensity Exceeded (mm/h) (Refer to Fig. 16.8)

Example 16.1 A ground station located at 45°N and 90°W is receiving the transmission from a DBS at 101°W. The [EIRP] is 55 dBW, and the downlink frequency may be taken as 12.5 GHz ($\lambda = 0.024$, *m*) for calculations. Transmission at the full capacity of 40 Mbps may be assumed. An 18-in (0.46 m) diameter parabolic dish antenna is used, and an efficiency of 0.55 may be assumed. Miscellaneous transmission losses of −2 dB also may be assumed. For the IRD, the equivalent noise temperature at the input to the LNA is 100 K, and the antenna noise temperature is 70 K. Calculate the look angles for the antenna, the range, and the $[E_b/N_0]$ at the IRD and comment on this.

Solution Use a value of 6378.14 km for the mean earth radius, and 42,164.20 km for the geostationary radius.

From Eq. (3.7):

$$B = \phi_E - \phi_{SS} = -90° - (-101°) = 11°$$

From Eq. (3.8):

$$b = a\cos(\cos B \cdot \cos \lambda_E) = a\ \cos(\cos 11° \cdot \cos 45°) = 46.04°$$

From Eq. (3.9):

$$A = a\sin\left(\frac{\sin |B|}{\sin b}\right) = 15.37°$$

By inspection, $\lambda_E > 0$; therefore, Fig. 3.3*d* applies and the required azimuth angle is:

$$A_Z = 180° + A = 195.37°$$

From Eq. (3.10):

$$d = \sqrt{R^2 + a_{GSO}^2 - 2 \cdot R \cdot a_{GSO}\cos b}$$

$$= \sqrt{6378.14^2 + 42,164.2^2 - 2 \times 6378.14 \times 42,164.2 \times \cos 46.04} = 38,015.32 \text{ km}$$

Equation (3.11) gives the required angle of elevation as

$$El = a\cos\left(\frac{a_{GSO}}{d}\sin b\right) = a\cos\left(\frac{42,164.2}{38,015.32}\sin 46.04°\right) \cong 37°$$

As noted in Sec. 3.2, the calculated values for azimuth and elevation provide a guide. Practical adjustments are made to maximize the received signal.

Equation (12.5b), with d and wavelength given in m, gives:

$$[Lp] = 10 \log \left[\left(\frac{1}{4\pi} \right) 2 \right] + 10 \log \left[\left(\frac{\lambda}{d} \right)^2 \right]$$

$$= [-21.98] + 10 \log \left[\left(\frac{0.024}{0.46} \right)^2 \right] = -205.98, \text{ dB}$$

Adding in the miscellaneous losses, Eq. (12.12) gives the total losses as

$$[\text{LOSSES}] = [Lp] + 2 = -207.98, \text{ dB}$$

The system noise temperature, from Eq. (12.23) is

$$T_S = T_{eq} + T_{ant} = 100 + 70 = 170 \text{ K}$$

In decibels relative to 1°K this is

$$[T_S] = 10 \log T_S = 22.31 \text{ dBK}$$

From Eq. (6.28), the gain of the receiving dish antenna is

$$G = \eta \left(\frac{\pi D}{\lambda} \right)^2 = 0.55 \times \left(\frac{0.46\pi}{0.024} \right)^2 = 1994 \Rightarrow 33.00, \text{ dB}$$

From Eq. (12.39):

$$\left[\frac{G}{T} \right] = [G] + [1/T_S] = 10.69 \text{ dBK}^{-1}$$

Equation (12.42) gives

$$\left[\frac{C}{N_0} \right] = [\text{EIRP}] + \left[\frac{G}{T} \right] - [\text{LOSSES}] - [K]$$

$$= 55 + 10.69 - 207.98 + 228.6 = 86.32 \text{ dBHz}$$

The downlink bit rate in decibels relative to 1 bps is

$$[R_b] = 10 \log (40 \times 10^6) \cong 76 \text{ dBbps}$$

Equation (10.32) gives:

$$\left[\frac{E_b}{N_0} \right] = \left[\frac{C}{N_0} \right] - [R_b] \cong 10.3 \text{ dB}$$

As noted in Sec. 16.8, a $[E_b/N_0]$ of at least 6 dB is required. The value obtained provides a margin of 4.3 dB under clear-sky conditions.

Example 16.2   Table 16.2 and Fig. 16.8 show the rainfall intensity in mm/h exceeded for given percentages of time. Calculate the upper limit for $[E_b/N_0]$ set by the rainfall for the percentage of time equal to 0.01 percent. The earth station is at mean sea level, and the rain attenuation may be assumed entirely absorptive, and the apparent absorber temperature may be taken as 272 K.

Solution   It is first necessary to calculate the attenuation resulting from the rain. The given data are shown below. Because the CCIR formula contains hidden conversion factors, units are not attached to the data, and it is understood that all lengths and heights are in km, and rain rate is in mm/h. From

Fig. 16.8, the earth station is seen to be located within region K. From the accompanying Table 16.2, the rainfall exceeds 42 mm/h for 0.01 percent of the time. Table 4.2 does not give the coefficients for 12.5 GHz; therefore, the values must be found by linear interpolation between 12 and 15 GHz. Denoting the 12-GHz values with subscript 12 and the 15 GHz values with subscript 15, then and using the values from Table 4.2,

$$a_h = a_{h12} + \frac{a_{h15} - a_{h12}}{15 - 12} \times (12.5 - 12)$$

$$= 0.0188 + \frac{0.0367 - 0.0188}{3} \times 0.5 = 0.0218$$

$$b_h = b_{h12} + \frac{a_{h15} - a_{h12}}{15 - 12} \times (12.5 - 12)$$

$$= 1.217 + \frac{1.154 - 1.217}{3} \times 0.5 = 1.207$$

$$a_v = a_{v12} + \frac{a_{v15} - a_{v12}}{15 - 12} \times (12.5 - 12)$$

$$= 0.0168 + \frac{0.0335 - 0.0168}{3} \times 0.5 = 0.0196$$

$$b_v = b_{v12} + \frac{b_{v15} - b_{v12}}{15 - 12} \times (12.5 - 12)$$

$$= 1.2 + \frac{1.128 - 1.2}{3} \times 0.5 = 1.188$$

Since circular polarization is used, the coefficients are found from Eq. (4.8):
From Eq. (4.8a):

$$a_c = \frac{a_h + a_v}{2} = \frac{0.0218 + 0.0196}{2} = 0.0207$$

From Eq. (4.8b):

$$b_c = \frac{a_h \cdot b_h + a_v \cdot b_v}{2 \cdot a_c} = \frac{0.0218 \times 1.207 + 0.0196 \times 1.188}{2 \times 0.0207} = 1.198$$

Using the method 3 curves in Fig. 4.4 for $p = 0.01$ percent and earth-station latitude 45°, the rain height is approximately 3.5 km, and as stated in the problem, at mean sea level $h_o = 0$
From Eq. (4.7):

$$L_S = \frac{h_R - h_0}{\sin El} = \frac{3.5}{\sin 37°} = 5.82 \text{ km}$$

From Eq. (4.9)

$$L_G = L_S \cos El = 5.82 \times \cos 37° = 4.64 \text{ km}$$

From Table 4.3:

$$r_{01} = \frac{90}{90 + 4L_G} = 0.829$$

From Eq. (4.8):

$$L = L_S \cdot r_{01} = 5.82 \times .829 = 4.82$$

From Eq. (4.2):

$$\alpha = a_c R_{01}^{b_c} = 0.0207 \times 42^{1.198} = 1.819$$

From Eq. (4.6):

$$[A_{01}] = -\alpha L = -1.819 \times 4.82 = -8.76 \text{ dB}$$

The effect of rain is calculated as shown in Sec. 12.9.2. This requires the attenuation to be expressed as a power ratio

$$A = 10^{[A_{01}]/10} = 10^{-0.876} = 7.52$$

For Eq. (12.60), the noise-to-signal ratios are required. From Example 16.1, $[C/N_0] = 86.3$ dBHz for clear sky, hence,

$$\left(\frac{C}{N_0}\right)_{CS} = 10^{-86.3/10} = 2.34 \times 10^{-9}$$

The system noise temperature under clear-sky conditions is just T_S, but the subscript is changed to conform with Eq. (12.66): $T_{SCS} = T_S = 170$ K
given as

$$\left(\frac{N_0}{C}\right)_{rain} = \left(\frac{N_0}{C}\right)_{CS}\left(A + (A-1)\cdot\frac{T_a}{T_{SCS}}\right)$$

$$= 2.34 \times 10^{-9}\left(7.52 + 6.52 \times \frac{272}{170}\right) = 4.21 \times 10^{-8}$$

Hence,

$$\left[\frac{C}{N_0}\right]_{rain} = 10\log(4.21 \times 10^{-8})^{-1} = 73.76 \text{ dBHz}$$

Recalculating the $[E_b/N_0]$ ratio, Eq. (10.24) gives:

$$\left[\frac{E_b}{N_0}\right]_{rain} = \left[\frac{C}{N_0}\right]_{rain} - [R_b] = -2.26 \text{ dB}$$

Thus the rain completely wipes out the signal for 0.01 percent of the time. It is left as an exercise for the reader to find the size of antenna that provides an adequate signal under these rain conditions.

16.12   Uplink

Ground stations that provide the uplink signals to the satellites in a DBS system are highly complex systems in themselves, utilizing a wide range of receiving, recording, encoding, and transmission equipment. Signals originate from many sources. Some are analog TV received from satellite broadcasts. Others originate in a studio, others from video cassette recordings, and some are brought in on cable or optical fiber. Data signals and audio broadcast material also may be included. All of these must be converted to a uniform digital format, compressed, and TDM. Necessary service additions that must be part of the multiplexed stream are the program guide and conditional access. FEC is added to the bit stream, which is then used to QPSK modulate the carrier

Figure 16.9 Uplink facilities for Echostar's DISH network. (*Courtesy of Echostar, at http://www.dishnetwork.com/content/aboutus/presskit/print_satellites/index.shtml*)

Company	Location
AlphaStar	Oxford, Connecticut
DirecTV	Castle Rock, Colorado
EchoStar	Cheyenne, Wyoming
U.S. Satellite Broadcasting	Oakdale, Minnesota

Table 16.3 Uplink Facilities in Operation as of 1996

for a given transponder. The whole process, of course, is duplicated for each transponder carrier.

Because of the complexity, the uplink facilities are concentrated at single locations specific to each broadcast company. The uplink facilities for Echostar's DISH network are shown in Fig. 16.9. The four uplink facilities in operation as of 1996, as given in Mead (2000), are shown in Table 16.3.

16.13   High Definition Television (HDTV)

Table 16.1 shows the 18 *advanced television systems committee* (ATSC) formats for digital television, which includes HDTV. Some of the early HDTV systems were analog, and a description of these is found in Kuhn (1995). However, all analog TV transmissions in the United States were terminated in January 2009. However, it is possible to get "set-top boxes" that convert HDTV signals to a format suitable for analog sets, but the high resolution, and many of the other advantages of digital TV are lost in this process.

In Europe, Astra is the major satellite provider, and they, along with 60 European broadcasters have agreed to standardize on two HDTV formats: 720p50 meaning 720 lines (or pixels) of vertical resolution, with progressive scan, at a 50 Hz refresh rate and 1080i25, meaning 1080 lines (or pixels) of vertical resolution, with interlace scan, at a 25 Hz refresh rate. The refresh rates are normally tied in with the frequency of the domestic electricity supply, 60 Hz in the United States and most of north America, and 50 Hz in Europe. In Table 16.1 a refresh rate of 24 Hz is listed, to make this format compatible with film projection at 24 frames per second.

DirecTV plans to use H.264/AVC (MPEG-4 Part 10, see Sec. 16.7) in its HDTV satellite broadcasts and all HDTV services in Europe are expected to use this rather than the MPEG-2. Comparing H.264/AVC to MPEG-2 the bit rate reduction can vary between about 50 (Mark, 2005) to 65 percent (Wipro, 2004). Two high-definition channels require a bit rate of 16 to 18 Mbps with MPEG-2 and 6 to 8 Mbps with H.264/AVC. As explained in Sec. 16.7, H.264/AVC is not backward compatible with MPEG-2, and this may mean considerable expense for the consumer who wishes to receive HDTV along with SDTV. The MPEG is working on making H.264/AVC compatible with MPEG-2 (Koenen, 1999).

16.13.1 HDTV Displays

Among the competing technologies suitable for HDTV are plasma displays, *liquid crystal displays* (LCD) and *digital light processing* (DLP) displays (and others). Plasma displays are made up of tiny cells coated with red, green, and blue phosphors. The video signal stimulates a gas inside the cells, which impacts the phosphors causing them to glow. Plasma flat-panel displays are around 3 to 5 in. thick and screen sizes up to 60 in. are available. In a LCD, light passes through a thin sheet of the liquid crystal material which forms the viewing screen. A thin-film transistor array carrying the video signal produces varying degrees of polarization of the liquid crystal allowing more or less light to pass through for each of the colors red, green, and blue. LCDs are thin, flat panel displays that can be made with screen sizes up to about 50 in. DLP displays utilize what is known as a *digital micromirror device* (DMD) invented by Texas Instruments. The DMD contains approximately 1.3 million micro-mirrors, each micro-mirror representing one pixel. The micro-mirrors can be mechanically pivoted up to 5000 times a second, the pivoting being activated by the video signal. The degree of pivoting is determined by the signal level, and light reflected by the DMD is passed through a "color wheel" consisting of red, blue, and green filters, which rotates at speeds of about 120 revolutions per second. Each filter projects an image for a brief period onto the screen and the eye integrates these to "see" the composite picture. The DLP display is a rear projection unit and is about 12 to 14 in. deep. These units are available in large screen size. LCD and DLP displays require a light source; they can also be made as front projection units, which provide corresponding larger screens than rear projection displays. Apart from the CRT, all these displays can be activated by digital signals, thus avoiding the need for digital to analog conversion.

Although the vision aspects of HDTV are by nature the most noticeable, sound quality is also very high. HDTV provides for Dolby 5.1 sound, which means there are five channels: left; right; center; left rear; and right rear; the .1 stands for a subwoofer, a very low frequency, or bass channel. The audio signal format is Dolby Digital/AC-3.

16.14 Video Frequency Bandwidth

An estimate of the highest frequency in the analog video signal can be found as follows. Let L_{act} be the number of active lines per frame, and L_{supp} the number of lines suppressed during picture flyback, then the total number of lines per frame is $L = L_{act} + L_{supp}$. Each pixel is the width of a line, so the number of lines also represents the number of pixels along the picture viewing height. Let h be the (viewing) picture height. The height of a pixel is therefore $h_{pix} = h/L_{act}$. Let w be the (viewing) picture width, then the number of active pixels per line is $PL_{act} = w/h_{pix} = (w/h)L_{act}$. But part of the total line is suppressed to allow for line flyback. Let the ratio of total line scan time to active scan time be F_{lfb}, then the total number of pixels per line is $PL_{tot} = PL_{act}F_{lfp} = (w/h)L_{act}F_{lfp}$. The total number of pixels per frame is, therefore, $PF_{tot} = L(w/h)L_{act}F_{lfp}$. An estimate of the highest frequency can be made by assuming that it occurs when a frame consists of alternate black and white pixels such that two pixels make up one cycle. Thus multiplying half the total number of pixels by the frame rate (in frames per second) gives the highest frequency. Subjective tests have shown that a reduction in the highest frequency obtained in this way can be tolerated, the reduction being accounted for by the *Kell factor K*. $K = 0.7$ is often assumed in practice, although different values are sometimes encountered. In a table of values published by Evans Associates (http://www.evansassoc.com) the Kell factor is given as 0.7 for interlaced scanning and 0.9 for progressive scanning.

The aspect ratio is defined as $a = w/h$ and denoting the number of frames per second by F, the highest frequency is given by

$$f_{max} = \frac{KPF_{tot}F}{2} = \frac{K \cdot a \cdot L \cdot L_{act} \cdot F_{lfb} \cdot F}{2} \tag{16.8}$$

As shown by Eq. (16.8) the highest frequency is directly proportional to frame rate. The frame rate is tied to half the frequency of the domestic electricity supply, 60 Hz in the United States and most of North America, and 50 Hz in Europe. At 30 complete frames per second (and more noticeably at 25 frames per second in European systems) flicker was apparent in analog TV. To overcome this, without increasing the frame rate (and hence the highest frequency) frames were divided into two fields, one field consisting of odd-numbered lines, the other of even-numbered lines. This is termed *interlaced* scanning. Displaying the odd and even fields alternately at 60 fields per second (50 in Europe) keeps the number of frames at 30 per second (25 in Europe), which eliminates the flicker without any increase in the video frequency. With *progressive* scanning the lines are scanned in sequence which provides a sharper picture.

Example 16.3 For NTSC analog TV (see Sec. 9.5) $L_{act} = 483$, $L = 525$, a 4/3 $F_{lfb} = 1.19$ and $F = 30$. With $K = 0.7$. Calculate the highest video frequency.

Solution Substituting the given values in Eq. (16.8) gives:

$$f_{max} = \frac{.7}{2} \times \frac{4}{3} \times 525 \times 483 \times 1.19 \times 30 \cong 4.2 \text{ MHz}$$

Values applicable to digital TV are shown in Table 16.4.

No.	Format Type	L	L_{act}	F_{lfb}	Aspect Ratio a	F, Frames per Second
1	SDTV	525	480i	1.13	4:3	30
2	EDTV	525	480p	1.13	4:3	24
3	EDTV	525	480p	1.13	4:3	30
4	EDTV	525	480p	1.13	4:3	60
5	EDTV	525	480i	1.13	4:3	30
6	EDTV	525	480p	1.13	4:3	24
7	EDTV	525	480p	1.13	4:3	30
8	EDTV	525	480p	1.13	4:3	60
9	EDTV	525	480i	1.22	16:9	30
10	EDTV	525	480p	1.22	16:9	24
11	EDTV	525	480p	1.22	16:9	30
12	EDTV	525	480p	1.22	16:9	60
13	HDTV	750	720p	1.29	16:9	24
14	HDTV	750	720p	1.29	16:9	30
15	HDTV	750	720p	1.29	16:9	60
16	HDTV	1125	1080i	1.15	16:9	30
17	HDTV	1125	1080p	1.15	16:9	24
18	HDTV	1125	1080p	1.15	16:9	30

Notes: EDTV, enhanced definition television; HDTV, high-definition television; i, interlaced scanning; p, progressive scanning; SDTV, standard definition television.

Sources: Booth (1999); www.timefordvd.com, 2004; Maxim/Dallas, 2001.

TABLE 16.4   Digital TV Parameters as Used in Eq. (16.8)

Example 16.4   Calculate the highest video frequency for HDTV 1080*i*. Assume a Kell factor of 0.7.

$$f_{max} = \frac{0.7}{2} \times \frac{16}{9} \times 1125 \times 1080 \times 1.15 \times 30 \cong 26 \text{ MHz}$$

16.15   Problems and Exercises

16.1.   Referring to Fig. 16.1, calculate (*a*) the total number of transponders broadcasting from each of the orbital positions shown and (*b*) the total number of transponders in use by each service provider.

16.2.   The [EIRP] of a 240-W transponder is 57 dBW. Calculate the approximate gain of the antenna.

16.3.   The transponder in Prob. 16.2 is switched to 120 W. Calculate the new [EIRP], assuming that the same antenna is used.

16.4. Draw accurately to scale the transponder frequency plan for the DBS transponders 5, 6, and 7 for uplink and downlink.

16.5. Calculate the total bit rate capacity available at each of the orbital slots for each of the service providers listed in Fig. 16.1. State any assumptions made.

16.6. Calculate the bandwidth required to transmit a SDTV format having a resolution of 704×480 pixels at 30 frames per second.

16.7. The R, G, and B colors in Eq. (16.2) are restored by the matrix multiplication

$$\begin{bmatrix} R \\ G \\ B \end{bmatrix} = \begin{bmatrix} 1 & 0 & 1.402 \\ 1 & -0.344136 & -0.714136 \\ 1 & 1.772 & 0 \end{bmatrix} \begin{bmatrix} Y \\ Cr \\ Cb \end{bmatrix}$$

where the 3×3 matrix is $\mathbf{M}^{-1}$ [the inverse of $\mathbf{M}$ in Eq. (16.2)]. Verify that this matrix multiplication is correct.

16.8. Briefly describe the video compression process used in MPEG-2.

16.9. Explain what is meant by *masking* in the context of audio compression. Describe how MPEG-1 utilizes the phenomenon of masking to achieve compression.

16.10. Using the overall codes rates (Mead, 2000) given in Sec. 16.8 and assuming a 40-Mbps transponder, calculate the payload bit rate and the overhead for the low-power and high-power transponders.

16.11. Plot antenna gain as a function of diameter for a paraboloidal reflector antenna for antenna diameters in the range 45 to 80 cm. Use a frequency of 12.5 GHz.

16.12. Assuming that the rms tolerance can be held to 0.2 percent of the diameter in antenna manufacture, calculate the reduction in gain that can be expected with antennas of diameter (*a*) 46 cm, (*b*) 60 cm, and (*c*) 80 cm.

16.13. Calculate the gain and -3-dB beamwidth for antennas of diameter 18, 24, and 30 in. at a frequency of 12.5 GHz. Assume that the antenna efficiency is 0.55 and that the rms manufacturing tolerance is 0.25 percent of diameter.

16.14. A DBS home receiver is being installed at a location $40°N, 75°W$ to receive from a satellite cluster at $61.5°$. Calculate the look angles for the antenna. It is hoped to use an 18-in. antenna, the antenna efficiency being 0.55, and the effect of surface irregularities may be ignored. The system noise temperature is 200 K. The downlink frequency may be taken as 12.5 GHz, the [EIRP] as 55 dBW, and the transponder bit rate as 40 Mbps. Miscellaneous transmission losses may be ignored. Calculate the received clear sky $[E_b/N_0]$, and state whether this results in satisfactory reception.

16.15. Following the procedure given in Example 16.2, calculate the rain attenuation for 0.01 percent of time and the corresponding $[E_b/N_0]$ for the system in Prob. 16.14. The ground station may be assumed to be at mean sea level. State if satisfactory reception occurs under these conditions.

16.16. A DBS home receiver is being installed at a location $60°N, 155°W$ to receive from a satellite cluster at $157°$. Calculate the look angles for the antenna. It is hoped to use an 18-in. antenna, the antenna efficiency being 0.55, and the effect of surface irregularities may be ignored. The system noise temperature is 200 K. The downlink frequency may be taken as 12.5 GHz, the [EIRP] as

55 dBW, and the transponder bit rate as 40 Mbps. Miscellaneous transmission losses may be ignored. Calculate the received clear sky $[E_b/N_0]$, and state whether this results in satisfactory reception.

16.17. Following the procedure given in Example 16.2, calculate the rain attenuation for the 0.01 percent of time, and the corresponding $[E_b/N_0]$ for the system in Prob. 16.16. The ground station may be assumed to be at mean sea level. State if satisfactory reception occurs under these conditions.

16.18. A DBS home receiver is being installed at a location 15°S, 50°W to receive from a satellite cluster at 61.5°. Calculate the look angles for the antenna. Also, calculate the diameter of antenna needed to provide a 5-dB margin approximately on the received $[E_b/N_0]$ under clear-sky conditions. An antenna efficiency of 0.55 may be assumed, and the effect of surface irregularities may be ignored. The antenna noise temperature is 70 K, the equivalent noise temperature at the input of the LNA is 120 K, and miscellaneous transmission losses are 2 dB. The downlink frequency may be taken as 12.5 GHz, the [EIRP] as 55 dBW, and the transponder bit rate as 40 Mbps.

16.19. For the system described in Prob. 16.18, determine the rainfall conditions that cause loss of signal. The ground station may be assumed to be at mean sea level.

16.20. For HDTV the R, G, and B colors in Eq. (16.6) are restored by the matrix multiplication

$$\begin{bmatrix} R \\ G \\ B \end{bmatrix} = \begin{bmatrix} 1 & 0 & 1.402 \\ 1.218399 & -0.217953 & -0.507777 \\ 1 & 1.772 & 0 \end{bmatrix} \begin{bmatrix} Y \\ Cr \\ Cb \end{bmatrix}$$

where the 3×3 matrix is $\mathbf{M}^{-1}$ [the inverse of $\mathbf{M}$ in Eq. (16.6)]. Verify that this matrix multiplication is correct.

16.21. Using the data from Table 16.4 calculate the highest video frequency for format Nos. 12 and 15. Use a Kell factor of 0.7 for No. 12 and 0.9 for No. 15.

References

Baylin, F., and B. Gale. 1991. *Ku-Band Satellite TV.* Baylin Publications, Boulder, Colorado.

Bhatt, B., David B., and D. Hermreck. 1997. "Digital Television: Making It Work." *IEEE Spectrum*, vol. 34, no. 10, October, pp. 19–28.

Booth, S. A. 1999. "Digital TV in the U.S." *IEEE Spectrum*, vol. 36, no. 3, March, pp. 39–46.

Breynaert, D., 2005. Analysis of the bandwidth efficiency of DVB-S2 in a typical data distribution network. CCBN, Beijing, March, at www.newtec.com.sg/news/articles/Beijing2005%20DVBS2%20whitepaper%202feb05%20MDOWaylon.pdf-29

Chaplin, J. 1992. "Development of Satellite TV Distribution and Broadcasting." *Electron. Commun.*, vol. 4, no. 1, February, pp. 33–41.

Dishnetwork Web page, at http://www.dishnetwork.com/

Koenen, R. 1999. "MPEG-4, Multimedia for Our Time." *IEEE Spectrum*, vol. 36, no. 2, pp. 26–33. February, at http://www.chiariglione.org/mpeg/tutorials/papers/koenen/mpeg-4.htm

Kuhn, K. J. 1995. HDTV—An Introduction, at http://www.ee.washington.edu/conselec/CE/kuhn/hdtv

Mark, S. 2005. HDTV Pushes Telcos Toward MPEG-4 Light Reading. June 24, at http://www.lightreading.com/document.asp?doc_id=76252

Maxim/Dallas. 2001. Application Note 750, Bandwidth Versus Video Resolution, at http://www.maxim-ic.com/appnotes.cfm/appnote_number/750.

Mead, D. C. 2000. *Direct Broadcast Satellite Communications.* Addison-Wesley, Reading, MA.

MPEG Web page, at http://www.mpeg.org/.

Netravali, A., and A. Lippman. 1995. "Digital Television: A Perspective." *Proc. IEEE,* vol. 83, no. 6, June, pp. 834–842.

Prichard, W. L., and M. Ogata. 1990. "Satellite Direct Broadcast." *Proc. IEEE,* vol. 78, no. 7, July, pp. 1116–1140.

Reinhart, E. E. 1990. "Satellite Broadcasting and Distribution in the United States." *Telcommun. J.,* vol. 57, no. V1, June, pp. 407–418.

Roddy, D., and J. Coolen. 1995. *Electronic Communications,* 4th ed. Prentice-Hall, Englewood Cliffs, NJ.

Sullivan, G. J., P. Topiwala, and Ajay Luthra, 2004. "The H.264/AVC Advanced Video Coding Standard: Overview and Introduction to the Fidelity Range Extensions." *SPIE Conference on Applications of Digital Image Processing XXVII, Special Session on Advances in the New Emerging Standard H.264/AVC,* August. Denver, Colorado.

Sweet, W. 1997. "Chiariglione and the Birth of MPEG." *IEEE Spectrum,* vol. 34, no. 9, pp. 70–77.

talk Satellite. 2004. Spectra Licensing Group Announces Gilat's Entrance into the Turbo Code Licensing Program for DVB-RCS Satellite Applications. December 15, at www.talksatellite.com/EMEAdoc.

Timefordvd.com. 2004. "Digital TV & HDTV Tutorial." (Last updated on 3.9.2004).

Wipro. 2004. MPEG-4 AVC/H.264 Video Coding. White paper. September, at http://www.wipro.com/insights/embedded.htm.

Yoshida, J. 2003. Hughes goes retro in digital satellite TV coding. EETimesUK, November, at www.eetuk.com/tech/news/OEG20031110S0081.

CHAPTER 17

Satellite Mobile and Specialized Services

17.1 Introduction

The idea that three geostationary satellites can provide communications coverage for the whole of the Earth, apart from relatively small regions at the north and south poles, is generally credited to Clarke (1945). The basic idea has been expanded to include many more satellites operated by various nations and companies. By spacing geostationary satellites in the orbital plane, the same frequencies can be reused. The limitation in spacing is primarily due to the beamwidth of ground-based antenna gain patterns in selecting between the satellites. If an average of spacing of 2° is assumed, there are 180 "slots" or orbital positions available. In practice satellites are often employed in clusters, with more than one satellite at an assigned orbital position. The populated orbital positions are concentrated to serve regions, where services are most in demand. In addition to geostationary satellites, nongeostationary orbits (especially *low earth orbits* (LEOs) and *medium earth orbits* (MEOs)) are becoming more common. As described in Chap. 15, satellite services may utilize a combination of geostationary and nongeostationary satellites.

Although past developments in satellites have led to larger satellites, recent interest in small satellites, often referred to as *microsats or smallsats* (see Sweeting, 1992, and Williamson, 1994), has increased with the number of smallsats in LEO or MEO outnumbering geostationary satellites.

In this chapter, brief descriptions are given of some of these services to illustrate the range of applications now found for satellites.

17.2 Satellite Mobile Services

Although many countries are well served by global communications, there remain large areas and population groups that have very limited access to telecommunications services. In the United States and some European countries, telephone landline density, measured by the number of phone lines per 100 people, is as much as 30 times higher than in China, India, Pakistan, and the Philippines, and an estimated 3 billion people worldwide have no phone at home (Miller, 1998). Developing a telephone network on the ground, whether wired or cellular, is time-consuming and expensive. Civil infrastructure may need to be installed or upgraded, including roads and utilities such as water and electricity. In contrast, satellites can provide wide area service for telephone and Internet, on an as-needed basis, without the need for extensive ground facilities.

An Internet search for mobile satellite companies yields a confusing array of data information with both new and outdated information. Newer systems, StarLink, for example, provide high-speed Internet while some older systems are limited telephone and packet services. Most of the systems that offer telephone services provide the users with dual-mode phones that operate to GSM standards. GSM stands for *global system for mobile communications* (originally Groupe Spe'cial Mobile) developed by the European Telecommunications Standards Institute (ETSI). It is the most widely used standard for cellular and personal communications. The user frequencies are in the L (1–2 GHz) and S (2–4 GHz) bands, and where geostationary satellites are employed, these require large on-board antennas in the range 100 to 200 m². To illustrate these, two specific systems are described below.

Thuraya.   The Thuraya satellite system is a United Arab Emirates-based regional mobile satellite services provider. It has two geosynchronous satellites located at 28.5°E (Thuraya 2) and 98.5°E (Thuraya 3) longitude, at an inclination of 6.3°, presumably to dispense with N-S station-keeping maneuvers as described in Sec. 7.4. An additional satellite (Thuraya 4-NGS) is being built to replace Thuraya 2. The Thuraya satellite system serves an area between about 20°W to 100°E longitude and 60°N to 2°S latitude, covering more than 110 countries with a combined population of 2.3 billion. The antenna footprint spans Europe, North and Central Africa, and large parts of Southern Africa. A 12.25 × 16 m elliptical antenna is employed providing 351 spot beams, with onboard beam-switching. An uplink beacon signal provides beam adjustment to compensate for the nongeostationary path of the satellites. The system operates with a 10 dB fade margin to allow for shadowing of handheld units. The network capacity is about 13,750 telephone channels. Quaternary phase-shift keying modulation is used, with *frequency-division multiple access* (FDMA)/*time-division multiple access* (TDMA). Dual-mode handsets are used that can be switched between GSM mode and satellite mode. Service features include voice telephony, fax, data, short messaging, location determination, emergency services, and high-power alerting. The mobile links operate at frequencies in the L-band (uplink 1625.5–1660.5 MHz; downlink 1525–1559 MHz), and the gateway frequencies are in the C band (uplink 6.425–6.725 GHz; downlink 3.400–3.700 GHz). Further information can be obtained from the Web site at https://www.thuraya.com/

MSAT.   The mobile satellite (MSAT) system was developed by the National Research Council of Canada to provide mobile telephone service. MSAT has two geostationary satellites, MSAT-1 and MSAT-2, which provide services across North and Central America, northern South America, the Caribbean, Hawaii, and in coastal waters. A variety of services by companies using this system are offered, including tracking, and managing trucking fleets, wireless phone, data and fax, dispatch radio services, and differential GPS. L-band frequencies are used for the satellite services and 1530 to 1559 MHz for downlink and 1631.5 to 1660.5 MHz for uplink. The feeder links to the ground segment (which connects with public and private telephone and data networks) are in the Ku-band 10.75 to 10.95 GHz for downlink and 13.0 to 13.15 GHz and 13.2 to 13.25 GHz for uplink. Two transponders are used for the Ku-band to L-band forward link, and one transponder is used for the L-band to Ku-band return link. Transmit power output at L-band is 600 W. Two L-band mesh reflector antennas, 5.7 m × 4.7 m provide an EIRP of 57.3 dBW at the edge of the service area. Transmit power output at Ku band is 110 W.

17.3 VSAT Networks

VSAT stands for *very small aperture terminal* system. The identifying feature, of a VSAT system, is that the Earth-station antennas are typically less than 2.4 m in diameter (Rana et al., 1990). The trend is toward even smaller dishes, not more than 1.5 m in diameter (Hughes et al., 1993). In this sense, the small TVRO terminals described in Sec. 16.9 for direct broadcast satellites could be labeled as VSATs, but the appellation is usually reserved for private networks, mostly providing two-way communications facilities. Typical user groups include banking and financial institutions, airline and hotel booking agencies, and large retail stores with geographically dispersed outlets. In a broad sense, VSAT networks operated as an early version of the Internet.

The basic structure of a VSAT network consists of a hub station, which provides a broadcast facility to all the VSATs in the network, and the VSATs themselves, which access the satellite in some form of multiple-access mode. The hub station is operated by the service provider, and it may be shared among many users, but of course, each user organization has exclusive access to its own VSAT network. The normal downlink mode, of transmission from hub to the VSATs, is TDMA. The transmission can be broadcast for reception by all the VSATs in a network, or address coding can be used to direct messages to selected VSATs.

Access the other way, from the VSATs to the hub, is more complicated, and several different methods are in use, many of them proprietary. A comprehensive summary of methods is given in Rana et al. (1990). The most popular access method is FDMA, which allows the use of comparatively low-power VSAT terminals (see Sec. 14.7.12). TDMA also can be used, but is not efficient for low-density uplink traffic from the VSAT. The traffic in a VSAT network is mostly data transfer of a burst nature, examples being inventory control, credit verification, and reservation requests occurring at random and possibly infrequent intervals, so allocation of time slots in the normal TDMA mode can lead to low channel occupancy. A form of *demand assigned multiple access* (DAMA) is employed in some systems, in which channel capacity is assigned, in response to the fluctuating demands of the VSATs in the network. DAMA can be used with FDMA as well as TDMA, but the disadvantage of the method is that a *reserve channel* must be instituted through which the VSATs can make requests for channel allocation. As pointed out by Abramson (1990), the problem of access then shifts to how the users may access the reserve channel in an efficient and equitable manner. Abramson presents a method of *code-division multiple access* (CDMA) using spread spectrum techniques, coupled with the Aloha protocol. The basic Aloha method is a random-access method in which packets are transmitted at random in defined time slots. The system is used where the packet time is small compared with the slot time, and provision is made for dealing with packet collisions, which can occur with packets sent up from different VSATs. Abramson calls this method *spread Aloha* and presents theoretical results, which show that the method provides the highest throughput for small Earth stations.

VSAT systems operate in a star configuration, which means that the connection of one VSAT to another must be made through the hub. This requires a double-hop circuit with a consequent increase in propagation delay, and twice the necessary satellite capacity is required compared with a single-hop circuit. In Hughes et al. (1993), a proposed VSAT system is presented, which provides for *mesh connection*, where the VSATs can connect with one another through the satellite in a single hop.

Most VSAT systems have operated in the Ku-band, although there are some C-band systems (Rana et al., 1990). For fixed-area coverage by the satellite beam, the system performance is essentially independent of the carrier frequency. For fixed-area coverage, the beamwidth and hence the ratio λ/D is a constant (see Eq. 6.29). The satellite antenna gain is therefore constant (see Eq. 6.28), and for a given high-power amplifier output, the satellite EIRP remains constant. As shown in Sec. 12.3.1, for a given size of antenna at the Earth station and a fixed EIRP from the satellite, the received power at the Earth station is independent of frequency. This ignores the propagation margins needed to combat atmospheric and rain attenuation. As shown in Hughes et al. (1993), the necessary fade margins are not excessive for a Ka-band VSAT system, and the performance otherwise is comparable with a Ku-band system. (From Table 1.1, the K-band covers 18 to 27 GHz and Ka-band covers 27 to 40 GHz. In Hughes et al. (1993), results are presented for frequencies of 18.7 and 28.5 GHz.

As summarized in Rana et al. (1990), the major shortcomings of prior VSAT systems are the high initial costs, the tendency toward optimizing systems for large networks (typically more than 500 VSATs), and the lack of direct VSAT-to-VSAT links. Technological improvements, especially in the areas of microwave technology and digital signal processing (Hughes et al., 1993), are expected to result in VSAT systems that ameliorate these shortcomings.

17.4   Satellite Internet

Two-way Internet traffic via satellite is a relatively new development based on large constellations of satellites operating in LEO. At LEO, the short signal travel time facilitates short lag times, whereas the much longer signal paths for satellites operating at higher altitudes, is a fundamental limit for two-way Internet traffic. Nevertheless, there are several geostationary satellite Internet providers, e.g., HughesNet (https://www.hughesnet.com/) and Viasat (https://www.viasat.com). These provide higher speed (up to 30 Mbps) downlink capability, but only low-speed (3 Mbps) uplink capability. An advantage of geostationary satellite Internet is that the antenna, once pointing is set, remains fixed. While a geostationary satellite is very expensive, only one is required. In contrast, similar coverage for LEO systems requires large constellations of satellites, which must, therefore, have much lower cost satellites to be economically viable. An example of the latter is Starlink.

Starlink

The Starlink is a constellation of small LEO satellites to provide global internet and future mobile phone service. Besides having shorter time lags, another advantage of LEO Internet is that the satellite transmitters can be lower power due to the shorter ranges, which helps to facilitate smaller and less expensive satellites. However, the ground terminal antenna must track rapidly moving satellites (up to ~0.3°/sec) and be able to switch quickly from one satellite to another. To satisfy this need, Starlink employs *phased array antennas* on its ground terminals.

Starlink is being developed and operated by a US aerospace company, SpaceX, and by the end of 2023, almost 5,000 satellites will be operating on-orbit. A key part of the constellation deployment is the reusable SpaceX falcone-9 booster that carries large batches (50–60) of small satellites in a single launch (see Fig. 17.1a). The Starlink satellite configuration is a flat panel with a deployable solar panel as shown in Fig. 17.1b. The communication systems are Ku/Ka-band: 10.7 GHz to 17.7 GHz and 37.5 GHz to 42.5 GHz that uses *orthogonal frequency division multiplex* (OFDM) modulation.

(a)

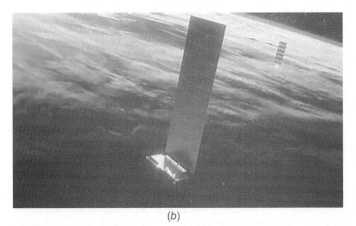

(b)

FIGURE 17.1 Artist rendition of: (a) sixty Starlink satellites stacked before on-orbit deployment, and (b) two Starlink satellites on-orbit. (*Courtesy of SpaceX, 2023.*)

Currently, the *inter-satellite links* (ISLs) operate at Ka-band, but next generation satellites will have infrared laser ISLs, which have higher link capacity (increased data rates) with less intra-satellite RFI.

Using onboard ion thrusters to raise their orbit, the satellites spread out in a dense constellation providing services to over 60 countries between ±50° latitudes. In April 2019, the US Federal Communications Commission (FCC) granted approval to operate nearly 12,000 Starlink satellites in three orbital shells with multiple inclinations. Initially about 1,600 satellites operate at 550 km altitude, and the remainder of the constellation will be approximately 7,500 at 340 km and 2,800 at 1,150 km. Satellites at the 340 km altitude are negatively impacted by atmospheric drag and re-enter the Earth's atmosphere within a few years; thereby producing a continuing need to replenish the constellation. For more information on Starlink, visit https://www.starlink.com/

17.5   Global Navigation Satellite System (GNSS)

GNSS services have become common place in everyday life. They provide global navigation, positioning, and timing information that have wide application. Multiple independent GNSS systems have been developed by different countries including the U.S. Global Positioning System (GPS) (previously known as NavStar GPS), the European Union Galileo system, the Russian Federation GLONASS, the Indian NavIC (previously known as IRNSAA), the Chinese BeiDou Navigation Satellite System (BDS) (previously known as Compass), and the Japanese QZSS. While each provides accurate information, precision is improved when they are combined (Misra and Enge, 2011). GPS, Galileo, and GLONASS provide global coverage with 24 satellites each. NavIC has seven satellites, while BDS has 35. QZSS currently has four satellites covering East Asia and Japan with plans to expand to seven satellites.

Global Positioning Satellite System (GPS)   In the GPS system, a constellation of 24 satellites circles the Earth in near-circular inclined orbits. By receiving signals from at least four of these satellites, the receiver position (latitude, longitude, and altitude) can be determined accurately using a triangulation technique that includes timing known as *trilateration* (Misra and Enge, 2011). In effect, the satellites substitute for the geodetic position markers used in terrestrial surveying. In terrestrial surveying, it is only necessary to have three such markers to determine the three unknowns of latitude, longitude, and altitude by means of triangulation. With the GPS system, a time marker is also required, which necessitates obtaining simultaneous measurements from (at least) four satellites.

The GPS system uses one-way transmissions, from satellites to users, so that the user does not require a transmitter, only a GPS receiver. The only quantity, that the receiver must measure, is time, from which propagation delay, and hence the range, to each satellite, can be determined. Each satellite broadcasts its ephemeris (which is a table of the orbital elements as described in Chap. 2), from which its position can be calculated. Knowing the range to three of the satellites and their positions, it is possible to compute the position of the observer (user). The geocentric-equatorial coordinate system (see Sec. 2.9.6) is used with the GPS system, where it is called the *earth-centered, earth-fixed* (ECEF) coordinate system. Denoting the coordinates for satellite n by (x_n, y_n, z_n)

and those for the observer by (x_0, y_0, z_0) the range from observer to satellite ρ_{On} is obtained from

$$\rho_{On}^2 = (x_n - x_0)^2 + (y_n - y_0)^2 + (z_n - z_0)^2 \tag{17.1}$$

As Eq. (17.1) shows, there are three unknowns (x_0, y_0, z_0) and in theory only three equations (for $n = 1$, 2, and 3) are needed. In practice, measurements for four satellites must be made (n ranges from 1 to 4) and this is explained shortly. For the moment, knowing the positions of three satellites and the measured range to each, the position of the observer relative to the coordinate system can be calculated. Of course, the satellites are moving, so their positions must be tracked accurately. The satellite orbits can be predicted from the orbital parameters (as described in Chap. 2). These parameters are continually updated by a master control station which transmits them up to the satellites, where they are broadcast as part of the navigational message from each satellite.

Just as in a land-based system, better accuracy is obtained by using reference points well separated in space. For example, the range measurements made to three reference points clustered together yield nearly equal values. Position calculations involve range differences, and where the ranges are nearly equal, any error is greatly magnified in the difference. This effect, brought about because of the satellite geometry, is known as *dilution of precision* (DOP). This means that range errors occurring from other causes, such as timing errors, are magnified by the geometric effect. With the GPS system, dilution of position is evaluated through a factor known as the *position dilution of precision* (PDOP) *factor*. This is the factor, by which the range errors are multiplied, to estimate the position error. The GPS system is designed to insure that the PDOP factor is less than 6 most of the time (Langley, 1991c).

The GPS constellation consists of 24 satellites in six near-circular orbits, at an altitude of approximately 20,000 km (Daly, 1993). The ascending nodes of the orbits are separated by 60°, and the inclination of each orbit is 55°. The four satellites in each orbit are irregularly spaced to keep the PDOP factor within the limits referred to earlier.

Time enters the position determination in two ways. First, the ephemerides must be associated with a particular time or epoch (as described in Chap. 2). The standard timekeeper is an atomic standard, maintained at the U.S. Naval Observatory, and the resulting time is known as *GPS time*. Each satellite carries its own atomic clock. The time broadcasts from the satellites are monitored by the control station, which transmits back to the satellites any errors detected in timing relative to GPS time. No attempt is made to correct the clocks aboard the satellites; rather, the error information is rebroadcast to the user stations, where corrections can be implemented in the calculations. It can be assumed, therefore, that the satellite position relative to the ECEF coordinate system (the geocentric-equatorial coordinate system, Sec. 2.9.6) is accurately known. The coordinates for satellite n are denoted by (x_n, y_n, z_n).

Second, time markers are needed to show, when transmissions leave the satellites. Thus, by measuring propagation times and knowing the speed of propagation, the ranges can be calculated. Therein lies a problem since the user stations have no direct way of knowing when a transmission from a satellite commenced. The problem is overcome by having the satellite transmit a continuous-wave carrier, which is modulated by a pseudo-random code, timing for the carrier and the code being derived from the atomic clock aboard the satellite. At a user station, the receiver generates a replica of the modulated signal from its own (nonatomic) clock, which is correlated with the received signal in a correlator. A delay is introduced into the replica signal path and is adjusted until the two signals show maximum correlation. If the receiver clock started at the same time as the satellite clock, the delay in the replica path is equal to the propagation delay. Unfortunately, there is an unknown difference in the start times. Let t_n be the true propagation time from satellite n to the observer, and let t_{dn} be the delay time as measured by the correlator. Let Δt be the unknown difference between the clock start times then $t_n = t_{dn} - \Delta t$ (Δt can be + or −). The satellite clocks are synchronized to a master atomic clock, so Δt is the same for all satellites. The range to the satellite n is therefore

$$\rho_{On} = ct_n = c(t_{dn} - \Delta t) \tag{17.2}$$

where c is the speed of light. Substituting this in Eq. (17.1) gives:

$$O = (x_n - x_0)^2 + (y_n - y_0)^2 + (z_n - z_0)^2 - c^2(t_{dn} - \Delta t)^2 \tag{17.3}$$

The unknowns here are the location (x_0, y_0, z_0) and the timing difference Δt. These are the trilateration equations. For satellite n the position (x_n, y_n, z_n) is known and delay time t_{dn} is measured by the receiver. Since there are four unknowns, the receiver must be capable of measuring t_{dn} for four satellites simultaneously ($n = 1, 2, 3, 4$) to yield four simultaneous equations of the form (17.3). These can be solved to find the unknowns Δt and (x_0, y_0, z_0). The latter, of course, are the required position coordinates for the receiver, which are converted into local coordinates (latitude, longitude, and altitude). All this requires quite sophisticated processing in the receiver. In practice, the composition of the GPS signal is much more complex than indicated here, utilizing spread-spectrum techniques that improve the timing accuracy.

The free-space value for c is used where high precision is not required. However, the free-space value cannot be used for radio waves traveling through the ionosphere and the troposphere. Although the change in propagation velocity is small in absolute terms, it can introduce significant timing errors in certain applications. Also, the satellites clocks, although highly accurate have their own timing errors. The dilution of position errors, described previously, combined with these timing errors set a limit on the accuracy of location determination. Where very high accuracy is required, differential GPS (DGPS) can be used. Two receivers are used, one of which is placed at an accurately known location. Thus, the reference receiver makes a measurement in the usual way, but since the coordinates for the location are known, the only unknown in Eq. (17.1) is the range ρ_{On}, which can now be calculated. Comparing the reference value with the value obtained from the receiver correlator enables the errors to be determined. The reference receiver is linked by radio to the receiver at the unknown location, which can now correct for the errors. The two receivers may be up to a few hundred kilometers apart, but this is insignificant in comparison with the distances to the satellites which are in thousands of kilometers, and it is often assumed that the signal paths through the ionosphere and troposphere are the same.

Further details of the GPS system are found in: Kaplan and Hegarty (2015), Miso and Enge (2011), Langley (1990a, 1991b, 1991c), Kleusberg and Langley (1990), and Mattos (1992, 1993a, 1993b, 1993c, 1993d, 1993e) and at https://www.trimble.com/

17.6 Iridium

The Iridium concept was originated by engineers at Motorola's Satellite Communications Division in 1987. Originally envisioned as consisting of 77 satellites in LEO, the name Iridium was adopted by analogy with the element *Iridium* which has 77 orbital electrons. Further studies led to a reduced constellation of only 66 satellites (Leopold, 1992). The original company, Iridium LLC went bankrupt in 2000. A group of private investors organized Iridium Satellite LLC, which acquired the operating assets of the bankrupt Iridium LLC including: the satellite constellation, the terrestrial network, Iridium real property, and intellectual capital to continue operation of the Iridium system.

The 66 Iridium satellites are grouped in six orbital planes each containing 11 active satellites. The orbits are circular, at a height of 780 km (485 miles). Prograde orbits with inclination angles of 86.4° are used. The 11 satellites in any given plane are uniformly spaced with a nominal spacing of 32.7°. An in-orbit spare is available for each plane at 130 km lower in the orbital plane.

The satellites travel in corotating planes, that is, they travel up one side of the Earth, cross over near the north pole, and travel down the other side. With 11 equispaced satellites in each plane, both sides of the Earth are covered continuously. The satellites in adjacent planes travel out of phase, meaning that adjacent planes are rotated by half the satellite spacing relative to one another. This is sketched in Fig.17.2, which shows the view from above the north pole. Collision avoidance is built into the orbital planning, and the closest approach between satellites is 223 km. Satellites in planes 1, 3, and 5 cross the equator in synchronization, while satellites in planes 2, 4, and 6 also cross in synchronization, but out of phase with those in planes 1, 3, and 5.

Although the planes are corotating, the first and last planes must be counter-rotating where they are adjacent as illustrated in this figure. The separation between corotating planes is 31.6° which allows 22° separation between the first and last planes.

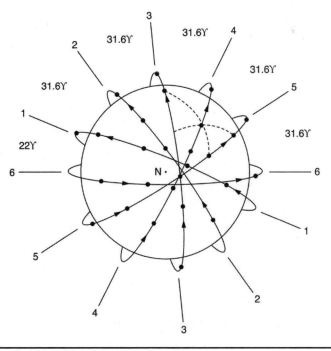

FIGURE 17.2 A polar view of the Iridium satellite orbits.

The closer separation is needed because Earth coverage under the counter-rotating "seam" is not as efficient as it is under the corotating seams. Two-way communication links exist between each satellite and its nearest neighbors ahead and front, and to the satellites in the adjacent planes, as shown by the dotted lines in this figure.

Figure 17.3 shows a system overview. The up/down links between subscribers and satellites take place in the L-band. A 48-beam antenna pattern is used from each satellite with each beam under separate control. At the equator, for instance, overlap of patterns is minimal and all beams may be on, while at high latitudes, considerable overlap occurs, and certain beams are switched off. Also, in regions where operation is prohibited by the telecommunications administration, the beams can be switched off. The switching of beams is referred to as *cell management*. The beams are similar to the cells encountered in cellular mobile, but with the fundamental difference that the beams move relative to the subscriber, whereas in cellular mobile, the cells are fixed. Note that the satellites are traveling at a much greater velocity than that any encountered with terrestrial vehicles, which can be considered stationary relative to the satellites. The orbital period is approximately 100 min, and using an average value of 6371 km for the Earth's radius, the surface speed is $(2 \times 6371\pi)/100 \approx 400$ km/min or just over 15,000 mph. A typical 48-cell pattern is illustrated in Fig. 17.4.

Intersatellite links, and the up/down links between satellites and gateway stations operate in the Ka-band. The link between the Iridium system control station and the satellites is also in the Ka-band. Circular polarization is used on all links (Leopold 1992). The multiple access utilizes a combination of TDMA and FDMA. Each subscriber unit operates in a burst-mode using a single carrier transmission (Motorola 1990). The TDMA frame format is shown in Fig. 17.5 and the uplink/downlink RF frequency plan in Fig. 17.6 (Motorola 1992). Finally, Fig. 17.7 shows an artist's conception of satellites.

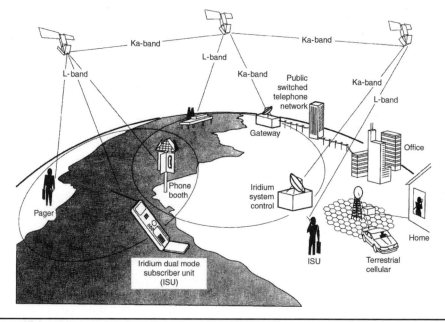

FIGURE 17.3   Iridium system overview. (*Courtesy of Motorola.*)

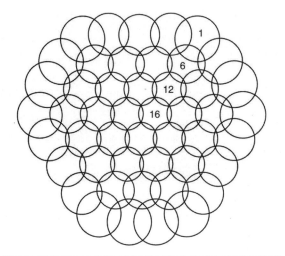

FIGURE 17.4 48-cell L-band integrated antenna pattern architecture. (*Courtesy of Motorola 1992.*)

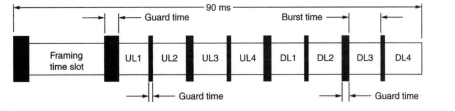

FIGURE 17.5 TDMA frame format. (*Courtesy of Motorola 1992.*)

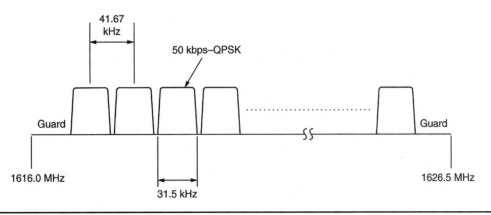

FIGURE 17.6 L-band uplink/downlink RF frequency plan. (*Courtesy of Motorola 1992.*)

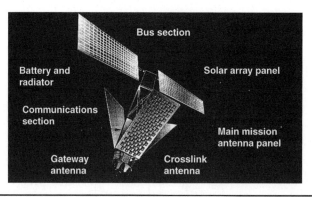

Figure 17.7   Motorola SV (Space Vehicle) deployed configuration. (*Courtesy of Motorola.*)

The original application of Iridium was an extension of wireless telephony. The major advantages listed by the company are:

- Iridium is less likely to be disabled in disaster situations (earthquakes, fire, floods, and the like), and therefore is available as an emergency service if terrestrial cellular services are knocked-out.

- Iridium can provide mobile services to those areas (e.g., remote and sparsely populated areas) that are not reached by terrestrial cellular services.

- With its LEO operation Iridium can offer more channels with shorter delays than mobile services through geostationary satellites.

17.7   Remote Sensing

With their high vantage points, satellites are ideal platforms for *remote sensing*. Earth remote sensing satellites carry sensors which observe the Earth at a variety of methods. Passive sensors include *imaging systems* (cameras) operating at visible, infrared, and ultraviolet frequencies (Chuvieco, 2016) and microwave sensors such as *radiometers* and *sounders* (Ulaby and Long, 2014). Active remote sensors include microwave radar sensors such as *synthetic aperture radars* (SARs), *scatterometers*, and *altimeters* (Ulaby and Long, 2014) and laser radars (*lidar*) (Winker et al., 2003; Hancock et al., 2021). Data from these sensors is used for weather forecasting, shipping and fisheries, farming, forestry, mapping and land use, cloud and aerosols, and climate research, among others. Such systems have been primarily developed and operated by space agencies in various countries, but commercial satellite remote sensing services are now being developed and operated by private companies.

Geostationary remote sensing satellites such as the U.S. *National Oceanographic and Atmospheric Administration* (NOAA) *Geostationary Satellite Server* (GOES) satellites described in Chap. 1 primarily use optical cameras at visible light and infrared wavelength. Microwave sensors and cameras are both employed on LEO remote sensing satellites that operate in polar orbits.

17.8 Problems and Exercises

17.1. Write brief notes on the advantages and disadvantages of using satellites in LEOs, MEOs, and GEOs for mobile satellite communications.

17.2. Write brief notes on the advantages and disadvantages of onboard switching and routing compared to the "bent pipe" mode of operation for satellite mobile communications.

17.3. Describe the operation of a typical VSAT system. State briefly where VSAT systems find widest application.

17.4. Explain why a minimum of four satellites must be visible at an Earth location utilizing the GPS system for position determination. What does the term *dilution of position* refer to?

17.5. For Iridium, calculate the uplink values of $[C/N_0]$ for cell-1 and cell-16 under no shadowing conditions. Calculate the corresponding $[E_b/N_0]$ values.

17.6. Repeat the calculations in Prob. 17.5 for the downlink.

17.7. For the shadowing situation with Iridium, calculate the uplink values of $[C/N_0]$ for cell-1 and cell-16. Calculate the corresponding $[E_b/N_0]$ values.

17.8. Repeat the calculations in Prob. 17.7 for the downlink.

17.9. Calculate for the Iridium SV–Gateway downlink the $[E_b/N_0]$ values for (*a*) rain conditions and (*b*) clear conditions.

17.10. Repeat the calculations of Prob. 17.9 for the Iridium SV–Gateway uplink.

17.11. Calculate for the Iridium east-west intersatellite link the $[C/N_0]$ values for (*a*) no sun, and (*b*) sun conditions.

17.12. Repeat the calculations of Prob. 17.11 for the north–south intersatellite link.

References

Abramson, N. 1990. "VSAT Data Networks." *Proc. IEEE*, vol. 78, no. 7, July, pp. 1267–1274.
Chuvieco, E. 2016. *Fundamentals of Satellite Remote Sensing: An Environmental Approach.* CRC Press, Boca Raton, FL.
Clarke, A. C. 1945. "Extraterrestrial Relays." *Wireless World*, vol. 51, October. pp. 305–308.
Daly, P. 1993. "Navstar GPS and GLONASS: Global Satellite Navigation Systems." *Electron. Commun. Eng. J.*, vol. 5, no. 6, December, pp. 349–357.
Hancock, S., C. McGrath, C. Lowe, I. Davenport, and I. Woodhouse. 2021. "Requirements for a global lidar system: spaceborne lidar with wall-to-wall coverage." *Royal Society Open Science*, https://doi.org/10.1098/rsos.211166
Hughes, C. D., C. Soprano, F. Feliciani, and M. Tomlinson, 1993. "Satellites Systems in a VSAT Environment." *Elect. Comm. Engr. J.*, vol. 5, no. 5, October, pp. 285–291.
Kleusberg, A., and R. B. Langley. 1990. "The Limitations of GPS." *GPS World*, vol. 1, no. 2, March–April, pp. 50–52.
Langley, R. B. 1991a. "Why Is the GPS Signal So Complex?" *GPS World*, vol. 1, no. 3, May–June, pp. 50–59.

Langley, R. B. 1991b. "The GPS Receiver: An Introduction." *GPS World*, vol. 2, no. 1, January, pp. 50–53.

Langley, R. B. 1991c. "The Mathematics of GPS." *GPS World*, vol. 2, no. 7 July–August, pp. 45–50.

Leopold, R. J. 1992. The Iridium Communication system, TUANZ'92, "Communications for Competitive Advantage," *Conference and Trade Exhibition*, Aotea Centre, Auckland, New Zealand, August 10–12.

Mattos, P. 1992. "GPS." *Electronics World + Wireless World*, December, No. 1681, pp. 982–987.

Mattos, P. 1993a. "GPS. 2: Receiver Architecture." *Electronics World + Wireless World*, January, No. 1682, pp. 29–32.

Mattos, P. 1993b. "GPS. 3: The GPS Message on the Hardware Platform." *Electronics World + Wireless World*, February, No. 1683, pp. 146–151.

Mattos, P. 1993c. "GPS. 4: Radio Architecture." *Electronics World + Wireless World*, March, no. 1684, pp. 210–216.

Mattos, P. 1993d. "GPS. 5: The Software Engine." *Electronics World + Wireless World*, April, No. 1685, pp. 296–304.

Mattos, P. 1993e. "GPS. 6: Applications." *Electronics World + Wireless World*, May, no. 1686, pp. 384–389.

Miller, B. 1998. "Satellites Free the Mobile Phone." *IEEE Spectrum*, vol. 35, no. 3, March, pp. 26–35.

Misra, P., P. Enge. 2011. Global Positioning System: Signals, Measurements, and Performance. Ganga-Jamuna Press, ISBN 9780970954428, 0970954425.

Motorola Communications Inc. 1990. *Application of Motorola Satellite Communications Inc. for IRIDIUM, a Low Earth Orbit Satellite System, Before the Federal Communications Commission.* Motorola Communications Inc., Washington D.C., December.

Motorola Communications Inc. 1992. Minor Amendment to Application Before the FCC to Construct and Operate a Low Earth Orbit Satellite System in the RDSS Uplink Band File No. 9-DSS-P91 (87) CSS-91-010.

Rana, H. A., J. McCoskey, and W. Check. 1990. "VSAT Technology, Trends, and Applications." *Proc. IEEE*, vol. 78, no. 7, July, pp. 1087–1095.

Sweeting, M. N. 1992. "UoSAT Microsatellite Missions." *Electron. Commun. Eng. J.*, June, Vol. 4, No. 3, pp. 141–150.

Ulaby, F., and D. G. Long. 2014. *Microwave Radiometric and Radar Remote Sensing.* University of Michigan Press, Ann Arbor, Michigan.

Williamson, M. 1994. "The Growth of Microsats." *IEE Review*, May, vol. 40, no. 3, pp. 117–120.

Winker, D.M., J.R. Pelon, M.P. McCormick. 2003. "The CALIPSO mission: spaceborne lidar for observation of aerosols and clouds." Proceedings Volume 4893, Lidar Remote Sensing for Industry and Environment Monitoring III, https://doi.org/10.1117/12.466539.

Answers to Selected Problems

Chapter 2

2.5. 11,016 km

2.7. 70.14°

2.9. (*a*) 3.842 km/s; (*b*) 12,605.85 ft/s; (*c*) 8594.89 mi/h

2.11. 42,165 km; 0.00018973

2.17. (*a*) 13.994 rev/day; (*b*) 0; (*c*) −3.158 deg/day

2.19. 2,451,248.5 days; 2,451,598.0 days; 2,452,700.1875 days; 2,455,382.125 days

2.21. (*a*) Jan 3, 0 h; (*b*) July 4, 03:00 h; (*c*) Oct 27, 2 h 55 min 4.21 s; (*d*) Jan 3, 7 h 3 min 57.75 s; (*e*) January 31, 2 h 26 min 9.6 s

2.23. (*a*) 27,027 km; (*b*) 0.74718; (*c*) 12.277 rad/day; (*d*) 12.276 rad/day; (*e*) 737.024 min; (*f*) −0.149 deg/day; (*g*) 0.007 deg/day

2.25. 0.856°W

2.27. (*a*) 54.53 rad/day; (*b*) −125.425°; (*c*) 10,509.1 km; (*d*) 36.66°S

2.29. 101.83°W

2.31. 249.8°; 84.96°; 290.7°; 289.52°; 6614.9 km

2.33. −0.707 deg/day; −2.502 deg/day; 82.778 rad/day

2.35. −0.206°; 110.063 min

2.37. (*a*) 7234.6 km; (*b*) 0° Latitude approximately.

2.39. 73.521°

2.41. (*a*) 07:00 h; (*b*) 05:00 h; (*c*) 15:00 h; (*d*) 20:00 h

Chapter 3

3.3. Satellite on same longitude as the earth station, and latitude = 76.3°.

3.5. Latitude zero; maximum possible longitudinal separation 76.3°.

3.7. 302.6°; 64.8°

551

3.9. 44.3°; 45°

3.15. 2.1°

3.21. $\phi_{SS} = 60.025°$; $\phi_{SSmean} = 59.989°$

3.23. 35,799.86 km; 35,772.87 km

Chapter 4

4.3. −0.29 dB

4.7. Horizontal −0.0067 dB; vertical −0.0055 dB

4.11. From Table 4.6: (*a*) $R_{0.01} = 42$ mm/h and $A_{0.01} = -1.12$ dB; (*b*) $R_{0.1} = 12$ mm/h and $A_{0.1} = -1.19$ dB

4.13. −0.006 dB/km

4.15. −0.015 dB/km

4.17. From Table 4.6: (*a*) $R_{0.01} = 42$ mm/h and $A_{0.01} = -11.54$ dB

Chapter 5

5.3. For E_{xmax} unity and $E_{ymax} = 3E_{xmax}$, the power relationship gives $E_{xmax} = 10$ V

5.5. 0.5 W

5.7. LH elliptical

5.9. LH elliptical

5.13. 59°

5.17. 19°

5.21. PL = −0.11 dB; XPD = 16 dB

5.25. 50.4 dB

5.27. 36.6 dB

Chapter 6

6.1. 26.78 dBW

6.3. (1.41, 0.513, 2.598)

6.5. −36.87° relative to the +*y* axis

6.7. 23.87×10^{-15} W/m²

6.9. 51.8 dB

6.13. 7.96×10^{-6} m²

6.19. $HPBE_H = 22°$; $HPBW_E = 25.2°$; 18.72 dB

6.25. 0.9 m

6.27. 12.76 m²; 48.07 dB

6.28   (theta, deg/norm_gain, dB): 0.20/−0.13; 0.40/−0.52; 0.60/−1.18; 0.80/−2.13; 0.944/−3.0; 1.0/−3.38; 1.2/−5.0; 1.4/−7.03; 1.6/−9.58

6.38.   Currents are in-phase

6.42.   (*a*) 1.172 cm; (*b*) 536.3 rad/m; 307.3 deg/cm

6.44. $\Delta l = 1.13$ mm

Chapter 9

9.1.   300–3400 Hz; −14.4 dBm

9.9. $Y = 0.3R + 0.59G + 0.11B$; $Q = 0.21R - 0.52G + 0.31B$; $I = 0.6R - 0.28G - 0.32B$

9.13.   500 Hz

9.17.   152 kHz

9.21.   (*a*) 1.465; (*b*) 2.322; (*c*) 7.545

9.23.   (*a*) 12.6 dB; (*b*) 13.9 dB; (*c*) 5.5 dB

9.25.   (*a*) 62.6 dB; (*b*) 2686.9:1

Chapter 10

10.5.   (*a*) 11100001, 01100001; (*b*) 11011111, 01011111; (*c*) 11010100, 01010100; (*d*) 10011000, 00011000

10.7.   (*a*) 51.7 dB; (*b*) 16

10.9.   (*a*) 0.079; (*b*) 3.87×10^{-6}; (*c*) approximately 0

10.15.   BER ~ 1.91×10^{-4}, <Te> ~ 5.24 sec

10.17.   BER ~ 1.6×10^{-4}, BET ~ $2.5 \times 10^{+2}$, <Te> ~ 4.05 millisec

10.19.   BER ~ 1.1×10^{-6}, BET ~ 55.0

Chapter 11

11.3.   codewords 256; datawords 128

11.5.   (*a*) (0000000); (*b*) (1111111); (*c*) (0010101)

11.7.   s = 001

11.9.   0.839

11.11.   3

11.17.   2.117 MHz; 1.059 MHz

11.19. 2.78×10^{-4}

11.21.   −3 dB

11.23.   For 111, the Euclidean metric is 3.942, and for 000, it is 4.143. This is closest to 111; therefore, the soft decision decoding produces a binary 1. For hard decision, majority vote results in a binary 0.

11.25.   0.201

Chapter 12

12.1. (*a*) 14.6 dB; (*b*) 23.6 dBW; (*c*) 75.6 dBHz; (*d*) 28.2 dB; (*e*) 23 dBK; (*f*) 43 dBi

12.3. (*a*) 42.9 dB; (*b*) 50.3 dB

12.5. 1.98 m^2

12.7. (*a*) 4.80e-12 W/m^2; (*b*) 214.9e-12 W

12.9. −98.18 dBW or 152.1e-12 W

12.12. (*a*) 20 dB; (*b*) 220 K

12.14. (*a*) 8.45 dB; (*b*) 1740 K

12.16. 940.8 K

12.20. 108.1 dBHz; 32.5 dB

12.22. 6 dB

12.24. (*a*) 178.6 pW/m^2; (*b*) −97.5 dBW/m^2

12.26. 45.6 dBW

12.28. 78.1 dBHz

12.30. 73.5 dBW

12.32. 92.1 dBHz; 86.1 dBHz

12.34. 14.6 dB

12.36. 76.1 dBHz

12.38. 14.5 dB

12.39. 7.62 dB

Chapter 13

13.3.P-12.1.3 For earth station 1, 2.83°, and for earth station 2, 2.85°

13.5. 0.5°; 314.2 km

13.7. 4.5 dB

13.9. 45.4 dB

13.11. 25.1 dB

13.13. 21.1 dB

13.15. 51.35 dB

13.17. Yes for 13.12; no for 13.13

13.19. (*a*) 22.4 dB; (*b*) 25 dB

Chapter 14

14.9. 6

14.11. 26.5 dB

14.13. (*a*) −146.8 dBWK⁻¹; (*b*) 15.7 dBW

14.15. (*b*)(1) 2.2 MHz; (*b*)(2) 11.2 MHz

14.21. 0.85

14.23. 0.51

14.25. 0.93; 872

14.27. 0.962; 1797

14.33. 60 Mbps

14.35. (*a*) 75.89 dBHz; (*b*) 49.3 dBW

14.37. 50.5 dBW

14.39. 24 modes

14.41. 8 mV; −21 dB

14.49. 199 (rounded down)

Chapter 15

15.11. 33.92 Mbps

15.23. 40 kbps

15.25. 251.7 Mbps

15.31. 1904 kilobytes

15.33. 0.327 s; 0.267 s

Chapter 16

16.1. (*a*) 22 at 148°, 10 unassigned; 30 at 166°, 2 unassigned; 27 at 157°, 5 unassigned; 32 at 119°, 110°, and 101°; 30 at 61.5°, 2 unassigned. (*b*) Continental 22; DBSC 22; DirecTV 54; Dominion 16; Echostar 24; Echostar/Directsat 44; MCI 28; TIC/Tempo 11; USSB 16; unassigned 19.

16.3. 54 dBW

16.5. **175°:** DBSC 440 Mbps; Echostar/Directsat 440 Mbps

166°: Continental 440 Mbps; Dominion 320 Mbps; Echostar/Directsat 440 Mbps

157°: DirecTV 1080 Mbps

148°: Echostar 960 Mbps; USSB 320 Mbps

119°: Echostar/Directsat 840 Mbps; TCI/Tempo 440 Mbps

110°: Echostar/Directsat 40 Mbps; MCI 1120 Mbps; USSB 120 Mbps

101°: DirecTV 1080 Mbps; USSB 200 Mbps

61.5°: Continental 440 Mbps; DBSC 440 Mbps; Dominion 320 Mbps

16.13. 31 dB; 33 dB; 34 dB; 3.67°; 2.76°; 2.21°

16.15. $[E_b/N_0]$ = −0.46 dB; therefore, signal is completely lost.

16.17. 6.9 dB. This is just satisfactory.

16.19. With an antenna diameter of 20 in., a rain rate of 18 mm/h will just lose the signal. This corresponds to a time percentage of about 0.49 percent (see Table 16.2).

16.21. 11.48 MHz; 33.44 MHz

Chapter 17

(Values rounded off to 1 decimal place).

17.7. 144.6 dB

17.9. 71.1 dBHz; 23.5 dB

17.11. 57.3 dBHz; 20.5 dB

17.13. 56.2 dBHz and 9.2 dB for both cells

17.15. Cell-1: 51.4 dBHz, 4.4 dB; cell-16: 52.1 dBHz, 5.1 dB

17.17. Rain conditions 4.8 dB; clear sky conditions 7.8 dB

17.19. No sun: 78.9 dBHz, 4.9 dB; sun: 78.5 dBHz, 4.6 dB.

APPENDIX B

Conic Sections

A _conic section_, as the name suggests, is a planar section taken through a cone. At the intersection of the sectional plane and the surface of the cone, curves having many different shapes are produced, depending on the inclination of the plane, and it is these curves which are referred to generally as conic sections. Although the origin of conic sections lies in solid geometry, the properties are readily expressed in terms of plane geometrical curves. In Fig. B.1a, a reference line for conic sections, known as the _directrix_, is shown as Z-D. The _axis_ for the conic sections is shown as line Z-Z'. The axis is perpendicular to the directrix. The point S on the axis is called the _focus_. For all conic sections, the focus has the particular property that the ratio of the distance SP to distance PQ is a constant. SP is the distance from the focus to any point P on the curve (conic section), and PQ is the distance, parallel to the axis, from point P to the directrix. The constant ratio is called the _eccentricity_, usually denoted by e. Referring to Fig. B.1a,

$$e = \frac{SP}{PQ} \tag{B.1}$$

Conic sections are given names according to the value of e, summarized here:

Curve	Eccentricity, e
Ellipse	$e < 1$
Parabola	$e = 1$
Hyperbola	$e > 1$

Conic sections are encountered in a number of situations. In this book they are used to describe:

1. The path of satellites orbiting the earth

2. The ellipsoidal shape of the earth

3. The outline curves for various antenna reflectors

A polar equation for conic sections with the pole at the foci can be obtained in terms of the fixed distance p, called the _semilatus rectum_. The polar equation relates the point P to the radius r and the angle θ (Fig. B.1b). From Fig. B.1b,

$$\begin{aligned} SN &= ZN - ZS \\ &= QP - WM \\ &= \frac{r}{e} - \frac{p}{e} \end{aligned} \tag{B.2}$$

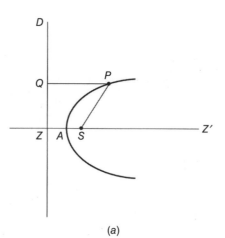

(a)

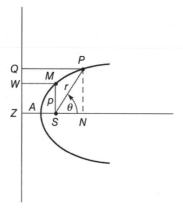

(b)

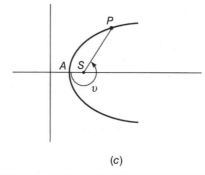

(c)

FIGURE B.1

Also,

$$SN = r \cos \theta \qquad (B.3)$$

Combining Eqs. (B.2) and (B.3) and simplifying give the polar equation as

$$r = \frac{p}{1 - e \cos \theta} \qquad (B.4)$$

If the angle is measured from SA, shown as v in Fig. B.1c, then, since $v = 180° + \theta$, the polar equation becomes

$$r = \frac{p}{1 + e \cos v} \qquad (B.5)$$

The Ellipse

For the ellipse, $e < 1$. Referring to Eq. (B.4), when $\theta = 0°$, $r = p/(1 - e)$. Since $e < 1$, r is positive. At $\theta = 90°$, $r = p$, and at $\theta = 180°$, $r = p/(1 + e)$. Thus r decreases from a maximum of $p/(1 - e)$ to a minimum of $r = p/(1 + e)$, the locus of r describing the closed curve $A'BMA$ (Fig. B.2). Also, since $\cos(-\theta) = \cos\theta$, the curve is symmetrical about the axis, and the closed figure results (Fig. B.2a).

The length AA' is known as the *major axis* of the ellipse. The semimajor axis $a = AA'/2$ and e are the parameters normally specified for an ellipse. The semilatus rectum p can be obtained in terms of these two quantities. As already shown, the maximum value for r is $SA' = p/(1 - e)$ and the minimum value is $SA = p/(1 + e)$. Adding these two values gives

$$AA' = AS + SA'$$

$$= \frac{p}{1 + e} + \frac{p}{1 - e}$$

$$= \frac{2p}{1 - e^2}$$

Since $a = AA'/2$, it follows that

$$p = a(1 - e^2) \qquad (B.6)$$

Substituting this into Eq. (B.5) gives

$$r = \frac{a(1 - e^2)}{1 + e \cos v} \qquad (B.7)$$

which is Eq. (2.23) of the text.

Equation (B.7) can be written as

$$r = \frac{a(1 + e)(1 - e)}{1 + e \cos v}$$

When $v = 360°$,

$$r = a(1 - e) \qquad (B.8)$$

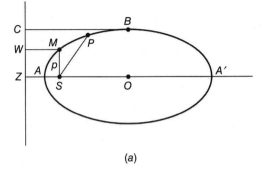

(a)

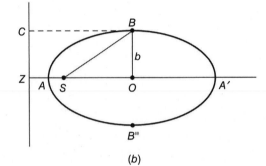

(b)

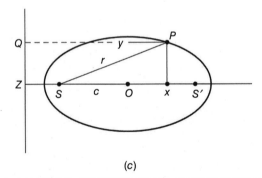

(c)

FIGURE B.2

which is Eq. (2.6) of the text. When $v = 180°$

$$r = a(1 + e) \qquad (B.9)$$

which is Eq. (2.5) of the text.

Referring again to Fig. B.2a, and denoting the length $SO = c$, it is seen that $AS + c = a$. But, as shown above, $AS = p/(1 + e)$ and $p = a(1 - e^2)$. Substituting these for AS and simplifying give

$$c = ae \qquad (B.10)$$

Point O bisects the major axis, and length BO is called the *semiminor axis*, denoted by b (BB'' is the minor axis) (Fig. B.2b).

The semiminor axis can be found in terms of a and e as follows: referring to Fig. B.2b, $2a = SA + SA' = e(AZ + A'Z) = e(2OZ)$, and therefore,

$$OZ = \frac{a}{e} \tag{B.11}$$

But $OZ = BC = SB/e$, and therefore, $SB = a$. SB is seen to be the radius at B. From the right-angled triangle so formed,

$$a^2 = b^2 + c^2$$
$$= b^2 + (ae)^2$$

From this it follows that

$$b = a\sqrt{1 - e^2} \tag{B.12}$$

and

$$e = \frac{\sqrt{a^2 - b^2}}{a} \tag{B.13}$$

This is Eq. (2.1) of the text.

The equation for an ellipse in rectangular coordinates with origin at the center of the ellipse can be found as follows: referring to Fig. B.2c, in which O is at the zero origin of the coordinate system,

$$r^2 = (c + x)^2 + y^2$$

But $r = ePQ = e(OZ + x) = e(a/e + x) = a + ex$. Hence,

$$(a + ex)^2 = (c + x)^2 + y^2$$

Multiplying this and simplifying give

$$a^2(1 - e^2) = x^2(1 - e^2) + y^2$$

Hence,

$$1 = \frac{x^2}{a^2} + \frac{y^2}{a^2(1 - e^2)} \tag{B.14}$$

$$\frac{x^2}{a^2} + \frac{y^2}{b^2}$$

This shows the symmetry of the ellipse, since for a fixed value of y the x^2 term is the same for positive and negative values of x. Also, because of the symmetry, there exists a second directrix and focal point to the right of the ellipse. The second focal point is shown as S' in Fig. B.2c. This is positioned at $x = c = ae$ (from Eq. B.10).

In the work to follow, y can be expressed in terms of x as

$$y = \frac{b}{a}\sqrt{a^2 - x^2} \tag{B.15}$$

To find the area of an ellipse, consider first the area of any segment (Fig. B.3). The area of the strip of width dx is $dA = y\,dx$, and hence the area ranging from $x = 0$ to x is

$$A_x = \int_0^x y\,dx$$

$$= \int_0^x \frac{b}{a}\sqrt{a^2 - x^2}\,dx \qquad (B.16)$$

This is a standard integral which has the solution

$$A_x = \frac{xy}{2} + \frac{ab}{2}\emptyset \qquad (B.17)$$

where $\phi = \arcsin(x/a)$. In particular, when $x = a$, $\phi = \pi/2$, and the area of the quadrant is

$$A_q = \frac{ab\pi}{4} \qquad (B.18)$$

It follows that the total area of the ellipse is

$$A = 4A_q = \pi ab \qquad (B.19)$$

Note the similarity of the equation to the area of a circle, which has $a = b$ is the circle's radius.

In satellite orbital calculations, time is often measured from the instant of perigee passage. Denote the time of perigee passage as T and any instant of time after perigee passage as t. Then the time interval of significance is $t - T$. Let A be the area swept out in this time interval, and let T_p be the periodic time. Then, from Kepler's second law,

$$A = \pi ab\,\frac{t - T}{T_p} \qquad (B.20)$$

The mean motion is $n = 2\pi/T_p$ and the mean anomaly is $M = n\,(t - T)$. Combining these with Eq. (B.20) yields

$$A = \frac{Mab}{2} \qquad (B.21)$$

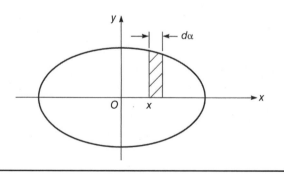

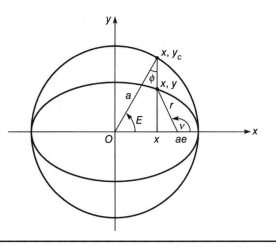

The auxiliary circle is the circle of radius a circumscribing the ellipse as shown in Fig. B.4. This also shows the *eccentric anomaly*, which is angle E, and the true anomaly v (both of which are measured from perigee). The true anomaly is found via the eccentric anomaly.

Some relationships of importance which can be seen from the figure are

$$E = \frac{\pi}{2} - \phi \qquad (B.22)$$

$$y_c = a \sin E \qquad (B.23)$$

The equation for the auxiliary circle is $x^2 + y_c^2 = a^2$. Substituting for x^2 from this into Eq. (B.15) and simplifying give

$$y = \frac{b}{a} y_c \qquad (B.24)$$

Combining this with Eq. (B.23) gives another important relationship:

$$y = b \sin E \qquad (B.25)$$

The area swept out in time $t - T$ can now be found in terms of the individual areas evaluated. Referring to Fig. B.5, this is

$$A = A_q - A_x - A_\Delta$$

$$= \frac{\pi a b}{4} - \left(\frac{xy}{2} + \frac{ab}{2}\phi \right) - \frac{(ae - x)}{2} y$$

$$= \frac{ab}{2} \left(\frac{\pi}{2} - \phi - e \sin E \right) \qquad (B.26)$$

$$= \frac{ab}{2} (E - e \sin E)$$

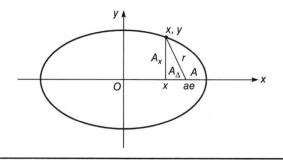

Comparing this with Eq. (B.21) shows that

$$M = E - e \sin E \qquad\qquad (B.27)$$

This is Kepler's equation, given as Eq. (2.27) in the text.

The orbital radius r and the true anomaly v can be found from the eccentric anomaly. Referring to Fig. B.4,

$$r \cos (180° - v) = ae - x$$

But $x = a \cos E$, and hence,

$$r \cos v = a (\cos E - e) \qquad\qquad (B.28)$$

Also, from Fig. B.4,

$$r \sin (180° - v) = y$$

and as previously shown, $y = b \sin E$ and $b = a\sqrt{1-e^2}$. Hence,

$$r \sin v = a\sqrt{1-e^2} \sin E \qquad\qquad (B.29)$$

Squaring and adding Eqs. (B.29) and (B.28) give

$$r^2 = a^2 (\cos E - e)^2 + a^2 (1 - e^2) \sin^2 E$$

from which

$$r = a (1 - e \cos E) \qquad\qquad (B.30)$$

This is Eq. (2.30) of the text.

One further manipulation yields a useful result. Combining Eqs. (B.28) and (B.30) gives

$$\cos v = \frac{\cos E - e}{1 - e \cos E}$$

and hence

$$\frac{1 - \cos v}{1 + \cos v} = \frac{1 - \dfrac{\cos E - e}{1 - e \cos E}}{1 + \dfrac{\cos E - e}{1 - e \cos E}}$$

$$= \frac{(1 + e)(1 - \cos E)}{(1 - e)(1 + \cos E)}$$

Using the trigonometric identity for an angle α that

$$\tan^2 \frac{\alpha}{2} = \frac{1 - \cos \alpha}{1 + \cos \alpha}$$

yields

$$\tan \frac{v}{2} = \sqrt{\frac{1 + e}{1 - e}} \, \tan \frac{E}{2} \tag{B.31}$$

This is Eq. (2.29) of the text.

An elliptical reflector has focusing properties, which may be derived as follows. From Fig. B.6,

$$OZ = a + AZ$$

$$= a + \frac{AS}{e}$$

But $AS = a(1 - e)$, as shown by Eq. (B.8); hence,

$$OZ = \frac{a}{e}$$

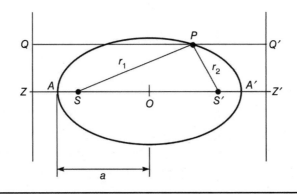

FIGURE B.6

Referring again to Fig. B.6, $SP = eQP$, and $S'P = ePQ'$. Hence,

$$SP + S'P = e\,(QP + PQ')$$
$$= e\,(ZZ')$$
$$= e\,(2 \times OZ)$$
$$= 2a$$

But $SP = r_1$ and $S'P = r_2$, and hence,

$$r_1 + r_2 = 2a \tag{B.32}$$

This result reveals that the sum of the focal distances is constant, or in other words, the ray paths from one focus to the other which go via an elliptical reflector are equal in length. Thus, electromagnetic radiation emanating from a source placed at one of the foci has the same propagation time over any reflected path, and therefore, the reflected waves from all parts of the reflector arrive at the other foci in phase. This property is made use of in the Gregorian reflector antenna described in Sec. 6.15.

The Parabola

For the parabola, the eccentricity $e = 1$. Referring to Fig. B.7, since $e = SA/AZ$, by definition it follows that $SA = AZ$. Let $f = SA = AZ$. This is known as the *focal length*. Consider a line LP' drawn parallel to the directrix. The path length from S to P' is $r + PP'$. But $PP' = AL - f - r\cos\theta$, and hence the path length is $r(1 - \cos\theta) + AL - f$. Substituting for r from Eq. (B.4) with $e = 1$ yields a path length of $p + AL - f$. This shows that the path length of a ray originating from the focus and reflected parallel to the axis (ZZ') is constant. It is this property which results in a parallel beam being radiated when a source is placed at the focus.

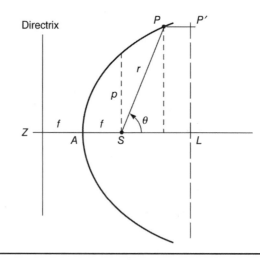

It can be seen that when $\theta = 90°$, $r = SP = p$. But $SP = eSZ = 2f$ (since $e = 1$), and therefore, $p = 2f$. Thus, the radius as given by Eq. (B.4) can be written as

$$r = \frac{2f}{1 - \cos\theta} \tag{B.33}$$

This is essentially the same as Eq. (6.27) of the text, where ρ is used to replace r, and $\Psi = 180° - \theta$, and use is made of the trigonometric identity $1 + \cos\Psi = 2\cos^2(\Psi/2)$. Thus

$$\frac{\rho}{f} = \sec^2\frac{\Psi}{2} \tag{B.34}$$

For the situation shown in Fig. B.8, where D is the diameter of a parabolic reflector, Eq. (B.34) gives

$$\frac{f}{\rho_0} = \cos^2\frac{\Psi_0}{2} \tag{B.35}$$

But from Fig. B.8 it is also seen that

$$\rho_0 = \frac{D}{2\sin\Psi_0} = \frac{D}{4\sin\frac{\Psi_0}{2}\cos\frac{\Psi_0}{2}}$$

where use is made of the double angle formula $\sin\Psi_0 = 2\sin(\Psi_0/2)\cos(\Psi_0/2)$. Substituting for ρ_0 in Eq. (B.34) and simplifying give

$$\frac{f}{D} = \frac{1}{4}\cot\frac{\Psi_0}{2} \tag{B.36}$$

This is Eq. (6.29) of the text.

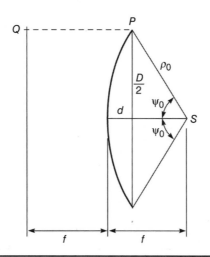

Figure B.8

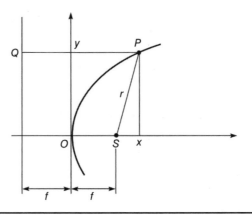

Figure B.9

It is sometimes useful to be able to locate the focal point, knowing the diameter D and the depth d of the dish. From the property of the parabola, $\rho_0 = PQ = f + d$. From Fig. B.8 it is also seen that $\rho_0^2 = (D/2)^2 + (f-d)^2$. Substituting for ρ_0 and simplifying yield

$$f = \frac{D^2}{16d} \tag{B.37}$$

With the zero origin of the xy coordinate system at the vertex (point A) of the parabola, the line PQ becomes equal to $x + f$ (Fig. B.9), and $y = r^2 - (x-f)^2$. These two results can be combined to give the equation for the parabola:

$$y^2 = 4fx \tag{B.38}$$

The Hyperbola

For the hyperbola, $e > 1$. The curve is sketched in Fig. B.10. Equation (B.4) still applies, but it can be seen that for $\cos \theta = 1/e$ the radius goes to infinity; in other words, the hyperbolic curve does not close on itself in the way that the ellipse does but lies parallel

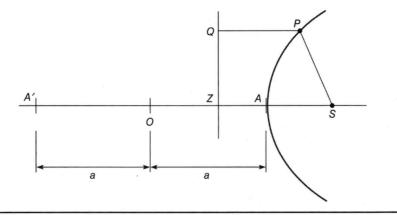

Figure B.10

to the radius at this value of θ. In constructing this ellipse, because e is less than unity, it is possible to find a point A' to the right of S for which the ratio $SA'/A'Z = e$ applied in addition to the ratio $e = SA/AZ$. With the hyperbola, because $e > 1$, a point A' to the left of S can be found for which $e = SA'/A'Z$ in addition to $e = SA/AZ$. By setting $2a$ equal to the distance $A'A$ and making the midpoint of $A'A$ equal to the x, y coordinate origin O as shown in Fig. B.10, it is seen that $2OS = SA + SA'$. But $SA' = eA'Z$ and $SA = eZA$. Hence,

$$2OS = e(ZA + A'Z)$$
$$= 2ae$$

Hence,

$$OS = ae \tag{B.39}$$

Also, $SA' - SA = 2a$, and hence

$$2a = e(A'Z - ZA)$$
$$= e2OZ$$

Therefore,

$$OZ = \frac{a}{e} \tag{B.40}$$

Point P on the curve can now be given in terms of the xy coordinates. Referring to Fig. B.11, S is at point ae, and

$$SP^2 = (ae - x)^2 + y^2$$

Also,

$$SP = ePQ = e\left(x - \frac{a}{e}\right)$$

Combining these two equations and simplifying yields

$$\frac{x^2}{a^2} - \frac{y^2}{b^2} = 1 \tag{B.41}$$

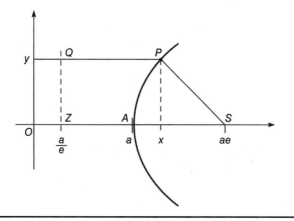

Figure B.11

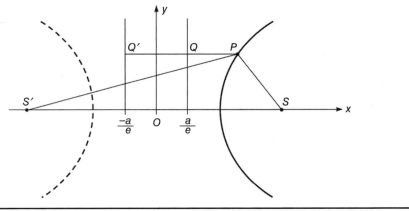

where in this case

$$b^2 = a^2 (e^2 - 1) \tag{B.42}$$

By plotting values it can be seen that a symmetrical curve results, as shown in Fig. B.12, and that there is a second focus at point S'. An important property of the hyperbola is that the difference of the two focal distances is a constant. Referring to Fig. B.12, $S'P = ePQ'$ and $SP = ePQ$. Hence,

$$S'P - SP = e(PQ' - PQ)$$

$$= e\left(\frac{2a}{e}\right) \tag{B.43}$$

$$= 2a$$

An application of this property is shown in Fig. 6.23a of the text which is redrawn in Fig. B.13. By placing the focus of the parabolic reflector at the focus S of the hyperbola

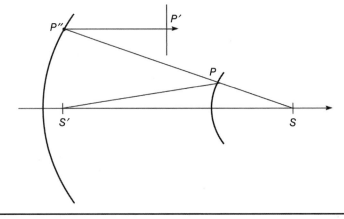

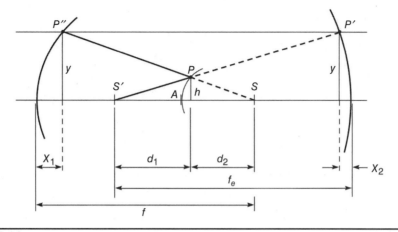

Figure B.14

and the primary source at the focus S', the total path length $S'P + PP''$ is equal to $2a + SP + PP''$ or $2a + SP''$. But, as shown previously in Fig. B.7, the focusing properties of the parabola rely on there being a constant path length $SP'' + P''P'$, and adding the constant $2a$ to SP'' does not destroy this property.

The double-reflector arrangement can be analyzed in terms of an *equivalent parabola*. The equivalent parabola has the same diameter as the real parabola and is formed by the locus of points obtained at P' which is the intersection of $S'P$ produced to P' and $P''P'$ which is parallel to the x axis, as shown in Fig. B.14. The focal distance of the equivalent parabola is shown as f_e and of the real parabola as f. Looking from the focus S to the real parabola, one sees that

$$\frac{h}{d_2} = \frac{y}{f - X_1}$$

Looking to the right from focus S' to the equivalent parabola, one sees that

$$\frac{h}{d_1} = \frac{y}{f_e - X_2}$$

Hence, equating h/y from these two results gives

$$\frac{d_1}{f_e - X_2} = \frac{d_2}{f - X_1}$$

In the limit as h goes to zero, X_1 and X_2 both go to zero, and d_1 goes to $S'A$ and d_2 to AS. Hence,

$$\frac{S'A}{f_e} = \frac{AS}{f}$$

From Fig. B.12 and Eq. (B. 39),

$$SS' = 2OS = 2ae$$

From Fig. B.11,

$$AS = ae - a = a(e - 1)$$

From Fig. B.14,

$$S'A = SS' - AS = a(e + 1)$$

Hence,

$$\frac{a(e + 1)}{f_e} = \frac{a(e - 1)}{f}$$

From which

$$f_e = \frac{e + 1}{e - 1} f \tag{B.44}$$

This is Eq. (6.35) of the text.

APPENDIX C

NASA Two-Line Orbital Elements

ORAD *two-line elements* (TLEs) are widely used for tracking satellites. These can be found at several Web sites (see Appendix D). The *National Oceanographic and Atmospheric Administration* (NOAA) Web site, at http://www.noaa.gov/, is probably the most useful to start with because it contains a great deal of general information on polar orbiting satellites as well as weather satellites in the geostationary orbit. An explanation of the two-line elements can be found in the FAQs section by Dr. T. S. Kelso at http://celestrak.com/. Figure C.1 shows how to interpret the TLEs or:

1 22969U 94003A 94284.57233250 .00000051 00000-0 10000-3 0 1147
2 22969 82.5601 334.1434 0015195 339.6133 20.4393 13.16724605 34163A

description of each line follows:

Line Number 1

Name	Description	Units	Example	Field Format
LINNO	Line number of element data (always 1 for line 1)	None	2	X
SATNO	Satellite number	None	22969	XXXXX
U	Not applicable	None		X
IDYR	International designator (last two digits of launch year)	Launch year	94	XX
IDLNO	International designator (launch number of the year)	None	3	XXX
EPYR	Epoch year (last two digits of the year)	Epoch year	94	XX
EPOCH	Epoch (Julian day and fractional portion of the day)	Day	284.57233250	XXX.XXXXXXXX
NDTO2 or BTERM	First time derivative of the mean motion or ballistic coefficient (depending on the ephemeris type)	Revolutions per day^2 or m^2/kg	0.00000051	±.XXXXXXXX*

Line Number 1 (*Continued*)

Name	Description	Units	Example	Field Format
NDDOT 6	Second time derivative of mean motion (field will be blank if NDDOT6 is not applicable)	Revolutions per day^3	00000-0	±XXXXX-X[†]
BSTAR or AGOM.	BSTAR drag term if GP4 general perturbations theory was used. Otherwise it will be the radiation pressure coefficient		10000-3	±XXXXX-X
EPHTYP	Ephemeris type (specifies the ephemeris theory used to produce the elements)	None	0	X
ELNO	Element number	None	1147	XXXX

*If NDOT2 is greater than unity, a positive value is assumed without a sign.
[†]Decimal point assumed after the ± signs.

Line Number 2

Name	Description	Units	Example	Field format
LINNO	Line number of element data (always 2 for line 2)	None	2	X
SATNO	Satellite number	None	22969	XXXXX
II	Inclination	Degrees	82.5601	XXX.XXXX
NODE	Right ascension of the ascending node	Degrees	334.1434	XXX.XXXX
EE	Eccentricity (decimal point assumed)	None	00151950	XXXXXXXX
OMEGA	Argument of perigee	Degrees	339.6133	XXX.XXXX
MM	Mean anomaly	Degrees	20.4393	XXX.XXXX
NN	Mean motion	Revolutions per day	13.16724605	XX.XXXXXXXX
REVNO	Revolution number at epoch	Revolutions	34163	XXXXX

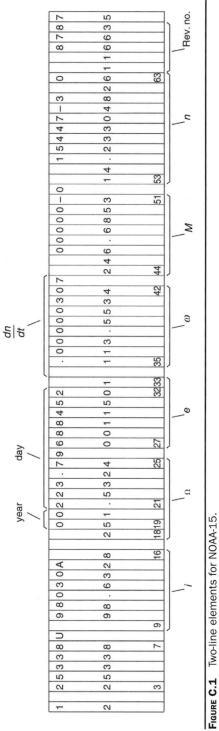

Figure C.1 Two-line elements for NOAA-15.

APPENDIX D
Listings of Artificial Satellites

There are many sources of current satellite orbit data. One of the best is provided by Dr. T. S. Kelso at CelesTrak, at https://celestrak.com/. The source of CelesTrak's data is NORAD two-line element sets (TLEs). TLEs are available at https://www.space-track.org/, but only approved registered users can access this site. Registration is quite straightforward.

Other sites with useful information are:

https://www.amsat.org/
https://www.lyngsat.com/

A Web search by satellite name or searching for TLEs can provide a list of useful Web sites on a particular satellite.

Nonlinear Transfer Functions and Third-Order Intermodulation Products

In general, the nonlinear voltage transfer characteristic for a general nonlinear device such as a diode, mixer, or TWT can be written as a power series expansion

$$e_o = ae_i + be_i^2 + ce_i^3 + \cdots$$

where e_o is the output and e_i is the input voltage. The first term is the linear transfer function. Higher-order terms give rise to the intermodulation products. To illustrate this, consider an input consisting of two unmodulated carriers

$$e_i = A\cos\omega_A t + B\cos\omega_B t$$

and let the carrier spacing be $\Delta\omega$ as shown in Fig. E.1. The terms on the right-hand side of the transfer characteristic equation can be interpreted as follows:

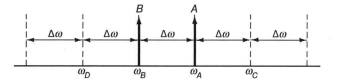

First term, ae_i. This gives the linear relationship between e_o and e_i.

Second term, be_i^2. For e_i as selected, this term can be expanded into the following components:

a zero-frequency dc component, a component at frequency $\Delta\omega$, second harmonic components of the carriers, second harmonic + $\Delta\omega$ components, plus higher order terms. A filter after the device can select or remove these components. For example, using a bandpass filter center at the second harmonic results in frequency doubler. In a mixer, a filter is used to select the + $\Delta\omega$ signals.

The third-order term, ce_i^3 can be expanded as

$$c(A^3 \cos^3 \omega_A t + B^3 \cos^3 \omega_B t + 3A^2 B \cos^2 \omega_B t \cos \omega_A t + 3AB^2 \cos \omega_A t \cos^2 \omega_B t)$$

The cubed terms in this expression can be further expanded as

$$\cos^3 \omega t = \frac{1}{4}(3\cos \omega t + \cos 3\omega t)$$

which is the fundamental frequency component plus a third harmonic component. The third harmonics can be removed by filtering.

The *intermodulation products* are contained in the cross-product terms $3A^2 B \cos^2 \omega_A t \cos \omega_B t$ and $3AB^2 \cos \omega_A t \cos^2 \omega_B t$. On further expansion, the first of these will be seen to contain a $\cos(2\omega_A - \omega_B)t$. With the carriers spaced equally by amount $\Delta \omega$, the $2\omega_A - \omega_B$ frequency is equal to $\omega_A + (\omega_A - \omega_B) = \omega_A + \Delta \omega$. This falls exactly on the adjacent carrier frequency at ω_C, as shown in Fig. E.1. Likewise, the expansion of the second cross-product term contains a $\cos(2\omega_B - \omega_A)t$ which yields an intermodulation product at frequency $\omega_B - \Delta \omega$. This falls exactly on the carrier frequency at ω_D.

APPENDIX F
Acronyms

AAL	ATM adaptation layer
ABR	Available bit rate
ACK	Acknowledgment
ACSSB	Amplitude-companded single-sideband
AFSC	Air force space command
ALA	ATM link accelerator
AM	Amplitude modulation
AMPS	Analog mobile phone service
AMSC	American Mobile Satellite Consortium
ANSI	American National Standards Institute
AOS	Acquisition of signal
APM	Amplitude phase modulation
ATM	Asynchronous transfer mode; asynchronous transmission mode
ATSC	Advanced Television Systems Committee
AVC	Advanced video coding
BAPTA	Bearing and power transfer assembly
BDP	Bandwidth delay product
BER	Bit error rate (errors/bit)
BET	Bit error time (errors/s)
BOL	Beginning of life
BPSK	Binary phase-shift keying
BSS	Broadcasting satellite service
BSU	Burst synchronization unit
BT	Burst tolerance
CAN	Campus area network
CBR	Constant bit rate
CCIR	Comite Consultatif International de Radio
CCITT	Comite Consultatif International de Telephone et Telegraph
CDV	Cell delay variation

CER	Cell error ratio
CFM	Companded frequency modulation
CLP	Cell loss priority
CLR	Cell loss ratio
COMSAT	Communications Satellite (Corporation)
CONUS	Contiguous United States
CPFSK	Continuous phase frequency shift keying
CS	Convergence sublayer
CSSB	Companded single sideband
CTD	Cell transfer delay
DAMA	Demand assignment multiple access
DBS	Direct broadcast satellite
DCS	Digital cross-connect switch
DCT	Discrete cosine transform
DM	Delta modulation
DOMSAT	Domestic satellite
DPCM	Differential PCM
DSI	Digital speech interpolation
DSS	Digital satellite services
DTV	Digital television
DVB	Digital video broadcast
DVD	Digital versatile disk
ECS	European communications satellite
EDTV	Enhanced definition television
EHF	Extremely high frequency
EIRP	Equivalent isotropically radiated power
EOL	End of life
ERS	Earth resources satellite
ESA	European Space Agency
ETSI	European Telecommunications Standards Institute
EUMETSAT	European Organization for the Exploration of METSATs
FCC	Federal Communications Commission
FDM	Frequency-division multiplex
FDMA	Frequency-division multiple access
FEC	Forward error correction
FFSK	Fast FSK
FM	Frequency modulation

FSK	Frequency-shift keying
FSL	Free-space loss
FSS	Fixed satellite service
GEO	Geostationary earth orbit
GEOSAT	Geostationary earth orbit satellite
GFC	Generic flow control
GHz	GigaHertz
GOES	Geostationary operational environmental satellite
GPS	Global positioning satellite
GSM	Global system for mobile communications (originally Groupe Spe'cial Mobile)
GSO	Geo-synchronous orbit
HAN	Home area network
HDTV	High-definition television
HEC	Header error control
HEO	Highly elliptical orbit
HP	Horizontal polarization
HPA	High-power amplifier
HPBW	Half-power beamwidth
IDU	Indoor unit
IEC	International Electrotechnical Commission
IEE	Institution of Electrical Engineers
IEEE	Institute of Electrical and Electronic Engineers
IMMARSAT	International Maritime Satellite (Organization)
INTELSAT	International Telecommunications Satellite (Organization)
INTERSPUTNIK	International Sputnik (Satellite Communications)
IOT	In-orbit testing
IP	Internet protocol
IRD	Integrated receiver-decoder
ISBN	Integrated satellite business network
ISL	Intersatellite link
ISO	International Standards organization
ISP	Internet service provider
ISU	Iridium subscriber unit
ITU	International Telecommunications Union
ITU-T	ITU-telecommunication standardization sector
IWU	Interworking unit

JPEG	Joint photographic experts group
kHz	KiloHertz
LAN	Local area network
LANDSAT	Land survey satellite
LDPC	Low-density parity check
LEO	Low earth orbit
LEOSAT	Low earth orbit satellite
LHCP	Left-hand circular polarization
LLR	Log likelihood ratio
LNA	Low-noise amplifier
LNB	Low-noise block
LOS	Line of Sight
LSAT	Large satellite (European)
MA	Multiple access
MAC	Monitoring alarm and control
MAC	Multiplexed analog compression
MAN	Metropolitan area network
MARISAT	Marine satellite (communications)
MEO	Medium earth orbit
MEOSAT	Medium earth orbit satellite
METOP	Meteorological operations
METSAT	Meteorological satellite
MHz	MegaHertz
MPEG	Motion pictures expert group
MSAT	Mobile satellite (for mobile communications)
MSK	Minimum shift keying
MUX	Multiplexer
NACK	Negative acknowledgment
NASA	National Aeronautics and Space Administration
NCC	Network control center
NCS	Network control station
NESS	National Earth Satellite Service
NNI	Network nod interface (also network network interface)
NOAA	National Oceanic and Atmospheric Administration
NORAD	North American Aerospace Defense Command
NPOESS	National polar operational environmental satellite system

NRT	Non real time
NSP	Network service provider
NTSC	National Television System Committee
OBP	On-board processing
ODU	Outdoor unit
OIG	Orbital information group
PAL	Phase alternation line
PC	Personal computer; personal communications
PCM	Pulse-code modulation
PCR	Peak cell rate
PLL	Phase-lock loop
PNNI	Private NNI
POES	Polar operational environmental satellite
POTS	Plain old telephone service
PSK	Phase-shift keying
PSTN	Public service telephone network
PTI	Payload type indicator
PVC	Permanent virtual circuit
PVP	Permanent virtual path
QoS	Quality of service
QPSK	Quaternary phase-shift keying
RADARSAT	Radar satellite (for remote sensing of earth's surface and atmosphere)
RDSS	Radio determination satellite service
RFC	Request for comments
RHCP	Right-hand circular polarization
RT	Real time
RTT	Round-trip time
SACK	Selective acknowledgment
SAR	Synthetic aperture radar
SARSAT	Search and rescue satellite
SATCOM	Satellite communications
SATM	Satellite ATM
SBS	Satellite Business Systems
SCPC	Single channel per carrier
SCR	Sustained cell rate

SDTV	Standard definition television
SECAM	Sequential couleur à memoire
SFD	Saturation flux density
SHF	Super high frequencies
SI	Systeme International (d' Unites) (International System of Units)
SPADE	SCPC PCM multiple-access demand-assignment equipment
SPEC	Speech predictive encoded communications
SRB	Synchronization reference burst
SS	Satellite switched
SS	Spread spectrum
SSB	Single sideband
STM	Synchronous transmission mode
SV	Space vehicle
SVC	Switched virtual circuit
TASI	Time assignment speech interpolation
TCP	Transmission control protocol; transport control protocol
TDM	Time-division multiplex
TDMA	Time-division multiple access
TDRSS	Tracking Data Radio Satellite System
TELESAT	Telecommunications satellite
TIROS	Television and infrared observational satellite
TLE	Two-line elements
TMUX	Trans-MUX
TT&C	Tracking telemetry and command
TV	Television
TVRO	TV receive only
TWT	Traveling-wave tube
TWTA	Traveling-wave tube amplifier
UBR	Unspecified bit rate
UDP	User datagram protocol
UHF	Ultra high frequencies
UNI	User network interface
VBR	Variable bit rate
VCEG	Video coding experts group
VCI	Virtual channel identifier
VHF	Very high frequencies
VoIP	Voice over internet protocol

VP	Vertical polarization
VPI	Virtual path identifier
VSAT	Very small aperture terminal
WAN	Wide area network
WARC	World Administrative Radio Conference
WWW	World Wide Web
XPD	Cross-polarization discrimination

Logarithmic Units

Decibels

A power ratio of P_1/P_2 expressed in *bels* is

$$\log_{10}\left(\frac{P_1}{P_2}\right)$$

The bel is a logarithmic unit that enables addition and subtraction to replace multiplication and division. A more commonly used logarithmic unit is the *decibel*, abbreviated as dB. The same power ratio expressed in decibels is

$$10\log_{10}\left(\frac{P_1}{P_2}\right)$$

Because the base 10 is understood, it is often not shown explicitly.

The abbreviation dB is often modified to indicate a reference level. Although referred to as a *unit*, the decibel is a dimensionless quantity since it is the *ratio* of two values—usually power values. Power by itself cannot be expressed in decibels. However, by selecting unit power as the denominator in the ratio, power can be expressed in decibels relative to this. A power expressed in decibels relative to 1 W is denoted as dBW. For example, 50 W expressed in decibels relative to 1 W is

$$10\log\frac{50 \text{ W}}{1 \text{ W}} \cong 17 \text{ dBW}$$

Another commonly used reference is the milliwatt, and decibels relative to this are denoted by dBm. Thus 50 W relative to 1 mW is

$$10\log\frac{50}{10^{-3}} \cong 47 \text{ dBm}$$

By definition, the ratio of two voltages V_1 and V_2 squared expressed in decibels is

$$10\log\left(\frac{V_1}{V_2}\right)^2 = 20\log\frac{V_1}{V_2} \text{ dB}$$

The multiplying factor of 20 comes about because power is proportional to voltage squared. Note carefully, however, that the definition of voltage decibels stands on its own and can be related to the corresponding power ratio only when the ratio of load resistances is also known. A unit voltage may be selected as a reference, which allows voltage to be expressed in decibels relative to 1 V. For example, 0.5 V expressed in decibels relative to 1 V is

$$20 \log \frac{0.5 \text{ V}}{1 \text{ V}} \cong -6 \text{ dBV}$$

Also −6 dB is a power ratio of $0.5^2 = 0.25$.

However, because of the chance for error in using the wrong formula, it is highly recommended that the voltage be squared before converting to dB. **This way a single formula (10 log (X)) is always used, and the argument "X" is considered a power ratio.**

Decilogs

The decibel concept is extended to allow the ratio of any two like quantities to be expressed in decibel units, these being termed *decilogs*.

The multiplying factor for decilogs is always 10. Decilogs may also be referred to a unit value. In practice, the name decilog is seldom used and common practice is to use decibels referred to the selected unit. For example, a temperature of 290 K (degrees kelvin) is

$$10 \log \frac{290 \text{ K}}{1 \text{ K}} \cong 24.62 \text{ dBK}$$

Another example that occurs widely in practice is that of frequency referred to 1 Hz. A bandwidth of 36 MHz is equivalent to

$$10 \log \frac{36 \times 10^6 \text{ Hz}}{1 \text{ Hz}} = 75.56 \text{ dBHz}$$

In this book, brackets are used to denote decibel quantities, using the basic power definition. A ratio X in decibels is thus

$$[X] \equiv 10 \log X$$

The three-bar symbol indicates an identity. The abbreviation dB is used, with letters added where convenient to signify the reference. For example, a bandwidth $b = 3$ kHz expressed in decibels is

$$[b] = 10 \log \frac{3000 \text{ Hz}}{1 \text{ Hz}} = 34.77 \text{ dBHz}$$

In many cases, the reference unit is understood and may not be shown explicitly. For example, the temperature of 290 K in decilogs may appear as:

$$10 \log 290 \cong 24.62 \text{ dBK}$$

A quantity that occurs frequently in link-budget calculations is Boltzmann's constant, which has the dimensions of J/K:

$$k = 1.38 \times 10^{-23} \text{ J/K}$$

Expressed in decibels relative to 1 J/K, this is

$$10 \log k = [k] = -228.6 \text{ dB}$$

Strictly, the dB units should be written as dBJ/K for decibels relative to 1 J/K, but this is awkward, and usually it is simply shown as −228.6 dB.

Note that when the argument of the log is <1, then the value of the logarithm is negative. In this text, we treat losses as gains < 1, and in this manner, the transfer function of cascaded blocks is the product of gains. We believe that this is more intuitive than treating the losses as power ratios > 1 and then dividing by the loss. Further, our approach is better for implementing in spreadsheets, where cells containing gains in ratio are multiplied, and the corresponding cells with dB values are added. So by constructing our formulas as products of gains, then the dBs always add. There are examples of this given throughout the text.

If a voltage ratio V_r is to be expressed in decibels using the bracket notation, it is written as $[V_r^2]$. A similar argument applies to electrical current ratios. Thus, it must be kept in mind that the brackets are identified always with $[X] = 10 \log (X)$ in this book.

The advantage of decibel units is that they can be added directly, even though different reference units may be used. For example, if a power of 34 dBW is transmitted through a circuit, which has a loss of −20 dB, the received power is

$$[P_R] = 34 + (-20) = 14 \text{ dBW}$$

In some instances, different types of ratios occur, an example being the G/T ratio of a receiving system described in detail in Sec. 12.6. G is the antenna power gain, and T is the system noise temperature. Now G by itself is dimensionless, so taking the log of this ratio requires that the unit reciprocal temperature K^{-1} be used. Thus

$$\left[\frac{G}{T}\right] = 10 \log\left(\frac{G}{\left(\dfrac{T}{1\,\text{K}}\right)}\right)$$

$$= 10 \log G + 10 \log\left(\frac{1\,\text{K}}{T}\right)$$

$$= [G] + [1/T] \text{ dBK}^{-1}$$

Note that the dBs add, but $[1/T]$ is negative dB. **In performing numerical calculations, the reader must be careful to include the sign of the logarithm.**

The units are often abbreviated to dB/K, but this must not be interpreted as decibels per kelvin. In practice $[G/T]$ is often a known parameter of the system with its value stated in dBK^{-1}.

A table of common dB values is provided in Table G.1. Values are given out to many significant digits, but in common use, only the leading values (indicated in bold) are often used.

Linear	dB		Linear	dB	
Value	Nominal	Accurate	Value	Nominal	Accurate
0.001	−30	−30.00000	1.10	0.4	0.41393
0.010	−20	−20.00000	1.25	1	0.96910
0.020	−17	−16.9897	1.33	1.25	1.24830
0.050	−13	−13.0103	1.50	1.75	1.76091
0.100	−10	−10.00000	1.75	2.4	2.43038
0.125	−9	−9.03090	2.00	3	3.01030
0.250	−6	−6.02060	3.00	4.8	4.77121
0.500	−3	−3.01030	pi	5	4.97150
0.750	−1.25	−1.24939	5.00	7	6.98970
0.900	−0.5	−0.45758	6.00	6	6.02060
1.000	0	0.00000	10.00	10	10.00000

TABLE G.1 Table of Common dB Values

The Neper

The *neper* is a logarithmic unit based on natural, or naperian logarithms. Originally it was defined in relation to currents on a transmission line which decayed according to an exponential law. The current magnitude may decay from a value I_1 to I_2 over some distance x (e.g., a length of transmission line) so that the ratio is

$$\frac{I_1}{I_2} = e^{-\alpha x}$$

where α is the attenuation constant. The attenuation in nepers is defined as

$$N = -\ln \frac{I_1}{I_2}$$

The same ratio expressed as an attenuation in decibels is

$$D = -10\log\left(\frac{I_1}{I_2}\right)^2 = -20\log\frac{I_1}{I_2}$$

Since the same ratio is involved in both definitions, it follows that

$$\frac{I_1}{I_2} = e^{-N} = 10^{-D/20}$$

Thus,

$$D = 20N\log_{10}e$$

$$= 8.686N$$

It follows that for $N = 1$, $D = 8.686$, that is, one neper is equivalent to 8.686 dB.

APPENDIX H

Optimum Bit Decisions and Probability of Bit Errors

In a communication link the received signal includes both the desired signal and additive noise. When demodulating the signal to extract the transmitted bit sequence a decision must be made for each bit based on the filtered received signal value. The presence of noise can cause incorrect decisions resulting in bit errors. For a given bit decision, the probability of error depends on the probability distribution of the noise as well as the probability of the particular bit being a 1 or a 0. The following considers a two-valued modulation scheme, but the ideas can be easily multi-bit modulation schemes.

Let x represent the filtered signal value. Let $p_n(x)$ be the *probability distribution function* (pdf) of the noise in the filtered received signal. We assume that the noise is independent of the signal and that the noise pdf is the same regardless of which bit is sent. When a particular bit, a 1 or 0, is received without noise, the filtered received value is denoted by x_i where i is the bit (0 or 1). The pdf of x depends on which bit was transmitted, which can be written as a *conditional pdf*, i.e., as $p_n(x - x_i \,|\, i)$ as illustrated in Fig. H.1. The decision is based on whether the value of x exceeds a threshold T. If the observed $x \geq T$, the decision is that a 1 was transmitted. If the $x < T$, the decision is that a 0 was transmitted. We note that in communication applications the threshold T is chosen to be halfway between x_0 and x_1, i.e., $T = (x_0 + x_1)/2$, assuming equally likely 1s and 0s. A bit error occurs if (a) the observed $x > T$ due to the noise, but the actual transmitted bit was a 0 or (b) if $x < T$ but a 1 was transmitted.

The probability $P_0(T)$ of $x > T$ for 0 transmitted depends on the noise distribution, $p_n(x)$, which is typically assumed to be a zero-mean *Gaussian pdf* that has variance s_n^2. This is the probably of a bit error for a 0 transmitted. The probability $P_0(T)$ is the integral of the pdf $p_n(x - x_0 \,|\, 0)$ for $x - x_0 > T$, i.e. (Kay, 1998)

$$P_0 = \int_T^\infty p_n(x - x_0 | 0)\, dx$$

which is the dark shaded area in Fig. H.1. Using the change of variables $z = (x - x_0)/s_n$ and $T_z = (T + x_0)/s_n$ this can be written as

$$P_0(T_z) = \int_{T_z}^\infty p(z)\, dz = \int_{T_z}^\infty \frac{1}{\sqrt{2\pi}} e^{-\frac{1}{2}x^2}\, dz = Q(T_z) = 1 - \mathrm{erfc}(T_z)$$

593

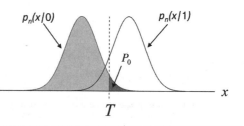

FIGURE H.1   Binary decision illustration for Gaussian noise.

where $p(z)$ is a zero-mean, unit-variance Gaussian pdf, $Q(T_z)$ is the *tail distribution function* of the *standard normal distribution*. $Q(T_z)$ is related to the *error function* $\text{erf}(T_z)$ and the *complementary error function* $\text{erfc}(T_z)$ is used in Sec. 10.6.4 by

$$Q(T_z) = \frac{1}{2} - \frac{1}{2}\text{erf}\left(\frac{T_z}{\sqrt{2}}\right) = \frac{1}{2}\text{erfc}\left(\frac{T_z}{\sqrt{2}}\right)$$

The probability $P_1(T)$ of $x < T$ for 1 transmitted is similarly computed. The total probability of error P_{err} is the sum of the probability of each bit transmitted ($P(0)$ and $P(1)$). Further, because the number of 0's and 1's transmitted are usually approximately equal, then it follows that $P(0)$ and $P(1)$ are usually equal to ½ of the total probability of error, i.e.,

$$P_{\text{err}} = P(0)P_0(T_z) + P(1)P_1(T_z)$$

Note that with $T_z = (x_1 - x_0)/2$ and $P(0) = P(1) = ½$, the effective signal-to-noise ratio (SNR) is

$$\text{SNR} = \frac{x_1 - x_0}{s} = \sqrt{\frac{E_b}{N_0}}$$

References

Kay, S.M., *Fundamentals of Statistical Signal Processing: Detection Theory*, Vol. 2, Prentice-Hall, Upper Saddle River, NJ, 1998.

Index